PRINCIPLES OF ENVIRONMENTAL, GROUNDWATER, AND ENGINEERING GEOLOGY

Principles of Environmental, Groundwater, and Engineering Geology

Editor:

Quan Cui, Ph.D.

Delve Publishing

www.delvepublishing.com

Principles of Environmental, Groundwater, and Engineering Geology

Editor: **Quan Cui, Ph.D.**

708 3rd Avenue, 6th Floor
New York, NY 10017
United States of America
www.delvepublishing.com

ISBN: 978-1-68095-822-5

Library of Congress Control Number: 2016951564

About the Editor

Quan Cui, Ph.D.

Quan Cui obtained her PhD from Beijing Normal University in 2015. She worked as environmental impact assessment engineer for years. Aiming at linking science to policy, her study interest spans from ecosystem modelling to water resources management. She published peer-reviewed articles about using machine learning models in inflow forecasting, CO2 management and watershed ecosystem health assessment.

Contents

List of Contributors

Gergely Tóth
European Commission, Joint Research Centre, Institute for Environment and Sustainability, Via E. Fermi 2749, 21027, Ispra, Italy

Tamás Hermann
University of Pannonia, Georgikon Faculty, Department of Crop Production and Soil Science, Keszthely, Hungary

Gábor Szatmári
Institute for Soil Science and Agricultural Chemistry, Centre for Agricultural Research, Hungarian Academy of Sciences, Budapest, Hungary

László Pásztor
Institute for Soil Science and Agricultural Chemistry, Centre for Agricultural Research, Hungarian Academy of Sciences, Budapest, Hungary

Oliver Kuras
British Geological Survey, Keyworth, Nottingham NG12 5GG, UK

Paul B. Wilkinson
British Geological Survey, Keyworth, Nottingham NG12 5GG, UK

Philip I. Meldrum
British Geological Survey, Keyworth, Nottingham NG12 5GG, UK

Lucy S. Oxby
British Geological Survey, Keyworth, Nottingham NG12 5GG, UK

Sebastian Uhlemann
British Geological Survey, Keyworth, Nottingham NG12 5GG, UK
ETH-Swiss Federal Institute of Technology, Institute of Geophysics, Sonneggstr. 5, 8092 Zurich, Switzerland

Jonathan E. Chambers
British Geological Survey, Keyworth, Nottingham NG12 5GG, UK

Andrew Binley
Lancaster Environment Centre, Lancaster University, Lancaster LA1 4YQ, UK

James Graham
National Nuclear Laboratory, Central Laboratory, Sellafield, Seascale, Cumbria CA20 1PG, UK

Nicholas T. Smith
National Nuclear Laboratory, Central Laboratory, Sellafield, Seascale, Cumbria CA20 1PG, UK
School of Earth, Atmospheric and Environmental Sciences, Williamson Building, University of Manchester, Oxford Road, Manchester M13 9PL, UK

Nick Atherton
Sellafield Ltd, Albion Square, Swingpump Lane, Whitehaven CA28 7NE, UK

J. Vatcher
Division of Mining and Geotechnical Engineering, Luleå University of Technology, SE-971 87 Luleå, Sweden

S.D. McKinnon
The Robert M. Buchan Department of Mining, Goodwin Hall, 25 Union St. Queen's University, Kingston, Ontario, K7L 3N6, Canada

J. Sjöberg
Itasca Consultants AB, Aurorum 2, SE-977 75 Luleå, Sweden

Che-Ming Yang
Graduate Institute of Applied Geology, National Central University, Taiwan

Wei-Lun Yu
Graduate Institute of Applied Geology, National Central University, Taiwan

Jia-Jyun Dong
Graduate Institute of Applied Geology, National Central University, Taiwan

Chih-Yu Kuo
Research Center for Applied Sciences, Academia Sinica, Taiwan

Toshihiko Shimamoto
State Key Laboratory of Earthquake Dynamics, Institute of Geology, China Earthquake Administration, China

Chyi-Tyi Lee
Graduate Institute of Applied Geology, National Central University, Taiwan

Tetsuhiro Togo
Seismotectonics Research Group, Institute of Earthquake and Volcano Geology, National Institute of Advanced Industrial Science and Technology, Japan

Yuki Miyamoto
Kimitsu Works, Nippon Steel & Sumitomo Metal Corpration, Japan

Mark W. McClure
Department of Petroleum Geosystems Engineering, University of Texas at Austin, 200 E. Dean Keeton, C0300, Austin, TX 78704, United States

Roland N. Horne
Energy Resources Engineering, Stanford University, 367 Panama Street, Green Earth Sciences 065, Stanford, CA 94305-4007, United States

Andrew Barkwith
British Geological Survey, Environmental Science Centre, Keyworth, Nottingham, NG12 5GG, UK

Martin D. Hurst
British Geological Survey, Environmental Science Centre, Keyworth, Nottingham, NG12 5GG, UK

Christopher R. Jackson
British Geological Survey, Environmental Science Centre, Keyworth, Nottingham, NG12 5GG, UK

Lei Wang
British Geological Survey, Environmental Science Centre, Keyworth, Nottingham, NG12 5GG, UK

Michael A. Ellis
British Geological Survey, Environmental Science Centre, Keyworth, Nottingham, NG12 5GG, UK

Tom J. Coulthard
Department of Geography, Environment and Earth Science, University of Hull, Cottingham Road, Hull, HU6 7RX, UK

Michael Lawson
School of Earth, Atmospheric and Environmental Sciences and Williamson Research Centre for Molecular Environmental Science, University of Manchester, Oxford Road, Manchester M13 9PL, UK

David A. Polya
School of Earth, Atmospheric and Environmental Sciences and Williamson Research Centre for Molecular Environmental Science, University of Manchester, Oxford Road, Manchester M13 9PL, UK

Adrian J. Boyce
Scottish Universities Environmental Research Centre, East Kilbride G75 0QF, UK

Charlotte Bryant
NERC Radiocarbon Facility, East Kilbride G75 OQF, UK

Christopher J. Ballentine
School of Earth, Atmospheric and Environmental Sciences and Williamson Research Centre for Molecular Environmental Science, University of Manchester, Oxford Road, Manchester M13 9PL, UK
Department of Earth Sciences, University of Oxford, South Parks Road, OX1 3AN Oxford, UK

Eirini Politi
Geography, School of Social Sciences, University of Dundee, Dundee DD1 4HN, UK

John S. Rowan
Geography, School of Social Sciences, University of Dundee, Dundee DD1 4HN, UK

Mark E.J. Cutler
Geography, School of Social Sciences, University of Dundee, Dundee DD1 4HN, UK

Paul S. Goodman
School of Civil Engineering and Geosciences, Newcastle University, Newcastle upon Tyne, Tyne and Wear NE1 7RU, UK

Fabio Galatioto
School of Civil Engineering and Geosciences, Newcastle University, Newcastle upon Tyne, Tyne and Wear NE1 7RU, UK

Neil Thorpe
School of Civil Engineering and Geosciences, Newcastle University, Newcastle upon Tyne, Tyne and Wear NE1 7RU, UK

Anil K. Namdeo
School of Civil Engineering and Geosciences, Newcastle University, Newcastle upon Tyne, Tyne and Wear NE1 7RU, UK

Richard J. Davies
School of Civil Engineering and Geosciences, Newcastle University, Newcastle upon Tyne, Tyne and Wear NE1 7RU, UK

Roger N. Bird
School of Civil Engineering and Geosciences, Newcastle University, Newcastle upon Tyne, Tyne and Wear NE1 7RU, UK

Jean-Christophe Comte
University of Aberdeen, School of Geosciences, Aberdeen, UK
Previously at Queen's University Belfast, Groundwater Research Group, Belfast, UK

Rachel Cassidy
Agri-Food and Biosciences Institute, Belfast, UK
Previously at Queen's University Belfast, Groundwater Research Group, Belfast, UK

Joy Obando
Kenyatta University, Department of Geography, Nairobi, Kenya

Nicholas Robins
Previously at Queen's University Belfast, Groundwater Research Group, Belfast, UK
Previously at British Geological Survey, Wallingford, UK

Kassim Ibrahim
University of Comoros, Faculty of Sciences, Moroni, Comoros

Simon Melchioly
University of Dar Es Salaam, Department of Geology, Dar Es Salaam, Tanzania

Ibrahimu Mjemah
Sokoine University of Agriculture, Department of Physical Sciences, Morogoro, Tanzania

Halimu Shauri
Pwani University, Department of Social Sciences, Kilifi, Kenya

Anli Bourhane
University of Reunion Island, Laboratory Geosciences Reunion, Saint-Denis, France
Water Office Reunion, Saint-Denis, France

Ibrahim Mohamed
University of Comoros, Faculty of Sciences, Moroni, Comoros

Christine Noe
University of Dar Es Salaam, Department of Geology, Dar Es Salaam, Tanzania

Beatrice Mwega
Kenyatta University, Department of Geography, Nairobi, Kenya

Mary Makokha
Kenyatta University, Department of Geography, Nairobi, Kenya

Jean-Lambert Join
University of Reunion Island, Laboratory Geosciences Reunion, Saint-Denis, France

Olivier Banton
University of Avignon, Laboratory of Hydrogeology, Avignon, France

Jeffrey Davies
Previously at British Geological Survey, Wallingford, UK
University College London, London, UK

Eric D. Morway
Nevada Water Science Center, U.S. Geological Survey, Carson City, NV 89701, USA

Richard G. Niswonger
National Research Program, U.S. Geological Survey, Menlo Park, CA 94025, USA

Enrique Triana
MWH Global, Inc., 3665 JFK Parkway Building 1, Suite 206, Fort Collins, CO 80525, USA

Agust Gudmundsson
Department of Earth Sciences, University of London Royal Holloway, Egham, UK

Francesco Vuolo
Institute of Surveying, Remote Sensing and Land Information, University of Natural Resources and Life Sciences, Vienna, Austria

Laura Essl
Institute of Surveying, Remote Sensing and Land Information, University of Natural Resources and Life Sciences, Vienna, Austria

Clement Atzberger
Institute of Surveying, Remote Sensing and Land Information, University of Natural Resources and Life Sciences, Vienna, Austria

Steven L. Forman
Department of Geology, Baylor University, Waco, TX, USA

Gary E. Stinchcomb
Department of Geosciences and Watershed Studies Institute, Murray State University, Murray, KY, USA

Katherine R. Royse
[a]British Geological Survey, Keyworth, Nottingham NG12 5GG, UK

Mike de Freitas
First Steps Ltd, Unit 17 Hurlingham Studios, London SW6 3PA, UK
Department of Civil and Environmental Engineering, Imperial College London, London SW7 2AZ, UK

William G. Burgess
Department of Earth Sciences, University College London, WC1E 6BT, UK

John Cosgrove
Department of Earth Science and Engineering, Imperial College London, London SW7 2AZ, UK

Richard C. Ghail
Department of Civil and Environmental Engineering, Imperial College London, London SW7 2AZ, UK

Phil Gibbard
Cambridge Quaternary, Department of Geography, University of Cambridge, CB2 3EN, UK

Chris King
16A Park Road, Bridport, Dorset, UK

Ursula Lawrence
Crossrail Ltd. 25 Canada Square, Canary Wharf, London E14 5LQ, UK

Rory N. Mortimore
University of Brighton & ChalkRock Ltd, 32 Prince Edwards Road, Lewes BN7 1BE, UK

Hugh Owen
Department of Palaeontology, the Natural History Museum, Cromwell Road, London SW7 5BD, UK

Jackie Skipper
Geotechnical Consulting Group (GCG), 52A Cromwell Road, London SW7 5BE, UK

Preface

In an interdisciplinary field, geology science is related to groundwater related issues, environmental and engineering problems. So understanding geology is critical to effectively manage water resources and solve related environmental or engineering issues precisely. Geotechnical methodologies are used interdisciplinary subjects such as elementary investigation and mapping, movement pattern understanding and modelling, groundwater management and regional development policies and plans. The focus of this book includes applied geomorphology, applied geophysics and geochemistry, environmental geology and hydrogeology, natural hazards, remote sensing techniques, applied geotechnical engineering, regional geographic review and policy-make. This book also contains up-to-bottom discussions about research challenges, economy analysis and assessment the utility of geographic methodologies.

Editor
Quan Cui, Ph.D.

CHAPTER 1

MAPS OF HEAVY METALS IN THE SOILS OF THE EUROPEAN UNION AND PROPOSED PRIORITY AREAS FOR DETAILED ASSESSMENT

Gergely Tóth [a], Tamás Hermann [b], Gábor Szatmári [c], László Pásztor [c]

[a] European Commission, Joint Research Centre, Institute for Environment and Sustainability, Via E. Fermi 2749, 21027, Ispra, Italy

[b] University of Pannonia, Georgikon Faculty, Department of Crop Production and Soil Science, Keszthely, Hungary

[c] Institute for Soil Science and Agricultural Chemistry, Centre for Agricultural Research, Hungarian Academy of Sciences, Budapest, Hungary

ABSTRACT

Soil contamination is one of the greatest concerns among the threats to soil resources in Europe and globally. Despite of its importance there was only very course scale (1/5000 km^2) data available on soil heavy metal concentrations prior to the LUCAS topsoil survey, which had a sampling density of 200 km^2. Based on the results of the LUCAS sampling and auxiliary information detailed and up-to-date maps of heavy metals (As, Cd, Cr, Cu, Hg, Pb, Zn, Sb, Co and Ni) in the topsoil of the European Union were produced. Using the maps of heavy metal concentration in topsoil we made a spatial prediction of areas where local assessment is suggested to monitor and eventually control the potential threat from heavy metals. Most of the examined elements remain under the corresponding threshold values in the majority of the land of the EU. However, one or more of the elements exceed the applied threshold concentration on 1.2 M km^2, which is 28.3% of the total surface area of the EU. While natural backgrounds might be the reason for high concentrations on large proportion of the affected soils, historical and recent industrial and mining areas show elevated concentrations (predominantly of As, Cd, Pb and Hg) too, indicating the magnitude of anthropogenic effect on soil quality in Europe.

INTRODUCTION

Soil contamination is one of the greatest concerns among the threats to soil resources in Europe and globally (CEC, 2006 and Kong, 2014). Heavy metal, together with mineral oils, is the most frequent contaminant in European soil. Recognizing the importance of this problem and consequent need to halt further contamination and start cleaning the soil of the European Union (EU), the 7th Environment Action Programme (OJEU, 2013) of the EU sets the aim to ensure that by 2020 "soil is adequately protected and the remediation of contaminated sites is well underway". A recent study by van Liedekerke et al. (2014) estimates the number of potentially contaminated sites in Europe to sum up to 2.5 Million, illustrating the extent of this challenge. While the new inventory of polluting sites by van Liedekerke et al. (2014) reflects on the magnitude of the problem with regards to risk from point sources, to date no detailed information was available on the size of the area affected by soil contamination in Europe. The FOREGS data produced by the EuroGeoSurvey (Salminen et al., 2005) and the derived continuous map sheets (Lado et al., 2008) have been the most comprehensive sources of

Citation: Gergely Tóth, Tamás Hermann, Gábor Szatmári, László Pásztor, Maps of heavy metals in the soils of the European Union and proposed priority areas for detailed assessment, Science of The Total Environment, Volume 565, 15 September 2016, Pages 1054-1062, ISSN 0048-9697, http://dx.doi.org/10.1016/j.scitotenv.2016.05.115.

information in this respect so far. However, the low sampling density (1 site/5000 km^2) of the FOREGS study (Demetriades et al., 2010) limits the possibilities of interpretation of the results to continental scale overview only.

The LUCAS Topsoil Survey of the European Union (Tóth et al., 2013) opened new possibilities to acquire detailed information on the soil cover in Europe, including heavy metal content. With its 1 site/200 km^2 sampling density it allows modeling and monitoring of soil resources on a finer scale than any previous attempts. This detail of sampling is adequate to create continuous maps for reliable spatial representation at 1 km resolution (Hengl, 2006) of heavy metals in topsoil of Europe. Comparing this detail in information with the sampling density of the geochemical and mineralogical survey of the United States Geological Survey, which was 1 site/1600 km^2 and resulted maps of heavy metals in the US (Smith et al., 2014), the current data can be regarded as a major step in the appraisal of heavy metal in European topsoil. Further to providing information for regional comparison, the LUCAS topsoil data offers the first appraisal for countries where national inventories did not exist before, or supplement the information in countries where monitoring sites are in low density (Jones et al., 2005).

Using this new survey data and applying fully validated comprehensive geostatistical method our aim was to produce the most detailed and up-to-date coverage of heavy metal in the topsoil of the European Union. Based on the maps of heavy metal concentration in topsoil we made a spatial prediction of areas where local assessment is suggested to monitor and eventually control the potential threat from heavy metals.

While heavy metal concentrations are predicted for topsoil of the whole territory of the European Union, differentiation between natural backgrounds and anthropogenic pollution was not the aim of the current research.

MATERIALS AND METHODS

Databases used

The LUCAS topsoil database

Over 23,000 topsoil samples (upper 20 cm) were collected from land of the European Union (EU) Member States (EU-28 except Croatia) with the aim to produce the first coherent baseline topsoil database for continental scale monitoring (Tóth et al., 2013). Sampling was performed in two campaigns, in 2009 and in 2012. The soil sampling was undertaken within the frame of the Land Use/Land Cover Area Frame Survey (LUCAS), an EU wide project to monitor changes in the management and character of the land surface (Eurostat, 2015). The survey, which represents the first effort to build a consistent spatial database of soil properties for environmental assessments is applied standard sampling and analytical procedures.

A stratified sampling scheme was applied and samples were taken from all land cover classes with unique georeferenced locations (Tóth et al., 2013). Basic soil properties, such as particle size distribution, pH, organic carbon, carbonates, NPK, cation exchange capacity (CEC) and multispectral signatures were determined from each sample, which were also analysed for heavy metal content. Analysis of soil parameters followed standard procedures. Tóth et al. (2013) provided detailed description on the methodology and data of the LUCAS Topsoil Survey.

After the analysis of the basic soil parameters - which project concluded in 2012 - soil tests for heavy metal content, including As, Cd, Co, Cr, Cu, Hg, Ni, Pb, Sb and Zn were carried out. Elements were analysed by inductively coupled plasma-optical emission spectrometry. Two certified reference materials (BCR 141R, Calcareous Loam Soil, and NIST 2711, Montana Soil) were used to compare the accuracies of the two digestion procedures. In the first phase of the HM analysis comparative tests were performed using two digestion methods on a subset of 500 samples (Comero et al., 2015). The standard method (ISO, 1995) using aqua regia as an extracting agent was matched with one using microwave-assisted acid digestion (ECS, 2010) and the same detection methods, employing ICP-OES (Inductively Coupled Plasma Optical Emission Spectrometer) for the above listed elements. Based on the reliable correspondence between the measured concentrations by the two methods and considering the advantages of the microwave assisted approach (Comero et al., 2015), all samples were analysed using the prEN16174 (ECS, 2010) procedure. The unit of measurement was mg/kg for As, Cd, Cr, Cu, Hg, Pb, Zn, Sb. Co and Ni, with detection limits 2.84, 0.07, 0.32, 0.26, 0.00005, 1.16, 2.12, 0.81, 0.15 and 0.27 mg/kg respectively.

As a result of the analytical procedure we obtained the concentrations of the studied elements. These are expressed by their elemental weight in milligram per 1 kg of soil. No elemental speciation was measured.

Auxiliary data

The applied environmental co-variables consisted of topography, geology, vegetation, meteorology, as well as anthropogenic factors.

Topography

Topography was taken into consideration by the EU-DEM dataset, which is one of the most detailed, freely available Digital Elevation Models. The EU-DEM is a 3D raster dataset with elevations captured at 1 arc second postings or about every 30 m. From EU-DEM we got the following 18 derivatives by SAGA GIS (Bock et al., 2007): (1) Aspect, (2) Channel Network Base Level, (3) Closed Depressions, (4) Convergence Index, (5) Cross Sectional Curvature, (6) Longitudinal Curvature, (7) Diurnal Anisotropic Heating, (8) Elevation, (9) LS Factor, (10) Mass Balance Index, (11) Multiresolution Index of Ridge Top Flatness (MRRTF), (12) Multiresolution Index of Valley Bottom Flatness (MRVBF), (13) Slope, (14) Topographic Wetness Index, (15) SAGA Wetness Index, (16) Total Direct Solar Insolation, (17) Valley Depth, and (18) Vertical Distance to Channel Network.

Geology

On a spatial basis, the lithogenic source (geological minerals in the soil parent material, on which the soil has developed) is the dominant factor determining the total concentration of heavy metals in world soils (Alloway, 2013). Geology data originated from the 'Digital data of the 1:5 Million International Geological Map of Europe and Adjacent Areas' (IGME 5000,Asch, 2005). The raw map with its 100 different rock types would be too detailed for the mapping, therefore the types were merged by Bakacsi (2016) into 16 categories according to their chemical properties: (1) igneous acidic, (2) igneous intermediate, (3) igneous basic, (4) igneous ultrabasic, (5) sedimentary calcareous, (6) sedimentary siliclastics, (7) sedimentary undifferentiated, (8) metamorphic acidic, (9) metamorphic intermediate, (10) metamorphic basic, (11) metamorphic ultrabasic, (12) metamorphic calcareous, (13) metamorphic siliclastics, (14) metamorphic undifferentiated, (15) volcano-sediments intermediate, and (16) volcano-sediments undifferentiated.

Vegetation

State of vegetation can especially be related to the heavy metal concentration in soils. Like all living organisms, plants are often sensitive both to the deficiency and to the excess availability of some heavy metal ions as essential micronutrient, while the same at higher concentrations and even more ions such as Cd, Hg, and As are strongly poisonous to the metabolic activities (Nagajyoti et al., 2010). Agricultural areas are particularly affected by heavy metal contamination due to the direct applications of organic or inorganic fertilisers, and agrichemicals (Alloway, 2013). Vegetation was taken into consideration by remotely sensed images, since MODIS enhanced vegetation index (EVI) effectively characterizes bio-physical/ biochemical states and processes from vegetated surfaces. Principal Components (2005–2006), furthermore mean value of the monthly MODIS EVI time series data (2001 – 2012) were used in the mapping process (MODIS, 2014).

Climate

Natural and man-made processes have been shown to result in metal containing airborne particulates. Depending on prevailing climatic conditions, these particulates may become wind-blown over great distances; nonetheless, they are subjected to the fate that they are ultimately returned to the lithosphere as precipitations by rain or snowfall (Nagajyoti et al., 2010). Climatic parameters can be described by various different spatial datasets. In the mapping process we used the following 10 variables from two different databases: (1) Annual mean temperature, (2) Mean diurnal range, (3) Isothermality, (4) Temperature seasonality, (5) Maximum temperature of the warmest month, (6) Minimum temperature of the coldest month, (7) Temperature annual range, (8) Mean temperature of the warmest quarter, (9) Mean temperature of the coldest quarter (EuroLST BIOCLIM: Metz et al., 2014), and (10) Sum of average monthly precipitation (WorldClim – Global Climate Data).

Human impact

Anthropogenic factors cannot be omitted from mapping heavy metal distribution in soil. Human impact – aside from agriculture, which was mentioned previously – is concentrated near cities, roads and industrial areas. Spatial distribution of the persistent lights at night trace well out these techno-areas. The layer is obtained from the Defense Meteorological Satellite Program.

Methods applied

Data preparation

Exploratory data analysis was performed on the LUCAS dataset to identify the outliers and examine the marginal distribution of the heavy metals, respectively. All of the marginal distributions have shown a positively skewed distribution, which is common in the case of earth and environmental science data (Manchuk et al., 2009). In fact, most of the geostatistical techniques rely on the MultiGaussian (or multivariate normal) approach. Hence, all of the observations on heavy metals have been transformed to their natural logarithm. The transformed values show a symmetric marginal distribution close to normal. The outliers were identified and then removed from the dataset.

A control dataset was created, which was independent from the mapping procedure. The control dataset incorporated 4396 randomly selected observation points, which is approximately 20% of the LUCAS dataset. The size of the control dataset is adequate to make statistical inferences for the accuracy of the compiled heavy metals maps.

Covariate data layers were resampled in SAGA GIS into a common grid system with 1 km resolution and all further processing was performed on this dataset.

Geostatistical prediction method

Regression kriging (RK) spatial prediction method was used to map the heavy metals spatial distributions. The RK technique combines the regression of the dependent variable on environmental covariates with kriging of the regression residuals. The estimation for z variable at an unvisited location $\mathbf{s}_0$ is:

$$z(\mathbf{s}_0) = \mathbf{q}_0^{\mathbf{T}} \cdot \boldsymbol{\beta} + \boldsymbol{\lambda}_0^{\mathbf{T}} \cdot (\mathbf{z} - \mathbf{q} \cdot \boldsymbol{\beta})$$

where $\boldsymbol{\beta}$ is the vector of the regression coefficients, $\mathbf{q}_0$ is the vector of the covariates at the unvisited location $\mathbf{s}_0$, $\boldsymbol{\lambda}_0$ is the vector of the kriging weights, $\mathbf{z}$ is the vector of the observations and $\mathbf{q}$ is the matrix of the covariates at the sampling locations. The prediction error variance of RK at $\mathbf{s}_0$ is given by:

$$\sigma^2(\mathbf{s}_0) = c(0) - \mathbf{c}_0^{\mathbf{T}} \cdot \mathbf{C}^{-1} \cdot \mathbf{c}_0 + \\ + \left(\mathbf{q}_0 - \mathbf{q}^{\mathbf{T}} \cdot \mathbf{C}^{-1} \cdot \mathbf{c}_0\right)^{\mathbf{T}} \cdot \left(\mathbf{q}^{\mathbf{T}} \cdot \mathbf{C}^{-1} \cdot \mathbf{q}\right)^{-1} \cdot \left(\mathbf{q}_0 - \mathbf{q}^{\mathbf{T}} \cdot \mathbf{C}^{-1} \cdot \mathbf{c}_0\right),$$

where $c(0)$ is the variance of the residuals, $\mathbf{c}_0$ is the vector of covariances between the residuals at the observed and unvisited locations and $\mathbf{C}$ is the covariance matrix of the residuals.

Validation method

Locally the accuracy of the prediction was estimated by the prediction error variance. Overall accuracy of the created heavy metals maps was evaluated using the control dataset. The following indicators were derived using the predicted $\hat{z}(\mathbf{s}_i)$ and observed $z(\mathbf{s}_i)$ values of the heavy metals at the control points $\mathbf{s}_i$:

$$\mathrm{ME} = \frac{1}{n} \cdot \sum_{i=1}^{n} [z(\mathbf{s}_i) - z(\mathbf{s}_i)]$$

$$\mathrm{MAE} = \frac{1}{n} \cdot \sum_{i=1}^{n} |z(\mathbf{s}_i) - z(\mathbf{s}_i)|$$

$$RMSE = \sqrt{\frac{1}{n} \cdot \sum_{i=1}^{n} [z(s_i) - z(s_i)]^2}$$

where n is the number of the control points (in this case 4396). ME (mean error) is often referred to as the bias. RMSE (root mean square error) quantifies the spread of the errors distribution, where the error is defined as the difference between the predicted and observed values (Isaaks and Srivastava, 1989). A reasonable goal for any mapping procedure is to produce unbiased predictions with small spread. MAE (mean absolute error) provides information about the average absolute error, which is more rigorous than ME, according to its definition.

Principal component analysis of heavy metal maps

In order to analyze complex linkages between different concentrations of heavy metals in their geographical context, principal component analysis (PCA) was performed. Because the concentration of some elements have gradual changes while that of others have sharper changes throughout the landscape of Europe the results of the PCA represent the sum of conditions, expressed by the main factors in the system. Results of the PCA analyzes are also mapped to indicate areas with these established interactions.

Application of threshold values to identify risk areas

Results of the mapping exercise performed in our study enables the identification of areas where heavy metal accumulation presents a hazard to health or the environment. To delineate those areas which are threatened by high heavy metal concentrations, we applied assessment thresholds on each map.

Threshold values proposed by the Finnish Ministry of Environment (MEF, 2007) were used, which indicate concentrations where soil contamination and remediation needs must be further assessed. The threshold values of the FEM (2007) represent a good approximation of the mean values of different national systems in Europe (Carlon et al., 2007), have been used in an international context for agricultural soils (UNEP, 2013) and were also successfully applied in regional assessment in Europe (Tóth et al., 2016). Soil covers with element concentration above the thresholds are displayed regardless of the origin of the given substance. Naturally occurring normal concentrations (background concentrations) or raised concentrations of hazardous substances that are found in topsoil which are suspected to be contaminated are treated equally in our delineation.

RESULTS AND DISCUSSION

Results of the geostatistical predictions

Nested models were used throughout the variogram modeling, which combine two or more basic variograms to model the spatial variability (Goovaerts, 1997). In this study three basic structures were used, i.e. a nugget effect and two exponential models. The first one arose from the measurement errors and/or small-scale heterogeneity, while later ones were applied to model the short range (10–100 km) and long range (300–1000 km) spatial variability. The short range variability can be considered as local and regional scale heavy metals distributions (e.g. industrial or mining activities, geological formations), while the long range spatial variability can represent continental scale tendencies.

Topsoil heavy metal concentrations show skewed distribution for all of the elements examined, with outliers of high concentrations. The number of outliers, which probably indicate polluted sites with point source pollution was the highest in the cases of Cd, Cu, Cr, Pb, Hg and Sb. An overview of the main statistics on heavy metal content predicted by the maps produced is given in Table 1.

Table 1: Summary statistics of heavy metal concentrations

Element	Concentration (mg/kg)			
Name	Min	Max	Mean	Std. deviation
Cd	0.02	3.17	0.09	0.11
As	0.46	252.53	3.72	2.92
Co	0.32	91.89	6.35	4.3
Cr	1.57	273.94	21.72	15.7
Cu	0.91	159.07	13.01	9.4
Hg	0	1.59	0.04	0.04
Mn	9.62	2285.23	373.05	237.68
Ni	0.36	466.48	18.36	18.15
Pb	1.63	151.12	15.3	8.33
Sb	0.01	10.91	0.25	0.37

The maps of Fig. 1 (a–i) show the concentrations of various heavy metals in the topsoil of the European Union. Heavy metal in topsoil is either of natural origin or from anthropogenic sources. Human activities can result pollution of limited areas with well-defined boundaries from point sources or diffuse contamination of larger land surfaces. The recognition of the origin of heavy metal is often not very easy, if the point source cannot be identified at the location where high concentration of element is found. The maps of heavy metal content of topsoil with European coverage help to assess both the spatial tendencies and hotspots in the continent.

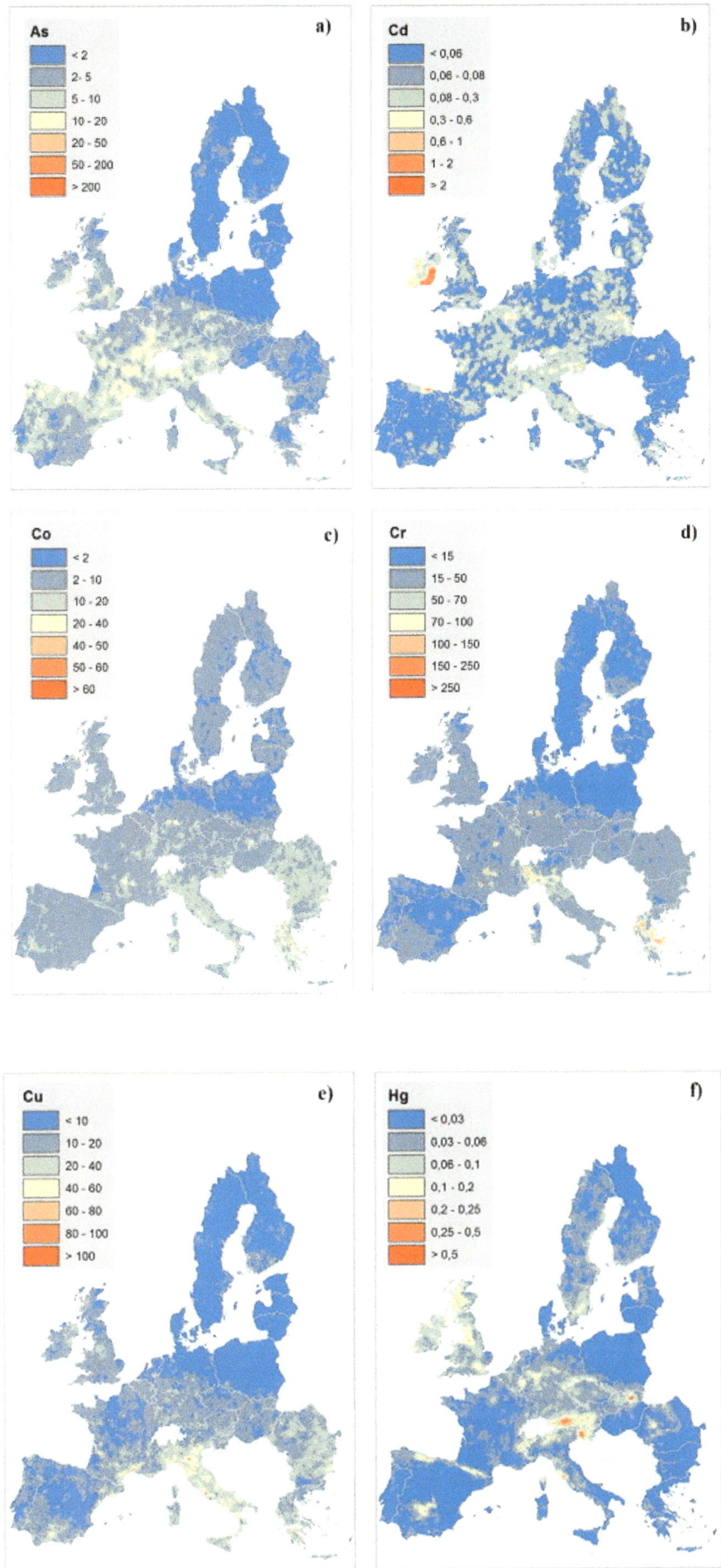
As
< 2
2 - 5
5 - 10
10 - 20
20 - 50
50 - 200
> 200
a)
Cd
< 0,06
0,06 - 0,08
0,08 - 0,3
0,3 - 0,6
0,6 - 1
1 - 2
> 2
b)
Co
< 2
2 - 10
10 - 20
20 - 40
40 - 50
50 - 60
> 60
c)
Cr
< 15
15 - 50
50 - 70
70 - 100
100 - 150
150 - 250
> 250
d)
Cu
< 10
10 - 20
20 - 40
40 - 60
60 - 80
80 - 100
> 100
e)
Hg
< 0,03
0,03 - 0,06
0,06 - 0,1
0,1 - 0,2
0,2 - 0,25
0,25 - 0,5
> 0,5
f)

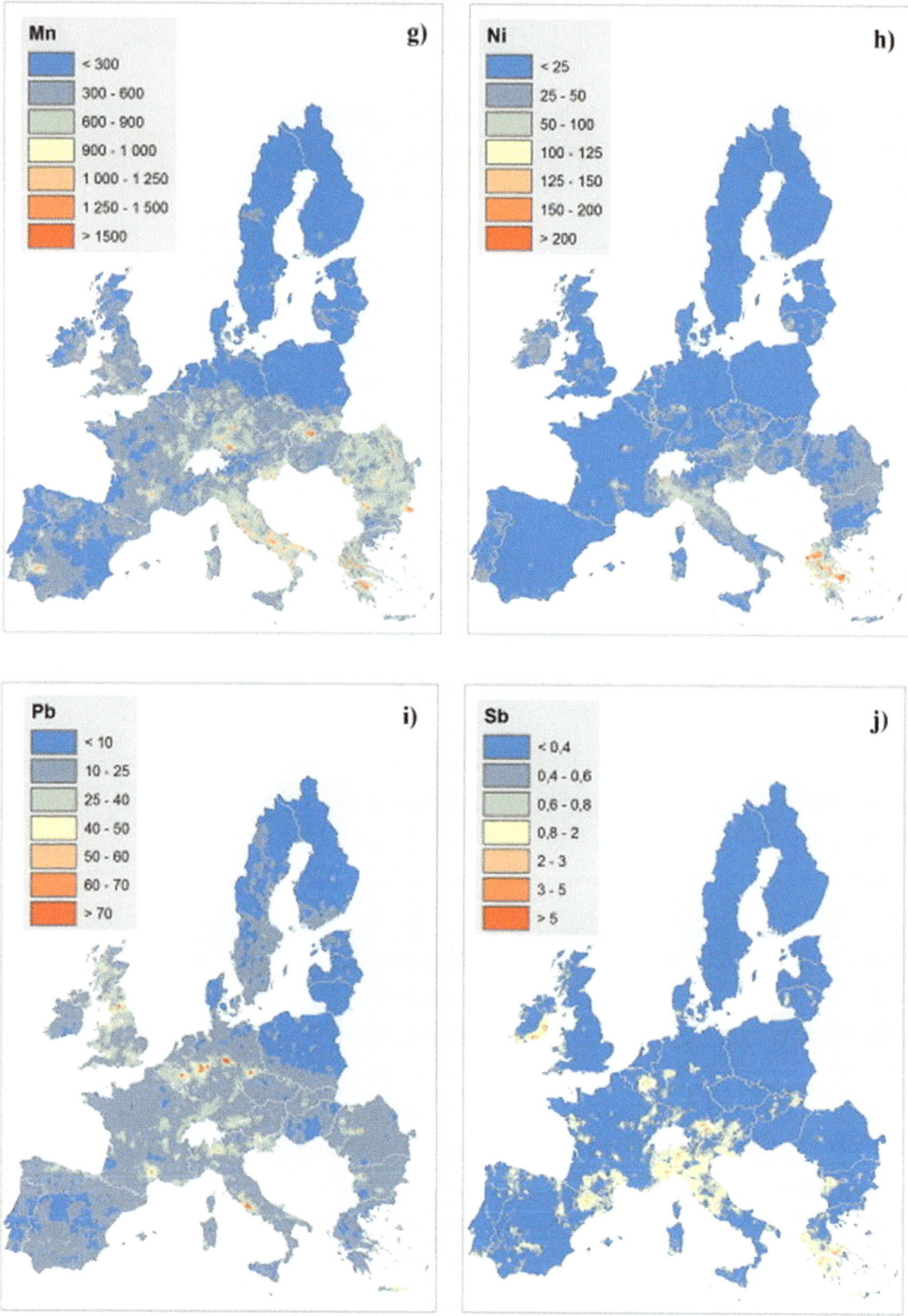

Figure 1: Concentration of heavy metals in topsoil of the European Union.

Geographical tendencies of heavy metals in topsoil of the EU

Results of the PCA analysis (Table 2.) indicate close correlation of Co, Cr, Cu, Mn, Ni and Sb, which are included to the first PC describing nearly half of the total variance and among which both siderophiles and chalcophiles elements can be found. PC-2, adds another 16% to the explanation of the variance in the system, by its main constituents, As, Cd, Hg and Pb, which are chalcophile elements. With PC-3, including Cd and Pb, two elements which likely to indicate anthropogenic pollution, the total explained variance reaches 75%.

Table 2. Results of the principal component analysis on 10 heavy metals (gray shadings indicate main constituents of the PCs).

	PC-1	PC-2	PC-3	PC-4	PC-5	PC-6
As	-0.255	-0.335	0.183	-0.662	0.215	0.464
Cd	-0.110	-0.446	-0.682	0.020	-0.530	0.108
Co	-0.398	0.245	0.097	0.055	-0.157	0.173
Cr	-0.394	0.252	-0.067	0.111	0.104	0.196
Cu	-0.375	0.113	0.021	-0.088	-0.047	-0.646
Hg	-0.131	-0.531	0.253	0.657	0.206	0.119
Mn	-0.376	0.153	0.159	-0.015	-0.461	0.075
Ni	-0.358	0.235	-0.250	0.291	0.252	0.289
Pb	-0.282	-0.377	0.442	-0.016	-0.232	-0.236
Sb	-0.320	-0.222	-0.373	-0.144	0.513	-0.358
Explained variance (%)	48.72	16.19	9.79	7.55	5.32	4.19
Explained cumulative variance (%)	48.72	64.91	74.71	82.26	87.58	91.77

When displaying the PCs on maps (Fig. 2a–c) we can see that the majority of interactions in PC-1 occur in the Central and Eastern Mediterranean region. Further contacts integrated to PC-1 are abundant also in other parts of the continents where glaciation did not interrupt soil genesis, except for the Carpathian basin and the Iberian peninsula, which regions have low concentration of these elements in their topsoils. PC-2, including four chalcophile elements is characteristic for the north-western part of the continent and PC-3, with Cd and Pb appears nearly all parts, although on limited areas, except for those abundant appearances in Ireland and Greece.

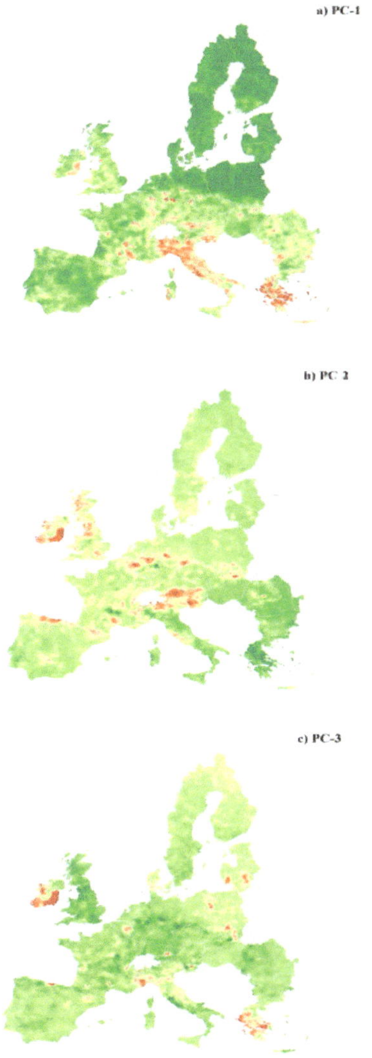

Figure 2: Spatial representation of results of the principal component analysis on heavy metals.

The examinations by elements show that < 20 mg/kg As was detected in > 95% of LUCAS topsoil samples and high portion of the samples contain Arsenic below the limit of detection (LOD). As is most abundant in topsoil of mountainous areas, such as the Alps, the Carpathians, the Massif Central or the Pyrenees (Fig. 1a). Alluvial As deposition can be observed in higher concentration in Holocene deposits at the foothills of younger mountains, where fluvial redistribution may also occur. The Po plain is a prominent example for such a process (Bianchini et al., 2014). On the other hand, post-glacial earth material of quaternary origin is practically free of As, as can be observed at the Northern European regions.

Cd levels in European soils can be characterized by irregular distribution throughout the continent (Fig. 1b.). The largest region with high concentration of Cd can be found in the limestone shale belt of Ireland, which finding is in accordance with the results of local survey (McGrath and Fleming, 2006). Bavarian Alps and the South Eastern Alps also show geographical tendencies of elevated Cd in topsoil. Based on the LUCAS topsoil data low Cd concentrations are predicted for Bulgaria, Portugal and the Scandinavian country. The Carpathian basin is practically free of Cd.

The major geological divide between topsoils of different Co concentration is the limit of the last glaciation (Fig. 1c.). The Co concentration in the postglacial surfaces of Northern Europe is rather low. Older surfaces south of the 55° latitude, where longer periods of soil formation allowed higher rates of mineral transformations exhibit higher Co concentrations. Higher Co concentrations can be observed also in areas associated with mountainous terrain, the Apennines and the Carpathians in particular, but Co concentrations in these areas are still not significantly different from other parts of the continent. High concentrations in the Massif Central in France, the Pindus Mountains in Greece and the Tatras in Slovakia are isolated cases where high concentrations can be observed on a regional scale, due to geochemical reasons.

The boundary of recent postglacial area north of the 50° latitude is the major divide also for Cr content in topsoil in different regions on a continental scale (Fig. 1d). There are six regions with predominantly geology-derived Cr accumulation, of which two has relatively large areas involving agricultural land uses. One in Central to North-Western Greece and the other in Piemonte and the Po plain in Northern Italy.

Cu, like many other elements can be found only in small, but still detectable concentrations in the postglacial plains and somewhat higher, but still low concentrations in the mountainous land of Northern Europe (Fig. 1e.). On the contrary, the Mediterranean hosts high concentrations of Cu in topsoil, the Apennine peninsula, in particular. Croplands contain Cu in significantly higher concentrations than woodland or in this region. When comparing our results with the map of Lado et al. (2008) we can see both similarities and differences. For instance Northern Italy and Andalusia are found to be affected by higher concentration in both maps, but the assessment based on the LUCAS shows generally higher Cu levels in these regions than what was previously predicted based on the FOREGS data.

Hg of higher concentrations in European topsoil can mainly be found in the Eastern ranges of the Alps and the Northern Carpathians and in Lazio province around Rome in Italy (Fig. 1f). The LUCAS based maps partly confirm the finding of the FOREGS analysis (Ottesen et al., 2013) in that the continental-scale distribution of the element is dominated by geology. On the other hand, some hotspots with high Hg concentrations of human release are clearly visible on the LUCAS based continental maps (for details seeSection 3.3). The climate-based comparison did not confirm significant difference between main climate zones of Europe.

Mn concentration in European topsoil shows irregular geographic distribution (Fig. 1g). Only the postglacial northern areas contain low concentration while on other parts of the continent Mn is generally more abundant in topsoil. The Southern Sub-Continental areas as well as Central- and Eastern Mediterranean have most diverse Mn contents, including regions with mean Mn concentrations above 1000 mg/kg soil.

Ni is among the elements of which the concentration in topsoil is geographically divided by the last glaciation (Fig. 1h). North of the 55° latitude its concentration is generally low, although in the Baltic States some 5% of the samples were found to have Ni concentration above 100 g/kg. This share is comparable with that of whole EU where over 95% of areas in the EU contain < 100 mg/kg Ni. It is worth mentioning that, according to the MEF (2007) this is the guideline value for ecological risk assessment. The highest concentration areas for Ni are in the Piemonte region in North-Western Italy and to Northern Greece, including the northern part of the Peloponnesus.

The major geological divide of the last glaciation has a smaller influence on topsoil lead concentration than in the cases of other heavy metals (Fig. 1i), probably because this element is mostly of anthropogenic origin in topsoil. High levels of Pb concentrations can be found in many regions of Europe, but those have the highest where historical industrial activities were combined with mining. Such cases are evident in southern Saxony, Central Saxony-Anhalt and North Rhine-Westphalia (Bergisches Land) in Germany, around Bristol and Manchester in England and Rome, Italy.

Sb is abundant in most European topsoil (Fig. 1j), especially in the southern and western part of the continent. Ireland and Greece have particular high concentrations of Sb in the upper layer of their soil cover but Sb accumulation ca also be seen in the North-Eastern Alps, in the region from Wallonia to the Ruhr land, in North-East Italy, Slovakia and the South Pyrenees.

Hotspots and delineation of priority areas for monitoring

After centuries of anthropogenic emission of heavy metals to the environment it is quite difficult to differentiate natural background contents of heavy metal from diffuse pollution in many cases. Background concentrations can be assessed against polluted sites where the source can be identified (FDEP, 2012), but the methodology to determine natural background concentrations of polluting substances are not straightforward and results might be biased especially in urban or industrial districts (Ander et al., 2013). The task to define areas where high concentrations of heavy metal in topsoil should be regarded as pollution was beyond the scope of the current study. However, those hotspots where the threat of high level of heavy metal is present can be identified based on the maps presented in this paper. Either. While the differentiation between natural and anthropogenic heavy metal and the detection of the magnitude of anthropogenic effect was out of the scope of our current study, some observations on areas with high concentration can be made based on the maps produced.

As seems to be one of the major threats to the topsoil in Southern Saxony which is one of the main historical mining areas in Germany (Fig. 1a). Cd is present in distinct areas in the Cantabria region in Northern Spain, a traditional Zn mining area where sandy conglomerates are combined with dolomite (ITGE, 1994) and also in the Ruhr region of Germany, where high concentration of Co and other heavy metals are also found, around Lyon and Nimes in France and southern Poland, which areas are all home of heavy industry and mining (Fig. 1b). The Mures county in Central Rumania was also an important mining district for centuries. Historically, mining for gold and mercury leads to high Hg concentrations in those mine areas. This may be the reason for the high Hg concentrations in some samples from Central Italy, North-West England and Eastern Slovakia (Fig. 1f).

All the above areas are included to the list of priority regions where more detailed assessment is proposed (Fig. 3). Based on the appraisal of risk areas under threat of soil contamination in Europe, most of the Western-Central European region, Greece, Central Italy and South East Ireland should be examined more in detail. Monitoring of the evolution of heavy metal contents in soil and related ecological systems, including crops is proposed on these lands, which cover more than a quarter of the land surface of the European Union (Table 3.). In order to identify anthropogenic pollution and natural background concentrations, subsoil and topsoil heavy metal concentrations should be compared. As currently such data is not available for reliable assumptions in the required density for Europe, we propose to extend the future LUCAS surveys towards the assessment of subsoil quality as well.

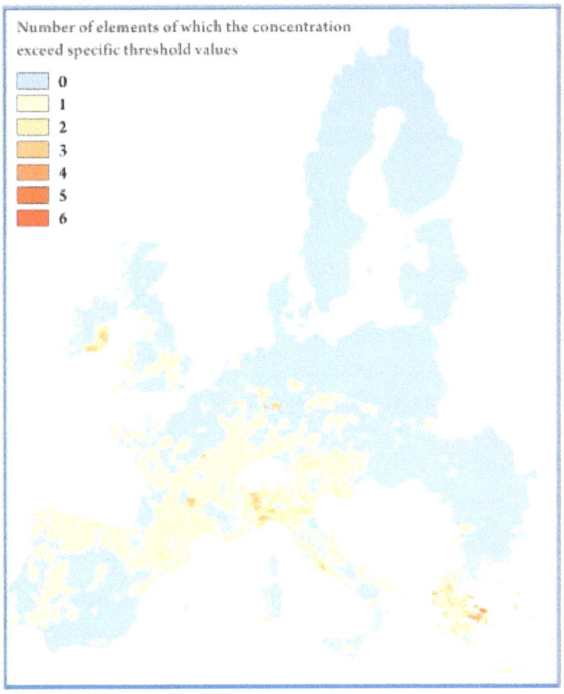

Figure 3: Priority areas of detailed assessment of soil heavy metals.

Table 3: Area extent of soil cover with heavy metal concentration for further assessment

	Area [km²]	Area [% of the total in the EU]
No threshold exceeded (no further assessment needed)	3,087,524	71.68%
Threshold exceeded for 1 element	1,091,013	25.33%
Threshold exceeded for 2 elements	95,372	2.21%
Threshold exceeded for 3 elements	27,036	0.63%
Threshold exceeded for 4 elements	5324	0.12%
Threshold exceeded for 5 elements	880	0.02%
Threshold exceeded for 6 elements	1	< 0.01%

CONCLUSION

Heavy metal contamination is among the main threats to soil-based ecosystem services, including food and feed production. Therefore a reliable information on the concentration of heavy metal in soil is essential. The LUCAS Topsoil Survey provides a unique opportunity for an appraisal of the situation of heavy metal levels in the soils of the European Union. Based on the data from the LUCAS samples we produced a series of maps which predict the concentrations of As, Cd, Cr, Cu, Pb, Zn, Sb. Co and Ni in the EU. Most of these elements remain under the corresponding threshold values in the majority of the land of the EU. However, one or more of the elements exceed the applied threshold concentration on 1.2 M km², which is 28.3% of the total surface area of the EU. In particular regions in Western Central Europe, Central Italy, Greece and South-East Ireland are in concern, where detailed assessment and monitoring is suggested. While natural background concentrations and anthropogenic pollution cannot be definitely separated on the basis of the maps presented, some tendencies and hotspots were identified. The main geological divide in the continent between regions with very low and higher concentration of most of the studied elements is the border of the last glaciation, running through around the 55° latitude. While isolated cases of highly polluted sites can be occasionally found in any regions of the continent, some of the larger historical and recent industrial and mining areas show elevated concentrations of As, Cd, Pb and Hg on most of their areas. These areas, together with those regions where one or more heavy metal show concentrations above the investigation thresholds are proposed to be priority areas for further detailed assessment.

REFERENCES

1. Alloway, B.J., 2013. Sources of heavy metals and metalloids in soils. In: Alloway, B.J. (Ed.), Heavy Metals in Soils. Trace Metals and Metalloids in Soils and their Bioavailability. Springer, Dordrecht, pp. 11–50.
2. Ander, E.L., Johnson, C.C., Cave, M.R., Palumbo-Roe, B., Nathanail, C.P., Murray Lark, R., 2013. Methodology for the determination of normal background concentrations of contaminants in English soil. Sci. Total Environ. 454–455, 604–618.
3. Asch, K., 2005. The 1:5 Million International Geological Map of Europe and Adjacent Areas. BGR (Hannover). Bakacsi, Zs, 2016. oral communication.

4. Bianchini, G., Cremonini, S., Di Giuseppe, D., Vianello, G., Antisari L.V. 2014. Multiproxy investigation of a Holocene sedimentary sequence near Ferrara (Italy): clues on the physiographic evolution of the eastern Padanian Plain. Journal of Soils and Sediments. Volume 14, Issue 1, pp 230–242
5. Bock, M., Böhner, J., Conrad, O., Köthe, R., Ringeler, A., 2007. Methods for creating functional soil databases and applying digital soil mapping with SAGA GIS. In: Hengl, T., Panagos, P., Jones, A., Toth, G. (Eds.), Status and Prospect of Soil Information in South-eastern Europe: Soil Databases, Projects and Applications. EUR 22646 EN Scientific and Technical Research series, Office for Official Publications of the European Communities, Luxemburg, pp. 149–162.
6. Carlon, C., D'Alessandro, M., Swartjes, F., 2007. Derivation methods of soil screening values in Europe: a review of national procedures towards harmonisation. Office for Official Publications of the European Communities EUR. Scientific and Technical Research Series 90 pp.
7. CEC, 2006. Thematic Strategy for Soil Protection. COM(2006)231 final. Commission of the European Communities, Brussels 22.9.2006.
8. Comero, S., Cristache, C., Fissiaux, I., Ruiza, A.A., Locoro, G., Tóth, G., Gawlik, B.M., 2015.
9. Comparative study on open vessel digestion versus microwave-assisted digestion for the determination of trace and major elements in agricultural soils by inductively coupled plasma-optical emission spectrometry. Anal. Chim. Acta. Under review.
10. Demetriades, A., Reimann, C., Birke, M., Salminen, R., De Vos, W., Tarvainen, T., and the EuroGeoSurveys Geochemistry Expert Group, 2010. Geochemical Atlases of Europe Produced by the EuroGeoSurveys Geochemistry Expert Group: State of Progress and Potential Uses. Bulletin of the Geological Society of Greece, Proceedings of the 12th International Congress, Patras, May, 2010.
11. Defense Meteorological Satellite Program - http://ngdc.noaa.gov/eog/ Retrieved from: http://esdac.jrc.ec.europa.eu/content/heavy-metals-topsoils ECS - European Committee for Standardization, 2010. European standard, Sludge, treated biowaste and soil – Digestion of aqua regia soluble fractions of elements, Draft prEN 16174 CEN/TC 400.
12. EU-DEM dataset, d. Produced using Copernicus data and information funded by the European Union - EU-DEM layers. http://www.eea.europa.eu/data-and-maps/data/eudem#tab-european-data (1km resolution EU-DEM was retrieved from: http:// esdac.jrc.ec.europa.eu/content/heavy-metals-topsoils).
13. Eurostat, 2015. LUCAS — Land use and land cover survey. http://ec.europa.eu/eurostat/ statistics-explained/index.php/LUCAS_-_Land_use_and_land_cover_survey (last accessed: October 9, 2015).
14. FDEP (Florida Department of Environmental Protection), 2012. Guidance for Comparing Background and Site Chemical Concentrations in Soil. Florida Department of Environmental Protection. Bureau of Waste Cleanup, Tallahassee, FL, p. 14.
15. Goovaerts, P., 1997. Geostatistics for Natural Resources Evaluation. Oxford University Press, New York.
16. Hengl, T., 2006. Finding the right pixel size. Comput. Geosci. 32 (9), 1283–1298.
17. ITGE (Instituto Tecnologico Geominero de Espana) 1994. Mapa Geologico de la Peninsula Iberica, Baleares e Canarias. 1:1.100.000 Madrid
18. Isaaks, E.H., Srivastava, R.M., 1989. An Introduction to Applied Geostatistics. Oxford University Press, New York.
19. ISO, 11466 International Standard, 1995. Soil Quality – Extraction of trace elements soluble in aqua regia. International Organization for Standardization, Genève, Switzerland.
20. Kong, X.B., 2014. China must protect high-quality arable land. Nature. 506, 7. http://dx. doi.org/10.1038/506007a (06 February 2014).
21. Lado, L.R., Hengl, T., Reuter, H.I., 2008. Heavy metals in European soils: A geostatistical analysis of the FOREGS Geochemical database. Geoderma 148, 189–199.
22. Manchuk, J.G., Leuangthong, O., Deutsch, C.V., 2009. The proportional effect. Math. Geosci. 41, 799–816.
23. McGrath, D., Fleming, G.A., 2006. Trace Elements and Heavy Metals in Irish Soils.
24. TEAGASC, Carlow, p. 216 (available at: http://www.teagasc.ie/publications/2011/ 823/Trace_Elements.pdf).
25. MODIS, 2014. Mean value of the monthly MODIS EVI time series data (2001−2012). http://worldgrids.org/doku.php?id=wiki:evmmod3 (Accessed in November 2014).
26. Metz, M., Rocchini, D., Neteler, M., 2014. Surface temperatures at the continental scale: tracking changes with remote sensing at unprecedented detail. Remote Sens. 6 (5), 3822–3840.
27. Ministry of the Environment, Finland, 2007. Government Decree on the Assessment of Soil Contamination and Remediation Needs. 214/2007 (March 1, 2007).
28. Nagajyoti, P.C., Lee, K.D., Sreekanth, T.V.M., 2010. Heavy metals, occurrence and toxicity for plants: a review. Environ. Chem. Lett. 8, 199–216.
29. Ottesen, R.T., Birke, M., Finne, T.E., Gosar, M., Locuturad, J., Reimann, C., Tarvainen, T., the GEMAS Project Team, 2013. Mercury in European agricultural and grazing land soils. Applied Geochemistry 33, 1–12.
30. Salminen, R., Batista, M.J., Bidovec, M., Demetriades, A., De Vivo, B., De Vos, W., Duris, M., Gilucis, A., Gregorauskiene, V., Halamic, J., Heitzmann, P., Lima, A., Jordan, G., Klaver, G., Klein, P., Lis, J., Locutura, J., Marsina, K., Mazreku, A., O'Connor, P.J., Olsson, S.Å., Ottesen, R.-T., Petersell, V., Plant, J.A., Reeder, S., Salpeteur, I., Sandström, H., Siewers, U., Steenfelt, A., Tarvainen, T., 2005. Geochemical atlas of Europe. Part 1: Background Information, Methodology and Maps. Geological Survey of Finland, Espoo, p. 526.
31. Smith, D.B., Cannon, W.F., Woodruff, L.G., Solano, F.,

Ellefsen, K.J., 2014. Geochemical and mineralogical maps for soils of the conterminous United States. U.S. Geological Survey Open-File Report 2014-1082 http://dx.doi.org/10.3133/ofr20141082 (386 pp.).

32. Soil resources of Europe. In: Jones, R.J.A., Houšková, B., Bullock, P., Montanarella, L. (Eds.), European Soil Bureau Research Report No.9, EUR 20559 EN, second ed. Office for Of- ficial Publications of the EuropeanCommunities, Luxembourg (420pp.).

33. System for Automated Geoscientific Analyses (SAGA GIS). Retrieved from http://www. saga-gis.org

34. Tóth, G., Jones, A., Montanarella, L., 2013. LUCAS topsoil survey – methodology, data and results. EUR 26102 EN. Office for Official Publications of the European Communities, Luxembourg, p. 141 (http://eusoils.jrc.ec.europa.eu/esdb_archive/eusoils_docs/other/EUR26102EN.pdf).

35. Tóth, G., Hermann, T., Da Silva, M.R., Montanarella, L., 2016. Heavy metals in agricultural soils of the European Union with implications for food safety. Environment International 88, 299–309.

36. OJEU - Official Journal of the European Union, 2013. Living well, within the limits of our planet. General Union Environment Action Programme to 2020. Off. J. Eur. Union 354, 171–200.

37. UNEP, 2013. Environmental Risks and Challenges of Anthropogenic Metals Flows and Cycles. In: van der Voet, E., Salminen, R., Eckelman, M., Mudd, G., Norgate, T., Hischier, R. (Eds.), A Report of the Working Group on the Global Metal Flows to the International Resource Panel pp231.

38. van Liedekerke, M., Prokop, G., Rabl-Berger, S., Kibblewhite, M., Louwagie, G., 2014. Progress in the Management of Contaminated Sites in Europe. EUR 26376. Publications, Office of the European Union. Luxembourg (68 pp).

39. WorldClim – global climate data http://www.worldclim.org/current.

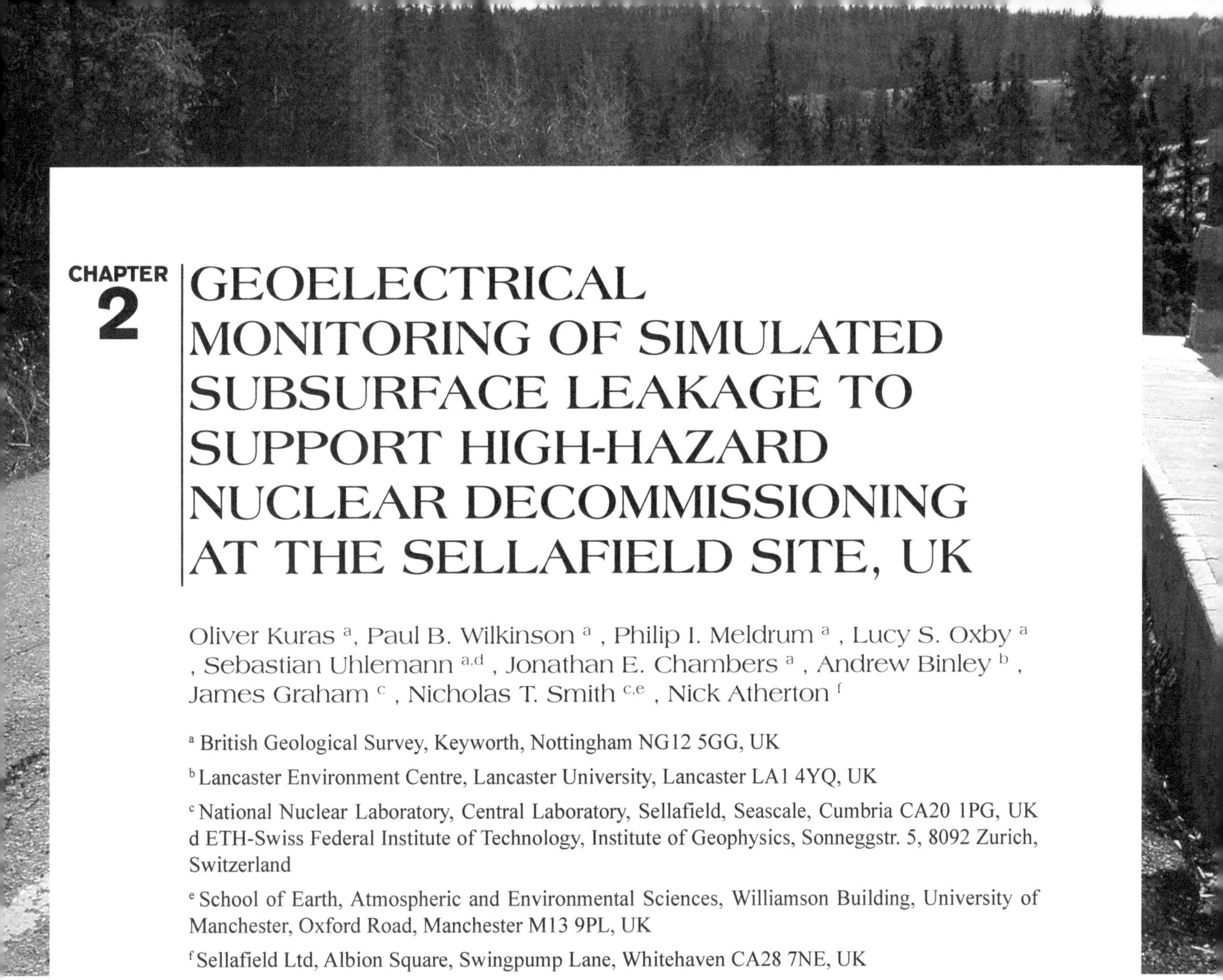

CHAPTER 2

GEOELECTRICAL MONITORING OF SIMULATED SUBSURFACE LEAKAGE TO SUPPORT HIGH-HAZARD NUCLEAR DECOMMISSIONING AT THE SELLAFIELD SITE, UK

Oliver Kuras [a], Paul B. Wilkinson [a] , Philip I. Meldrum [a] , Lucy S. Oxby [a] , Sebastian Uhlemann [a,d] , Jonathan E. Chambers [a] , Andrew Binley [b] , James Graham [c] , Nicholas T. Smith [c,e] , Nick Atherton [f]

[a] British Geological Survey, Keyworth, Nottingham NG12 5GG, UK

[b] Lancaster Environment Centre, Lancaster University, Lancaster LA1 4YQ, UK

[c] National Nuclear Laboratory, Central Laboratory, Sellafield, Seascale, Cumbria CA20 1PG, UK
d ETH-Swiss Federal Institute of Technology, Institute of Geophysics, Sonneggstr. 5, 8092 Zurich, Switzerland

[e] School of Earth, Atmospheric and Environmental Sciences, Williamson Building, University of Manchester, Oxford Road, Manchester M13 9PL, UK

[f] Sellafield Ltd, Albion Square, Swingpump Lane, Whitehaven CA28 7NE, UK

ABSTRACT

A full-scale field experiment applying 4D (3D time-lapse) cross-borehole Electrical Resistivity Tomography (ERT) to the monitoring of simulated subsurface leakage was undertaken at a legacy nuclear waste silo at the Sellafield Site, UK. The experiment constituted the first application of geoelectrical monitoring in support of decommissioning work at a UK nuclear licensed site. Images of resistivity changes occurring since a baseline date prior to the simulated leaks revealed likely preferential pathways of silo liquor simulant flow in the vadose zone and upper groundwater system. Geophysical evidence was found to be compatible with historic contamination detected in permeable facies in sediment cores retrieved from the ERT boreholes. Results indicate that laterally discontinuous till units forming localized hydraulic barriers substantially affect flow patterns and contaminant transport in the shallow subsurface at Sellafield. We conclude that only geophysical imaging of the kind presented here has the potential to provide the detailed spatial and temporal information at the (sub-)meter scale needed to reduce the uncertainty in models of subsurface processes at nuclear sites.

GRAPHICAL ABSTRACT

3D fractional resistivity change (resistivity change Δρ divided by baseline resistivity ρ_0) image showing results of Stage 1 silo liquor simulant injection. The black line delineates the preferential flow path; green cylinders show regions of historic contamination found in sediment cores from ERT boreholes.

Citation: Oliver Kuras, Paul B. Wilkinson, Philip I. Meldrum, Lucy S. Oxby, Sebastian Uhlemann, Jonathan E. Chambers, Andrew Binley, James Graham, Nicholas T. Smith, Nick Atherton, Geoelectrical monitoring of simulated subsurface leakage to support high-hazard nuclear decommissioning at the Sellafield Site, UK, Science of The Total Environment, Volumes 566–567, 1 October 2016, Pages 350-359, ISSN 0048-9697, http://dx.doi.org/10.1016/j.scitotenv.2016.04.212.

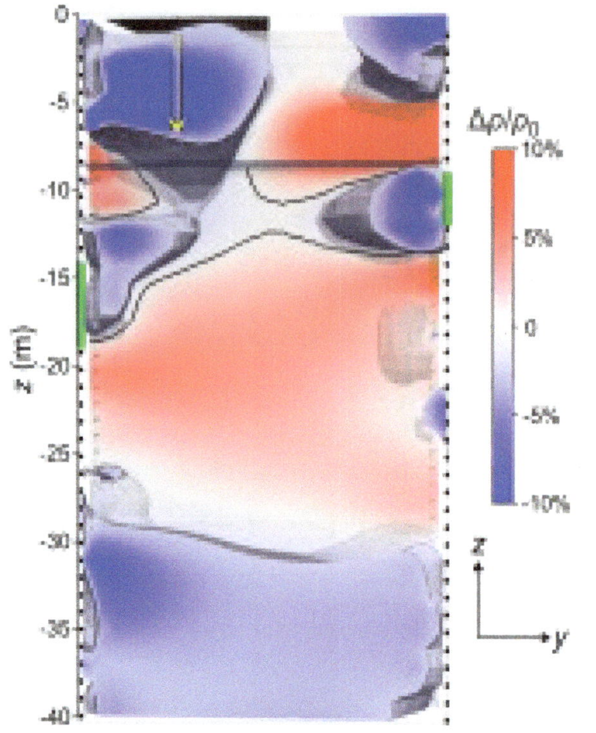

INTRODUCTION

Whilst modern geophysical methodology is increasingly applied at nuclear legacy sites worldwide, the full potential of near-surface geophysics has yet to be fully recognized by the nuclear industry and its regulators. Recent studies have more frequently focused on characterizing historic contamination and hydrological processes (Johnson et al., 2012a,Johnson et al., 2012b and Slater et al., 2010), rather than directly supporting nuclear decommissioning activities (Calendine et al., 2011 and Rucker et al., 2008). At the Sellafield Site in Cumbria, United Kingdom, major long-term decommissioning and remediation projects of national priority and international significance are currently underway that can demonstrably benefit from non-invasive subsurface monitoring approaches. The reduction of risk and hazard at four Sellafield plants known as the Legacy Ponds and Silos is a strategic priority for the UK›s Nuclear Decommissioning Authority (NDA), but presents unique technical challenges. The safe emptying and decommissioning of the Magnox Swarf Storage Silos (MSSS) is a flagship project that Sellafield Ltd. are currently undertaking on behalf of the NDA. The complex and congested industrial setting of this facility demand the development and use of innovative decommissioning and monitoring technologies, both to prepare and to execute the retrieval of wastes from the silos (Nuclear Decommissioning Authority, 2011). The present study demonstrates the value of 4D subsurface geoelectrical monitoring in support of those efforts, and highlights its potential for underpinning risk management and environmental assurance for nuclear decommissioning programs.

Site history, leak scenarios and liquor properties

As nuclear generating capacity within the UK increased in the late 1950s, new Magnox reactors were built and enhanced waste management facilities were brought into service at Sellafield in the early 1960s. Fuel cladding and other solid wastes were stored in water-filled concrete silos (Baldwin, 2003), which over their 25-year operational life received Magnox swarf (debris from fuel decanning) and other wastes from nuclear sites across the UK. The original building was a simple reinforced concrete construction. Three later extensions increased capacity and introduced progressively improved containment standards. All material in the silos is submerged in water and much of the Magnox swarf has degraded into a sludge (BNFL, 1998).

During the 1970s, silo liquor is known to have leaked out of the original building, entering the ground below and creating a plume of contamination (BNFL, 2006). A rate of 3 m^3/day has generally been accepted as the maximum measured rate of historic leakage. While none has been measured since, there is an increased risk that new leakage may occur during waste retrievals due to changes in structural loading caused by the installation and operation of waste retrieval equipment. The historic leak was originally discovered when contamination was identified in an adjacent excavation. Liquor balances have since been closely monitored and measurable leakage is thought to have occurred until 1980/81, when the estimated rate of liquor loss fell to the levels expected for evaporation only. A negative liquor balance of 5 m^3/month (~ 0.17 m^3/day) is currently regarded as the threshold for investigation.

Little information is available about the physical properties of the silo liquor and their variation over time. The electrical properties of the liquor are a function of the aqueous chemistry and temperature of the wastes and are expected to be controlled predominantly by the ionic concentrations of dissolved solids, however few samples have in the past been obtained from the silos, and those only at the liquor surface. Electrical conductivities determined by lowering a sensor in the compartments fell in a range between 865 and 1940 μS/cm (J Graham, pers. Comm.), but these may not be representative of the liquor properties at the base of the silos or at the point of entry into the geosphere.

Environmental monitoring at nuclear sites - a role for geophysics?

Environmental monitoring forms a key component of Sellafield Ltd.'s Ground Environment Management Scheme (GEMS) to assess the impact on groundwater contamination and risk to offsite receptors of contaminants from past and potential future leaks at the facility (Dewey et al., 2014). Prior research within the UK nuclear regulatory framework had highlighted the advantages of geophysical methodologies for leak mitigation (Cummings RLK, 2012 and Emptage et al., 2013). In particular, Electrical Resistivity Tomography (ERT) was identified as a preferential technology for in-ground detection and volumetric monitoring of potential subsurface leakage.

At an international level, there is also a growing body of research that has shown success with geoelectrical techniques for investigating leakage and contaminant transport at nuclear legacy sites, such as the Department of Energy›s Hanford Nuclear Reservation in the United States (Daily et al., 2004, Johnson et al., 2012a, Johnson et al., 2012b,Ramirez et al., 1996, Rucker et al., 2008, Slater et al., 2010 and Truex et al., 2013). In the UK however, ERT had never been used for leak detection at a nuclear licensed site before, and the much greater complexity of the site layout and geological setting at Sellafield compared to Hanford made a successful application of the method appear less likely. The near-surface geological setting at Hanford is dominated by a substantial vadose zone comprising coarse grained, Pleistocene-age cataclysmic flood deposits (the Hanford Formation, containing pebble-to-boulder-size gravels and interbedded sands) that are hydraulically permeable and electrically resistive (Slater et al., 2010). Typical contaminants at Hanford include liquids leaking from underground storage tank farms; unlike MSSS liquors, they are typically fully saturated saline solutions with very high electrical conductivity (Ramirez et al., 1996). The high property contrast between contaminants and geological background therefore lends itself well to the geophysical detection of plumes at Hanford.

Given the much more subtle contrasts expected at Sellafield, it was decided that a full-scale controlled field experiment should be undertaken at the facility, with the aims of assessing the suitability of ERT for leak detection, and monitoring and determining its ‘Technology Readiness’ (Mankins, 1995) for future permanent deployment at Sellafield and comparable sites. Predictive numerical simulations of possible subsurface leak scenarios had demonstrated the feasibility of the approach in principle (Kuras et al., 2011).

Geoelectrical monitoring with ERT

Direct Current (DC) resistivity (specifically ERT) is a key technique in the field of hydrogeophysics, which over the past 25 years has transformed ‘our ability to see into the very fabric of the subsurface environment, and monitor the dynamics of fluids [...] that occur within it’ (Binley et al., 2015). ERT monitoring methodology in particular has seen radical improvements (Loke et al., 2013a); without the most recent advances in data acquisition and inverse geophysical modelling the present study would have been inconceivable. ERT relies upon multiple and repeated measurements of bulk electrical resistance of the soil and subsurface deposits in the region of interest; these are carried out with a buried array of electrodes (‘sensors’) that must cover (as a minimum) the perimeter of the volume of ground to be monitored. Through inverse modelling, ERT then transforms these measurements into volumetric images of ground resistivity. Resistivity is a fundamental geophysical property that is linked to hydrogeological properties and states through property relationships (Binley et al., 2015). Since it has a strong dependence on pore fluid saturation and dissolved contamination, changes in resistivity can be used to study the evolution of leakage plumes both in the vadose zone and in saturated groundwater bodies, provided the aqueous chemistry of the plume is significantly different to that of the native groundwater.

We present results for ERT monitoring of a series of controlled injection experiments, during which simulated silo liquor tracers were released into the vadose zone at locations, depth and release rates commensurate with those that might be expected of a real leak from the facility. The aim of this study is to demonstrate the effectiveness of 4D geoelectrical imaging as a flexible and powerful in-ground detection and volumetric monitoring tool that can provide subsurface information unavailable from conventional techniques, thus aiding the wider nuclear decommissioning effort at critical sites.

MATERIALS AND METHODS

Hydrogeological setting and implications for geophysical leak detection

A detailed understanding of the hydrogeology beneath the storage facility was essential for meaningful conceptualization and quantitative interpretation of tracer migration during the experiment. Much of the present level of geological knowledge about the site stems from the Sellafield Contaminated Land Study (SCLS) (Smith and Cooper, 2004) and follow-on investigation under the Sellafield Contaminated Land and Groundwater Management Project (Cruickshank, 2012). The conceptual ground model is shown inFig. 1. Shallow bedrock beneath the site is of Triassic age and comprises the two uppermost lithostratigraphic units of the Sherwood Sandstone Group. Below the silos, bedrock is encountered at depths of ~ 40–45 m below ground level (bgl); at this location however, borehole information suggests the presence of a ‘trough’ in the bedrock surface topography, which is buried beneath superficial deposits of a comparatively greater thickness relative to adjacent areas of the Sellafield site. Superficial geology comprises Quaternary-aged deposits of variable thickness. A layer of made ground typically extends to depths of 1–3 m bgl, but greater thicknesses are encountered in the direct vicinity of the silos where, during construction of the building, excavations exceeded a depth of 6 m bgl (the base of the silo foundations). Both the sandstone bedrock and the superficials are of hydrogeological significance and classified as aquifers (El-Ghonemey, 2004). Superficial deposits are the principal medium through which contaminant transport occurs in the ground (Smith and Cooper, 2004). Partially saturated conditions prevail in the vadose zone, which is expected to be the primary region of entry into the geosphere for any leakage from the building. The water table below the silos is located at approximately 9–10 m bgl and shows only moderate seasonal variation (typically ± 1 m).

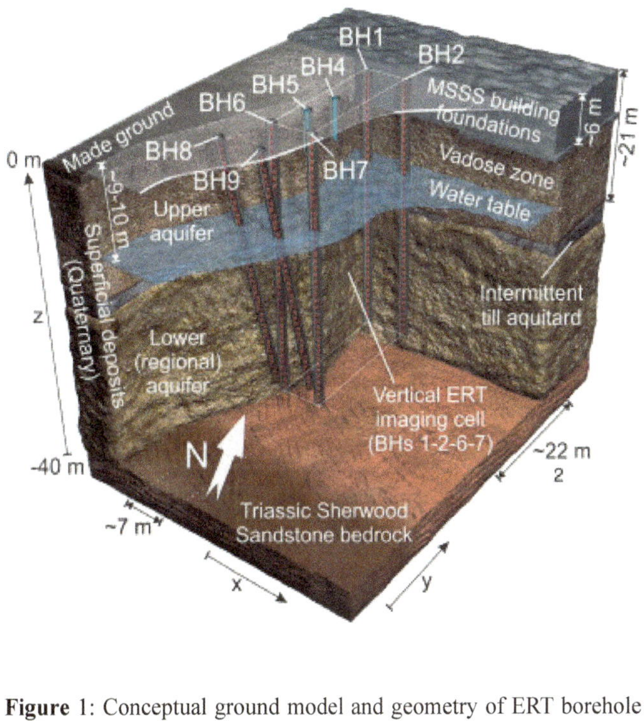

Figure 1: Conceptual ground model and geometry of ERT borehole sensor arrays deployed at the silo facility.

The surviving Quaternary sediments can be attributed to the last major glaciation that affected West Cumbria (the Devensian glaciation: 26–10 ka BP (Merritt and Auton, 2000)), and comprise a complex sequence of glacial and postglacial deposits including tills, alluvial sands and lacustrine sediments. The system has been divided into sixteen hydrostratigraphic units, whose relationship with lithological units (mappable lithofacies as well as lithostratigraphic units used in the 3D geological model of the site) is summarized in Table S1 (Supplementary information). It should be noted that the '3D model lithostratigraphy' referred to here was created purely for modelling purposes using definition of lithofacies units, and in separation to the regional Quaternary lithostratigraphy of Merritt and Auton (2000). Of key hydrological significance is the depth and lateral extent of clay- and/or silt-dominated lithologies that are suspected to restrict vertical permeability at least locally, and, where present, act as barriers to the downward migration of groundwater and associated contaminant transport.

A low-permeability layer ('Glacial till extensive aquitard unit' below ~ 20 m bgl) is responsible for a high degree of hydraulic separation between upper and lower groundwater systems in the superficial deposits (El-Ghonemey, 2004) (Fig. 1). Groundwater in the lower Quaternary system is generally in connection with groundwater in the underlying sandstone, and together they are referred to as 'lower (regional) aquifer' in Fig. 1. Piezometric surfaces constructed for the upper and lower (regional) groundwaters using data from monitoring boreholes surrounding the experimental site had suggested contrasting patterns in average (site-scale) groundwater flow, with flow in the upper aquifer approximately to the east and flow in the regional aquifer approximately to the south-west.

However, the Quaternary deposits at Sellafield are highly variable and not laterally continuous, thus, it is generally difficult to predict hydraulic properties and resulting flow regimes in the shallow subsurface at the scale of an individual building. Simplified predictive models of the groundwater flow and tracer injection that incorporated the available geological knowledge at the time, suggested that a plume was likely to be detectable in the majority of scenarios considered (Kuras et al., 2011). However, that study used a generic translation of hydrological state (tracer concentrations) to geophysical properties (bulk resistivity), whose practical validity is likely to be limited.

The desire to robustly assess the performance of a geoelectrical leak detection system therefore made an in-situ experiment imperative. A central objective of the latter was to enhance our conceptualization of the hydrogeology at the silo facility and to improve our understanding of contaminant transport and likely plume behavior in the shallow subsurface at relevant (meter to sub-meter) scales. Given the difficulty, and hence scarcity, of intrusive investigations in this environment, the ERT experiment offered a unique opportunity to incorporate new borehole lithological information into the geological model of the site.

Experimental design

At the core of our experiment were three controlled injections into the vadose zone of environmentally benign, electrically conductive simulant fluids, continually monitored by automated cross-borehole ERT measurements at daily intervals. The simulants comprised saline tracer solution as a proxy for the silo liquor, and were developed to replicate the range of measured conductivities. The three injections simulated leaks from the base of the silo foundations of moderate electrolytic conductivity contrast (3:1 compared to groundwater conductivity) at a typical leak rate of ~ 0.7 m^3/day (Stage 1); low contrast (2:1) at ~ 0.7 m^3/day (Stage 2); and higher contrast (5:1) at a high (hypothetical) leak rate of ~ 10.8 m^3/day (Stage 3). The Stage 1 injection was designed to last for an intermediate period (38 days) and occur from a location nearer the edge of the building (BH5); the Stage 2 injection lasted a longer time (97 days) and occurred from the building center (BH4); and the Stage 3 injection simulated a more significant leak over a short period (3 days, BH5). A detailed overview of the experimental parameters is given in Table S2 (Supplementary information).

Sellafield process water was used as the basis for the solutions. This process water was dosed with NaCl and NaBr to specific conductivities of 850 μS/cm, 550 μS/cm and 1500 μS/cm at 25 °C, respectively. A target molar ratio of Cl:Br of 10:1 (mass ratio of 4.4:1) was used in order to make the solution sufficiently distinct from the known Sellafield groundwater chemistry.

The experimental procedure covered six key practical steps, including (1) the drilling of boreholes and the installation of the ERT equipment, (2) baseline ERT monitoring to assess the ambient noise conditions at the site, (3) Stage 1 simulant injection and continual ERT monitoring, (4) ERT monitoring to study the decay of the Stage 1 plume and to re-baseline the system, (5) Stage 2 simulant injection, with any remnant Stage 1 plume remaining in the ground, and (6) Stage 3 simulant injection ('high flow'). Finally, extended ERT baseline monitoring was undertaken to capture plume decay and assess seasonal variation of external parameters (noise, sensor health, and hydrology).

Borehole ERT sensor array geometry

Leak detection and monitoring at the silo facility required a focus on regions at or below the base of the building foundations, and ideally a non-destructive capability to detect and characterize potential leakage plumes beneath the building itself. A cross-borehole configuration of ERT sensors fulfills these criteria, with sensors located along boreholes situated on both sides of the building. However, sensitivity and resolution in crosshole ERT imaging are closely linked to and constrained by the separation between boreholes. The aspect ratio for a crosshole panel (separation between boreholes divided by length of electrode string in each borehole) should not exceed 0.75 in order to provide satisfactory image resolution (LaBrecque et al., 1996b). The most favorable imaging conditions are expected for ratios below 0.5 (Wilkinson et al., 2010). The limiting distance for borehole deployment at the silos is the width of the building that is ~ 18 m. Boreholes could not be installed within ~ 2 m of the silo walls, so a minimum 'cross-building' borehole separation of 22 m had to be considered for our experiment. A lateral borehole spacing along the building walls of 7 m was chosen based on realistic numbers of boreholes required for a future monitoring installation covering the original building and first extension.

Below the older sections of the silo facility that are expected to have a greater potential for leakage due to their outdated design and building standards, the silo foundations extend to approximately 6 m bgl, which is well above the water table. Monitoring variations in fluid saturation, determining likely pathways and tracking the direction and timing of fluid flow between this depth and the water table is of particular interest for leak characterization at the facility. Due to the lack of detailed hydrogeological information, it was unclear how quickly any potential contaminant plume would migrate and how deep it would eventually sink. For the purposes of the ERT experiment it was therefore decided to include the entire sequence of superficial deposits in the initial monitoring strategy. Boreholes were therefore extended down to bedrock at ~ 40 m bgl, which also helped satisfy the geometric condition for imaging discussed above (aspect ratio achieved ~ 0.55).

Downhole ERT sensor arrays in six boreholes adjacent to the MSSS structure, but offset to the south, were arranged in an approximately rectangular fashion, simulating deployment on opposite sides of the building (Fig. 1 and Fig. S1, Supplementary information). Four of the boreholes (BH1, 2, 6, 7) were vertical and represented the fundamental ERT imaging cell. Note that we use the term 'imaging cell' to refer to the parameter space bounded by four boreholes (one in each corner). Two further boreholes (BH8, 9) were installed at ~ 12° inclination and oriented perpendicularly towards the silo wall, in order to test deployment in areas of the original building where surface access was more restricted. These, together with the two vertical boreholes adjacent to the Silo (BH1, 2) formed a separate 'inclined imaging cell', for which the geometry was known from simulations to be less favorable. Each sensor array comprised 40 low-carbon grade austenitic stainless steel (UNS S31603) tubular electrodes with an exposed surface area of ~ 44 cm^2, spaced at 1 m separations, resulting in a total of 240 buried sensors available for ERT imaging. Two shallow boreholes (6 m length) for simulant injection were placed between the four deep vertical boreholes. BH4 was located near the center of the area, approximately halfway between the borehole panels BH1–BH2 and BH6–BH7. BH5 was located closer towards the BH6–BH7 panel, at a distance of approximately 4.7 m from BH4. Simulant was injected at a depth of 6 m bgl in either borehole, simulating possible leak locations at the base of the silo foundations (see Table S2, Supplementary information).

Borehole and ERT system installation

The six ERT boreholes and two shallow injection boreholes were installed by rota-sonic drilling between August and October 2012 (Fig. S2, Supplementary information). Temporary steel casing was initially deployed in each borehole. ERT sensor cables mounted onto flexible PVC carrier tubing were carefully lowered into the open boreholes. The casings were then gradually withdrawn, whilst simultaneously backfilling the borehole annulus with extruded bentonite sealing pellets prepared from naturally occurring high purity montmorillonite (sodium activated bentonite). Given the ability of bentonite to swell and self-seal when hydrated by natural waters (here: the groundwater column in the open borehole), this procedure resulted in sensor emplacement that was mechanically stable and provided good galvanic contact between electrodes and the surrounding sediments. Moreover, the hydraulic sealing properties of the bentonite packer were essential to satisfy regulatory requirements. Performance tests showed that only three electrodes out of the 240 installed needed to be excluded from ERT measurements due to poor electrical contact.

Some concerns were raised over the impact of bentonite in the borehole column related to potential current shorting effects due to its low bulk resistivity (~ 2.5 Ωm when saturated with deionized water in the laboratory), and to potential bias in the ERT measurements as a result of induced polarization (IP; ground charging phenomenon) effects associated with the complex conductivity of clays (Okay et al., 2014). However, predictive ERT modelling showed that shorting effects were acceptable and the use of a switched-DC source signal, combined with suppression of early time windows for potential measurements, helped reduce the risk of IP-related bias.

After the boreholes were completed, sensor arrays were connected to BGS-designed geoelectrical monitoring instrumentation known as ALERT technology (Kuras et al., 2009, Ogilvy et al., 2009 and Wilkinson et al., 2010) that was installed in an environmental enclosure at the site. System batteries were recharged from grid power; this arrangement allowed automated scheduled collection of large electrical resistance datasets over long monitoring periods without operator intervention at site. A telemetric link via broadband internet to the BGS offices in Nottingham, UK, enabled fully remote operation and management of the ERT system, including upload of command schedules and regular (daily) download of datasets for analysis and interpretation.

ERT baseline measurements and variability assessment

Initial testing of the ERT system focused on an assessment of data quality that could be expected at the site. As was anticipated in the complex industrial environment at Sellafield, spurious electrical potentials were found to affect the measured data to varying degrees at different times. These potentials were suspected to be a combination of spontaneous potential (SP; electrical fields unrelated to the ERT current injection) and IP effects. In order to reduce their impact on monitoring the processes of interest, the measurement scheme was reorganized, and specific electrodes were excluded from further measurements (2 for vertical/inclined imaging cells and 2 further for inclined only).

After system commissioning, a 2-week period of quiescent baseline measurements was undertaken (20th Jan. – 2nd Feb. 2013), during which no further changes to the setup were made. Throughout the experiment, data acquisition on vertical and inclined imaging cells was scheduled separately on alternating days. A comprehensive measurement sequence was programmed for each cell, resulting in a total number of four-point resistance measurements per dataset of ~ 53,100. Crosshole bipole-bipole configurations were used (instead of configurations with current electrodes in a common hole and potential electrodes in a common hole), as they are known to be less sensitive to previously mentioned borehole effects (Chambers et al., 2007), for example the presence of an electrically conductive bentonite packer. Measurements contained reciprocal configurations (LaBrecque et al., 1996a), where the current and potential dipoles are swapped to assess the reciprocity of the resistance measurement. This allowed the calculation of ~ 13,400 averaged reciprocal measurements per dataset. The typical measurement duration per imaging cell was approximately 22.5 h; this time combined with data retrieval and transmission to the BGS servers meant that a routine 24-h measurement cycle could be achieved at the silo facility.

Error estimates from reciprocal measurements (timescale ~ several hours) were compared with apparent resistivity variations in the data over the baseline measurement period (timescale ~ several weeks), in order to understand short-term and long-term variability at the site under baseline ('no leak') conditions (Fig. 2). Short term variability (blue histograms) was found to be very low, with reciprocal error distributions peaking below 0.1%. Longer term variability of resistivities (orange histograms) was found to be significantly greater, but still acceptable compared with the levels of change caused by simulated leaks that were forecast by desk study numerical models (Kuras et al., 2011). Weather variations (air temperature and rainfall) for the ERT experiment are shown in Fig. S3 (Supplementary information). Over the 2-week quiescent baseline period, they included temperature variations typical for the time of year (average air temperature 3.9° ± 2.7°) and total rainfall of ~ 40 mm. These are not considered major changes, but sufficiently large to be at least partially responsible for the longer term variability observed. Likely effects on the vadose zone electrical response as a result of precipitation-driven soil moisture variations are discussed in Section 3.2.

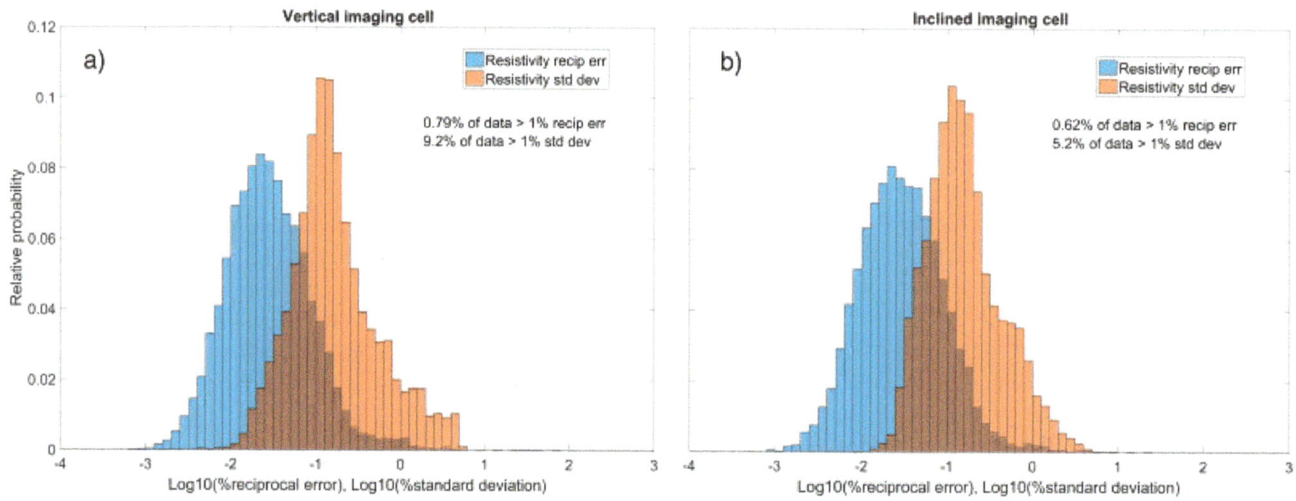

Figure 2: Distributions of the reciprocal error (blue) and apparent resistivity standard deviation (orange) during baseline acquisition, providing measures for short-term variability and for longer-term variability in the study data, respectively. Distributions are shown for (a) vertical and (b) inclined imaging cells (based on 7 ERT datasets each). The reciprocal error is defined as the standard error of the mean of forward (fwd) and reciprocal (rec) measurement, $RE = 100\ |(\rho fwd - \rho rec)/(\rho fwd + \rho rec)|$.

Simulant injection

Once baseline measurements were complete, controlled tracer injections were carried out from a purpose-built pump and tank skid installed near the boreholes. Details of injection timings, indicative duration and simulant properties are described in Table S2 (Supplementary information). Stage 1 and 2 injections were performed 'blind', i.e. the start date and time of the injections were controlled by Sellafield Ltd. and remained unknown to the ERT monitoring team. Continuously repeated ERT cross-borehole measurements using the vertical and inclined imaging cells were made before, during and after the injections allowing us to assess the information content of the ERT data with respect to both the occurrence of the simulated leak as well as the fate of the resulting plume.

ERT data processing

The relative complexity of the Quaternary geology at Sellafield, combined with the presence of clay-rich, electrically conductive sediments, result in limited electrical property contrasts between silo liquor, groundwater and site geology. This was seen as a major constraint for the success of geoelectrical methods at the silo facility. We also recognized that previous applications of ERT to nuclear waste leak detection reported in the literature were subject to significantly more favorable geological and geometrical conditions, for example at Hanford (Daily et al., 2004). Moreover, leak detection based on ERT is challenging in any circumstance as competing (but unrelated) environmental processes are known to affect bulk resistivity, including natural or man-made soil temperature variations, infiltration of surface water into the vadose zone, and electrical noise from plant operation and natural sources.

Nevertheless, initial efforts at MSSS included an attempt to perform leak detection based upon raw resistance data statistics alone, as such an approach had been successfully demonstrated at Hanford (Barnett et al., 2003 and Daily et al., 2004). However, the results of an equivalent analysis at Sellafield were deemed unreliable, and consequently we proceeded by carrying out full 4D ERT inversion (3D time-lapse) to generate volumetric, time-dependent resistivity images of the subsurface in the trial area for further analysis. A 4D spatially and temporally smoothness-constrained inversion algorithm (Res3DInvX64 code by Geotomo Software, Malaysia) was used, implementing the method proposed in (Kim et al., 2009 and Loke et al., 2013b). Our inversions used a model discretization based on a non-uniform rectangular grid to account for the borehole surface locations not being located on a regular grid pattern.

RESULTS AND DISCUSSION

Array sensitivity and baseline model

Given the constraints encountered with placing ERT boreholes at and around the silo building, the sensitivity for the inverse problem (McGillivray and Oldenburg, 1990) is known to vary significantly across the subsurface volume bounded by the ERT sensor array, with the central part of that volume having significantly lower sensitivity than the regions closer to the boreholes themselves. The cumulative sensitivity patterns for the entire field measurement set on the vertical and inclined imaging cells are shown in Fig. 3.

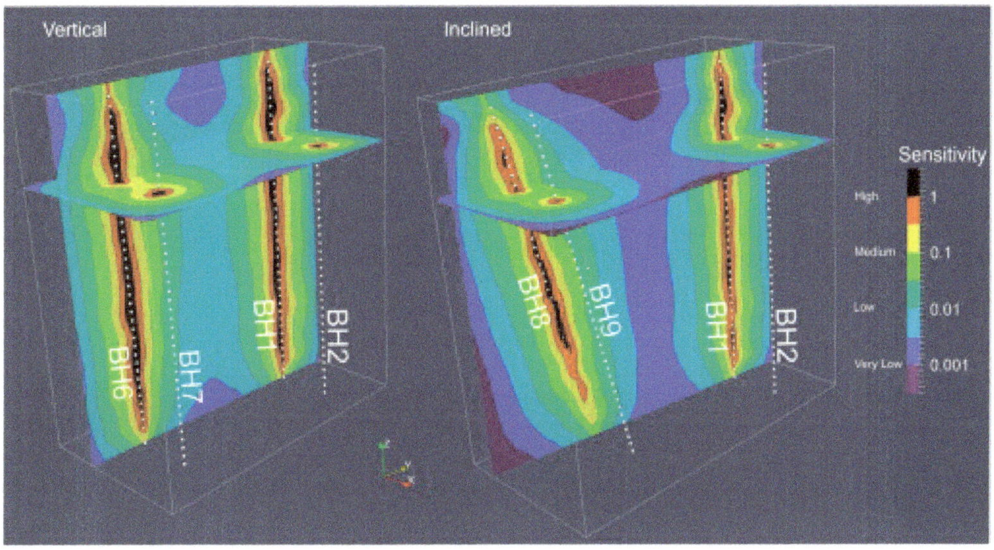

Figure 3: Cumulative sensitivity patterns for the vertical (left) and inclined ERT cross-borehole imaging cells (right).

Absolute images of resistivity obtained during the baseline period (Fig. 4) reflected the heterogeneity of the geological setting at the site, but recovered electrical property distributions broadly compatible with the conceptual ground model (Fig. 1), subject to sensitivity constraints (Fig. 3). The images were also found to correlate well with stratigraphic logs obtained from the inspection of sediment cores from ERT boreholes. Whilst absolute resistivity images can be expected to be sensitive to saturation state, detailed saturation profiles have not been obtained at Sellafield, and uncertainty remains over actual saturation levels within the vadose zone below MSSS. A sharp decrease in resistivity is observed at ~ 7–9 m bgl in both BH1–2 and BH6–7 crosshole panels (right-hand side of Fig. 4), likely associated with an increase in saturation levels near the water table. However, strong dependence of resistivity on sediment texture (particularly clay content) can also be expected. In the absence of a petrophysical model for Sellafield sediments, the arising ambiguity is best resolved by focusing on the analysis of resistivity change images.

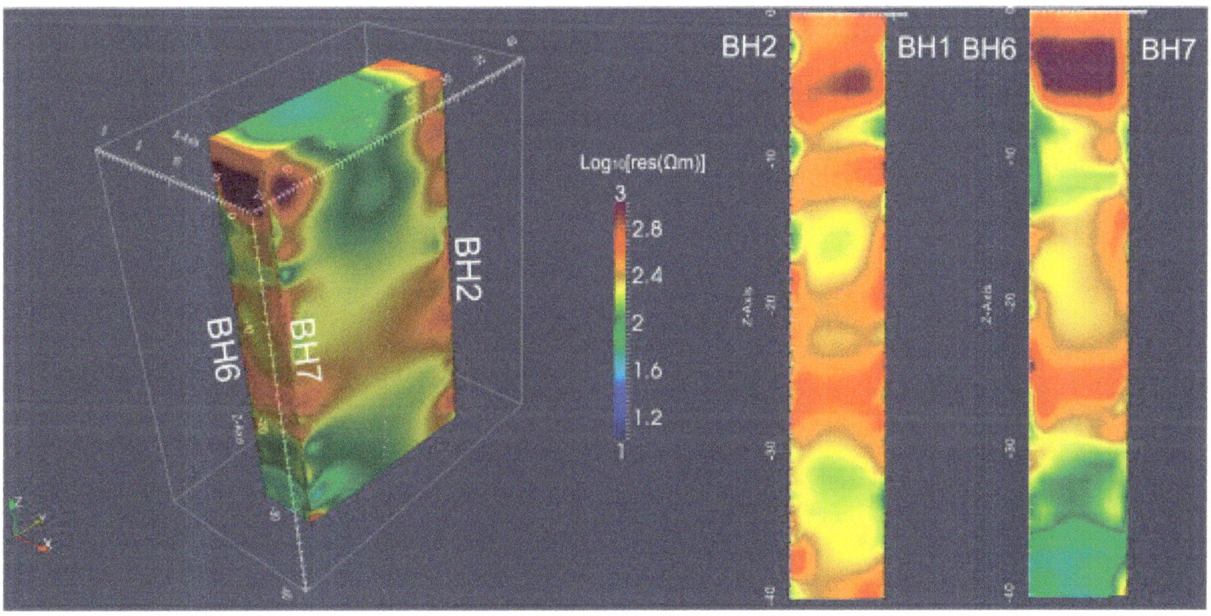

Figure 4: Baseline 3D resistivity image of the ERT Trial area (left), and 2D crosshole panels (right) on the building side (BH1–2) and away from the building (BH6–7).

Simulant injections and analysis by progressive stages

Quiescent baseline

3D fractional resistivity change during the quiescent baseline period and as a result of the Stage 1, 2 and 3 injections is shown in Fig. 5. 4D inverse modelling was performed at regular intervals, and the results were plotted as % changes relative to the inverted model at the start of the most recent injection stage (change in inverted resistivity at time tdivided by inverted resistivity at baseline time). The quiescent baseline results (Fig. 5a) show that resistivity changes over that period are confined to the very shallow subsurface (upper vadose zone). Virtually no changes are observed in the saturated zone, and few isolated changes in the lower vadose zone. It is therefore reasonable to assume that the changes are largely associated with weather variations during the quiescent baseline period, possibly caused by shallow infiltration of rainfall (see Section 2.5) in unpaved areas of the experimental site.

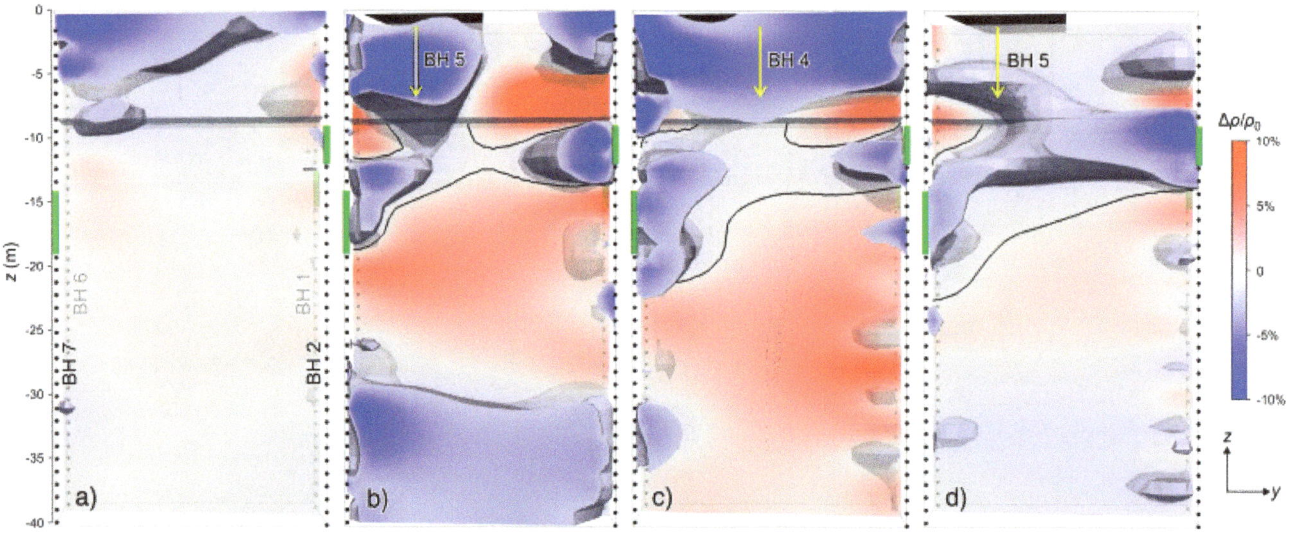

Figure 5: 3D fractional resistivity change (resistivity change Δρ divided by baseline resistivity ρ_0) images showing changes observed during the quiescent baseline period (a), and as a result of the Stage 1 (b), Stage 2 (c) and Stage 3 (d) injections. Results are obtained by performing 4D inverse modelling at regular intervals, and plots show % changes relative to the inverted model at the start of the most recent injection stage (change Δρ in inverted resistivity at time t divided by inverted resistivity ρ_0 at baseline time). Baseline dates are 13th Jan. 2013 for quiescent baseline, 5th Feb. 2013 for Stage 1, 28th May 2013 for Stage 2, 16th Sep. 2013 for Stage 3. Approximate locations of injections are shown by yellow arrows. Red indicates resistive changes, i.e. increases relative to baseline resistivity, while blue reflects conductive changes. The black line is the 0% change contour, which is assumed to delineate the inferred pathway; green cylinders show windows of historic contamination found in sediment cores from ERT boreholes. (For interpretation of the references to color in this figure legend, the reader is referred to the web version of this article.)

Stage 1 injection

Simulant with a moderate electrical conductivity contrast was injected via BH5, but at a low leak rate. After ~ 4 weeks of continued injection for Stage 1, a consistent picture emerged; regions of increased conductivity developing in the time-lapse ERT images had strengthened in contrast and formed a coherent volume in the vadose zone and upper saturated zone, bounded by the 0% change contour (Fig. 5b). Its location was consistent with the known injection point (base of BH5, yellow arrow) and the assumption that the simulant plume will experience gravitational sinking due to its greater density, whilst broadly responding to the ambient lateral hydraulic gradients of the shallow groundwater system. Whilst the ERT images will allow interpretation of flow patterns on the scale of the imaging cell, deduction of more regional (site-scale) flow directions from one cell alone is not possible – only additional ERT boreholes (e.g. multiple imaging cells) could provide a capability to determine the fate of the plume on a larger (site) scale.

The strongest features of conductive change in the upper saturated zone were imaged in the immediate vicinity of ERT boreholes 1, 2, and 7 within narrow depth ranges that were consistent with windows of historic contamination (green cylinders), which had been detected in gravel-dominated (and hence likely more permeable) facies within sediment core recovered from the boreholes. No changes had been observed in this region during the quiescent baseline. This led us to infer the presence of a preferential hydraulic flow path, which appears to have been occupied by contaminants leaking from the silo facility in the past, and which has subsequently been re-occupied by the Stage 1 simulant. No appreciable decay of the conductive plume signature was observed over the space of ~ 10 weeks of routine monitoring following the end of the Stage 1 injection, which indicates either long residence times of the plume itself or the retention of residual conductivity in the vadose zone pore spaces after the main plume had passed. The conductive changes towards the base of the images are unconnected; they are thought to be due to seasonal variations in the electrical conductivity of the regional groundwater body (lower Quaternary plus sandstone bedrock aquifer), which is separated from the upper aquifer by a low permeability till layer (Fig. 1).

We found that ERT inversion was essential for this interpretation, as the statistics of conductive changes in the raw resistance data relative to the baseline alone could not be attributed unambiguously to the injection. Instead, we observed much stronger correlation of such changes with unrelated (but clearly relevant) processes in the near surface such as rainfall and subsequent infiltration into the vadose zone (see quiescent baseline), surface runoff, and infiltration of dissolved road de-icing salt following snowfall, and variations in ground temperature driven by air temperature changes and natural gradients in the subsurface. Furthermore, statistical analysis alone is unable to provide spatial information; leak path geometry can only be recovered through inverse modelling of the subsurface.

Stage 2 injection

For this stage, simulant with the lowest conductivity contrast was injected into BH4 at the same low leak rate used in Stage 1, partly to assess the extent to which ERT could distinguish separate events in a 'leak-on-leak' scenario. However, expectations were modest as the Stage 2 plume behavior was likely to be complex, and leak-on-leak scenarios had not been simulated prior to the experiment. Once again, no significant response was noted for a period of at least ~ 4 weeks. Further negative changes (i.e. relative increases in conductivity) were imaged eventually (Fig. 5c), which, compared with Stage 1, had increased in strength and appeared in new regions, consistently enveloping the Stage 1 plume volume. This evidence demonstrates the successful detection of the lower contrast Stage 2 leak against the background of the Stage 1 leak, which was a welcome result, given the challenges of detecting a low-contrast tracer in a pathway already occupied by the stronger previous tracer.

Stage 3 injection

This final stage involved the introduction of simulant with a significantly higher conductivity and at a high leak rate, reflecting the most pessimistic leak scenario. Stage 3 was designed to corroborate the results of the previous two stages, as it was expected that the existing flow path would be reoccupied once more and that the cumulative effect of the more conductive simulant would overcome the limitations of the low-sensitivity array geometry, by consistently highlighting flow paths throughout the ERT model. The Stage 3 injection lasted three days, during which approximately 34 m^3 of simulant were released into the vadose zone via BH5. Resistivity ratio images for Stage 3 responded very quickly (~ 1 week) after the injection event, and the change evolution over the subsequent 6 weeks showed that the Stage 1 flow path was indeed occupied once again (Fig. 5d). The resistivity ratio isosurface indicating a 2% conductive change was now connected across the model and provided an intuitive visual representation (as well as a quantitative measure) of inferred fluid flow in the shallow subsurface beneath the experimental area. More detailed quantitative interpretation is difficult without knowledge of hydraulic properties of the pathway and the subsurface adjacent to (but outside) the imaging volume. The cumulative nature of the injections and uncertain residence times in the unsaturated zone add complexity. It is also likely that the higher injection rate of the Stage 3 injection will create stronger hydraulic gradients, forcing the simulant to dissipate more rapidly into the preferential pathway compared to Stages 1 and 2. This may also cause a larger proportion of the simulant to exit the volume imaged by the borehole sensor array and thus become invisible to the present experimental setup. A more comprehensive monitoring deployment comprising multiple adjacent imaging cells would address this shortfall.

Extended baseline monitoring and full trial analysis

After the staged injections, ERT monitoring continued for a further 12 months with the aim of tracking the fate of the simulant plume that was expected to decay and disperse over time. Our goal was to explore current practical limits of our inversion strategy that required simultaneous 4D inversion of a time series of consecutive 3D datasets. It was found that, given the density and nature of the ERT monitoring data, a sequence of not more than 48 full datasets, spread approximately evenly throughout the full duration of the trial, could be simultaneously inverted with the available algorithm and computational resources. Typical inversion times for full 48-dataset inversions were below 22 h. Given that data acquisition itself requires ~ 24 h per imaging cell and data pre-processing time was negligible, a typical time frame of < 48 h from start of data acquisition to an inverted result was achievable. This was deemed sufficient to provide adequate temporal resolution for leak events within the suspected range of leak rates, and at the same time allow detection of subtle changes in subsurface properties and structure over relevant time periods of many months. For a larger monitoring system comprising multiple imaging cells the overall numerical demands would increase significantly, but inversion strategies could be tailored to cover partial imaging volumes at a time, whilst computational resources could be increased.

An issue with noticeable distortion of the original Stage 2 image (not shown) in the vicinity of BH1, was found to be due to water ingress in an electrical connector on BH1 at the surface that had suffered mechanical damage from activity unrelated to the ERT experiment. It was not possible to repair the connector for the remainder of the experiment. Therefore to investigate the full duration of the experiment, including the three injection stages and the monitoring of the fate of the tracers, all data involving any BH1 electrode was removed from the inversion. A subset of the resulting change images for the entire monitoring period of ~ 2 years is shown in Fig. 6.

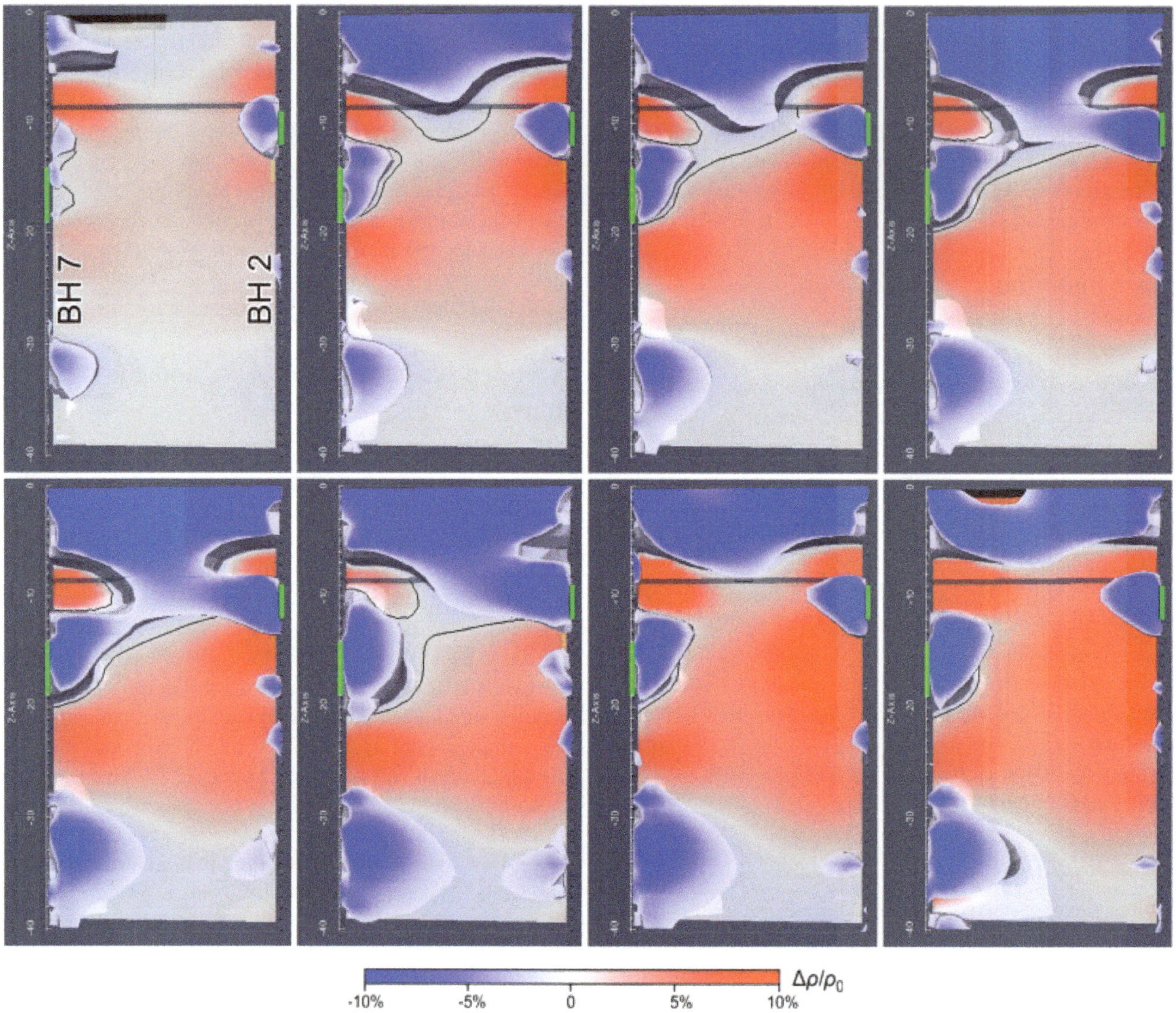

Figure 6: Entire monitoring sequence fractional resistivity changes in inverted vertical cell images for 28th Mar. 2013, 28th May 2013, 18th Jul. 2013, 13th Sep. 2013, 4th Nov. 2013, 10th Mar. 2014, 27th Jul. 2014, 2nd Dec. 2014 (left-to-right, top-to-bottom). Changes plotted relative to 5th Feb. 2013, no BH1 data used. Contours at the approximate boundaries of conductive changes highlight the inferred preferential pathway over time.

Revised geological model and enhanced conceptualization

A revised geological model of the MSSS area was built using existing stratigraphic information from around 190 boreholes in the wider Sellafield Separation Area, combined with new lithology data from the eight ERT trial boreholes. The model was generated using the GSI3D software and workflow (Kessler et al., 2009) and comprises around 30 geological cross-sections, which were defined using boreholes from the above dataset and manually correlated using a modelling lithostratigraphy based on the 16 separate geological units described earlier (Table S1, Supplementary information). The resulting model is a fully digital 3D spatial model of the lithostratigraphy in the study area; it can thus be interrogated at arbitrary locations to create synthetic borehole logs. We found that this form of structural correlation across the trial area offered enhanced spatial context and new conceptual understanding of the relationships between sand-dominated and gravel-dominated facies and the clay-rich till layers present in the area. Crucially, it also allowed direct and interactive comparison between the interpreted geology, measured electrical properties and hydrogeological monitoring information. Fig. 7presents the % change ERT image following the Stage 1 injection, superimposed on a geological cross-section through the trial area, which shows the lithostratigraphic context in the region between boreholes BH7 and BH1. Amongst other features, the section shows the laterally coherent 'Glacial till extensive aquitard unit' ('Lower Till 1', light blue) extending across the entire section at approximately constant depth (~ 23 m bgl). It is thought that this low-permeability layer is responsible for providing a high degree of hydraulic separation between the upper and lower groundwater systems, as observed in previous studies (El-Ghonemey, 2004). Other till units are clearly not as extensive and appear to provide more localized hydraulic barriers, potentially diverting or retarding flow in a particular direction. One example is the 'Glacial till local aquitard unit' observed at ~ 12 m bgl in BH7 ('Upper Till 2'), which appears to locally divert flow in the inferred pathway downwards (Fig. 7).

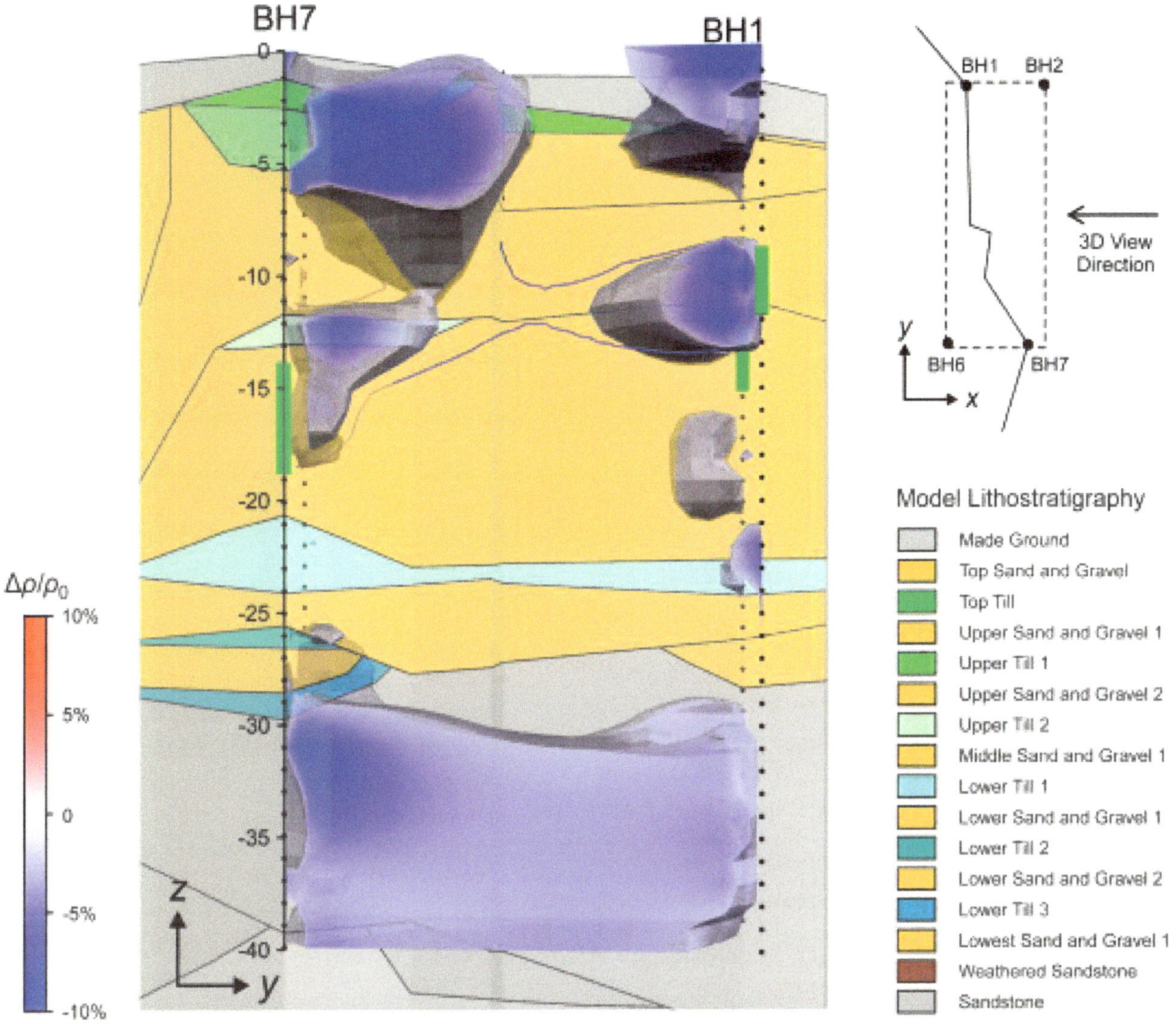

Figure 7: Spatial correlation of geoelectrical monitoring results with the revised 3D geological model of the experimental area. Fractional resistivity change image following Stage 1 injection superimposed on geological cross-section through BH9-BH710205-BH1. Green cylinders along the boreholes indicate regions of historic contamination detected in sediment cores. The solid blue contour line in the resistivity image reflects the 0% change boundary and serves to outline the inferred preferential pathway. (For interpretation of the references to color in this figure legend, the reader is referred to the web version of this article.)

CONCLUSIONS

Automated ERT monitoring was applied at a UK nuclear licensed site for the first time and a full-scale field experiment trialing the methodology at MSSS in Sellafield successfully detected, characterized, and tracked simulated leaks. Contrary to experience obtained by previous studies under more favorable conditions at US sites, at Sellafield we were unable to discriminate simulated leaks from unrelated processes using simple metrics based on raw data statistics alone. Instead, we have shown in this study that careful experimental design and full 4D ERT imaging can overcome the lack of sensitivity associated with challenging borehole sensor geometries and the poor property contrasts presented by UK Quaternary geology. Images of resistivity changes relative to a baseline date prior to a leak have revealed likely pathways of simulant flow in the vadose zone and upper groundwater system. The geophysical evidence was found to be compatible with historic contamination detected in permeable facies in borehole sediment cores, and with a geological model based on wider scale borehole stratigraphy. Results showed that laterally discontinuous till units forming localized hydraulic barriers can substantially affect flow patterns and contaminant transport in the shallow subsurface at Sellafield. Thus, when compared with traditional methods of leak detection at nuclear sites, only geophysical imaging (of the kind presented here, but not necessarily limited to ERT) can provide the detailed spatial and temporal information at the (sub-meter) scale needed to reduce uncertainty in models of subsurface processes, particularly where they underpin decision-making in nuclear risk management. Our experiment at one of the most challenging nuclear legacy facilities in the world has shown how routine operation of a permanent ERT monitoring system could support critical decommissioning activities over the coming decades. Full-scale permanent systems covering entire buildings or subsurface structures would require about an order-of-magnitude more boreholes and ERT sensors, and a proportionally higher effort in terms of data collection and processing, thus setting new challenges for the future direction of R&D in geoelectrical monitoring.

ACKNOWLEDGMENTS

This paper is published with the permission of the Executive Director of the British Geological Survey (NERC) and Sellafield Ltd. (on behalf of the Nuclear Decommissioning Authority). Arnaud Watlet and Peter Bergmann provided detailed reviews that helped improve an earlier version of the manuscript.

REFERENCES

1. BNFL, 1998. Engineering developing Sellafield drypack process with world in mind. Nuclear Engineering International, Global Trade Media, London Available from http:// www.neimagazine.com/features/featurebnfl-engineering-developing-sellafielddrypack-process-with-world-in-mind/ (accessed 29 March 2016).
2. BNFL, 2006. Decommissioning B38. New Civil Engineer, Emap, London, Available from http://www.newcivilengineer.com/decommissioning-b38/484772.article (accessed 29 March 2016).
3. Baldwin, N.D., 2003. Remediating Sellafield - A New Focus for the Site. Waste Management (WM) Conference. WM Symposia, Tucson, AZ.
4. Barnett, D.B., Gee, G.W., Sweeney, M.D., Johnson, M.D., Medina, V.F., Mendoza, D.P., et al., 2003. Results of Performance Evaluation Testing of Electrical Leak-Detection Methods at the Hanford Site Mock Tank – FY 2002–2003 Pacific Northwest National Laboratory.
5. Binley, A., Hubbard, S.S., Huisman, J.A., Revil, A., Robinson, D.A., Singha, K., et al., 2015. The emergence of hydrogeophysics for improved understanding of subsurface processes over multiple scales. Water Resour. Res. 51, 3837–3866.
6. Calendine, S., Rucker, D.F., Fink, J.B., Levitt, M.T., 2011. Automated leak detection of buried tanks using geophysical methods at the Hanford Nuclear site. Symposium on the Application of Geophysics to Engineering and Environmental Problems. EEGS, Charleston, SC
7. Chambers, J.E., Wilkinson, P.B., Weller, A.L., Meldrum, P.I., Gilvy, R.D., Caunt, S., 2007. Mineshaft imaging using surface and crosshole 3D electrical resistivity tomography: a case history from the East Pennine Coalfield, UK. J. Appl. Geophys. 62, 324–337.
8. Cruickshank, J., 2012. Findings of the Sellafield Contaminated Land & Groundwater Management Project and the Next Steps for the Land Quality Programme. Land Quality Report, Sellafield Ltd.
9. Cummings RLK, 2012. Options for Leak Management from Nuclear Legacy Facilities during Decommissioning. Royal Holloway University of London. p. 91.
10. Daily, W., Ramirez, A., Binley, A., 2004. Remote monitoring of leaks in storage tanks using electrical resistance tomography: application at the Hanford Site. J. Environ. Eng. Geophys. 9, 11–24.
11. Dewey, G., Atherton, N., Ball, T., Kuras, O., Wilkinson, P., Meldrum, P., 2014. The Ground Environment Management Scheme (GEMS): Development of Technologies for Detecting and Monitoring Subsurface Leakage and Contaminant Transport, Supporting the Decommissioning of Legacy Silos at the Sellafield Site, UK. Waste Management (WM) Conference. WM Symposia, Phoenix, AZ, p. 14160.
12. El-Ghonemey, H., 2004. SCLS Groundwater Conceptual Model. Nuclear Sciences and Technology Services. BNFL.
13. Emptage, M., Hepworth, S., Winspear-Roberts, V., Cummings, R., 2013. The leak management hierarchy. Decommissioning Challenges: An Industrial Reality and Prospects. Société Française d'Energie Nucléaire, Avignon, France.
14. Johnson, T.C., Slater, L.D., Ntarlagiannis, D., Day-Lewis, F.D., Elwaseif, M., 2012a. Monitoring groundwater-surface water interaction using time-series and time-frequency analysis of transient three-dimensional electrical resistivity changes. Water Resour. Res. 48.
15. Johnson, T.C., Versteeg, R.J., Rockhold, M., Slater, L.D., Ntarlagiannis, D., Greenwood, W.J., et al., 2012b. Characterization of a contaminated wellfield using 3D electrical resistivity tomography implemented with geostatistical, discontinuous boundary, and known conductivity constraints. Geophysics 77, En85–En96.
16. Kessler, H., Mathers, S., Sobisch, H.G., 2009. The capture and dissemination of integrated
17. 3D geospatial knowledge at the British Geological Survey using GSI3D software and methodology. Comput. Geosci. 35, 1311–1321.
18. Kim, J.H., Yi, M.J., Park, S.G., Kim, J.G., 2009. 4-D inversion of DC resistivity monitoring data acquired over a dynamically changing earth model. J. Appl. Geophys. 68, 522–532.
19. Kuras, O., Pritchard, J.D., Meldrum, P.I., Chambers, J.E., Wilkinson, P.B., Ogilvy, R.D., et al, 2009. Monitoring hydraulic processes with automated time-lapse electrical resistivity tomography (ALERT). Compt. Rendus Geosci. 341, 868–885.
20. Kuras, O., Wilkinson, P.B., White, J.C., Chambers, J.E., Meldrum, P.I., Ogilvy, R.D., 2011. MSSS Leak Mitigation - Leak Detection Phase 3: Desk Study for ERT Technology. Commissioned Report CR/11/053. British Geological Survey.
21. LaBrecque, D.J., Miletto, M., Daily, W., Ramirez, A., Owen, E., 1996a. The effects of noise on Occam's inversion of resistivity tomography data. Geophysics 61, 538–548.
22. LaBrecque, D.J., Ramirez, A.L., Daily, W.D., Binley, A.M., Schima, S.A., 1996b. ERT monitoring on environmental remediation processes. Meas. Sci. Technol. 7, 375–383.
23. Loke, M.H., Chambers, J.E., Rucker, D.F., Kuras, O., Wilkinson, P.B., 2013a. Recent developments in the direct-current geoelectrical imaging method. J. Appl. Geophys. 95, 135–156.
24. Loke, M.H., Dahlin, T., Rucker, D.F., 2013b. Smoothness-constrained time-lapse inversion of data from 3D resistivity surveys. Near Surf. Geophys. http:// dx.doi.org/10.3997/ 1873-0604.2013025.

25. Mankins, J.C., 1995. Technology Readiness Levels - A White Paper. Office of Space Access and Technology, NASA.

26. McGillivray, P.R., Oldenburg, D.W., 1990. Methods for calculating Frechet derivatives and sensitivities for the nonlinear inverse problem - a comparative-study. Geophys. Prospect. 38, 499–524.

27. Merritt, J.W., Auton, C.A., 2000. An outline of the lithostratigraphy and depositional history of Quaternary deposits in the Sellafield district, west Cumbria. Proc. Yorks. Geol. Soc. 53, 129–154.

28. Nuclear Decommissioning Authority, 2011. NDA Strategy, Effective from April 2011. United Kingdom Nuclear Decommissioning Authority.

29. Ogilvy, R.D., Meldrum, P.I., Kuras, O., Wilkinson, P.B., Chambers, J.E., Sen, M., et al., 2009. Automated monitoring of coastal aquifers with electrical resistivity tomography. Near Surf. Geophys. 7, 367–375.

30. Okay, G., Leroy, P., Ghorbani, A., Cosenza, P., Camerlynck, C., Cabrera, J., et al., 2014. Spectral induced polarization of clay-sand mixtures: experiments and modeling. Geophysics 79, E353–E375.

31. Ramirez, A., Daily, W., Binley, A., LaBrecque, D., Roelant, D., 1996. Detection of leaks in underground storage tanks using electrical resistance methods. J. Environ. Eng. Geophys. 1, 189–203.

32. Rucker, D.F., Levitt, M.T., Myers, D.A., Henderson, C., 2008. Development of an Electrical Resistivity Imaging Program for Subsurface Characterization at Hanford. Waste Management (WM) Conference. WM Symposia, Phoenix, AZ, p. 8193.

33. Slater, L.D., Ntarlagiannis, D., Day-Lewis, F.D., Mwakanyamale, K., Versteeg, R.J., Ward, A., et al., 2010. Use of electrical imaging and distributed temperature sensing methods to characterize surface water-groundwater exchange regulating uranium transport at the Hanford 300 Area, Washington. Water Resour. Res. 46.

34. Smith, N., Cooper, S., 2004. SCLS Phase 1 - Sellafield Geological Conceptual Model. Nuclear Sciences and Technology Services. British Nuclear Fuels Limited (BNFL).

35. Truex, M.J., Johnson, T.C., Strickland, C.E., Peterson, J.E., Hubbard, S.S., 2013. Monitoring vadose zone desiccation with geophysical methods. Vadose Zone J. 12.

36. Wilkinson, P.B., Meldrum, P.I., Kuras, O., Chambers, J.E., Holyoake, S.J., Ogilvy, R.D., 2010. High-resolution electrical resistivity tomography monitoring of a tracer test in a con- fined aquifer. J. Appl. Geophys. 70, 268–276.

CHAPTER 3

DEVELOPING 3-D MINE-SCALE GEOMECHANICAL MODELS IN COMPLEX GEOLOGICAL ENVIRONMENTS, AS APPLIED TO THE KIIRUNAVAARA MINE

J. Vatcher [a], S.D. McKinnon [b], J. Sjöberg [c]

[a] Division of Mining and Geotechnical Engineering, Luleå University of Technology, SE-971 87 Luleå, Sweden

[b] The Robert M. Buchan Department of Mining, Goodwin Hall, 25 Union St. Queen's University, Kingston, Ontario, K7L 3N6, Canada

[c] Itasca Consultants AB, Aurorum 2, SE-977 75 Luleå, Sweden

ABSTRACT

An understanding of the relationship between the geological environment and rock mass behaviour induced by mining activities can lead to hazard reduction through knowledge-based design. However, characterisation of complex and heterogeneous rock masses that typify mining environments is difficult. A methodology to characterise these types of rock masses, based largely on classical statistics, geostatistics and an extension of previous quantitative structural domaining work, is presented and applied to the Kiirunavaara Mine, Sweden. In addition to a new perspective on intact rock strengths of geological units at the mine, a correlation was found between modelled volumes of clay, modelled RQD, newly identified structural domains and falls of ground. These relationships enabled development of a conceptual model of the role of geology in rock mass behaviour at the mine. The results demonstrate that the proposed methodology can be useful in characterisation of complex rock masses.

INTRODUCTION

Mining induced rock mass behaviour is a result of the interaction between the mining and geological environments. An understanding of the role of 3-D geomechanical characteristics in rock mass behaviour can lead to significant opportunities for risk-mitigating engineering design. Typically, geomechanical models of mining environments are developed by using classification methods such as Geological Strength Index (GSI) or Rock Mass Rating (RMR). These methods were originally developed and calibrated for civil engineering excavations, such as tunnels, and are well suited to homogeneously jointed rock masses. Alone, however, they are unsuited to heterogeneous rock masses in which the strength and stiffness is affected by characteristics in addition to discontinuities, which typify the geological environment of many mines. Adaptations of GSI for heterogeneous rock masses have been developed (Hoek and Karzulovic, 2001 and Marinos and Hoek, 2001); however, they are still mostly based upon planes of weakness, and as identified by Mandrone (2006), are less applicable when weakness exists in other forms, such as alteration. Over the life of a mine, geomechanically relevant data is often collected for different purposes, and there can be difficulty synthesizing these multiple sources of data into a coherent geomechanical model. Also, data availability and coverage issues specific to mining caused by economical and physical constraints (such as shotcreting limiting available mapping faces) often compound the difficulties associated with this process.

Citation: J. Vatcher, S.D. McKinnon, J. Sjöberg, Developing 3-D mine-scale geomechanical models in complex geological environments, as applied to the Kiirunavaara Mine, Engineering Geology, Volume 203, 25 March 2016, Pages 140-150, ISSN 0013-7952, http://dx.doi.org/10.1016/j.enggeo.2015.07.020.

This paper proposes a methodology to create a 3-D mine-scale geomechanical model for complex and heterogeneous rock masses using readily available data from mines, such as geologically logged core, Rock Quality Designation (RQD) and mapped discontinuities. Luossavaara-Kiirunavaara Aktiebolag's (LKAB) Kiirunavaara Mine, in northern Sweden, is used to illustrate the techniques. This geomechanical model created for the mine is then compared to fall of ground information, with the intention of developing a conceptual model of causes of rock mass behaviour and identifying indicators of problem areas. In 2008, the Kiirunavaara Mine became seismically active (refer to Dahnér et al. (2012) for a detailed description). Several studies were undertaken by LKAB to understand the underlying causes and nature of the behaviour (Sjöberg et al., 2011, Sjöberg et al., 2012 and Vatcher et al., 2014). Recent underground mapping, core logging and laboratory testing campaigns done in support of these projects have revealed that the rock mass is heterogeneous, with variable characteristics across both the 5 km length and > 1000 m depth of the orebody. An additional potential complication for geomechanical characterisation of the rock mass is that there are volumes of clay alteration, which are often meters to tens of meters in their longest dimension.

Previous studies at the Kiirunavaara Mine have been limited in scope with respect to development of a geomechanical model. The majority of these studies were limited in breadth of analysis, often focusing on portions of the available data at the time of each study (Henry and Marcotte, 2001, Lindgren, 2013, Mattsson et al., 2010 and Rådberg et al., 1989). Much additional data collection and analysis has been done since these studies with the intention of increasing the geomechanical understanding of the rock mass, providing a good opportunity to enhance knowledge.

The data analysis techniques presented to create large-scale 3-D geomechanical models were intentionally selected for their compatibility with characteristics of the geology and typical limitations associated with data from mining environments, such as sparse coverage and few samples. The methodology utilises standard statistical tests, geostatistics, and an extension of previously published techniques related to quantifiable identification of structural domains.

KIIRUNAVAARA MINE

The Kiirunavaara Mine is a large sublevel caving mine that produces approximately 28 million tonnes per annum of iron ore. With the deepest production levels currently at approximately 800 m below surface (Level 1051 m), the rock mass is seismically active and rockburst prone. The newest main haulage level is Level 1365 m (see Fig. 1), and sublevels currently have a 28.5 m separation. Mine coordinates are approximately aligned with the cardinal directions, and the Y coordinates (positive axis aligned with south), which crosscut the orebody, are used to divide the orebody into blocks for production and ore handling (see the numbered ore pass groups along the orebody inFig. 1).

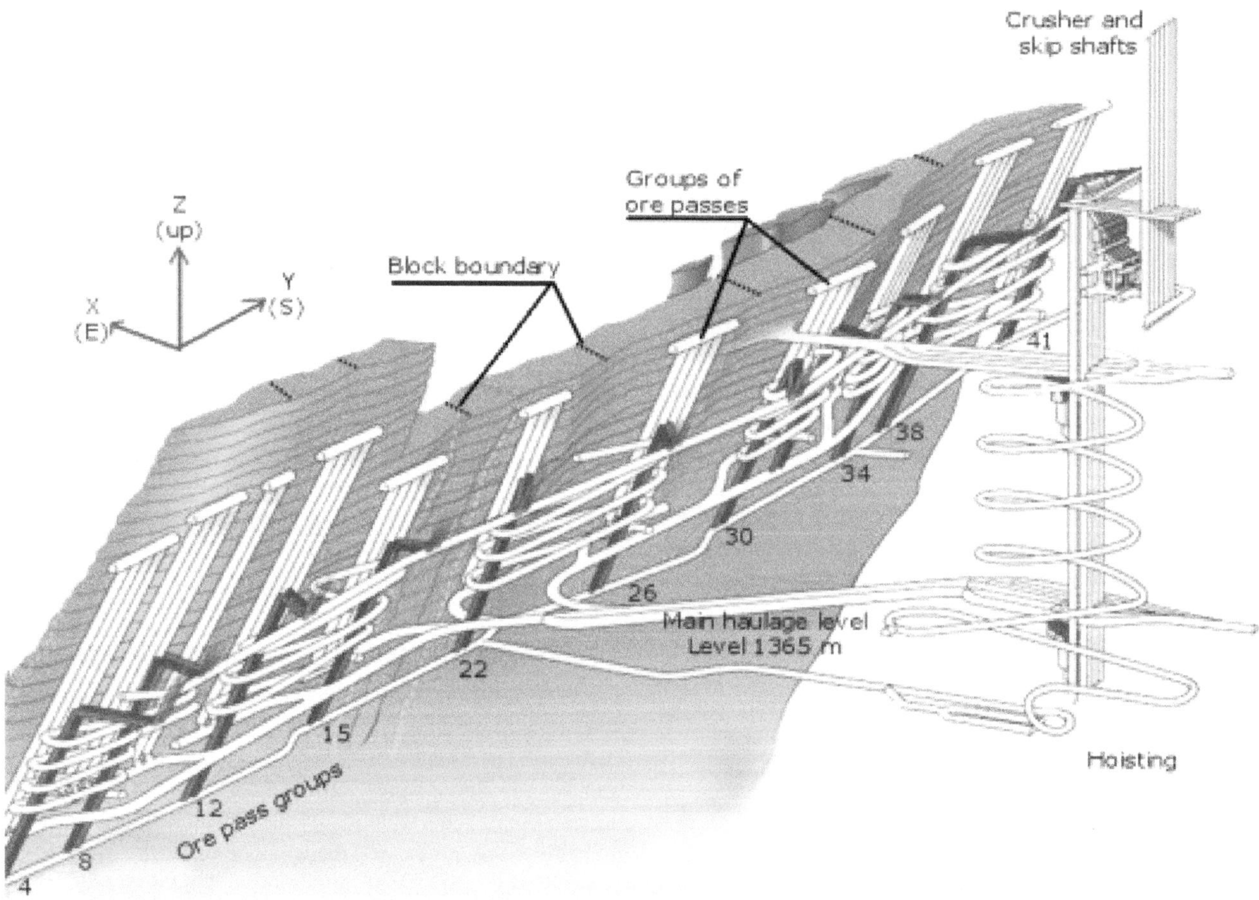

Figure 1: Sketch of the orebody and mine layout. Production blocks and associated ore pass groups are numbered based on their Y coordinate (numbered at the draw points in this image). Modified image from LKAB.

The magnetite orebody extends approximately 5 km along strike, which is nearly north–south, with a varying width from meters to over 150 m and a dip of 50-70° towards the east (positive X-axis in mine coordinates). For a detailed description of the geology, refer to Geijer (1910). The most specific geological classifications at the mine are done during core logging, whereas underground geological mapping uses more generic geological classifications. The mine distinguishes ore rock types based on their grade and contaminant concentration. Footwall rock types are divided into three major classifications; granite, skarn or trachyte-trachyandesite (referred to as syenite porphyries at the mine). The trachyte-trachyandesite is further divided into 5 subgroups based on their mineralogy, texture and/or alteration, referred to as Sp1 through Sp5. Porphyry dykes are also common in the footwall region. Hangingwall material, mainly rhyodacite (referred to as quartz-bearing porphyries at the mine), is similarly divided into 5 subcategories (Qp1 through Qp5) based on their mineralogy, texture and/or alteration. Underground geological mapping identifies the ore units based on grade and contamination (the same units as mapped during core logging), however footwall and hangingwall materials are mapped in less detail underground than they are during core logging. Fewer hangingwall data are available by either mapping or core logging due to underground access and ore-targeted drilling.

Some of the rock mass at the Kiirunavaara Mine has undergone significant alteration. This alteration is in the form of both replacement of minerals by clay as well as leaching, leaving the rock porous (Berglund and Andersson, 2013). The clay-altered lens-shaped volumes are visible underground and in core from diamond drilling (as clay and possibly as core losses), and their extent ranges from centimeters to tens of meters throughout the mine.

DATA AND METHODOLOGY

The selection of data was based on potential importance to geomechanical characterisation, availability, and reliability. Data analyzed in the development of the geomechanical model of the mine was selected because of their possible relations to intact rock strength, zones of strength and stiffness contrasts, rock mass quality and structural domains. The forms of these data sources available at the Kiirunavaara Mine are described below. Some of this data is routinely collected at the mine, and some was collected during specifically targeted campaigns in support of this and other projects at the mine.

The diamond drill core database consists of approximately 590 000 m of mostly 28 mm diameter core, from approximately 3000 boreholes (LKAB, 2014a). The extent of the data is shown by Volume A in Fig. 2. With few exceptions, all drilling is from underground, with fans of four holes on average at 50 m spacing along strike targeting orebody definition. The majority of drilling is done from the footwall due to access restrictions. On surface, geologists log these drill holes, identifying geological units, core losses and zones of poor material (such as clay), and RQD.

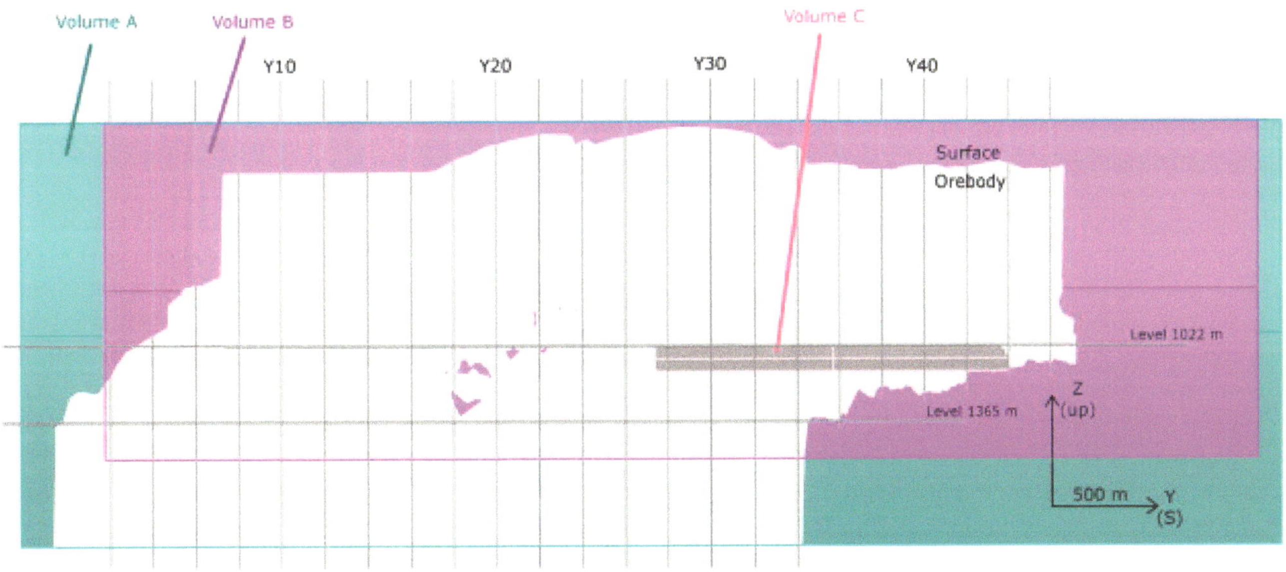

Figure 2: Data extents (not coverage) represented by boxes overlaid on an outline of the orebody (longitudinal view from the footwall towards the hangingwall). All volumes extend into the hangingwall and footwall. Mine coordinates along orebody strike (Y coordinates) are labeled. Volume A represents the data extents of the drill cores. Volume B shows the data extents of the discontinuities mapped underground. Volume C illustrates the location of extensive underground mapping campaigns to better identify geological features, such as structures and clay altered volumes.

A total of 56 unconfined compressive strength (UCS) tests were completed on 11 of the rock units, which are sourced from previous testing done by LKAB employees (such as Andersson and Israelsson, 2009), unpublished work (Westblom, 2014) and as a part of this study. Competent and representative samples were selected. Samples were extracted from many locations in the mine, mostly contained by Volume B in Fig. 2.

Approximately 740 Point Load Strength tests were completed on 20 of the rock units on 28 mm core as a part of this study. The majority of these samples were from Volume C in Fig. 2. Weaker samples were preferentially eliminated due to sampling technique. An additional 141 Point Load Strength tests were completed by Lindgren (2013) on lump samples from varied locations throughout the mine. These 141 tests were not used in this study to reduce the introduction of potential sources of error caused by 1) the different sample type, and 2) the different source location.

Geological data exists from routine underground drift mapping, which focuses on definition of the orebody and the location of different types of ore rock units. The majority of mapping data is in the footwall and in the orebody. Extensive shotcreting limits mapping anywhere except working faces. In addition, a recent underground mapping campaign over the span of one year has provided a more detailed definition of clay zones in Volume C in Fig. 2.

Over 4000 discontinuities exist in the database from underground drift mapping (LKAB, 2014b). The extents of these data are represented by Volume B in Fig. 2. All mapped joints have information about their orientation.

Geological Strength Index (GSI) has been mapped during underground campaigns (LKAB, 2014c). Data is available between Level 1051 m and Level 1543 m within Volume B (Fig. 2), but data coverage is limited due to shotcreting.

A recent campaign acquired and analyzed data from underground mapping to provide an understanding of the kinematic regime of the rock mass (Berglund and Andersson, 2013). This is the first major underground campaign specifically related to faults at the mine, and was mostly conducted in Volume C (Fig. 2).

Rock falls occurring during production are individually documented in reports and stored in a database at the mine. All rock falls have been systematically recorded since circa 2010, with selected events recorded in 2008 and 2009.

The Kiirunavaara Mine has an extensive seismic network with related information stored in the database. Dahnér et al. (2012) provides a description of the network and mining-induced seismicity.

In complicated rock masses such as at the Kiirunavaara Mine, the importance of using statistics as a basis for analysis is heightened. With such large complexity and quantity of available data, there is a likelihood that geomechanical patterns become too convoluted to analyze without the assistance of statistics. As such, the methodology to build the geomechanical models is based on statistical analysis of physical data, using operational experience and separate data sources for verification where possible.

Geostatistical techniques have been used with success to assess rock masses from a geomechanical perspective for a variety of purposes (Ayalew et al., 2002, Coli et al., 2012, Ozturk and Simdi, 2014 and Stavropoulou et al., 2007). These techniques are well suited to data-rich environments that have large-scale changes in characteristics. Due to its robust and quantifiable nature and suitability with complex and heterogeneous geological environments, geostatistics was selected as one of the main spatial analysis techniques to develop the geomechanical model at the Kiirunavaara Mine.

The proposed methodology employs the use of standard geostatistical techniques with additional statistical analysis to understand the large-scale 3-D patterns of rock mass characteristics, such as structural domains, zones of high or low stiffness material and rock mass quality. These areas are first analyzed separately and then subsequently synthesised in 3-D space to create one geomechanical model. Geostatistical analysis is limited to identification of volumetric characteristics of the rock mass, and specifically excludes identification of discrete features such as faults, for which targeted underground mapping is better suited.

As with all modelling, control data for calibration and verification purposes is essential. Good candidates include data related to rock mass behaviour as well as seismicity, rock falls and drift mapping. Data sets which are too sparse for statistical analysis or which are in an unsuitable format (such as maps or photos) can be used as control data.

GEOMECHANICAL CHARACTERISATION

Intact rock strength

A complete understanding of the intact strength differences between geological units does not yet exist at the Kiirunavaara Mine. In the literature, petrographic data has been used to evaluate intact compressive rock strength through regression, expert systems, and artificial intelligence (Alvarez Grima and Babuška, 1999, Gokceoglu, 2002,Gokceoglu et al., 2009 and Singh et al., 2001), however, insufficient data was available for this study to use such an approach. As an alternative, UCS and Point Load Strength tests were selected to evaluate intact strength. Typically, the results from Point Load Strength tests are used to gain additional strength information by scaling the Point Load Strength Indices ($Is_{(50)}$) to UCS values. Numerous empirical relationships exist in literature for the scaling of $Is_{(50)}$ data to UCS values (Bieniawski, 1975, D'Andrea et al., 1965, ISRM Commission on Standardization of Laboratory and Field Tests, 1985,Kahraman, 2001, Kaya and Karaman, 2015 and Kohno and Maeda, 2012). Most of these relationships use linear scaling as suggested by ISRM (1985):

$$UCS=kIs_{(50)} \qquad (1)$$

where k is the scaling factor. The scaling factor suggested in the testing standards ranges from 20–25, however it is acknowledged in the standards and confirmed by others (previously listed) that this value may vary between 15 and 50 and is highly dependent upon rock type (ISRM, 1985). Some authors, such as Kaya and Karaman (2015), have found even lower scaling factors.

Since this scaling is dependent upon rock type, it is critical to first approach grouping of rock units by intact rock strength using statistical tests. The purpose of this analysis is to identify 1) the underlying distribution of strength parameters, 2) groups of geological units that share common strengths, and 3) the estimated mean strengths and standard deviations of these groups. There is insufficient data availability at the mine to determine if intact rock strength is dependent upon location, not only geological units.

Due to the large number of samples, the Kolmogorov-Smirnov test was used on the $Is_{(50)}$ data to evaluate all possible groupings of geological units. This method uses the cumulative distribution functions of two samples to evaluate if differences exist between the location and shape of the distribution functions, and was selected due to its robust, nonparametric nature. Each sample was represented by one $Is_{(50)}$ value; an average value for the sample was used if there was more than one acceptable data point. $Is_{(50)}$ data were then scaled to UCS data for each group of geological units individually to estimate the strength of each group according to the following equation:

$$k = \left(\frac{\mu_{UCS}}{\mu_{Is_{(50)}}}\right) \quad (2)$$

where k is the scaling factor, μUCS is the average UCS from the UCS tests and $\mu_{Is(50)}$ is the average $Is_{(50)}$.

Based on the results from the Kolmogorov-Smirnov tests and considerations of sample size, it is evident that some mapped geological units likely share common underlying distributions. The most likely grouping of geological units into strength groups based on the statistical analysis are shown in Fig. 3, where the distribution of the grouped and scaled $Is_{(50)}$ data is shown by box plots, providing information about the distributions of each group through the visualisation of their quartiles. No significant difference between the distributions of 1) the 5 types of hangingwall units or 2) the remaining ore types were found, however not all units were well represented. Units that have limited sample numbers are starred (*). Additional point load tests would improve the precision and accuracy of this analysis. Based on Kolmogrov-Smirnov tests and χ^2 tests, all data before and after grouping by multiple geological units indicated normal distributions, with the exception of the ore material with few samples (ore, higher contaminant levels). This may be a result of few samples or at least two unique distributions of strength within these geological units.

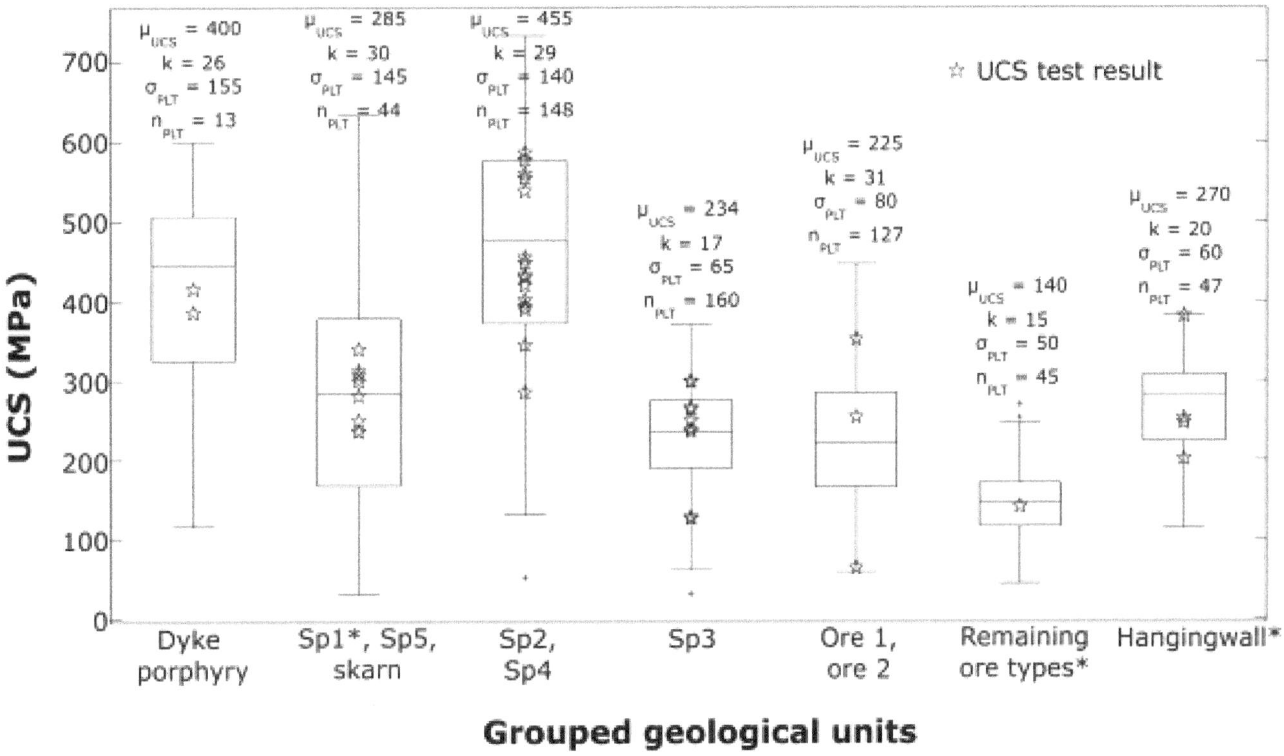

Figure 3: Scaled Point Load Strength test results (box plots) and UCS test results (stars) by grouped geological units. Boxplots of the point load tests for each grouped geological unit show outliers (considered as +/− 2.7 * standard deviation) as crosses. Average UCS from the UCS tests (μ_{UCS}), the scaling factor (k), the scaled standard deviation from the Point Load Strength tests (σ_{PLT}), and the number of Point Load Strength tests (n_{PLT}) for each group are presented over the box plots.

The UCS data for each group of geological units are overlaid on the scaled $Is_{(50)}$ data inFig. 3 as stars. Additional UCS tests would significantly improve the accuracy of the estimated strengths, in particular for groups with few UCS samples. Both the UCS data and the Point Load Test data had a relatively wide spread, in particular in the majority of the footwall material. The porphyritic dyke and Sp2/Sp4 groups showed higher strengths than the other groups. Kolmogorov-Smirnov tests on the scaled point load test data and the UCS data sets indicated that the distributions between the two data sets (Point Load Test data and UCS test data) for individual grouped geological units were consistent.

There is a significant spread of estimated intact rock strengths for the geological units present at the Kiirunavaara Mine. Lindgren (2013) showed that increasing the pervasive alteration into softer minerals, such as chlorite, carbonates and clay, lowers the intact rock strength through analysis of Point Load Strength tests conducted on lump samples; however, the current study does not have sufficient data to evaluate the role of alteration on intact strength at the mine. Although very clear during core logging and underground site investigations, the intact rock strength tests do not illustrate the significant strength and stiffness difference of some of the weaker units found at the mine, such as clay, due to preferential sampling.

Volumes with clay alteration

Due to the significantly different material properties of clay compared to the surrounding rock, and due to the total volume of clay present, the locations of volumes of clay alteration are an important consideration for the geomechanical model at the Kiirunavaara Mine. 3-D volumes of clay based on isotropic search parameters from mapped data were developed by Mattsson et al. (2010). However, this work has not been updated to reflect newly acquired data and knowledge.

Geostatistical techniques were applied to understand the underlying spatial distribution of these clay volumes. This includes an analysis of appropriate search parameters via semi-variograms and then interpolation of the data. The interpolated model from the drill core data was compared to drift mapping to evaluate accuracy of the technique for this rock mass.

A common difficulty in identifying clay material in drill cores is that clay is often washed away by the drilling water and does not remain as part of the core. To account for this, both mapped clay and core loss in the core logging data (LKAB, 2014a) were considered to be indications of clay altered zones.

Indicator semi-variograms were created to determine 1) if directional anisotropy exists, and 2) the appropriate search parameters for the composited clay data from core mapping. Sample compositing was used to reduce bias caused by variable sample length. The clay and core loss data were downhole composited on 0.5 m intervals, as the majority of sample lengths were shorter than this value. The semivariograms indicated that the major and minor search ellipse axes are horizontal (Fig. 4). The major axis is semi-perpendicular to orebody formation. The semi-major axis was steeply dipped towards the north. The major and semi-major axes had approximately the same range, however the range was significantly reduced in the minor axes direction. As evident inFig. 4, the minor axis, approximately aligned along orebody strike, has a significantly smaller range than the borehole spacing (circa 50 m). This indicates that the data from the drilling has insufficient density for the use of geostatistical techniques to estimate clay volumes at the Kiirunavaara Mine. This could be related the possible genesis method of the hydrothermal clay alteration at the mine; hydrothermal fluid travels through existing discontinuities until it reaches a volume of the rock mass which is susceptible to larger amounts of alteration. The preferential travel along existing discontinuities can result in locations of clay that have a discrete nature, rather than a volumetric one. Due to this discrete nature, this data a difficult candidate to consider for interpolation methods.

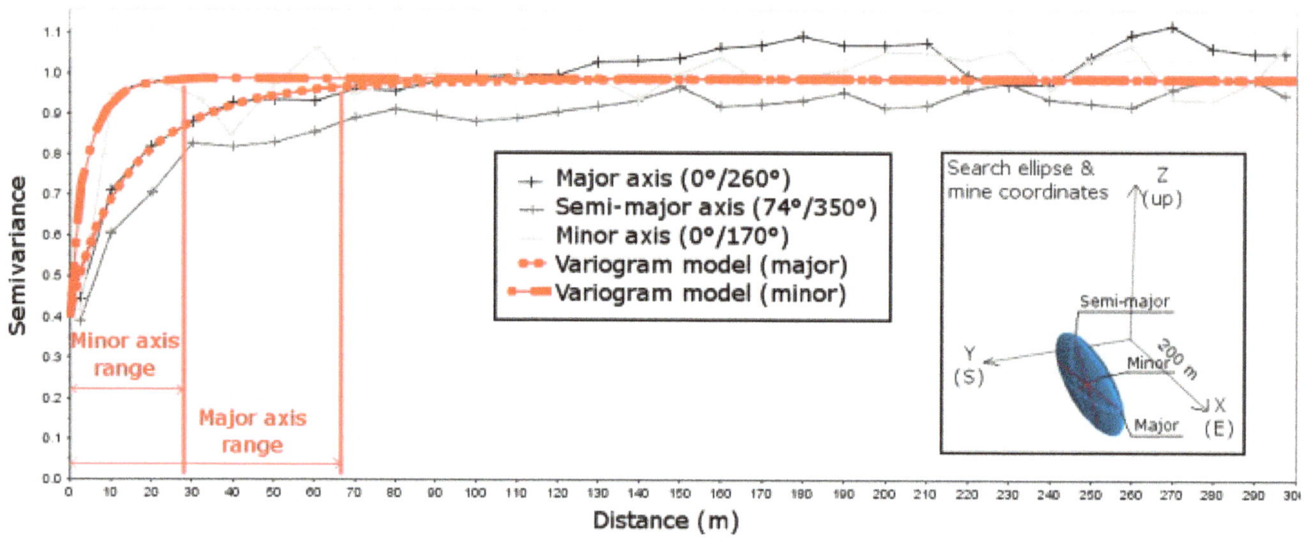

Figure 4: Semi-variograms in the major, semi-major and minor search ellipse directions. Search ellipsoid is visualised with mine coordinates.

However, a simple spatial analysis of the density of clay indicators from the core data in 3-D space using search radii identified from the semi-variograms yielded a model of the clay volumes (not clay alteration found in joints) that could be calibrated to clay mapped underground where data was available, see Fig. 5a) and b). Fig. 5c) shows the calibrated clay model with the underground mapping data. Calibration was done visually to determine which isosurface value in the clay model corresponded to the mapped data. Within the calibration volume, this isosurface represents the density at which the clay indicators represent a volume of the rock mass that has more clay. The only mapped clay volumes that were not represented in the model were volumes which had no associated drill core data. The calibrated model does not necessarily represent volumes that have entirely been altered to clay, rather volumes that have a higher concentration of clay than their neighbours. At the Kiirunavaara Mine, this simplistic form of spatial analysis of core data has the potential to assist with prediction of location and size of clay volumes prior to drifting. Tighter exploration drill spacing and additional underground mapping in other areas of the mine have the potential to improve future models.

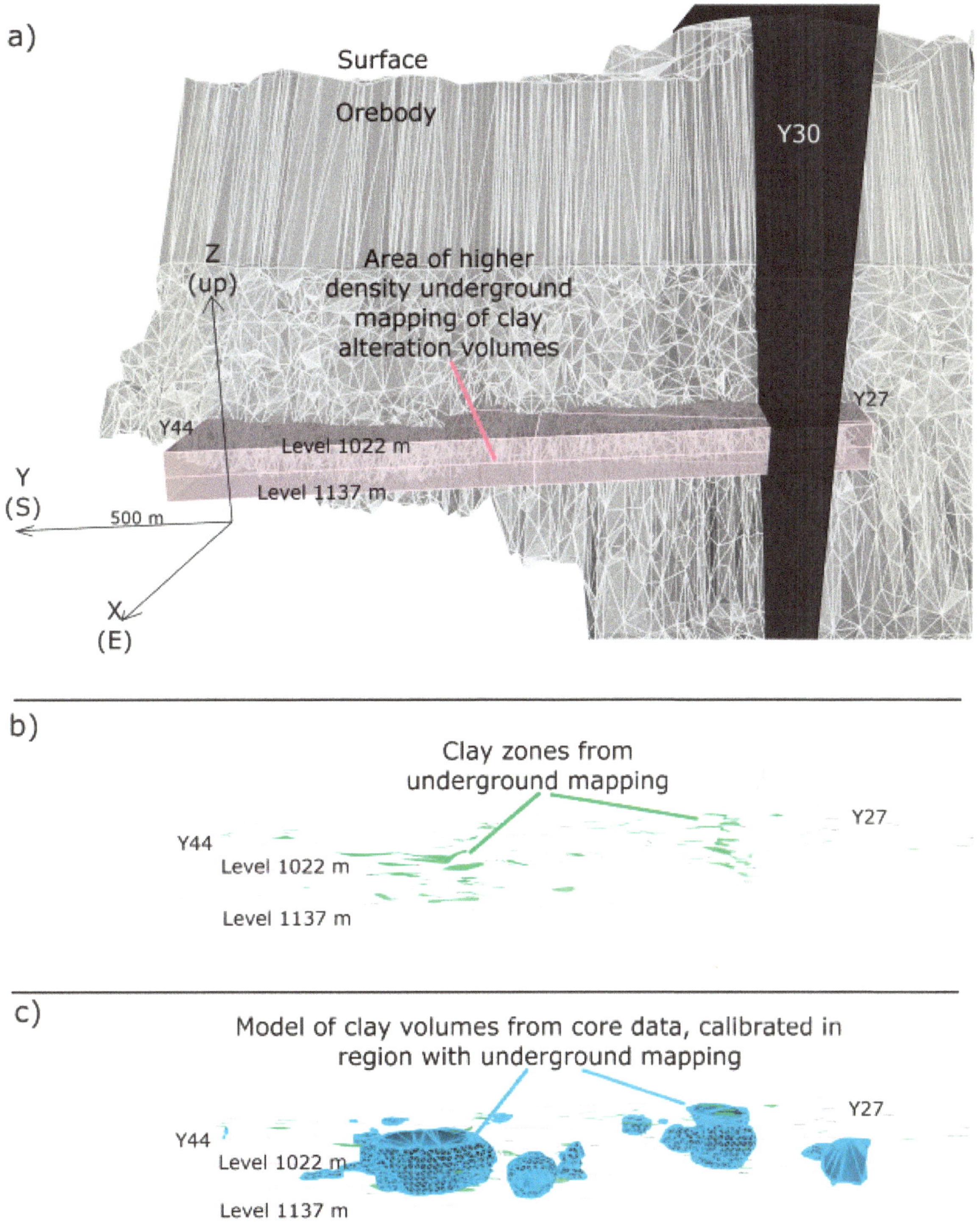

Figure 5: Calibration of model of clay volumes from core data with underground mapping data (view from the hangingwall towards the footwall). Area of higher density underground mapping shown in a), mapped clay volumes are presented in b) with the overlaid clay model from the core data in c).

Rock mass quality

When considering the data available over the entire mine, neither GSI (LKAB, 2014c) nor RMR data sets are good candidates for geostatistical analysis. The RMR data set at the mine is very sparse in both number of sample points as well as their physical locations. GSI has more data points than RMR, but still suffers from poor coverage. However, the RQD data set (LKAB, 2014a) is much more complete, and is amenable to geostatistical analysis.

Two previous works exist using geostatistics to estimate RQD at the Kiirunavaara Mine. The first was limited to a section of the rock mass located in the north (Henry and Marcotte, 2001), and the second used isotropic search parameters to develop a mine-scale model (Cotesta, 2011). To update the model with newly mapped data and to provide additional analysis, a new mine-scale model that used anisotropic search parameters was created.

Using similar techniques as previously discussed, anisotropy was found to exist through the analysis of semi-variograms. Despite results from Henry and Marcotte (2001) illustrating that the range of the semi-variograms indicated insufficient drill spacing for geostatistical analysis of the analyzed portion of the rock mass (only the northern portion), the semi-variograms of the current database for the entire mine indicated otherwise. Based on the ranges from the semi-variograms search radii were evaluated to be 100 m along strike, 100 m across strike, and 80 m vertically, which are greater than the maximum drill spacing of 50 m.

Models of RQD were created using inverse distance, as a preliminary evaluation technique. Kriging methods have the potential to improve these models. An example vertical slice through the mine-scale RQD model is illustrated in Fig. 6 in combination with a longitudinal projection of the orebody. Distinct zones of lower RQD crosscut the orebody, in particular in the southern end (Fig. 6). These zones are physically associated with the calibrated clay volume model; low RQD is found both inside and in the immediate vicinity of the volumes that are estimated to have more clay alteration (Fig. 6). Consistent with the mine-scale results found by Cotesta (2011), RQD increases with depth in the northern portion of the rock mass, with a sharp contrast at approximately the Level 950 m.

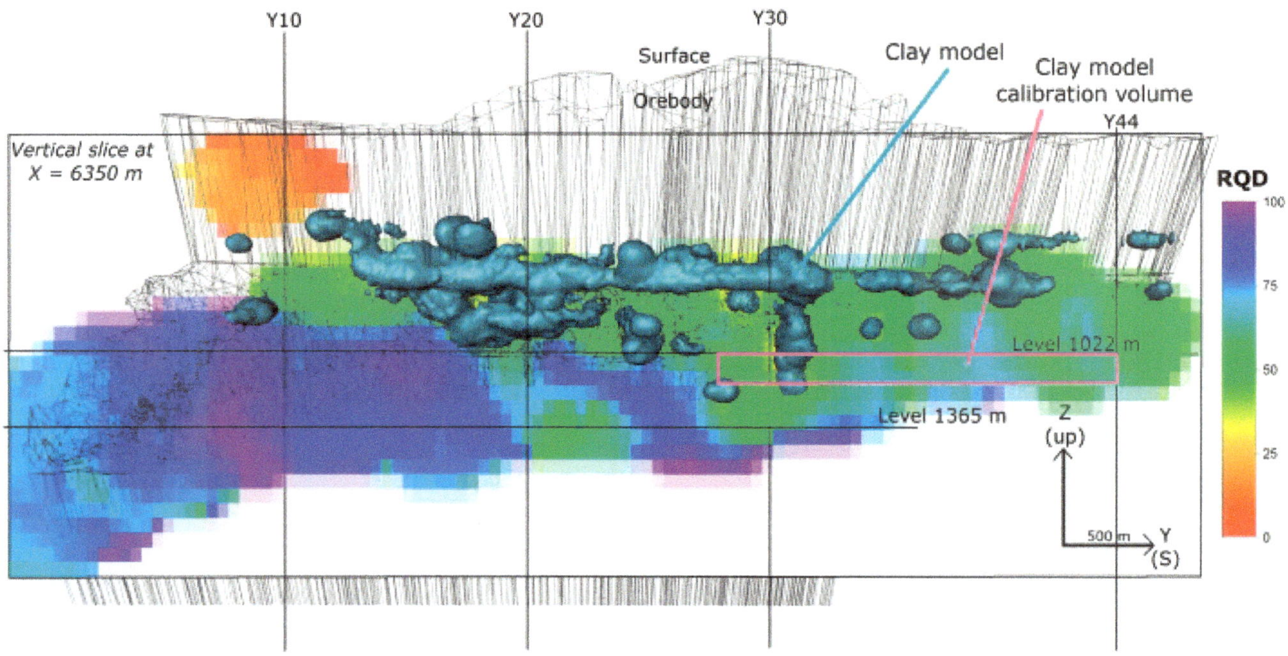

Figure 6: Example vertical slice through the mine-scale RQD model through X = 6350 m viewed from the footwall overlaid on a longitudinal projection of the orebody. The RQD model is represented by the block model, where values are scaled by colour. The mine-scale clay model (from Section 4.2) is superimposed on the RQD model, in addition to its calibration volume.

Structural domains

The last structural domaining done at the mine scale was completed in 1995 (Holmstedt, 1994 and Holmstedt, 1995), however the structural domains found by Rådberg et al. (1989) are more commonly used. Since that time, some additional work on identifying domains has been done sporadically for specific volumes of the mine. The current study updates the structural domain model by incorporating historical and newly acquired data (LKAB, 2014b).

Due to the importance of the identification of structural domains, multiple techniques have been developed to assess their boundaries (Escuder Viruete et al., 2001, Jimenez-Rodriguez and Sitar, 2006, Martin and Tannant, 2004, Piteau and Russell, 1971 and Zhou and Maerz, 2002). Most techniques have limitations, however, such as their reliance upon a priori geological knowledge, lack of evaluation in 3-D space, appropriate distance metrics for clustering, etc., which may misrepresent the underlying spatial distribution of joint sets.

Available discontinuity data at the mine scale in the Kiirunavaara Mine is limited to drift mapping of joint orientations at active faces (shotcrete limits mapping of completed excavation). The nature of this data eliminates the possibility of multivariate clustering as suggested by Zhou and Maerz (2002), as only dip and dip direction is readily available at the mine scale.

Piteau and Russell (1971) developed an effective and simple technique to identify the locations of structural domains. Their technique is based upon the mathematical definition of structural domains; within each domain, the slope of the cumulative sum of dip or dip direction is constant. The major limitation of their domain location technique is that it is based upon evaluating data along a 1-D ray or in a 2-D plane. Identifying the correct locations of domains using this technique assumes that the domains can be defined upon the user selected ray or plane, and that there are no other domains in other directions. The success of this approach is highly dependent upon the underlying spatial distribution of the joint sets.

An extension of Piteau and Russell (1971) has been developed to quantitatively identify domains in 3-D. Identified domains are then statistically evaluated using similar strategies as presented by Mahtab and Yegulalp (1984) and Martin and Tannant (2004). The proposed technique for quantitative identification of structural domains is detailed below.

1. Create a block model representing the span and extent of the original, cleaned data. Block dimensions are user specified; the intended scale of the domains, measurement spacing and geological environment are evaluated to determine this.

2. Within each block, for every joint measured, calculate the slope of the cumulative sum of 1) strike and 2) dip. Filtering of blocks based on user-defined minimum number of data points and linear fit parameters is recommended.
3. Using the slope of the cumulative sum and/or dip direction, apply geostatistical interpolation methods to supplement gaps in the data. The decision of which data to use should be based upon statistical analysis of the underlying data. With most data sets, the slope of the cumulative sum of dip is the most appropriate source to use, as when strike is used, there is a numerical discrepancy between 0° and 359° that does not represent a large physical direction difference. The methodology could be adapted for other mapped joint characteristics, such as joint roughness, however due to data availability at the mine this was not explored.
4. Using isosurfaces (3-D analog of an isoline or contour) of the block model, identify continuous volumes in 3-D space representing volumes of distinct structural domains. The selection of isosurfaces requires user input. Particular attention should be paid to large changes in shape/location of isosurfaces. Distinct and individual isosurfaces represent domains; however, the numeric values of the isosurfaces are irrelevant.
5. Statistically and visually compare the stereonets for the determined structural domains.

This technique was tested using the underground discontinuity mapping data at the Kiirunavaara Mine to evaluate the location and shape of the structural domains. Domains were identified using cubic blocks of 30 m edge length, that had greater than five data points and an r^2 value from linear regression greater than 0.95. A significant number of the structures measured at the mine had strikes that were orientated around 0° N, so the cumulative sum of dip was the most appropriate metric. Isotropic inverse distance was applied to the block model to interpolate the data. Alternative methods of interpolation have the potential to improve this model.

Example slices through the interpolated block model are shown in Fig. 7 in conjunction with a wireframe of the orebody, where each block is coloured based on its value of the slope of cumulative sum of dip. Changes in colour represent distinct 3-D volumes that share similar slopes of cumulative sum of dip, indicating distinct domains.

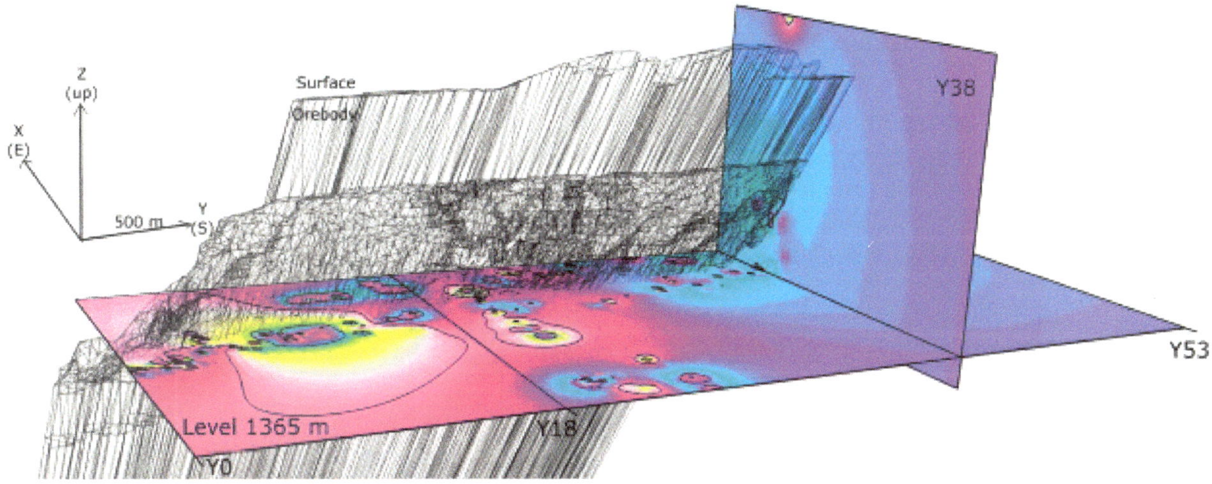

Figure 7: Example slices through the slope of the cumulative sum of dip block model (30 m cubes) used to identify structural domains (seen from the footwall towards the hangingwall). The colour of each block represents the slope of the cumulative sum of dip for all joint sets mapped inside of that block. The values themselves are inconsequential; it is the location of changes in the values that represents possible structural domain boundaries.

The 3-D isosurfaces from the domain block model were used to group and test the original data points. Mahtab and Yegulalp (1984) suggested a method of statistically comparing stereonets based on pole count in equal area windows, and then using the χ^2test to test if the distribution of the pole count for two stereonets is different, where a difference represents a new domain. Due to sample size limitations associated with the χ^2 test, Martin and Tannant (2004) adopted the window technique, however used the Pearson product–moment correlation coefficient (Pearson's r) as an indication of similarity between two stereonets. For the Kiirunavaara Mine, Pearson's r, Spearman's rank correlation coefficient (Spearman's rho (ρ)) as well as the Kendall rank correlation coefficient (Kendall's tau (τ) coefficient) were tested to evaluate statistical similarity between stereonets. For the joint sets in the Kiirunavaara Mine, the use of Spearman's ρ correlation coefficient and Kendall's τ were found to be most appropriate since the joint set data series after the count of poles in the windows are discrete and non-normal.

The results from the domaining analysis at the Kiirunavaara Mine highlight that the domains have 3-D shapes at the mine-scale, which cannot always be accurately represented in 2-D. The methodology is useful not only to identify domains and their shape, but also to explore rotations of joint sets within one domain. The magnitude of a joint set's rotation and intended use of the structural domains form the basis of the decision about if it constitutes a domain boundary. Statistical comparison of the stereonets can assist with this process via quantification of how different two stereonets are. At the Kiirunavaara Mine, one of the identified domains consists of a joint set that rotates within the volume. The three volumes presented in Fig. 8 are statistically from the same domain (see domain stereonet in the upper left corner of Fig. 8) using both Spearman's ρ correlation coefficient and Kendall's τ. However, as shown by the other stereonets in Fig. 8, the orientation of the main joint set varies slightly between the individual volumes.

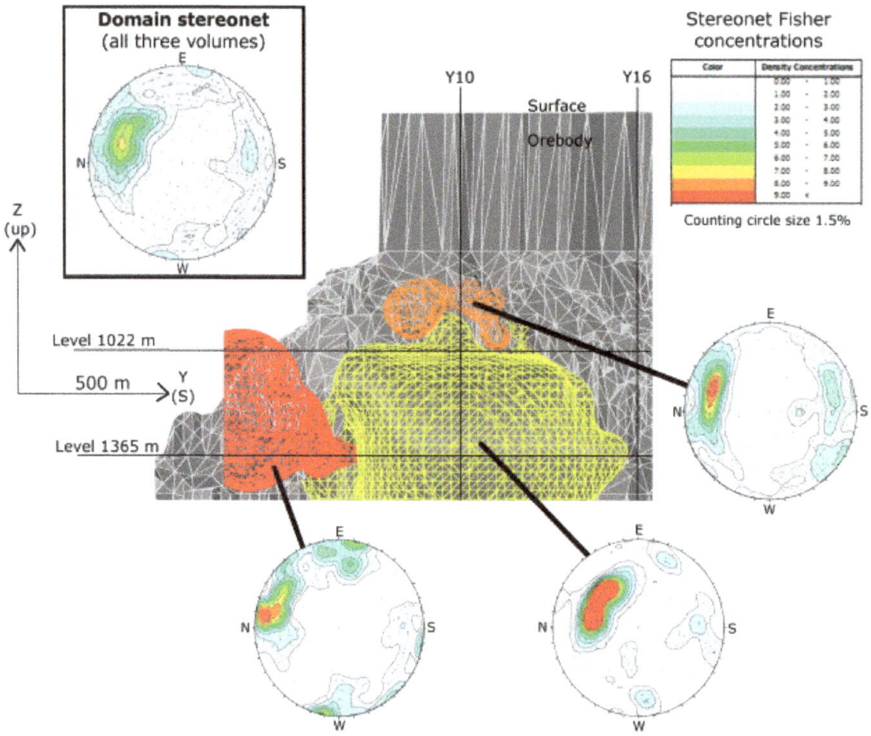

Figure 8: Example of one domain (domain stereonet in upper left) at the Kiirunavaara Mine with slight variation of orientation of the major joint set for different volumes within the domain. Fisher concentrations for the poles are shown on equal area lower hemisphere stereonets. Stereonets are rotated so they are aligned with the mine coordinate system.

At the Kiirunavaara Mine, transitions between domains at the mine-scale were rarely sharp. One example to the contrary, however, corresponds with a known, mapped geological feature. Despite not being used as a priori boundary during the analysis, where the rock mass transitions from north to south into a more clay altered volume (from approximately Y 28 and southwards), there is a sharp domain change. In this southern portion of the orebody, only two domains have been identified at the mine-scale (Fig. 9). They consist of many joints, and joint set identification is difficult (Fig. 9). The joint sets have a more random component in the heavily clay altered volumes of the rock mass.

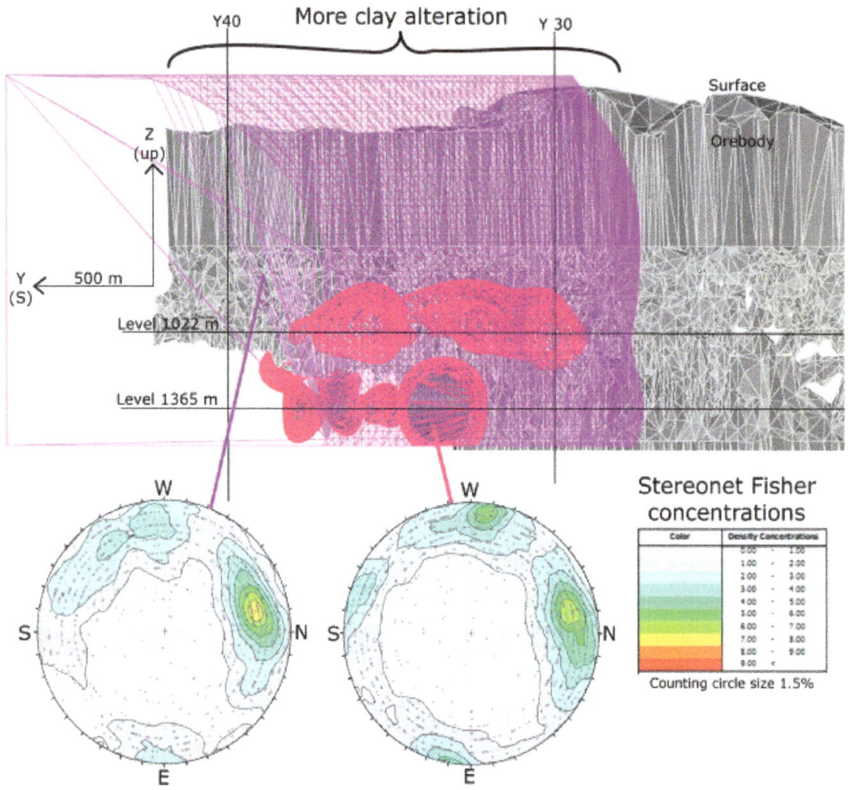

Figure 9: Domains in the southern portion of the rock mass overlaid on the orebody as seen from the hangingwall. Domains are represented by the shaded wireframes, and their associated lower hemisphere equal area stereonets are provided. Stereonets are rotated so they are aligned with the mine coordinate system.

The 3-D domains identified for the rock mass using the suggested technique offered a significant improvement when compared to vertical extension of domains previously analyzed as suggested by Rådberg et al. (1989). More unique domains were identified, in different locations, with clearer definition of joint sets and transition zones between domains. Statistical comparison of the old and new domains showed that the new domains were more distinct than the domains identified by Rådberg et al. (1989).

Faults

Numerous faults (discontinuities with identifiable shear displacement) are present at the mine. Berglund and Andersson's (2013) underground mapping work around Y 34 on the Level 1051 m, Level 1060 m, and Level 1079 m identified many faults. The majority of these faults are steeply dipping and follow the strike of the orebody (Berglund and Andersson, 2013). Currently, their length, spacing and persistence is unknown. Evidence of faults can also be seen during core logging in the form of slickenslides along discontinuity surfaces. Visible underground, many of these ore parallel structures, which exhibit slickenslides, are hematite coated with a mirror like surface and undulate on the scale of meters. Few orebody perpendicular faults have been mapped underground. Seismic clustering of events by the mine has identified possible fault structures, however only one structure has been confirmed by underground mapping, which was subsequently used in numerical stress analysis by Sjöberg et al. (2011). The currently available data does not enable a mine-scale characterisation to include faults, since mapped data is mainly limited to that done by Berglund and Andersson (2013).

CORRELATION TO FALLS OF GROUND

Based on the geomechanical characterisation, the portion of the rock mass which contains significant volumes of clay has different characteristics than the unaltered portion. The different characteristics are not limited to the clay volumes; rather the entire rock mass surrounding the clay. This is apparent from information related to jointing. Surrounding the clay volumes there are lower RQD values, few distinguishable structural domains, and a more random component to the discontinuities. Because of the seemingly controlling nature the clay zones have over the geomechanical characteristics of the rock mass, it is logical to evaluate if there is a correlation between the calibrated clay model and falls of ground.

The falls of ground at the Kiirunavaara Mine are often associated with seismic events, in the form of shakedown and/or bursting. With available data, it is difficult to accurately distinguish between these causes of the falls of ground, so all of these events were considered in this analysis. All fall of ground events at the mine are comparable since the support system is the same for almost the entire extent of the rock fall database. The direction of the major principal in situ (pre-mining) stress at the Kiirunavaara Mine is horizontal and approximately perpendicular to the strike of the orebody, whilst the intermediate and minor principal in situ stresses are nearly equal (Sandström, 2003). When viewing the data from the approximate direction of principal in situ stress, a correlation between the modelled clay zones and the falls of ground is evident (Fig. 10). The majority of the falls of ground are concentrated in the rock mass surrounding the clay zones.

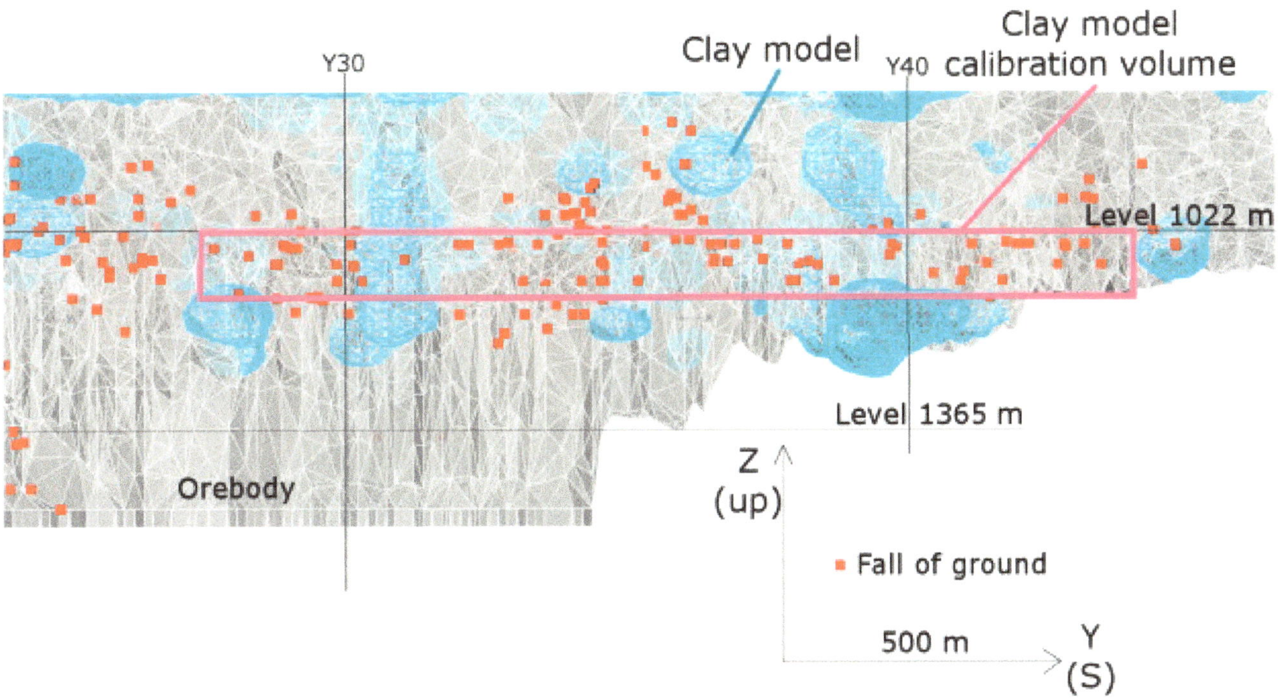

Figure 10: Falls of ground (squares) versus the calibrated clay volume model (shown from the footwall towards the hangingwall). Viewed from the direction of major principal in situ stress (pre-mining).

There were significantly fewer falls of ground from approximately Y 23 and northwards. This could be related to the lack of clay volumes in the area, although additional calibration of the clay model in that volume is required.

DISCUSSION

Analysis of intact rock strength at the Kiirunavaara mine showed large variations not only between the groupings of geological units, but within them. This spread has implications for the geomechanical model and future rock engineering analyses. Representing the rock mass with constant strengths and stiffnesses is a simplification that cannot fully capture the rock mass response related to intact strength variation. The causes of this variation are unclear; the role of alteration in intact rock strength needs to be further evaluated. Additional UCS values (from UCS tests or triaxial tests) for geological groups with few of these data points may improve the scaling of the Point Load Strength tests. At this stage, how to scale the intact rock strength to rock mass strength for this complex geological environment remains unresolved.

At the Kiirunavaara Mine, a link exists between modelled clay volumes and falls of ground. Considering how the in situ stresses in this region likely behave, a possible explanation of this mine-scale behaviour is suggested. Since the volumes with clay alteration have a lower stiffness than the surrounding rock mass, the stresses will flow around the clay zones and concentrate in the more competent rock. This increase in stress magnitude in the surrounding rock mass has the potential to initiate new failure not necessarily induced by mining, such as those possibly indicated by the RQD model and the random component of the joint sets in the region. Stress concentrations caused by the clay volumes combined with mining induced stress changes have the potential to explain some of the behaviour. Numerical stress analysis in combination with more detailed study of the failure modes of the ground falls will be carried out in a future stage of this investigation to further explore this hypothesis and its implications on rock mass behaviour at the mine.

The methodology presented, as applied to the Kiirunavaara Mine, resulted in an increased understanding of the geomechanical characteristics of the rock mass. It successfully improved upon many aspects of geomechanical characterisation, such as the identification of new structural domains and a new understanding of the distributions of intact rock strengths. The methodology worked particularly well with the large quantity of borehole data combined with separate data sets for calibration. As an additional benefit, the development of a geomechanical model using these techniques and comparison of that model to rock mass behaviour highlighted where supplementary data collection could add the most value.

The methodology is robust; it has the potential to be adapted to different types of data in other mining environments. However, a priori knowledge of the rock mass was used to select the analysis techniques. Due to the smoothing nature of geostatistics, consideration must be given to if the parameters of interest are good candidates for interpolation. Data sets for interpolation must show spatial correlation over a greater distance than the data spacing, which is evaluated through semi-variograms. The highest resolution of the interpolation is a function of the input data set; interpolation requires supporting data points and accuracy of the results cannot be improved without well-positioned data. Provided the data supports interpolation, the presented methodology is well suited to geological environments with heterogeneity and alteration, and/or environments that have domain boundaries with irregular shapes in 3-D.

CONCLUSION

Through enhancement of the understanding of the geomechanical characterisation of the Kiirunavaara Mine, a conceptual model of some aspects of the rock mass behaviour was developed. Clay volumes (represented by a model based on borehole data calibrated to underground mapping) correlated to the geomechanical characteristics and behaviour of the rock mass. The rock mass in the immediate vicinity of the volumes of clay alteration had lower RQD values, more random jointing, and a higher concentration of falls of ground than the surrounding rock mass.

Consideration of multiple sources of data in 3-D space through the application of geostatistics was a key component to the development of this new conceptual model. The methodology presented, based on statistics, geostatistics and an extension of previous quantitative domaining work, is of interest to other mines that have complex and heterogeneous geological environments, specifically those which have volumes of alteration, structural domains which possibly have complex shapes in 3-D, or other characteristics which are well suited to interpolation methods. The methodology is well suited to geological environments that are appropriate data for interpolation; data types which can logically be interpolated between data points and that are spatially correlated over a greater distance than the data spacing. The use of geostatistics on multiple data sets that are often available at mines (such as borehole data, structural mapping data, etc.) has the potential to give indications of problem areas, provided the interpolation is well calibrated against data that is distinct from the input data.

ACKNOWLEDGEMENTS

This work would not have been possible without the generous financial contributions from the Hjalmar Lundbohm Research Centre (HLRC) and LKAB. The importance of the resources provided by LKAB in the form of personnel, time, equipment, and data cannot be understated. The authors gratefully acknowledge Christina Dahnér (LKAB) for her continued support as our mine contact and insightful contributions to discussions. We thank Jimmy Töyrä (LKAB) for his assistance with yet another short-notice review. Ulf B. Andersson is thanked for his review of the manuscript. We are also immensely grateful to the following people, who have aided the formulation of portions of this manuscript via countless in-depth discussions and much assistance with data acquisition (employed by LKAB unless otherwise stated): Lars Malmgren, Carlos Quinteiro, Stig Fjellborg, Henrikki Rutanen, Karin Lindgren, Stina Hallinder, Anousheh Abdollahpour, Mirjana Bošković, Helena Sturk, Johan Berglund (Vattenfall), and Tomas Björnell (Norconsult).

REFERENCES

1. Alvarez Grima, M., Babuška, R., 1999. Fuzzy model for the prediction of unconfined compressive strength of rock samples. Int. J. Rock Mech. Min. Sci. 36 (3), 339–349. http:// dx.doi.org/10.1016/S0148-9062(99)00007-8.

2. Andersson, U.B., Israelsson, J., 2009. Resultat av bermekaniska materialtester (in Swedish) (LKAB Internal Report) .

3. Ayalew, L., Reik, G., Busch, W., 2002. Characterizing weathered rock masses—a geostatistical approach. Int. J. Rock Mech. Min. Sci. 39 (1), 105–114. http://dx.doi.org/10.1016/S1365-1609(02)00004-7.

4. Berglund, J., Andersson, U.B., 2013. Kinematic analysis of geological structures in Block 34, Kiirunavaara (LKAB Internal Report), (Ref no. 13–746, 46 pp.).

5. Bieniawski, Z.T., 1975. The point-load test in geotechnical practice. Eng. Geol. 9 (1), 1–11. http://dx.doi.org/10.1016/0013-7952(75)90024-1.

6. Coli, N., Berry, P., Boldini, D., Bruno, R., 2012. The contribution of geostatistics to the characterization of some bimrock properties. Eng. Geol. 137–138, 53–63. http://dx.doi. org/10.1016/j.enggeo.2012.03.015.

7. Cotesta, L., 2011. Development of an RQD Voxet Model for LKAB's Kiruna Mine (LKAB Internal Report)

8. D'Andrea, D.V., Fogelson, D.E., Fischer, R.L., 1965. Prediction of compressive strength from other rock properties. Report of investigations (Ref no. 6702). U.S. Dept. of the Interior, Bureau of Mines, Washington, D.C.

9. Dahnér, C., Malmgren, L., Bošković, M., 2012. Transition From Non-seismic Mine to a Seismically Active Mine: Kiirunavaara Mine. The 2012 ISRM European Rock Mechanics Symposium (Eurock 2012), Stockholm, Sweden.

10. Escuder Viruete, J., Carbonell, R., Jurado, M.J., Martí, D., Pérez-Estaún, A., 2001. Twodimensional geostatistical modeling and prediction of the fracture system in the Albala Granitic Pluton, SW Iberian Massif, Spain. J. Struct. Geol. 23 (12), 2011–2023. http://dx.doi.org/10.1016/S0191-8141(01)00026-8.

11. Geijer, P., 1910. Geology of the Kiruna district, Vetenskapliga och praktiska undersökningar i Lappland. P.A.Norstedt, Stockholm.

12. Gokceoglu, C., 2002. A fuzzy triangular chart to predict the uniaxial compressive strength of the Ankara agglomerates from their petrographic composition. Eng. Geol. 66 (1–2), 39–51.

13. Gokceoglu, C., Sönmez, H., Zorlu, K., 2009. Estimating the uniaxial compressive strength of some clay bearing rocks selected from Turkey by nonlinear multivariable regression and rule-based fuzzy models. Expert. Syst. 26 (2), 176–190.

14. Henry, E., Marcotte, D., 2001. Developments in ground control planning enhanced by RQD geostatistical analysis at LKAB's Kiruna mine. The 2001 ISRM European Rock Mechanics Symposium (Eurock 2001), Espoo, Finland.

15. Hoek, E., Karzulovic, A., 2001. Rock Mass Properties for Surface Mines. In: Hustrulid, W.A., McCarter, M.K., Van Zyl, D.J.A. (Eds.), Slope Stability in Surface Mining. Society for Mining, Metallurgy, and Exploraction, Inc. (SME), Littleton, pp. 59–69.

16. Holmstedt, A., 1994. Bergmasseklassificering av Kiirunavaara malmkropp (in Swedish) (LKAB Internal Report), (Ref no. 94–788).

17. Holmstedt, A., 1995. Bergmasseklassificering av liggvägen I KUJ ovanför nivå 740 m (in Swedish) (LKAB Internal Report), (Ref no. 95–782).

18. ISRM Commission on Standardization of Laboratory and Field Tests, 1985. Suggested method for determining point load strength. Int. J. Rock Mech. Min. Sci. Geomech. Abstr. 22, 51–60. http://dx.doi.org/10.1016/0148-9062(85)92327-7.

19. Jimenez-Rodriguez, R., Sitar, N., 2006. A spectral method for clustering of rock discontinuity sets. Int. J. Rock Mech. Min. Sci. 43, 1052–1061. http://dx.doi.org/10.1016/j.ijrmms. 2006.02.003.

20. Kahraman, S., 2001. Evaluation of simple methods for assessing the uniaxial compressive strength of rock. Int. J. Rock Mech. Min. Sci. 38, 981–994. http://dx.doi.org/10.1016/ S1365-1609(01)00039-9.

21. Kaya, A., Karaman, K., 2015. Utilizing the strength conversion factor in estimation of the uniaxial compressive strength from the point load index. Bull. Eng. Geol. Environ. http://dx.doi.org/10.1007/s10064-015-0721-1.

22. Kohno, M., Maeda, H., 2012. Relationship between point load strength index and uniaxial compressive strength of hydrothermally altered soft rocks. Int. J. Rock Mech. Min. Sci. 50, 147–157. http://dx.doi.org/10.1016/j.ijrmms.2012.01.011.

23. Lindgren, K., 2013. Punktlasttestning av olika bergartstyper i Kiirunavaaras liggvägg (in Swedish) (Bachelor's Thesis 41921476) Luleå University of Technology, Luleå, Sweden.

24. LKAB, 2014a. Core logging database. LKAB Internal Database (Accessed 2014-12-18).

25. LKAB, 2014b. Discontinuities database. LKAB Internal Database (Accessed 2014-12-18).

26. LKAB, 2014c. GSI database. LKAB Internal Database (Accessed 2014-12-18).

27. Mahtab, M.A., Yegulalp, T.M., 1984. Similarity Test For Grouping Orientation Data In Rock Mechanics. The 25th U.S. Symposium on Rock Mechanics (USRMS). American Rock Mechanics Association.

28. Mandrone, G., 2006. Engineering geological mapping of heterogeneous rock masses in the Northern Apennines: an example from the Parma Valley (Italy). Bull. Eng. Geol. Environ. 65 (3), 245–252. http://dx.doi.org/10.1007/s10064-006-0039-0.

29. Marinos, P., Hoek, E., 2001. Estimating the geotechnical properties of heterogeneous rock masses such as flysch. Bull. Eng. Geol. Environ. 60 (2), 85–92. http://dx.doi.org/10.1007/s100640000090.

30. Martin, M.W., Tannant, D.D., 2004. A technique for identifying structural domain boundaries at the EKATI Diamond Mine. Eng. Geol. 74 (3–4), 247–264. http://dx.doi.org/10. 1016/j.enggeo.2004.04.001.

31. Mattsson, H., Berglund, J., Magnor, B., 2010. Strukturgeologisk modell över Kiruna gruvområde (in Swedish) (LKAB Internal Report), (Ref no. 10–874).

32. Ozturk, C.A., Simdi, E., 2014. Geostatistical investigation of geotechnical and constructional properties in Kadikoy–

Kartal subway, Turkey. Tunn. Undergr. Space Technol. 41, 35–45. http://dx.doi.org/10.1016/j.tust.2013.11.002.

33. Piteau, D.R., Russell, L., 1971. Cumulative sums technique: A new approach to analyzing joints in rock. Rock Mech. 13, 1–29.

34. Rådberg, G., Liedberg, S., Holmstedt, A., 1989. Domänindelning av Kirunavaara efter sprickors orientering (in Swedish) (LKAB Internal Report), (Ref no. 89–737).

35. Sandström, D., 2003. State of stress at the Kiirunavaara mine: its significance for rock mechanics analysis (Licentiate Thesis (2003:02)) Luleå University of Technology, Luleå, Sweden.

36. Singh, V.K., Singh, D., Singh, T.N., 2001. Prediction of strength properties of some schistose rocks from petrographic properties using artificial neural networks. Int. J. Rock Mech. Min. Sci. 38 (2), 269–284. http://dx.doi.org/10.1016/S1365-1609(00)00078-2.

37. Sjöberg, J., Dahnér, C., Malmgren, L., 2011. Forensic analysis of a rock burst event at the Kiirunavaara Mine – results and implications for the future. In: Sainsbury, D., Hart, R., Detournay, C., Nelson, M. (Eds.), Proceedings of the 2nd International FLAC/DEM Symposium. Continuum and Distinct Element Numerical Modeling in Geomechanics–2011. Itasca International Inc., Melboune, Australia, pp. 67–74.

38. Sjöberg, J., Perman, F., Quinteiro, C., Malmgren, L., Dahnér-Lindkvist, C., Boskovic, M., 2012. Numerical analysis of alternative mining sequences to minimise potential for fault slip rockbursting. Min. Technol. 121 (4), 226–235.

39. Stavropoulou, M., Exadaktylos, G., Saratsis, G., 2007. A Combined Three-Dimensional Geological-Geostatistical-Numerical Model of Underground Excavations in Rock. Rock Mech. Rock. Eng. 40 (3), 213–243. http://dx.doi.org/10.1007/s00603-006- 0125-4.

40. Vatcher, J., McKinnon, S.D., Sjöberg, J., Dahnér, C., 2014. Modelling methodology: Structural geology and rock mass behaviour at Kiirunavaara Mine. Rock Engineering and Rock Mechanics: Structures in and on Rock MassesThe 2014 ISRM European Rock Mechanics Symposium (Eurock 2014). CRC Press, Vigo, Spain, pp. 643–648.

41. Westblom, M., 2014. Bergmekaniska tester av Syenitporfyrer, Leptiter och Granit (in Swedish). (Draft Master's Thesis) Luleå University of Technology, Luleå, Sweden.

42. Zhou, W., Maerz, N.H., 2002. Implementation of multivariate clustering methods for characterizing discontinuities data from scanlines and oriented boreholes. Comput. Geosci. 28 (7), 827–839. http://dx.doi.org/10.1016/S0098-3004(01)00111-X.

CHAPTER 4

INITIATION, MOVEMENT, AND RUN-OUT OF THE GIANT TSAOLING LANDSLIDE — WHAT CAN WE LEARN FROM A SIMPLE RIGID BLOCK MODEL AND A VELOCITY–DISPLACEMENT DEPENDENT FRICTION LAW?

Che-Ming Yang [a], Wei-Lun Yu [a], Jia-Jyun Dong [a], Chih-Yu Kuo [b], Toshihiko Shimamoto [c], Chyi-Tyi Lee [a], Tetsuhiro Togo [d], Yuki Miyamoto [e]

[a] Graduate Institute of Applied Geology, National Central University, Taiwan

[b] Research Center for Applied Sciences, Academia Sinica, Taiwan

[c] State Key Laboratory of Earthquake Dynamics, Institute of Geology, China Earthquake Administration, China

[d] Seismotectonics Research Group, Institute of Earthquake and Volcano Geology, National Institute of Advanced Industrial Science and Technology, Japan

[e] Kimitsu Works, Nippon Steel & Sumitomo Metal Corpration, Japan

ABSTRACT

Tsaoling landslide is the largest and best documented landslide among several large landslides induced by the 1999 Taiwan Chi-Chi earthquake. Pliocene sedimentary rocks of about 125 Mm^3 in volume slid along very flat bedding planes dipping by 14° with an average speed of 35–40 m/s for about 1650 m, before hitting the bank of the Chinshui River during the landslide. Detailed analysis of DTMs before and after the earthquake using a GIS software leads to an accurate determination of the locations of the centroids of landslide mass, revealing the horizontal and vertical displacements of the 2524 m and 524 m, respectively. Those displacements and landslide mass give an apparent friction coefficient of 0.21 and the release of the potential energy of 1.6 × 10^{15} J. We conducted rotary-shear high-velocity friction experiments on fault gouge from bedding-parallel faults under semi-wet conditions and at 3 MPa normal stress corresponding to the overburden pressure of the landslide mass. We also compiled reported data on the frictional properties on shale powders and fault gouge from the landslide site under both dry and wet conditions, and proposed a velocity–displacement dependent friction law that can describe most experimental data. Newmark analysis of landslide motion with six scenarios for different landslide materials and conditions, assuming a simple rigid block sliding and using measured frictional

Citation: Che-Ming Yang, Wei-Lun Yu, Jia-Jyun Dong, Chih-Yu Kuo, Toshihiko Shimamoto, Chyi-Tyi Lee, Tetsuhiro Togo, Yuki Miyamoto, Initiation, movement, and run-out of the giant Tsaoling landslide — What can we learn from a simple rigid block model and a velocity–displacement dependent friction law?, Engineering Geology, Volume 182, Part B, 19 November 2014, Pages 158-181, ISSN 0013-7952, http://dx.doi.org/10.1016/j.enggeo.2014.08.008.

parameters, revealed that the landslide did not occur with dry frictional properties, and that the landslide occurred at 38–39 s with accumulated displacements of 0.62 m–1.09 m and reached at the river bank at 82–87 s after the generation of Chi-Chi earthquake at its epicenter. Those timings are consistent with high-frequency signals at 32–40 s and at 76 s recorded at a nearby seismic station and with a survivor's witness that the landslide initiated 10 s after he felt strong ground motion, possible S wave arrival at 25.2 s. Slip-weakening is essential in initiating the landslide and low friction coefficient (0.08–0.1) allowed high-speed of the landslide possible. The landslide was caused by a few peaks of northeast-oriented strong accelerations of the ground motion. Frictional work during the sliding of the landslide mass was estimated to be of about 23% of potential energy, and the rest of the released energy is likely to have been consumed during the stopping phase of the landslide after hitting the river bank in complex processes such as fragmentation, heat dissipation, and spreading of the landslide deposits.

INTRODUCTION

Giant landslides frequently induce catastrophic hazards. A key issue for mitigating landslide hazards is how to predict the processes of triggering, movement, and deposition of a landslide mass. Numerical simulations are widely used approach to predict the kinematics of large landslides that have fast sliding velocities and long run-out distances (e.g., Campbell, 1989, Cleary and Campbell, 1993, Campbell et al., 1995,Hungr, 1995, Hungr et al., 2001, Legros, 2002, Crosta et al., 2003, Tang et al., 2009 and Wu and Chen, 2011). However, it is not easy, if not impossible, to determine the appropriate parameters for the numerical model. In the case of high-speed rock sliding, frictional properties along the sliding surface at high velocities and large shear displacements are the most important factor that dominates the kinematics of large landslides (e.g., Voight and Faust, 1982, Tika and Hutchinson, 1999 and Chang et al., 2005). Dong et al. (2009) showed that the kinematics of landslides can be depicted well using a simple rigid block model that incorporates a velocity-dependent friction law. This paper attempts to combine laboratory studies on the high-velocity frictional properties of sliding surface and a Newmark analysis of landslide motion, for a realistic prediction of the Tsaoling landslide. This landslide is the largest and best-documented landslide among several large landslides triggering by the 1999 Taiwan Chi-Chi earthquake (M_w = 7.6) with its volume of about 125 Mm^3 (Hung, 2000, Weissel and Stark, 2001,Hung et al., 2002, Chigira et al., 2003, Wang et al., 2003, Chen et al., 2004, Chang et al., 2005, Chen et al., 2005, Chen et al., 2006, Kuo et al., 2009, Miyamoto et al., 2009, Tang et al., 2009, Wu and Chen, 2009, Wu and Chen, 2011, Chang et al., 2013, Kuo et al., 2013, Chen et al., 2014 and Togo et al., 2014). In particular, a landslide mass of the Pliocene Cholan Formation (about 140 m in thickness) slid along very flat bedding planes within alternated fine sandstone and shale beds dipping by 14° on the southern slope of the Tsaoling Mountain, as we discuss Fig. 1, Fig. 2 and Fig. 3 later. Thus, the geometry and kinematics of the Tsaoling landslide are ideal for a simple rigid-block analysis with frictional properties, using the Newmark method (Newmark, 1965).

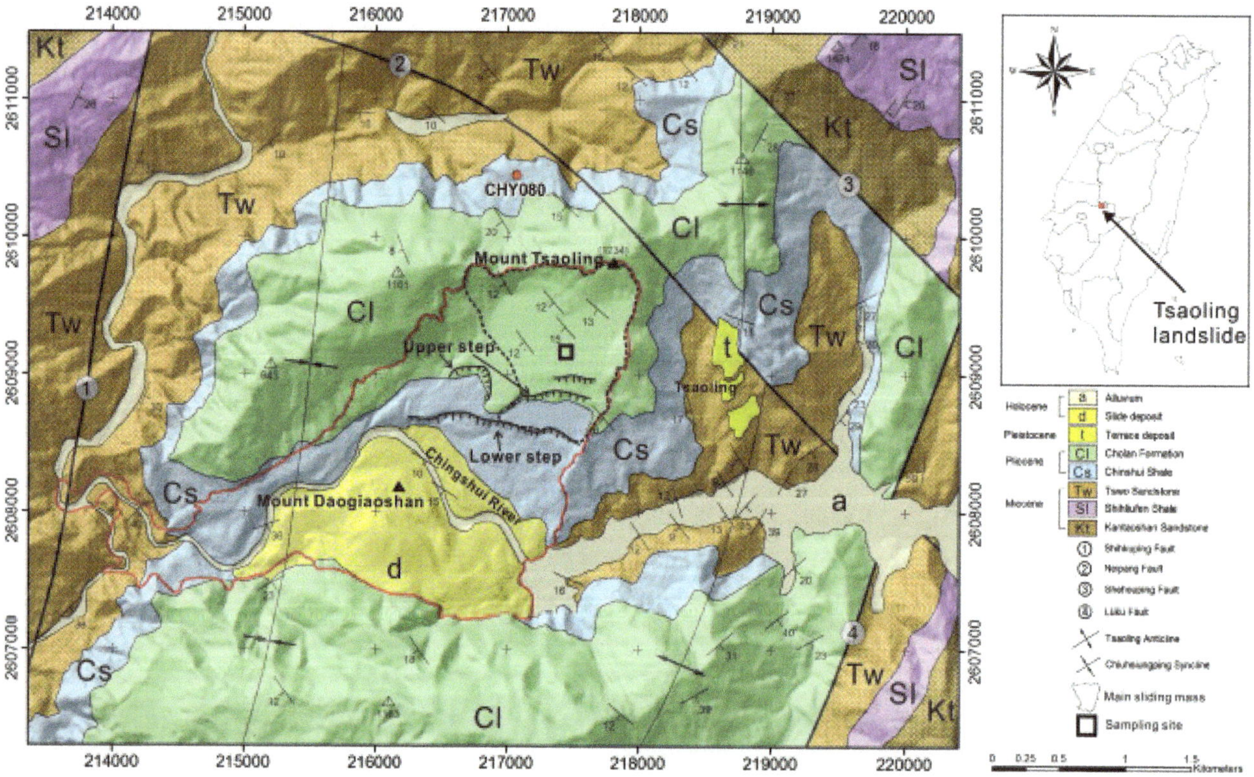

Figure 1: Geological map of the Tsaoling landslide site (red curve) and the neighboring area. The strong motion station CHY080 is located north of the Tsaoling landslide (red circle). The location of the landslide is shown by a red square in the inset map of Taiwan.

Figure 2: Photographs showing (a) a side view of the Tsaoling landslide surface, (b) Mt. Daogiaoshan on the SW side of the Chingshui River, where the Tsaoling landslide mass was deposited, (c) sampling location of the gouge, and (d) a bedding-parallel fault with gouge zone approximately 10 mm thick. During the Sept. 21, 1999 Chi-Chi earthquake, the landslide mass slid from right to left along the bedding plane (yellow dashed line in Figure 2a) above the upper step. The gouge was soft and easily removed using metal tools.

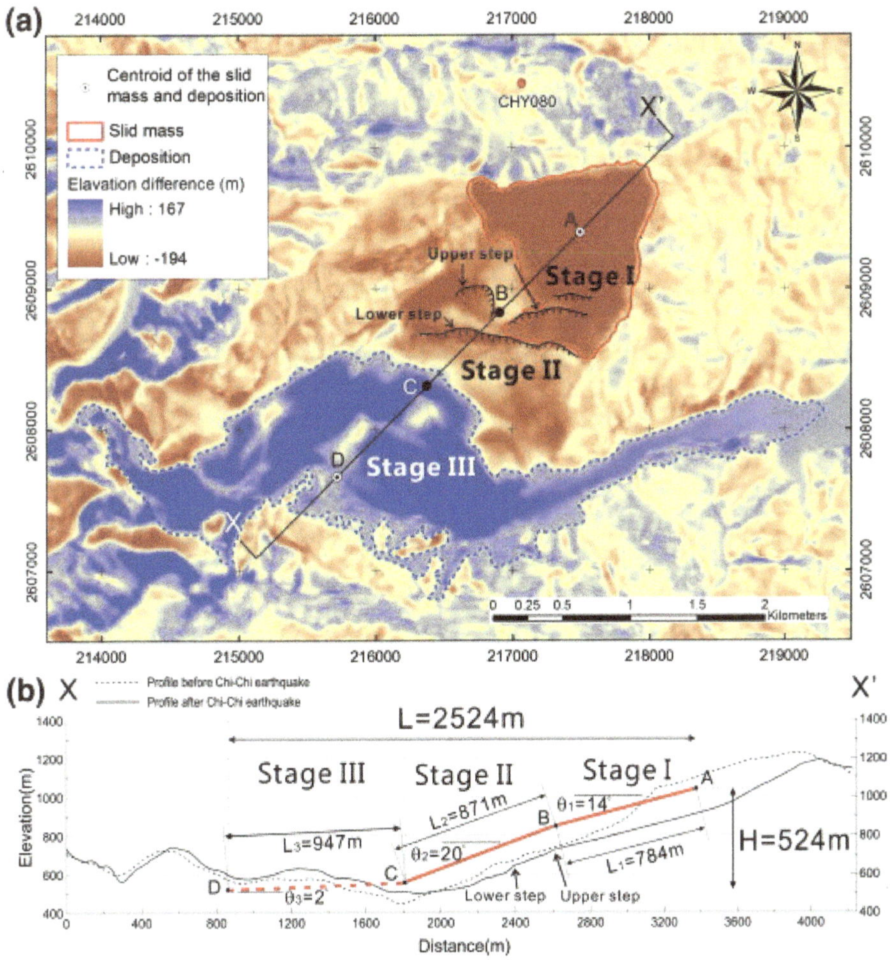

Figure 3: (a) A color contour map of the Tsaoling landslide site showing the differences in elevation before and after the landslide, derived from pre- and post-event DTMs. The warm and cool colors indicate areas of erosion and deposition, respectively. Stages I to III and the re-

lated boundary points are marked in the figure. (b) Profile XX′ with the dashed and solid lines showing the ground surfaces before and after the Chi-Chi earthquake (Tsaoling landslide). The red solid and dashed lines represent the trajectory of the moving mass in Stages I to III, using the centroids of the landslide mass that were determined from the DTMs.

After sliding for about 800 m along bedding planes dipping 14° and then for about the same distance along steeper slope dipping about 20°, the Tsaoling landslide mass hit a southwestern bank of Chinshui river and deposited over Mt. Daoqiaoshan of about 60 m in height (e.g., Chen et al., 2004; Chigira et al., 2003, Chen et al., 2006 and Tang et al., 2009). A strong motion seismometer at station CHY080 (TWD97: X = 217,073, Y = 2,610,450, EL. 840 m) located at about 800 m to the north from the center of the Tsaoling landslide mass recorded high-frequency signals starting at 75.8 s after the generation of Chi-Chi earthquake and lasting for about 1 s. Kuo et al. (2009) interpreted that the signals are due to the impact of the landslide mass to the river bank. Moreover, thirty-nine people in a village on the landslide mass slide with the landslide and seven of them survived (Chen et al., 2004). We found a survivor who witnessed that the landslide motion initiated about 10 s after he felt "strong shaking", possibly due to the S wave arrival (25.2 s after the Chi-Chi earthquake). Finally, Chen et al. (2014) found earlier high-frequency signals in the seismic record and proposed that the Tsaoling landslide started at 12–16 s after the S-wave arrival. Those information yields an average speed of landslide motion of 22–42 m/s, or a typical speed of around 30 m/s (Togo et al., 2014, Supplementary Information S1). Thus, the initiation and motion of the Tsaoling landslide are documented well and any model can be tested against those constraints from the seismic records and a survivor's witness.

Traditional direct shear tests at low slip rates yielded a peak friction angle of 39° (friction coefficient of 0.8) and an average residual friction angle of 23° (friction coefficient of 0.43) on the rocks from the Tsaoling landslide (WRA, 2000, Lee, 2001 and Towhata et al., 2001). Huang et al. (2001) conducted a Newmark analysis of the Tsaoling landslide using those values, but the landslide motion stopped upon the reduction of the seismic ground motion. Based on the pseudostatic analysis, Chen et al. (2004) inferred that the Tsaoling landslide could be triggered by the Chi-Chi earthquake with a peak friction angle of 30° (equivalent to a friction coefficient of 0.58). A serious problem in those analyses with the results from the direct shear tests is that the landslide motion stops upon a reduction of seismic ground motion because the residual friction angle (23°) is greater than the dipping angle of the slope (14°). On the other hand, Kuo et al. (2009) used the continuum model of hydraulic flow, SHALTOP2D, which is based on the equations ofBouchut and Westdickenberg (2004) to simulate the dynamic motion of the Tsaoling landslide and obtain a calibrated friction coefficient of 0.09 (friction angle of 5°). Using the code PFC2D and parameters constrained by the final geometry of the landslide deposits,Tang et al. (2009) obtained a calibrated friction coefficient of 0.15 (friction angle of 8.5°). Furthermore, the apparent friction coefficient was calculated from the vertical fall height and the horizontal run-out distance (Heim, 1932, Scheidegger, 1975 and Legros, 2002) to be approximately 0.21 (friction angle of 12°). Those friction angles allow runaway instability of the Tsaoling landslide along a slope with a dip angle of 14°. However, if the initial friction angle is as low as those values, the landslide mass could not stay along a flat surface in the first place. Hence, those analyses do not elucidate the triggering part of the Tsaoling landslide.

Thus, an essential process for triggering the Tsaoling landslide and for causing its high-speed motion is the marked slip-weakening from a high initial friction coefficient (high enough to keep the landslide mass on the slope) to a very low friction coefficient (low enough to allow high-speed sliding). Rotary-shear high-velocity friction experiments in the last two decades demonstrated that such dramatic slip-weakening indeed occurs for frictional sliding of rocks with or without fault gouge (e.g., Shimamoto and Tsutsumi, 1994, Tsutsumi and Shimamoto, 1996, Tsutsumi and Shimamoto, 1997, Hirose and Shimamoto, 2005, Di Toro et al., 2006, Di Toro et al., 2011, Han et al., 2007 and Mizoguchi et al., 2007), and for landslide materials too (Miyamoto et al., 2009,Yano et al., 2009, Ferri et al., 2010, Ferri et al., 2011 and Liao et al., 2011). Friction coefficient reduced to the order of 0.05 to 0.2 at seismic slip rates on the order of 1 m/s, allowing high speed landslide motion possible. Dong et al. (2013) conducted the Newmark analysis and successfully reproduced the Tsaoling landslide using a velocity- and displacement-dependent friction law to describe the results from high-velocity experiments on the Tsaoling landslide materials by Miyamoto et al. (2009). We employed the same approach in the study to refine further the analysis of the Tsaoling landslide motion, (1) by conducting high-velocity friction experiments on the gouge from bedding-parallel faults in the Cholan Formation close to the sliding surface during the Tsaoling landslide, (2) by compiling all existing data on the frictional properties of the Tsaoling landslide materials, and (3) by slightly modifying the velocity- and displacement-dependent friction law in the Newmark analysis of the landslide. We used a rotary-shear low- to high-velocity friction apparatus at the National Central University of Taiwan (LHVR-Taiwan) to study frictional behaviors of the fault gouge. The Tsaoling landslide mass was simplified as a rigid block (centroid of the moving block) and we idealized the landslide motion as the three stages of mass movement from the initiation to stop.

Recently, Togo et al. (2014) performed oscillatory velocity experiments on dry and wet shale gouge from the Tsaoling landslide site, by idealizing the Tsaoling landslide motion during the Chi-Chi earthquake as repeated accelerating/decelerating motion with a frequency ranging from 0.3 to 1.2 Hz. They observed small weakening and strengthening in association of the accelerating/decelerating motion, while undergoing overall slip-weakening. They successfully reproduced the Tsaoling landslide motion by the Newmark analysis with a simple exponential slip-weakening law. We use in this paper a more involved friction law, particularly by including the velocity dependence of the steady-state friction which can affect the high-velocity sliding portion of the Tsaoling landslide. We also compiled all existing friction data, including those by Togo et al. (2014). After our initial manuscript was submitted, an interesting paper was published byChen et al. (2014) who conducted the Newmark analysis of the Tsaoling landslide by combining the results from traditional direct shear tests cited above and from high-velocity friction experiments by Miyamoto et al. (2009). Their results are similar to ours for the high-speed sliding part because they used a very low steady-state friction coefficient (0.05) after Miyamoto et al. (2009). They called it "mobilized friction angle" and distinguished it from the residual friction angle which is derived from the direct shear tests, and assumed arbitrarily that the friction drops instantly from the peak to residual friction, and then friction drops abruptly to mobilized friction at a displacement of 0.5 m. Those assumptions on the changes in friction are oversimplified and Newmark analysis of the triggering portion should be testified with real experimental results.

1999 TSAOLING LANDSLIDE EVENT

Geological background

The Tsaoling landslide is located in central-western Taiwan in the Western Foothills geologic province (Figure 1). The Pliocene Chinshui Shale and the Cholan Formation crop out in the Tsaoling area. The Cholan Formation is more than 1000 m thick and consists of intercalated fine-grained sandstone and shale, whereas the Chinshui Shale consists of massive mudstone and shale having an average thickness of approximately 110 m. The Tsaoling anticline and the Chiusiungping syncline dominate the attitudes of the bedding planes at the landslide site, which have a dip direction and dip angle of 220–230° and 12–15°, respectively (Kuo et al., 2009, Tang et al., 2009 and Lee, 2011). Four historical rock slides have occurred on the southwest slope of Mount Tsaoling between 1862 and 1979 (Hung et al., 2002); the landslides in 1862 and 1941 were triggered by earthquakes, and those in 1942 and 1979 were triggered by heavy rainfall. In addition, a large rock slide with a volume of 125-million m^3 and an average thickness of 140 m slid down Mount Tsaoling during the Sept. 21, 1999 Chi-Chi earthquake.

The Tsaoling landslide is a typical dip-slope rock slide that moved along bedding surfaces (Figure 2a). The Chingshui River cuts into the toe of the dip-slope and eliminates obstacles that would prevent landslide motion. The colluvium of the Tsaoling landslide was deposited on the other (SW) side of the Chingshui River and on Mount Daogiaoshan (Figure 2b). Flexural-slip along bedding shears commonly occur in alternating beds of folded sedimentary rocks (Donath and Parker, 1964, Tanner, 1989 and Tanner, 1992). Bedding-parallel shears or faults that develop in moderate to tight folds or fault-propagation folds, which can facilitate dip slope failures (Hart, 2000 and Lee, 2012). Immediately below the exposed sliding surface in the Tsaoling area, we observed four bedding-parallel faults with outcrop lengths greater than 20 m that were spaced approximately 1 m apart (Figure 2c). We sampled gouge from the fault zones for high-velocity friction experiments (Figure 2d) at the site marked in Fig. 1. X-ray diffraction analysis revealed that the bedding-parallel fault gouge contains quartz, plagioclase, illite, kaolinite, chlorite, and montmorillonite (Miyamoto et al., 2009). The liquid limit and plastic limit of the fault gouges are 23% and 6%, respectively. The fault gouge materials can thus be categorized as low-plasticity clays (CL) according to the Unified Soil Classification System.

Topography and motion of the Tsaoling landslide

Based on the post-event topography, two steps are present (Figure 1; upper and lower steps) in the landslide source area, and the three outcropping planar sliding surfaces parallel to the bedding plane in alternating beds (Chigira et al., 2003). Fig. 2a shows the upper step and the main sliding surface (yellow line). The main sliding surface is located within the Cholan Formation, but the lower part extends into the Chinshui Shale (Chen et al., 2014). Approximately 20% of the slide mass dropped into the Chingshui River, which created a 5-km-long landslide dam and reservoir. The landslide deposit included rock blocks greater than 10 m in diameter. A village with 39 residents and visitors slid with the landslide mass, and some of the houses were moved to the top of Mount Daogiaoshan without being buried in debris (cf. Chen et al., 2004 and Chen et al., 2014). Surprisingly, 7 people survived the catastrophic landslide.

We determined the centroids of the sliding mass (Point A) and landslide deposit (Point D) in the following manner by determining the distribution of the elevation-difference using the pre- and post-event DTMs (Figure 3a). We made slices of landslide body and landslide deposits of 5 m in width using a GIS software and determined the mass distributions in the slices. Then the moment balance equations can be solved for the coordinates of the centroids as unknowns for accurate determinations of the centroids, plotted on a profile XX′ as points A and D. The horizontal and vertical displacements from A to D were 2524 m and 524 m, respectively, and the apparent friction angle is 0.21. The horizontal projection of the vector from Point A to Point D is almost identical to the sliding direction (dip direction, 226°). Fig. 3b shows profile XX′, which is located along the vector from A to D. Points B and C are the locations of the upper step and the Chingshui River Valley, respectively. The landslide process can be summarized as follows:

The initiation of movement of the mass (Point A) was triggered by the earthquake, and the mass slid in the dip direction of the bedding plane. The friction on the sliding plane controlled the mass movement. When the landslide mass passed through the upper (Point B) and lower steps (Fig. 1 and Fig. 3), it "dropped" slightly and continued sliding along the bedding planes. The mass then "moved through" in the Chingshui River Valley (Point C) and finally hit Mount Daogiaoshan (Point D). A large amount of energy would have been dissipated by the collision and disintegration of the fast-moving mass. After the collision, the mass flowed up- and down-stream along the river bed. Accordingly, three stages of the Tsaoling landslide can be identified based on the aforementioned processes (Figure 3a and b). Stages I, II, and III are identified as the motions from A to B, B to C, and C to D, respectively.

In view of the kinematics of the Tsaoling landslide, we idealize the Tsaoling landslide as follows. (1) In Stage I (from A to B in Figure 3b), the landslide movement is idealized as a rigid-body sliding of the landslide mass along a flat bedding plane dipping 14°. (2) In Stage II (from B to C), the landslide mass went over two steps on surface topography (Figure 2a) as it entered into Stage II and must have fragmented to some degree, but we treat it as a rigid-block sliding along a steeper plane dipping 20°. (3)In Stage III (from C to D), the landslide block must have broken into fragments when it collided against the southwestern bank of the Chingshui River (Point C), and the surface topography is quite irregular from point C to the Mt. Daogiaoshan (Figures 2b and 3b). Thus a simple rigid-block motion along a planar surface is no longer applicable to the stopping phase of the landslide in Stage III, and this paper will not attempt to analyze the landslide behavior in Stage III. With those idealization of landslide motion in Stages I and II, frictional properties along the sliding surfaces are critical in conducting the landslide analysis. Thus, the next two sections will be devoted to the high-velocity frictional properties of the landslide materials.

FRICTION EXPERIMENTS ON BEDDING-PARALLEL FAULT GOUGE — LOW- TO HIGH-VELOCITY ROTARY-SHEAR TESTS

A low- to high-velocity rotary-shear friction apparatus

A high-velocity rotary-shear friction apparatus developed by Shimamoto and Tsutsumi (1994) and Hirose and Shimamoto (2005) has been widely used in friction experiments on fault rocks (e.g., Di Toro et al., 2011; references therein). The fifth author (TS) developed a second machine that expanded the capability of the apparatus to a wide range of velocities from 5 mm/yr to almost 10 m/s (see Togo and Shimamoto (2012) for the details of the second apparatus). In this study, we used a third-generation low- to high-velocity rotary-shear friction apparatus (Marui Co. Ltd, Osaka, Japan, MIS-233-1-78), that was designed by TS and installed at the National Central University of Taiwan in 2011 (Figure 4). The apparatus is capable of producing slip rates of 30 mm/yr to 1.3 m/s on 25-mm-diameter solid cylindrical specimens at normal stresses up to 20 MPa. A unique feature of the apparatus is a large chamber in which a variety of specimen assemblies, such as a hydrothermal pressure vessel, can be developed and mounted easily.

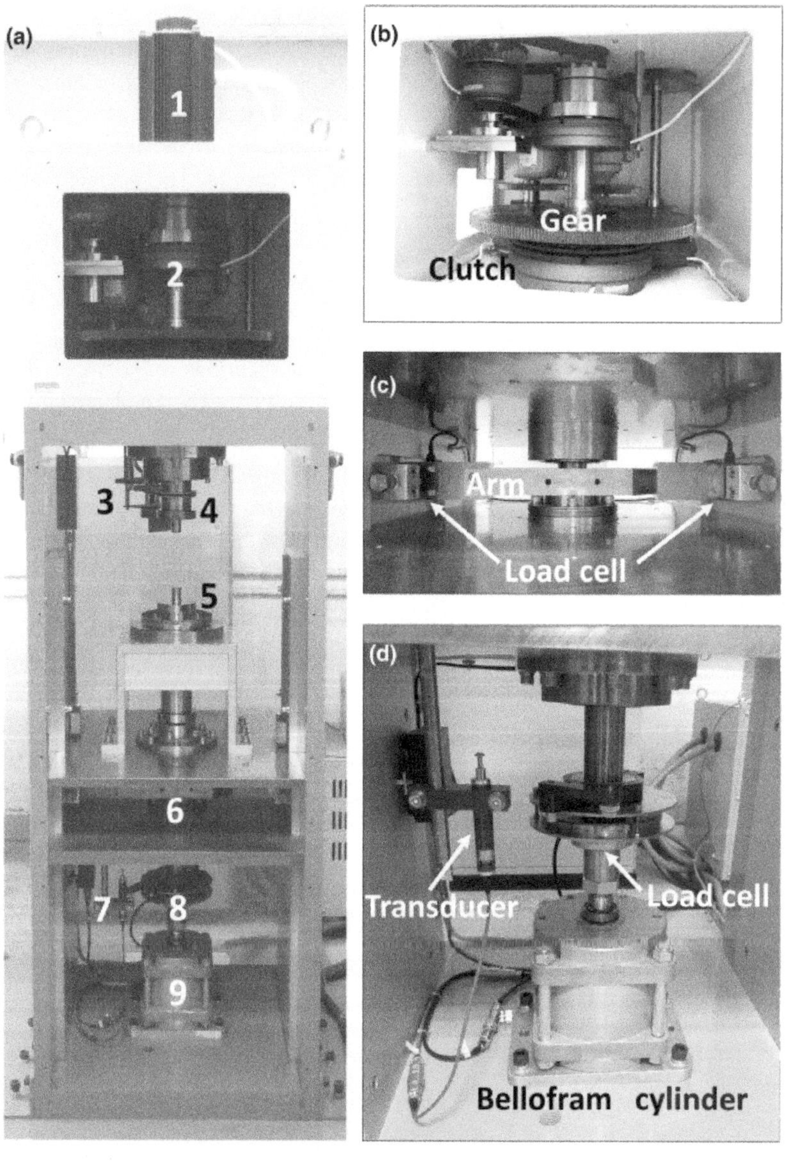

Figure 4: Photographs showing (a) the rotary-shear low- to high-velocity frictional testing apparatus at the National Central University of Taiwan (LHVR-Taiwan, built by Marui Co. Ltd., Osaka, Japan), (b) the gear/belt system for three ranges of velocities, (c) the cantilever-type

torque gauge, and (d) the bottom part with a Bellofram cylinder, an axial load cell, and an axial displacement transducer. The part numbers in (a) are used in the description of the apparatus in the text.

The apparatus consists of a 7.5 kW servomotor (1 in Figure 4a), a gear/belt system for low, intermediate and high velocities (2), a rotary encoder and potentiometer for monitoring the rotary motion (3), an upper rotary piston (4), a stationary lower piston (5), a cantilever-type torque gauge (6), an axial displacement transducer (7), an axial force gauge (8), and an air actuator (Fujikura Composites, Bellofram Cylinder) for applying axial loads up to 10 kN (9). The gear/belt system is composed of belts, gears, two gear boxes, and five electromagnetic clutches (Figure 4b) and can achieve a range of slip rates of almost 9 orders of magnitude by changing the motor speed (1.5–1500 rpm) and/or gear/belt lines. Five electromagnetic clutches allow velocity increases of 10^3 or 10^6 times by changing the gear/belt lines without stopping the motor. Torque is measured by a cantilever-type force gauge with an arm and two force gauges at both ends (Figure 4c). The normal stress on the sliding surfaces can be kept nearly constant even if the specimen shortens by controlling the air pressure in the Bellofram cylinder (Figure 4d) with a regulator. The axial force and axial displacements are measured with a strain-gauge-based force gauge and a displacement transducer, respectively (Figure 4a).

Hollow or solid cylindrical specimens with outer diameters of 25 mm are set between points 4 and 5 in Fig. 4a using an ETP bush from Miki Pulley Co. Ltd., Japan. This bush is a cylindrical flat-jack that can be tightened by turning a bolt with a torque wrench to hold a specimen tightly against the inner wall of the cylinder. The specimen has to be slightly less than 25 mm in diameter to prevent permanent deformation of the ETP bushes. A disadvantage of rotary shear apparatuses is the large gradient in slip rate across the sample from zero in the center to the maximum at the periphery. Because the shear velocity is not constant along the radius of the cylinder, an equivalent slip velocity (Veq) is defined such that VeqS gives the rate of frictional work over a sliding surface with an areaS, assuming a uniform shear stress (Shimamoto and Tsutsumi, 1994 and Hirose and Shimamoto, 2005). For hollow cylindrical specimens with outer and inner diameters r_1 and r_2, respectively, Veq is given by:

$$V_{eq} = \frac{4\pi R\left(r_1^2 + r_1 r_2 + r_2^2\right)}{3(r_1 + r_2)} \tag{1}$$

where R is the revolution rate of the motor. We used solid cylindrical specimens that were 25 mm in diameter ($r_1 = 0.0125$ m and $r_2 = 0$ m), and the maximum revolution rate of 1500 rpm gives a maximum equivalent slip rate of 1.3 m/s. Hereafter, the equivalent slip velocity is referred to simply as "velocity" or "slip rate" V.

Experimental procedures

Friction experiments were conducted on fault gouge collected from near the sliding surface of the Tsaoling landslide under a normal stress (σn) of 3 MPa, at slip rates of 0.0023 to 1.30 m/s and with displacements up to 53 m. The normal stress corresponds to the overburden pressure of the 140-m-thick sliding mass and the density of rock is 2430 kg/m^3 (Hung, 1982). The tests were performed at room temperature and room humidity conditions. We conducted the friction experiments on the fault gouge following the procedures of Mizoguchi et al. (2007). The fault gouges were sandwiched by a pair of solid cylinder of gabbro samples that were 25 mm in diameter and surrounded by a Teflon sleeve (Figure 5). The inner diameter of the Teflon sleeve was approximately 100 μm smaller than the specimen diameter to impose a pressure on the specimen and prevent the loss of fault gouge (see Appendix of Sawai et al. (2012) for a description of problems caused by Teflon sleeves). The sliding surfaces of the specimens were roughened by grinding them with #80 silicon carbide abrasive (180 μm in size) so the fault gouge would deform rather than slipping at the gouge-rock interfaces. The specimen alignment was adjusted to within several microns to avoid distortion of the Teflon sleeve that would lead to fault gouge leakage and erratic frictional behavior. Gouge particles smaller than 150 μm were selected using a #100 sieve to avoid mixing of large particles in the approximately 1-mm-thick gouge zone (1 g of gouge). Distilled water was added to the samples to keep the water content (w) nearly identical to the natural water content of the fault gouge (~ 25%). The details of preparing the samples can be found in Togo et al. (2011). After the gabbro specimens with fault gouge were placed in the apparatus, the axial stress was increased to 3 MPa within approximately 1 min, and the specimens were left at this normal stress until the axial displacement due to gouge compaction had stopped. It usually took approximately 10–15 min to fully consolidate the gouges, and some water was squeezed out during consolidation; i.e., the water content in the gouges during the experiments was less than 25%.

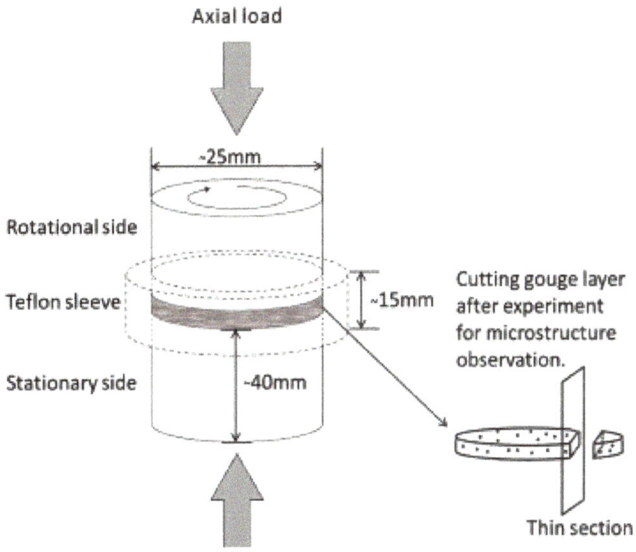

Figure 5: Schematic of the sample assembly used in the friction experiments. Thin sections of the deformed fault gouge were prepared for microstructural observations along a surface shown in the lower-right corner.

Experimental results on bedding-parallel fault gouge

The results from the friction experiments on the gouge from the bedding-parallel faults at a slip rate V of 0.0023 to 1.3 m/s reveal three types of frictional behavior: (1) slip-strengthening behavior to the peak friction followed by slight slip-weakening or nearly constant friction (Figure 6a; V = 0.0023, 0.0124, 0.0436, and 0.1 m/s); (2) transitional behavior with slip-strengthening to the peak friction followed by slip-weakening (Figure 6b; 0.22 m/s); and (3) marked slip-weakening behavior (Figure 6c to f; V = 0.43, 0.50, 0.87, and 1.30 m/s). The slip-strengthening behavior at low slip rates is similar to the results from traditional direct shear tests of normally consolidated clay. The ultimate friction coefficients μult, defined as the friction coefficient as the shear displacement approaches infinity, were within a narrow range of 0.4–0.5 (Figure 6a). We calculated μultby taking an average of the measured friction coefficients at shear displacements of greater than 2 m (Table 1). The value of μult increases with decreasing slip rate; this could be due to the loss of water from the gouge in the slow experiments. At shear velocities greater than 0.43 m/s, the peak friction was reached very quickly and was followed by marked slip-weakening (Figure 6c–f). The post-peak slip-weakening can be described by an empirical slip-weakening law after Mizoguchi et al. (2007):

$$\mu(d)=\mu ss+(\mu_p-\mu ss)\exp(\ln(0.05)d/D_c) \quad (2)$$

where μ(d) is the friction coefficient at a displacement d, μp, and μss are the peak and steady-state friction coefficients (friction coefficient when the displacement approach to infinity), respectively, and the slip-weakening distance Dc is defined as the displacement at which μ decreases by 95% (μp − μss). This definition can be confirmed by substitution of d = Dc into Eq. (2), which yields (μ(Dc) − μss) = 0.05 (μp − μss) (seeFigure 7 and the dashed lines in Figure 6c–f). The slip-weakening parameters μp, μss, and Dc are listed in Table 1.

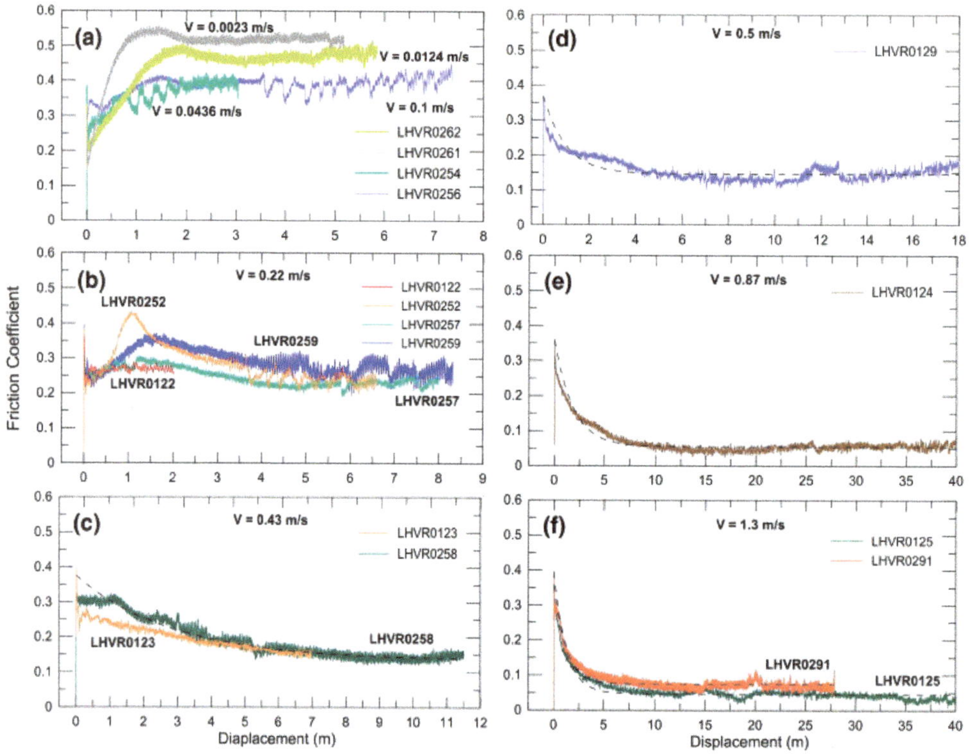

Figure 6: Curves of friction coefficient versus shear displacement for fault gouges from a bedding-parallel fault near the sliding surface of the Tsaoling landslide showing (a) slip-strengthening behavior, (b) transitional behavior, and (c) to (f) slip-weakening behavior. The shear velocities are given in the diagrams. The dashed lines show the least-squares fits to the experimental data using an empirical slip-weakening law (Eq. (2)).

Table 1: Experimental conditions and slip-weakening parameters in Eq. (2) for the bedding-parallel fault gouge.

	Slip velocity [m/s]	σn [MPa]	Test numbers	μult	μp	μss	Dc[m]
Slip-strengthening regime	0.0023	3	LHVR0261	0.52	–	–	–
	0.0436	3	LHVR0254	0.47	–	–	–
	0.0124	3	LHVR0262	0.39	–	–	–
	0.10	3	LHVR0256	0.39	–	–	–
Mean value				0.44			
Transition regime	0.22	3	LHVR0122	–	0.36	0.27	–
	0.22	3	LHVR0252	–	0.38	0.24	–
	0.22	3	LHVR0257	–	0.38	0.23	–
	0.22	3	LHVR0259	–	0.40	0.27	–

Slip-weakening regime	0.43	3	LHVR0123	–	0.40	0.18	8.55
	0.43	3	LHVR0258		0.36	0.13	
	0.50	3	LHVR0129	–	0.37	0.15	2.94
	0.87	3	LHVR0124	–	0.36	0.05	5.60
	1.30	3	LHVR0125	–	0.36	0.05	3.97
	1.30	3	LHVR0291	–	0.40	0.08	3.42
Mean value					0.38		4.90

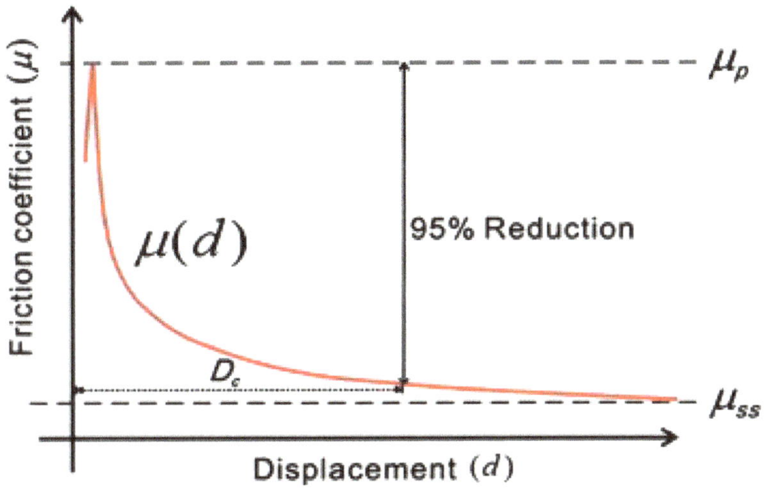

Figure 7: Schematic diagram showing a slip-weakening behavior described by an empirical law (Eq. (2)).

The behavior at V = 0.22 m/s is transitional between the slip-strengthening and the slip-weakening behaviors; i.e., the peak friction was reached at d = 1–2 m and it is followed by minor slip-weakening (Figure 6b). For the transitional behaviors, we read μp directly from the curves and determined μss by taking an average of the measured values of μ ford > 2 m for LHVR0122 and d > 4 m for LHVR0252, 0257, and 0259 (see Table 1 for the values). The water contents of the samples generally decreased from 25% to 18% after consolidation.

The derived ultimate and peak friction coefficients are compared with those that were obtained from direct shear tests. First, we compiled three documented strength datasets of shale samples from the Tsaoling landslide site, using traditional direct shear test at slip rates of 10^{-6} to 10^{-5} m/s (WRA, 2000, Lee, 2001 and Towhata et al., 2001), and compared with our experimental results (Figure 8). The peak friction coefficient of the tested shale with its natural water content (w = 2.2–5.6% after the direct shear test) was approximately 0.8 (Lee, 2001). The residual friction coefficients ranged from 0.3 to 0.6 (WRA, 2000, Lee, 2001 and Towhata et al., 2001) and had an average value of 0.43 (Figure 8). It is notable that the lowest residual friction coefficient of approximately 0.3 was obtained for a slickensided discontinuity in a shale sample.

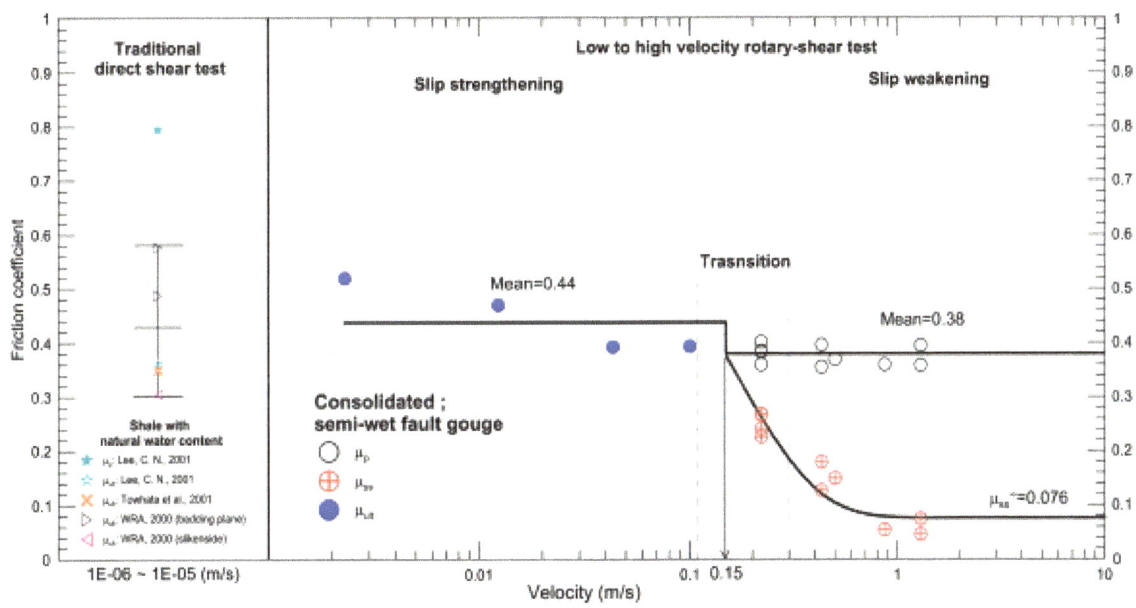

Figure 8: Friction coefficients of the tested bedding-parallel fault gouge collected from the Tsaoling landslide site and fitting with our idealized velocity–displacement friction law (black lines). μult is the friction coefficient at velocities less than Vcritical (= 0.15 m/s), μp and μss

are the peak and steady-state friction coefficient in the slip-weakening regime, respectively. The left part shows the results of traditional direct shear tests (WRA, 2000, Lee, 2001 and Towhata et al., 2001).

In the slip-strengthening regime (velocities from 0.001 to 0.1 m/s), the average ultimate friction coefficient (μult; blue circles) of the tested wet fault gouges is 0.44 (Table 1 andFigure 8), which is nearly identical to the friction coefficient obtained from traditional direct shear tests of wet shale (0.43). In the transitional and slip-weakening regimes (velocity > 0.1 m/s), the average peak friction coefficient (μp; hollow circles) is 0.38 (Table 1 and Figure 8), which is slightly less than the ultimate value in the slip strengthening regime. As mentioned previously, the friction coefficient decreased from a peak value to a steady-state value with increasing displacement in the slip-weakening regime. The steady-state friction coefficients (μss; red circles with crosses) decreased from approximately 0.27 to 0.05 with increasing velocity (Figure 8), which has been observed in numerous studies of high-velocity rotary shear tests (Miyamoto et al., 2009 and Togo et al., 2014). The slip-weakening distances (Dc) for our data range from 2.94 to 8.55 m with an average value of 4.90 m (Table 1).

Previous studies (e.g., Togo et al., 2011) proposed a velocity dependent steady-state friction coefficient that can be expressed by:

$$\mu_{ss}(V) = \mu_{ss}^{\infty} + \left(\mu_p - \mu_{ss}^{\infty}\right)\exp\left(-V/V_c\right) \quad (3)$$

where the friction coefficient μss$^{\infty}$ is defined as the steady-state friction coefficient when the shear velocity approaches infinity and Vc is a material constant. This equation was proposed to describe a change in μss only in the slip-weakening regime, so that we modified the equation to describe μss in both slip-weakening and strengthening regimes as follows. The frictional coefficient in the slip-strengthening regime was assumed to be a constant (=μult) when the shear velocity is less than a critical velocity Vcritical and μss was described by a similar equation as in Eq. (2) for velocity greater than Vcritical.Then the changes in friction coefficients are shown by thick solid curves in Fig. 8 and given by:

$$\begin{aligned} &\mu(V) = \mu_{ult} = 0.44 && \text{for } V < V_{critical} \\ &\mu_{ss}(V) = \mu_{ss}^{\infty} + \left(\mu_p - \mu_{ss}^{\infty}\right)\exp[-(V - V_{critical})/V_{critical}] && \text{for } V \geq V_{critical} \\ &\mu_p = 0.38 && \text{for } V \geq V_{critical} \end{aligned} \quad (4)$$

where μp is the peak friction coefficient in the slip-weakening regime. In this study, Eqs. (2) and (4) were used to calculate the velocity–displacement dependent friction coefficient for the Newmark analysis. The calibrated parameters are as follows:μult = 0.44, μp = 0.38, μss$^{\infty}$ = 0.076, Dc = 4.9 m (an average of Dc in Table 1), andVcritical = 0.15 m/s (cf. black curve in Figure 8).

Microstructural observations

Microstructural observations were made on thin sections of the deformed fault gouge samples impregnated with epoxy. The samples were cut parallel to the specimen axis between the center and margin of the specimens (Figure 5). A field emission environmental SEM (JEOL JSM-7000 F) with an accelerating voltage of 15 or 20 kV was utilized to observe the microstructures. Back-scattered electron images (BSIs) of carbon-coated thin sections are shown in Fig. 9, Fig. 10, Fig. 11, Fig. 12 and Fig. 13. The top of the BSIs are the rotational sides, and the arrows indicate the shear sense. The black areas represent the epoxy used to glue the sample.

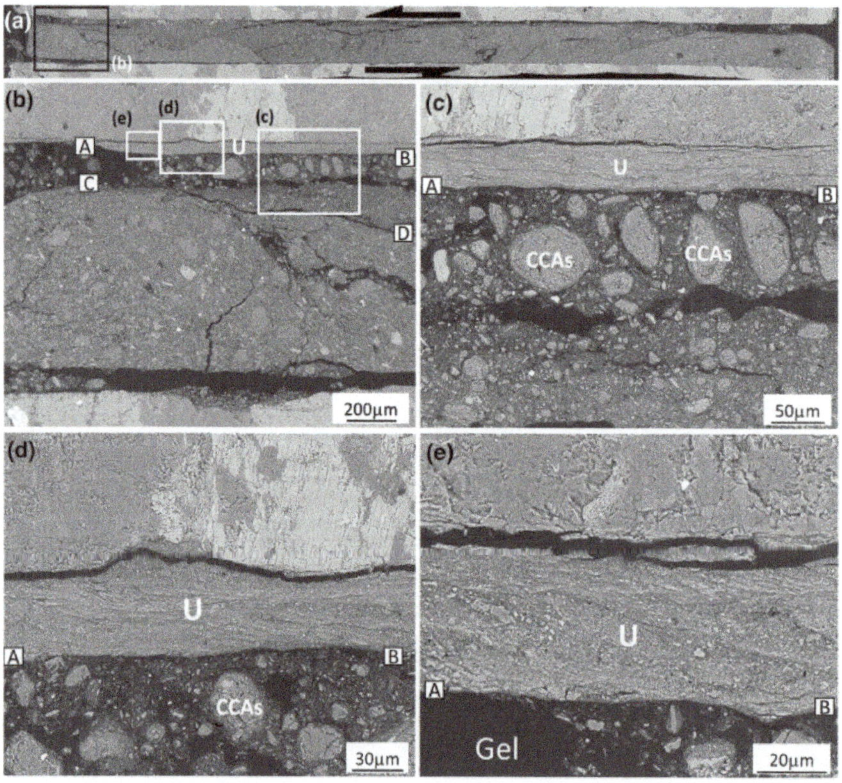

Figure 9: Back-scattered electron images of fault gouge deformed at a normal stress of 3 MPa and a slip rate of 1.3 m/s (LHVR0125). (a) Photomicrograph showing the textures of the deformed fault gouge. (b) a close-up photomicrographs of the rectangular frame in (a), and (c),

(d) and (e) are close-up photomicrographs of rectangular portions with corresponding marks in (b). The U layer and CCA layer are above and below AB, respectively. Although CD is a sharp surface, CCAs are present above and below CD. It mainly comprised of foliated mineral with few quartz grains.

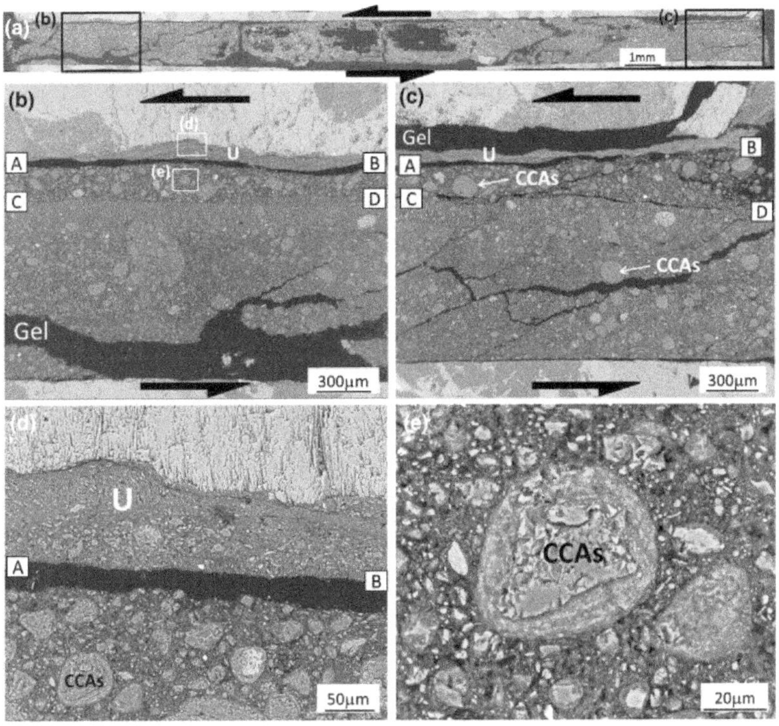

Figure 10: Back-scattered electron images of fault gouge deformed at a normal stress of 3 MPa and a slip rate of 0.87 m/s (LHVR0124). (a) Photomicrograph showing the textures of the deformed fault gouge. (b and c) Detailed photomicrographs of the stratified fault gouge zone. The U layer and CCA layer are above and below AB, respectively. Although CD is a sharp surface, CCAs are present above and below CD. (d) Detailed photomicrograph of the U layer. (e) Detailed photomicrograph of CCAs between AB and CD. The thickness of the cortex of this CCA is approximately 10 μm.

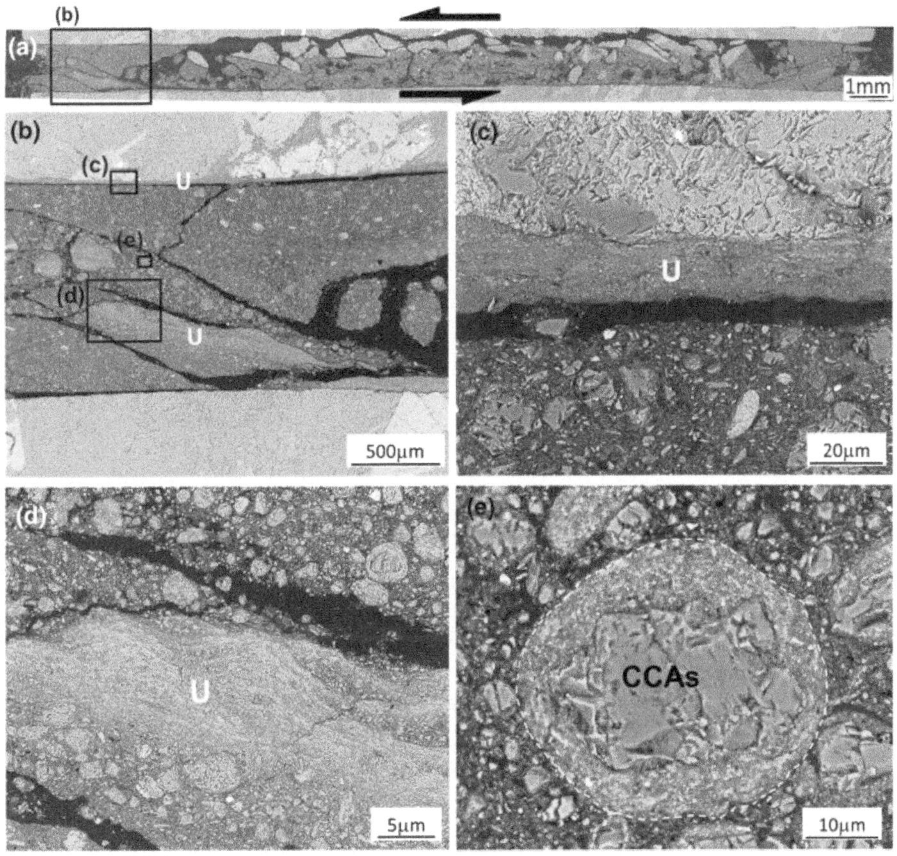

Figure 11: Back-scattered electron images of fault gouge deformed at a normal stress of 3 MPa and a slip rate of 0.5 m/s (LHVR0129). (a) Photomicrograph showing the textures of the deformed fault gouge. (b) Detailed photomicrograph of the stratified fault gouge zone, including

two U layers and spherical CCAs distributed throughout the fault gouge zone with the exception of the U layer. (c and d) The stratified fault gouge zone contains the upper and middle U layers and is mainly comprised of foliated minerals with few quartz grains. The middle U layer contacts the overlying CCA layer along a well-defined boundary, while the boundary between this fault gouge layer and the underlying CCA layer cannot be identified clearly. (e) Detailed photomicrograph of CCAs near the middle U layer.

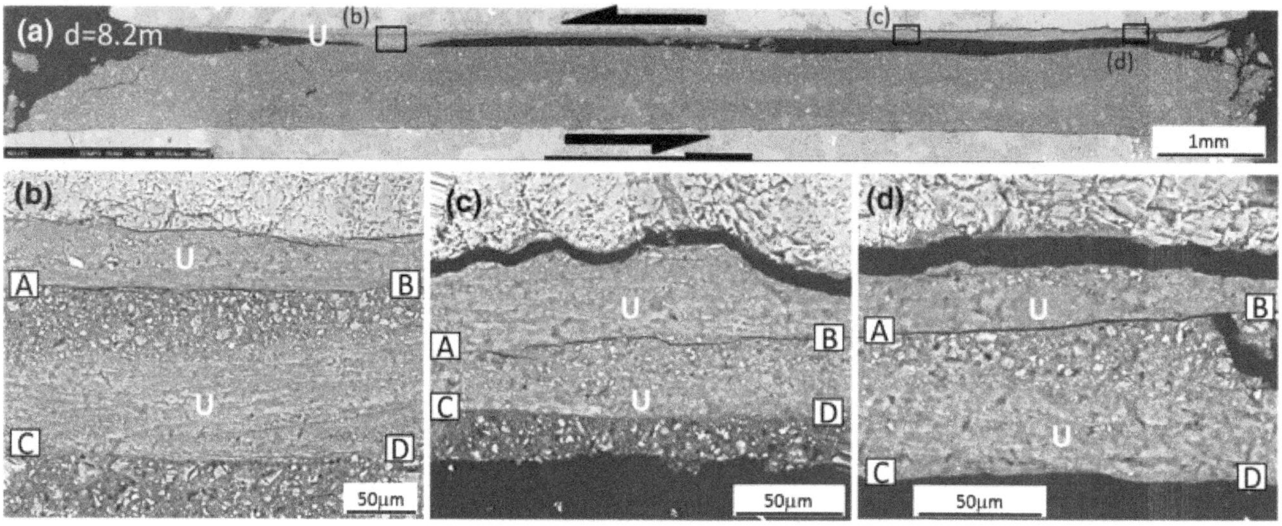

Figure 12: Back-scattered electron images of fault gouge deformed at a normal stress of 3 MPa and a slip rate of 0.22 m/s (LHVR0259, transitional regime). (a) Photomicrograph showing the texture of the deformed fault gouge. (b to d) Two U layers above AB and CD observed at this scale. The U layer was separated by several groups of fine quartz particles across the entire section.

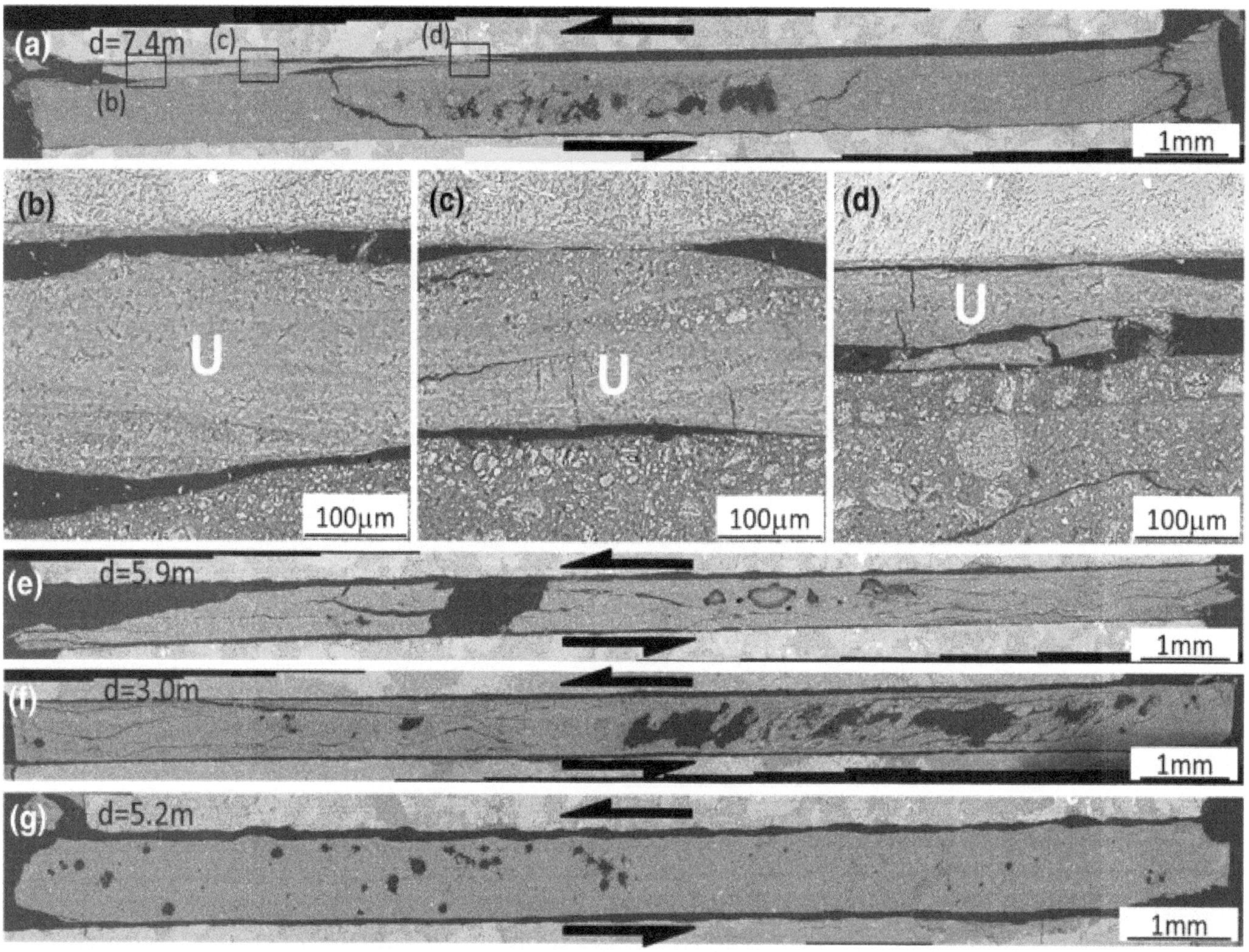

Figure 13: Photomicrographs of samples from the slip-strengthening regime. The fault gouge was deformed at a normal stress of 3 MPa and a slip velocity of (a) 0.1 m/s (LHVR256). (b to d) These textures are similar to those in Fig. 12b to d but are less extensive (approximately 4 mm wide). The fault gouges were deformed at a normal stress of 3 MPa and slip velocities of (e) 0.0023 m/s (LHVR261), (f) 0.0124 m/s (LHVR262), and (g) 0.0436 m/s (LHVR254).

The microstructures of the tested fault gouge in the slip-weakening regime are illustrated first. Fault gouge samples LHVR0125 and 124, in which the slip rates were 0.87 and 1.30 m/s, respectively, were slipped 40 m and 53 m, respectively. Highly deformed zones with ultrafine grains (the "U" layer) were observed near the rotational sides (Fig. 9 and Fig. 10). The U layers consisted of ultrafine grains that were mostly less than 1 μm in diameter and appeared to be welded or sintered. Similar textures were reported previously (Sawai et al., 2012, Togo and Shimamoto, 2012, Yao et al., 2013a and Yao et al., 2013b). These highly deformed zones consisting of ultrafine grains may be slip zones in which the shear deformation was concentrated. Clast-clay aggregates (CCAs) are commonly observed within the moderately deformed zones below the "U" layers. The moderately deformed zones and CCAs are shown in Fig. 9c and d and Fig. 10c and d. Similar to microstructures, the presence of the U layer and moderately deformed zones with CCAs can be observed in the fault gouge after a slip distance of approximately 18 m at a velocity of 0.5 m/s (LHVR 0129; Figure 11). Two U layers are shown in Fig. 11b, and detailed photomicrographs are shown in Fig. 11c and d.

The transitional and slip-strengthening regimes are described below. Fig. 12 shows BSIs of sample LHVR0259 (slip velocity of 0.22 m/s), and Fig. 12b–d shows the highly deformed zones, including the U layer. The U layer is well-developed, as was observed in the samples from the slip-weakening regime. Continuous U layers are absent when the shear velocity was less than 0.22 m/s (slip-strengthening regime). The U layer can be observed at a slip velocity of 0.1 m/s (Figure 13a) but is less extensive (only approximately 4 mm wide). The fault gouges from the low slip velocity tests (LHVR0261, 262, 254; slip rates of 0.0023–0.0436 m/s) having total slip of 3.0–7.4 m show uniform deformation textures with few low angle fractures (Figure 13b and c) or no observable fractures (Figure 13d). Interestingly, CCAs are absent in the transitional and slip-strengthening regimes.

COMPILATION OF EXISTING FRICTION DATA FROM THE TSAOLING LANDSLIDE MATERIALS

Friction behaviors of shale powder and fault gouge under dry and wet conditions

Two datasets of rotary shear tests using dry and wet samples from the Tsaoling landslide site have been reported (Miyamoto et al., 2009 and Togo et al., 2014). Togo et al. (2014)documented the friction coefficients of dry shale powders of samples from the lower Cholan Formation near the slide plane (squares in Figure 14), and Miyamoto et al. (2009)reported friction coefficients of dry bedding-parallel fault gouges (triangles in Figure 14). In the slip-strengthening regime (velocities from 0.001 to 0.1 m/s), the ultimate friction coefficient (μult; squares with crosses) of the tested dry shale powder is higher than those of the wet fault gouges obtained in this study (solid blue lines) and those from traditional direct shear tests of wet shale. In the transition and slip-weakening regimes (velocity > 0.1 m/s), the peak friction coefficients (μp) of dry shale powders (solid squares) and dry fault gouges (solid triangles) are nearly the same as the ultimate friction coefficients in the slip-strengthening regime. These values ranged from approximately 0.6 to 0.8 with a mean value of 0.7, which is similar to the typical Byerlee (1978) friction values that range from 0.6 to 0.85. Notably, these ultimate and peak values are much higher than those of the wet fault gouges (approximately 0.4 to 0.5) obtained in this study.

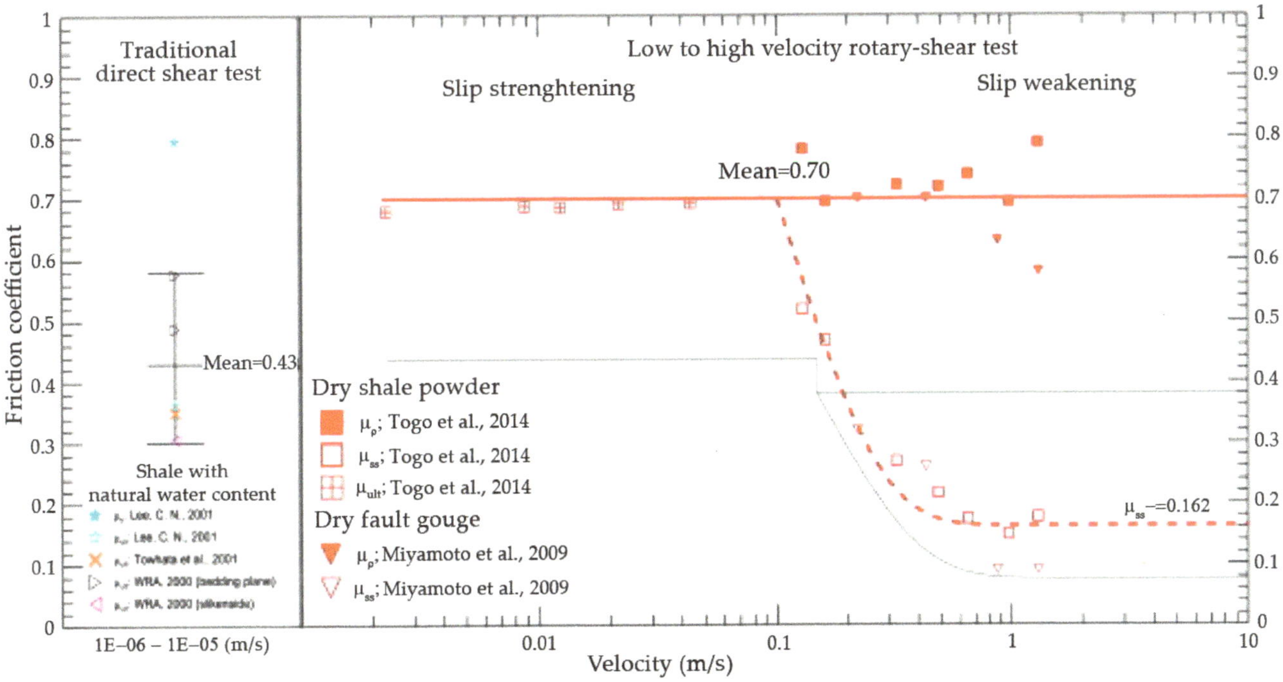

Figure 14: Friction coefficients of dry samples collected from the Tsaoling landslide site reported by Miyamoto et al. (2009) and Togo et al. (2014). The squares and triangles represent the results from dry shale powders and dry fault gouges, respectively. The solid gray lines represent the curves fitted to the results of this study (Figure 8). The left part shows the friction coefficients of wet shales from traditional direct shear tests.

As mentioned previously, the friction coefficient decreased from the peak to a steady-state value with increasing displacement in the slip-weakening regime. The steady-state friction coefficients (μss) of dry shale powders and dry fault gouges (hollow squares and triangles) decreased from approximately 0.5 to 0.1 with increasing velocity. As expected, the steady-state friction coefficients of the dry samples are higher than those of the wet fault gouges obtained in this study (approximately 0.27 to 0.05).

In addition to dry samples, Miyamoto et al. (2009) documented a dataset of friction coefficients in the weakening regime using wet shale powders and fault gouges sampled from the Tsaoling landslide site. The friction behaviors of the wet shale powders and wet fault gouges reported by Miyamoto et al. (2009) are shown in Fig. 15. The peak friction coefficients of the wet shale powders (average value of 0.61) are similar to those of dry powders reported by Togo et al. (2014) (Figure 14) while the steady-state values are similar with those of wet fault gouges derived from this study (solid gray lines).

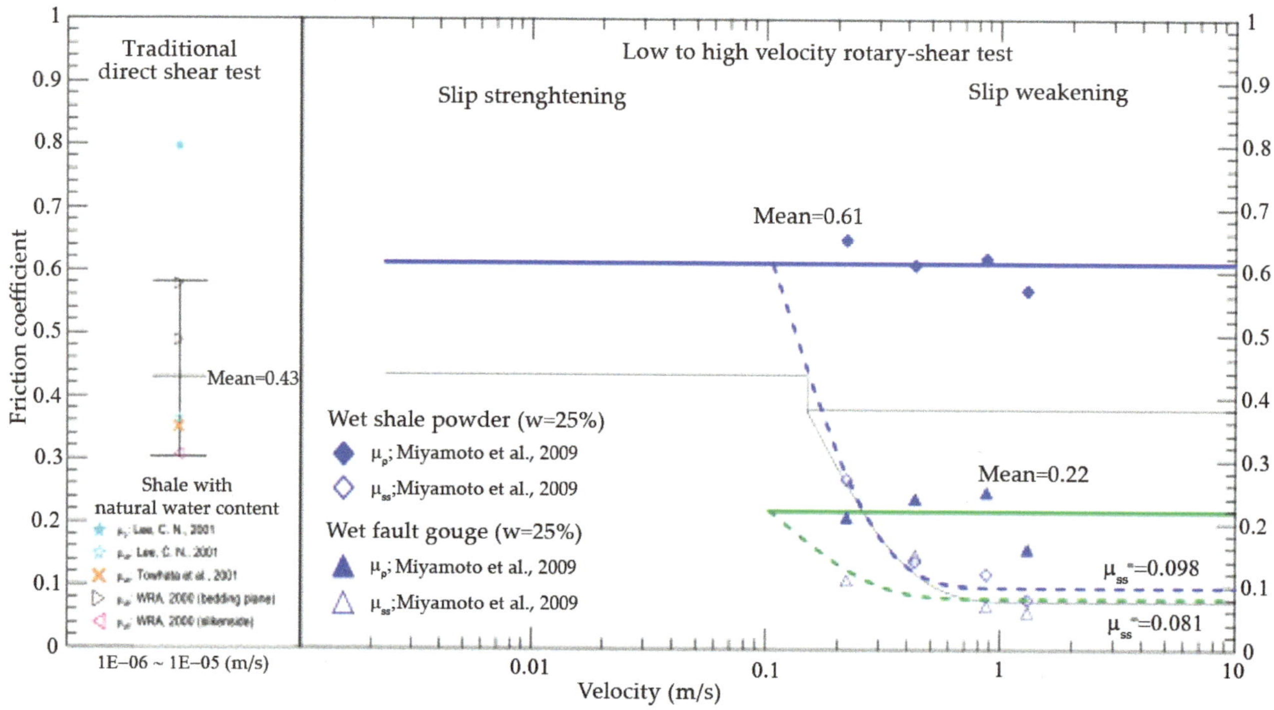

Figure 15: Friction coefficients of wet samples collected from the Tsaoling landslide site reported by Miyamoto et al. (2009). The diamonds and triangles represent the results of wet shale powders and wet fault gouges, respectively. The solid gray lines represent the curves fitted to the results of this study (Figure 8). The left part shows the friction coefficients of the wet shales from traditional direct shear tests.

The reported values of μss of the wet fault gouges are also similar to our experimental results (Figure 15). The average μp of the wet fault gouges (0.22) is the lowest of all of the datasets. Notably, although the fault gouge samples are identical to those in this study, the consolidation procedure was not adopted by Miyamoto et al. (2009). Therefore, the initial water contents of the fault gouge samples tested in this study are less than those used in Miyamoto et al. (2009). The excess pore pressure generated by the normal load is inferred to have dissipated during the consolidation procedure and resulted in higher peak strengths obtained in this study. The microstructures of the sheared samples with different water contents will be compared in Section 6.3. To discriminate the wet fault gouges, the term "semi-wet fault gouge" is used in the following discussion to represent the samples from this study, and "wet fault gouge" is used for the samples of Miyamoto et al. (2009). The influence of the frictional behaviors at different water contents on the initiation and kinematics of the earthquake-triggered landslide will be elucidated further inSection 5.3.

Velocity-dependence of the slip-weakening parameters

To evaluate the effect of the water content on the kinematics of the Tsaoling landslide, the parameters of the proposed friction law for all of the published datasets were calibrated. For simplicity, the ultimate and peak friction coefficients are assumed to be identical. The peak friction coefficients μp of the dry shale powder (Togo et al., 2014), wet shale powder, and wet fault gouge (Miyamoto et al., 2009) are 0.70, 0.61, and 0.22, respectively. Accordingly, μss^{∞} and VCritical in Eq. (4) of all of the datasets can be calibrated using the least-squares curve fitting technique. The velocity dependent steady-state friction coefficients that incorporate the calibrated parameters of all of the datasets are shown in Fig. 16a and the calibrated parameters are listed in Table 2. In general, the steady-state friction coefficients of the dry materials are slightly higher than those of the wet materials. The μss^{∞} value of the dry materials is 0.162, which is higher than that of the wet material (0.076–0.098). The critical velocities (VCritical) that separate the slip-weakening and slip-strengthening regimes are within a narrow range (0.1–0.15 m/s) for different water contents.

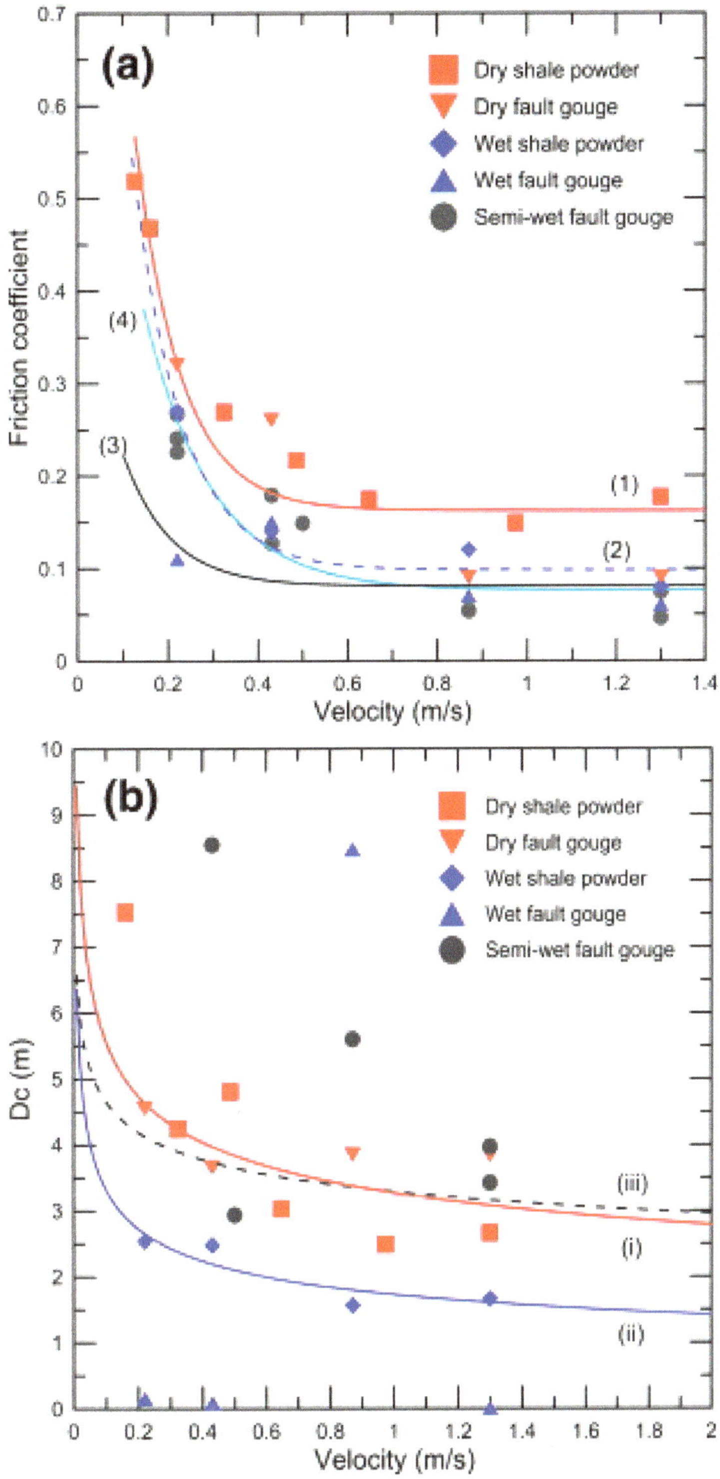

Figure 16: (a) Steady-state friction coefficients (μ_{ss}) of all of the datasets. The red line (1) represents the dry shale powder and fault gouge; the blue dashed line (2) represents the wet shale powder; the black line (3) represents the wet fault gouge; the blue solid line (4) represents the semi-wet fault gouge from this study. (b) Velocity dependency of slip-weakening distance. The red line (i) represents the dry samples; the blue line (ii) represents the wet shale powders; the black dashed line (iii) represents the fitted curve of all of the datasets (Togo et al., 2014: dry shale powder; Miyamoto et al., 2009: dry and wet fault gouge and wet shale powder; this study: semi-wet fault gouge).

Table 2: Parameters of the dry and wet samples in the velocity-dependent friction law.

Materials	μ_{ult}	μ_p	$V_{Critical}$ [m/s]	μ_{ss}^{∞}
(1) Dry shale powders and fault gouges (Togo et al., 2014)	0.70	0.70	0.10	0.162
(2) Wet shale powders (Miyamoto et al., 2009)	0.61	0.61	0.11	0.098
(3) Wet fault gouges (Miyamoto et al., 2009)	0.22	0.22	0.10	0.081
(4) Semi-wet fault gouges (this study)	0.44	0.38	0.15	0.076

Togo et al. (2014) reported the Dc values of the tested materials for different velocities. The Dc values of the tested samples of Miyamoto et al. (2009) were calibrated using the reported experimental results. Fig. 16b shows the slip-weakening distance (Dc) of all of the dry and wet samples from the Tsaoling landslide site. The values of Dc range from 1 to 9 m with the exception of the wet fault gouges reported by Miyamoto et al. (2009). The documented Dc values of the dry samples (fault gouges and shale powders) and wet shale powder decrease with increasing shear velocity. Togo et al. (2011) and Yao et al., 2013a and Yao et al., 2013b fit the dependence of Dc on the slip rate and the normal stress as a power law. We use their empirical equation in a form:

$$D_c(V)=a(V/V_0)-b \quad (5)$$

where a and b are constants (a has a dimension [length], and b is dimensionless), andV_0 was introduced for the dimensional consistency (we take $V_0 = 1$ m/s in this study). The values of a and b of the dry samples (shale powders and fault gouges) and wet shale powder were calibrated and the fit line (Eq. (5)) are shown in Fig. 16b (red and blue lines). In addition, the black dashed line represents the fitted curve of all of the data (dry and wet samples). The calibrated parameters are listed in Table 3. The trend of the velocity dependency on the slip-weakening distance is not clear for the wet fault gouges. The extremely low slip-weakening distance resulted from the decrease of the friction coefficient with increasing shear displacement. We speculate that the decrease in strength is due to the rapid generation of excess pore pressure. The tests of the semi-wet fault gouges (sheared after consolidation) yielded a larger Dc.

Table 3: Parameters of the dry and wet samples in the velocity dependent slip-weakening distance.

Materials	a [m]	b
(i) Dry shale powders and fault gouges (Togo et al., 2014)	3.27	0.23
(ii) Wet shale powders (Miyamoto et al., 2009)	1.73	0.28
(iii) All datasets	3.29	0.15

INITIATION AND KINEMATICS OF THE TSAOLING LANDSLIDE BASED ON A RIGID BLOCK MODEL THAT INCORPORATES A VELOCITY–DISPLACEMENT DEPENDENT FRICTION LAW

Newmark displacement method

This section reports our Newmark analysis of the Tsaoling landslide that incorporates the proposed velocity–displacement dependent friction law and the frictional properties summarized in the last two sections. We idealize the Tsaoling landslide as a rigid-block sliding on along a bedding plane dipping 14° in Stage I and along a slope dipping 20° in Stage II as summarized at the end of Section 2 (Figure 3b). Newmark (1965) proposed a procedure for evaluating the potential deformation of an earth dam due to earthquake shaking. The Newmark displacement analysis has been widely employed to investigate the initiation and kinematics of earthquake-induced landslides and to back-calculate strength parameters (Wilson and Keefer, 1983, Jibson and Keefer, 1993, Romeo, 2000,Huang et al., 2001, Lee, 2001, Sepúlveda et al., 2005, Wu and Tsai, 2011 and Chen et al., 2014). Movement of a block (or landslide mass) is assumed to start when the ground acceleration exceeds a critical acceleration Ac. But here we use an acceleration of the sliding block S derived by Huang et al. (2001):

$$S=(g\sin\theta-A_d)-\mu(g\cos\theta+A_n) \quad (6)$$

where g is the acceleration of gravity, θ is the dip angle of the sliding surface, μ is the friction coefficient along the sliding surface, Ad is the acceleration in the down-dip direction (positive downward), and An is the acceleration normal to the sliding surface (positive upward). We neglected the cohesion term based on the results from friction experiments in the last two sections. The down dip velocity of the rigid block can be calculated easily by integrating the positive value of the down-dip sliding accelerationS, and the permanent displacement of the sliding block can be obtained by integrating the velocity of the block.

Note that an upward movement of ground (or negative Ad) promotes landslide motion and increases S (the first term in Eq. (6)). Moreover, an upward ground motion (or positive An) increases the normal stress on the sliding surface, thereby increasing the friction and decreasing S (the second term). Traditionally, the critical acceleration Ac is treated as a constant which can be expressed as the function of the friction angle (or friction coefficient) on the sliding surface ($Ac = [(\mu - \tan\theta)/(1 + \mu\tan\theta)]g$). Dong et al. (2009) utilized a velocity-dependent friction coefficient to calculate the temporal variation of the critical acceleration which is a function of the rigid block velocity. In this study, we use a displacement- and velocity-dependent friction law, Eqs. (2), (4) and (5), for μ in Eq. (6) with measured parameters summarized in the last two sections. The critical acceleration of the simulated rigid block is therefore velocity–displacement dependent. Marked slip- and velocity-weakening (reduction in μ) increases S causing high-speed catastrophic landslides.

Fig. 17a exhibits acceleration of the ground motion record during Chi-Chi earthquake recorded at station CHY080 (see Figure 1 for its location) that was used in our Newmark analysis of the Tsaoling landslide. The total duration of the earthquake signals recorded by CHY080 was 159 s, but the figures show only records of 25 s since the arrival of P waves, using the time since the generation of the Chi-Chi earthquake at its epicenter on the horizontal axis. The peak ground accelerations (PGA) were 715.9, 792.4, and 841.5 gal for the vertical, E–W, and N–S, components, respectively (Figure 17a; Lee et al., 2001). The P-wave and S-wave arrivals were 20.0 s and 25.2 s, respectively (Kuo et al., 2013). The site effects in the landslide source area may have also influence the analysis. However, the observation site is located at about the same elevation as the original landslide site on the same mountain, and we did not consider an amplification of the recorded ground accelerations in our calculations. Fig. 17b shows normal and down-dip accelerations An and Ad calculated from the acceleration records for the time periods of 30 to 45 s that are critical to the analysis of the Tsaoling landslide and are discussed in detail in the next subsection.

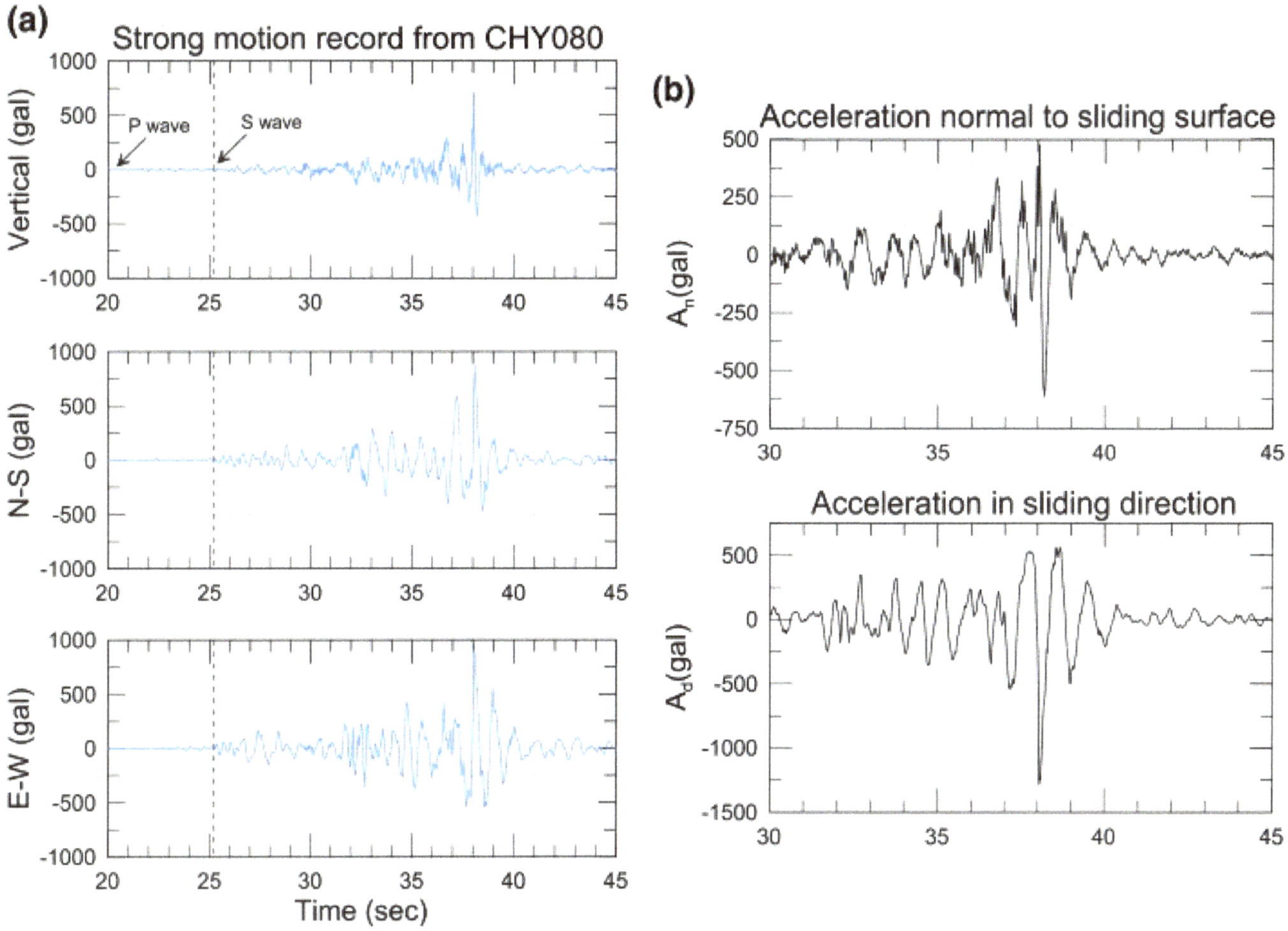

Figure 17: (a) Ground acceleration in the vertical, N–S, and E–W directions during the 1999 Chi-Chi earthquake recorded with a sampling rate of 200 Hz at station CHY080, located about 700 m to the north from the original landslide site. (b) Calculated acceleration in the normal direction to the sliding surface (An, positive upward; top diagram) and acceleration in the sliding direction (Ad, positive downward; bottom diagram), using Eq. (3) in Huang et al. (2001). Figures show acceleration data only for the time periods that are important in landslide analyses.

Rigid block modeling of the Tsaoling landslide using semi-wet fault gouge parameters

We now conduct a Newmark analysis of the Tsaoling landslide for Stages I and II, using the frictional parameters for the semi-wet fault gouge (Section 3). Parameters for the idealized velocity–displacement friction law are listed in Table 4 as Scenario 1, corresponding to our experimental data summarized in Fig. 8 (note that we used an average value of Dc (4.9 m) in Table 1). Eqs. (2) and (4) with those parameters were incorporated into the friction coefficient term μ in Eq. (6) to calculate the time history of the down dip acceleration S (gray curves), velocity V (green curves), cumulative displacement d (blue curves), and friction coefficient μ(V) (red curves) in Fig. 18, using the time in seconds since the generation of the Chi-Chi earthquake on the horizontal axes. The result for the initiation portion of the landslide in Fig. 18a indicates that the landslide motion started at 31.6 s (a small pulse in the green curve on the lower-left side of the bottom diagram) where the down-dip acceleration S became positive at this time to initiate the landslide motion (gray curve at 31.6 s in the figure). But the landslide motion stopped soon after the S went back to negative again. There were five similar pulse-like movements before the onset of continuous sliding at 38.0 s, and they caused small step-like increase in the cumulative displacement d (blue curve in Figure 18a). They correspond to positive peaks of S above the dashed line for S = 0 in Fig. 18a. There are two small peaks between 32 s and 33 s, but the positive S peaks are too small to cause corresponding velocity pulses visible in the figure. After 38 s the velocity V increased and decreased twice, but V did not go back to zero and the Tsaoling landslide was triggered. The accumulated displacement was 0.65 m. We thus take 38.0 s as the time for the onset of landslide since the generation of the Chi-Chi earthquake at its epicenter (denoted asTChi_Chi in Table 4) and 0.65 m is the critical slip distance for the occurrence of an earthquake induced landslide. TChi_Chi = 38.0 s corresponds to 12.8 s after the S wave arrival to the landslide site (TS in Table 4). Apparently, the landslide was triggered by a sharp spike in S exceeding 1000 gal at 38.0 s that corresponds to the sharp negative peaks of An and Ad in Fig. 17 b, both promoting the landslide motion (note the effects ofAn and Ad on S in Eq. (6)). Strong ground motion ended at around 41 s after which both normal and down-dip accelerations An and Ad decreased (Figure 17), but the down-dip acceleration S of the landslide mass remained positive even after that. This was caused by the velocity- and slip-weakening of the sliding surface, i.e., the reduction in μ in Eq. (6)which allowed the continued high-speed landslide possible. Then the landslide mass went into Stage II at 69.3 s after the generation of the Chi-Chi earthquake with a velocity of 50.7 m/s and into Stage III at 82.2 s with 84.6 m/s (Figure 18b; Table 4, two columns on the right). A knee in the velocity curve at the boundary between Stages I and II is caused by the difference in the slope angles (14° and 20°, Figure 3b).

Table 4: A summary of frictional parameters for five scenarios and results of the Newark analysis of the Tsaoling landslide. μult: an ultimate friction coefficient in the slip-strengthening regime, μp: peak friction coefficient, μ_{ss}^{∞}: steady-state friction coefficient at infinite velocity, Vcritical: critical velocity for the onset of marked slip-weakening, Dc: slip-weakening distance, a and b: parameters for the velocity dependence of Dc, TChi_Chi: time for the onset of Tsaoling landslide since the time of generation of the 1999 Chi-Chi earthquake at its epicenter, and TS: time for the onset of the landslide since the S-wave arrival to the landslide area. The two numbers on the two columns on the right side give the arrival time since the generation of the Chi-Chi earthquake, and final velocity of the landslide mass at the ending points of Stages I and II.

Scenario	Gouge types	μult	μp	μ_{ss}^{∞}	Vcritical [m/s]	Dc [m]		Landslide [s]TChi_Chi/TS	Accumulated displacement before initiation [m]	Arrival time [s]/ final velocity [m/s]	
						a[m]	b			End of Stage I	End of Stage II
1	Semi-wet fault gouge(this study)	0.44	0.38	0.076	0.15	4.9		38.0/12.8	0.65	69.3/50.7	82.2/84.6
2	Dry shale powders and dry fault gouges(Togo et al., 2014)	0.70	0.70	0.162	0.10	3.27	0.23	Landslide stopped.			
3	Wet shale powders(Miyamoto et al., 2009)	0.61	0.61	0.098	0.11	1.73	0.28	38.9/13.7	1.09	72.8/47.5	86.8/80.5
4	Semi-wet fault gouge(this study)	0.44	0.38	0.076	0.15	8.55		38.0/12.8	0.62	71.4/50.7	84.9/82.7
5	Semi-wet fault gouge(this study)	0.44	0.38	0.076	0.15	3.29	0.15	38.0/12.8	0.69	68.7/50.7	82.2/84.6
6	Wet fault gouges(Miyamoto et al., 2009)		0.22	0.081	0.10	< 1		Slope is unstable.			

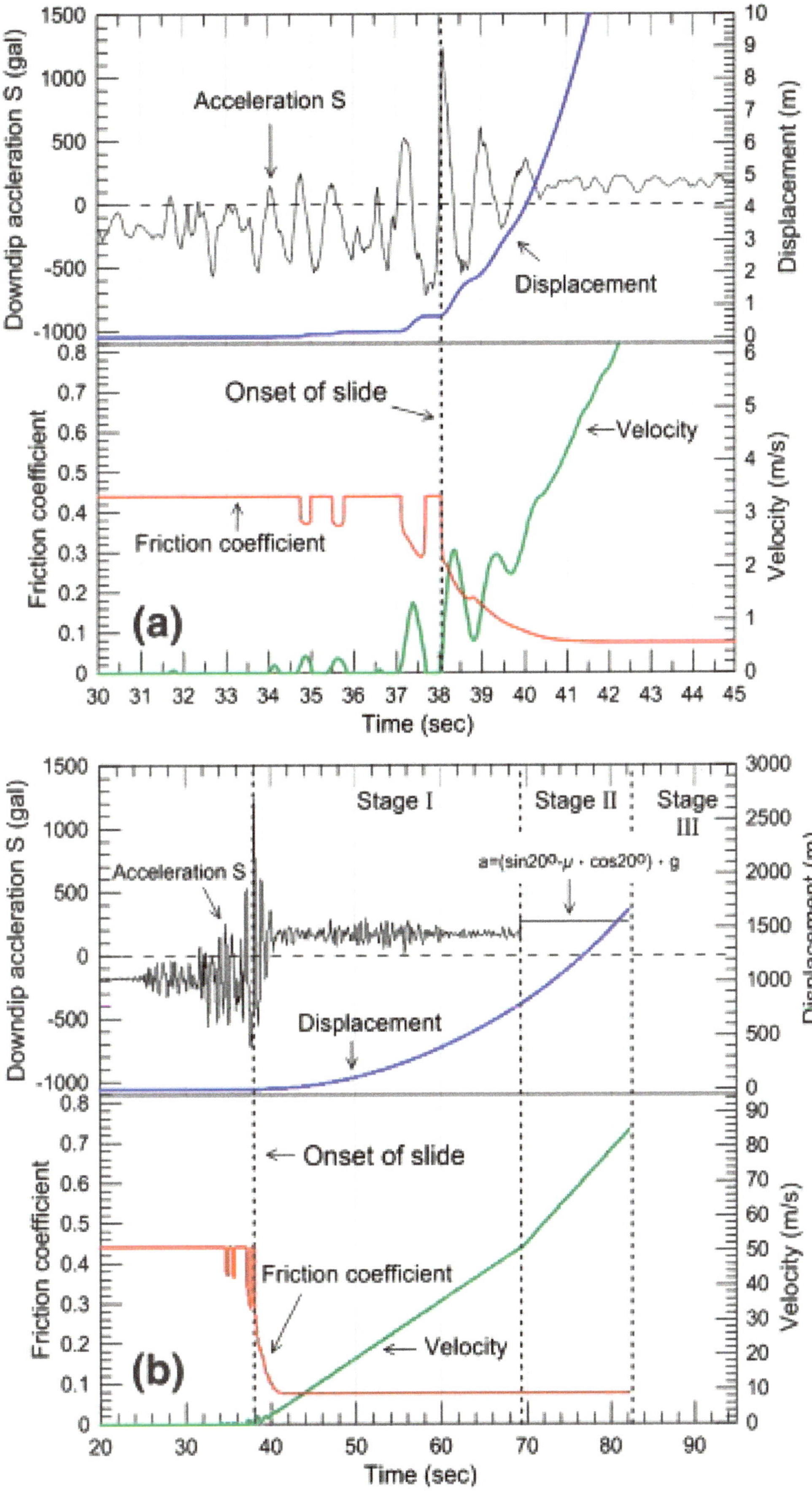

Figure 18: Results of the Newmark analysis with the frictional parameters for the semi-wet fault gouge (a) for the initiation portion and (b) for Stages I and II of the Tsaoling landslide, showing time histories of the down-dip sliding acceleration S (gray), velocity (green), and cumulative displacement (blue) of the moving block, and friction coefficient along the sliding surface (red). The horizontal axis gives time since the generation of the Chi-Chi earthquake at its epicenter. The horizontal dashed lines indicate S = 0 above which landslide motion takes place with an acceleration S > 0.

Friction coefficient µ along the sliding surface changes as the landslide velocity V and the cumulative displacement d changes as shown by red curves in Fig. 18. We assumedµ(V) = µult for sliding velocity V less than a critical velocity VCritical (Figure 8), so that two small pulses of movements at 34 s and 36.5 s did not cause any changes in µ(V). The three larger pulses of movements at 34.6 s, 35.4 s, and 37.1 s caused reduction in µ(V) following the velocity- and displacement-dependent friction specified by Eqs. (2) and (4), but µ(V) went back to µult as V became zero (see the red curve corresponding to the pulses in Figure 18a). The landslide velocity V increased to 2.3 m/s at 38.4 s and then decreased to 0.6 m/s at 38.9 s after the onset of landslide at 38 s, but this reduction in Vcaused only a slight increase in µ(V) because V was greater than VCritical = 0.15 m/s (green and red curves in Figure 18a). The friction coefficient µ(V) quickly changes toµ s s$^{\infty}$ in a few seconds and ~µ s s$^{\infty}$ after about 42 s (red curves in Figure 18a, b), andµ s s$^{\infty}$ essentially determines the landslide speed.

Rigid block modeling using existing friction data of shale powder and fault gouge

Based on the existing friction data summarized in Section 4 (Fig. 14, Fig. 15 and Fig. 16and Table 2 and Table 3), we set up Scenarios 2 to 6 with materials from the landslide site and their frictional parameters shown in Table 4. The critical velocity for the onset of marked slip-weakening Vcritical is fairly similar in all scenarios, but other frictional parameters vary considerably with different scenarios. Scenario 2 represents the case of a sliding surface that is composed of dry shale powders or dry fault gouges whose frictional properties (Miyamoto et al., 2009 and Togo et al., 2014) are summarized inFig. 14 and Fig. 16. The bedding-parallel fault gouges used in Miyamoto et al. (2009)were collected from the same fault zones as we collected our fault gouge samples. Shale is the weaker member of the alternated sandstone and shale, and is likely to determine the frictional strength of the bedding planes. Under dry conditions with room humidity, the shale powder and fault gouge exhibit similar frictional properties with the ultimate friction coefficient in the slip-strengthening regime µult = 0.70, peak friction coefficient µp = 0.70, steady-state friction coefficient µ s s$^{\infty}$ = 0.162, and slip-weakening distanceDc = $3.27(V/V_0)^{-0.23}$ (Table 3, second row; Figure 16b). Values of Dc, µp, and µ s s$^{\infty}$, are much larger than those in Scenario 1, and the Newmark analysis reveals three pulses of movements at 37.1 s, 38 s, and 38.9 s corresponding to three S peaks above S = 0 line (Figure 19a), and the pulse-like motion caused a cumulative displacement d of 0.66 m. The landslide motion stopped in Scenario 2 because the friction coefficient µ remained high and S did not stay positive after the movements. This result is consistent with the result by Togo et al. (2014) using the frictional properties for the dry shale powder.

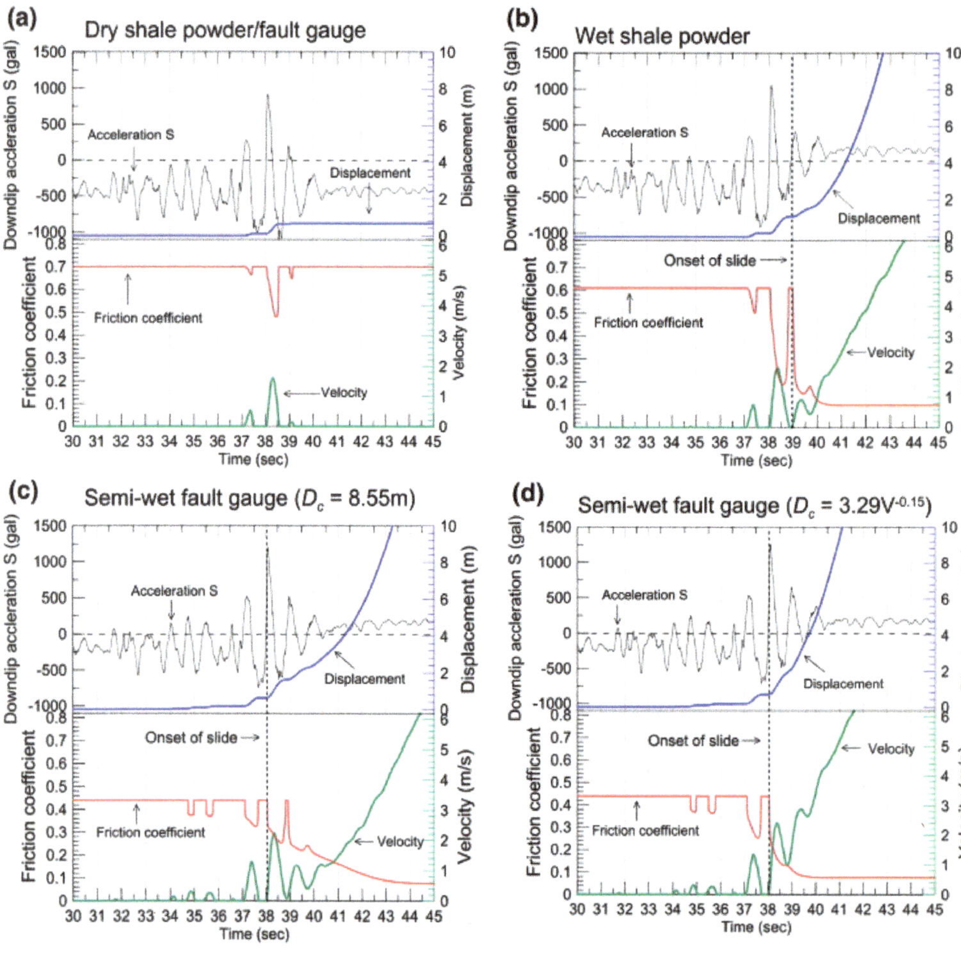

Figure 19: Results of the Newmark analysis of the Tsaoling landslide (a) for Scenario 2 with the dry shale powder/fault gouge, (b) for Scenario 3 with the wet shale powder, (c) for Scenario 4 with the semi-wet fault gouge assuming Dc = 8.55 m, and (d) for Scenario 5 with semi-wet fault gouge assumingDc = $3.29(V/V_0)^{-0.15}$. Frictional parameters for the four scenarios are given in Table 4, and all figures are constructed similar to Fig. 18a. The horizontal dashed lines indicate S = 0 above which landslide motion occurs.

Scenario 3 is the wet shale powder whose frictional properties (Miyamoto et al., 2009) can be described by μult = 0.61, μp = 0.61, μss^{∞} = 0.098, and Dc = $1.73(V/V_0)^{-0.28}$(Table 4, third row; Figure 16b). Those friction coefficients are intermediate between those in Scenarios 1 and 2, but Dc is smaller than those in the two scenarios (Figure 16b). The result of the Newmark analysis in Figure 19b revealed that two pulse-like movements with accumulated displacement d of 0.19 m and 1.09 m occurred at 37.1 s and 38.0 s corresponding to two positive S peaks and that the landslide initiated at 38.9 s. That is, the critical slip distance of this scenario is 1.09 m. Several pulse-like movements as recognized in Scenario 1 did not occur because μult was higher in Scenario 3 than in Scenario 1. This scenario appears to be very marginal in causing the Tsaoling landslide because the block of this case was not triggered by the sharp spike in S exceeding 1000 gal at 38.0 s. The friction coefficient μ recovered back to the peak value of 0.61 after the largest spike of S (red curve in Figure 19b). The block was triggered by the next spike follow the largest one and the μ(V) drops to μss^{∞} rapidly and a high-speed landslide occurs. The landslide motion after this is similar to that shown in Fig. 18a and we give only the time t and velocity V for the landslide mass to enter into Stage II (t = 72.8 s andV = 47.5 m/s) and into State III (t = 86.8 s and V = 80.5 m/s) in Table 4.

Scenarios 4 and 5 are for the semi-wet fault gouge as in Scenario 1, but we used a largest measured value to Dc (8.55 m) in Scenario 4 and Dc given by an equation that was fit to the whole data in Fig. 16b, Dc = $3.29(V/V_0)^{-0.15}$, for Scenario 5 (fourth row inTable 3). The landslide motions in both scenarios are similar to that in Scenario 1 with several pulses of motion between 31.7 s and 37 s, ending up with the onset of landslide at 38.0 s with accumulated displacements of 0.62 m and 0.69 m (cf. Figure 19c, d, andFigure 18a). A large Dc suppresses slip-weakening and the movement at 38 s stopped in Scenario 4, but it ran away at the next pulse of movement (Figure 19c). The landslide motions are quite similar between Scenarios 1 and 5 (cf. Figures 18a and 19d). Friction coefficient μ dropped to μss^{∞} = 0.076 at about 44 s and 40 s in Scenarios 4 and 5, reflecting the difference in Dc (cf. Figure 19c, d), but this much of difference does not change the high-speed landslide motion. The landslide mass went into Stage II att = 71.4 s and V = 50.7 m/s and into Stage III at t = 84.9 s and V = 82.7 m/s in Scenario 4, whereas the times and speeds for Scenario 5 were t = 68.7 s and V = 50.7 m/s for State II and t = 82.2 s and V = 84.6 m/s for State III (Table 4, two columns on the right).

Scenario 6 is for the frictional properties of wet fault gouge (Miyamoto et al., 2009) withμp = 0.22, μss^{∞} = 0.081, and Dc < 1 m (Table 1, Fig. 15 and Fig. 16). Although μult is not reported in their paper, the friction angle for μp is 12.4° and is less than the slope angle of 14°. Thus the friction is not high enough to keep the Tsaoling landslide on the slope and the Newmark analysis cannot be done with this Scenario.

DISCUSSION

Newmark analysis of the Tsaoling landslide

Previous studies on the Newmark analysis of landslides often assumed a critical slip distance of rock or soil slides (typically 50–100 mm), beyond which a slope instability might occur (Wieczorek et al., 1985, Keefer and Wilson, 1989 and Jibson and Keefer, 1993). However, it has been unclear what determines the critical distance and how high-speed landslide initiates. Although the mechanisms that dominate the rapid and long run-outs of large landslides have not been conclusively determined (Hungr and Evans, 2004), Dong et al. (2013) attempted a Newmark analysis of the Tsaoling landslide using a block-sliding model with a velocity- and displacement-dependent friction law, describing marked slip-weakening behaviors of the landslide materials at high velocities. In this paper, we have expanded their analysis by presenting new experimental data on the fault gouge collected from bedding-parallel faults close to the sliding surface of the Tsaoling landslide (Section 3), by compiling existing friction data on the Tsaoling landslide materials (Section 4), and by modifying their velocity- and displacement-friction law (Eqs. (2), (4) and (5)). The Tsaoling landslide is an ideal case for such a study because the landslide occurred along a very flat bedding surface, particularly in Stage I (Figure 3b), and because the seismic records and a survivor's witness give constraints on the landslide motion. A high-frequency signal lasting for about 1 s was recorded at a nearby station CHY080 at 75.8 s after the generation of the Chi-Chi earthquake at its epicenter, and Kuo et al. (2009) proposed that these signals were caused by the impact of the Tsaoling landslide mass to the bank of the Chinshui River (point C in Figure 3b).Chen et al. (2014) recognized high-frequency signals at 32 to 40 s in the same seismic records and proposed that those are due to the initiation of the Tsaoling landslide.

In view of the geometry of the landslide surface along the sliding direction in Fig. 3b, we conducted the Newmark analysis for the rigid-block sliding during Stages I and II, before the landslide mass hit the river bank. One can argue that the sliding mass may not remain intact in Stage II because the two steps between B and C in Fig. 3b might have fragmented the landslide block. However, high-frequency signals were not recorded except for the two time periods quoted above, and hence severe fragmentation of the landslide mass probably did not occur before it hit the river bank. Moreover, 39 people slid with the Tsaoling landslide mass over the Chinshui River and landed on top of the landslide deposits, and 7 people survived after sliding for a distance of 2250 m. Chen et al. (2014) reports an interview with a survivor who witnessed that "the sliding block remained more or less intact, especially in the rear part". Tang et al. (2009) also concluded that a chance for survival is low if the landslide mass disaggregated early in the sliding because of the extensive mixing of blocks, based on an analysis of the Tsaoling landslide using a discrete element method. Accordingly, the assumption of rigid body motion along the sliding plane in Stages I and II would be reasonable. However, the landslide mass must have broken into pieces after it hit the river bank to cover a wide area as shown in Fig. 3a. Alternative approaches such as an analysis using Voellmy rheology with a dynamic basal friction (Hungr and Evans, 1996, Hungr and Evans, 2004,Pirulli and Mangeney, 2008 and Sosio et al., 2008) may be applicable to the evaluation of the kinematics of disturbed landslide deposits, but the analysis of the stopping phase of the Tsaoling landslide is beyond the scope of this paper. However, we estimate later the loss of potential energy by the Tsaoling landslide and compare it with the frictional work during Stages I and II. This gives an estimate of the energy dissipated during the stopping phase of the landslide in Stage III.

We set up six scenarios using the frictional parameters determined for the experimental results on the shale powders and fault gouge under dry and wet conditions (Table 4). The landslide stopped after a few pulse-like movements in Scenario 2 for the dry shale powder and fault gouge because the ultimate friction coefficients in the slip-strengthening regime μult and the peak friction coefficient μp were high (Figure 19a). The sliding surface was about 140 m below the surface at the time of Tsaoling landslide and probably was not under dry conditions because of frequent heavy rains in Taiwan. The other extreme case is Scenario 6 for wet fault gouge with μp = 0.22 (or a friction angle of 12.4°), that is not high enough to keep the landslide mass on a slope dipping 14°. We will discuss a possible significance of this case in the next subsection in relation to rain-triggered landslides.

In all other scenarios, the landslide motion started either at 38.0 s or 38.9 s after the generation of the Chi-Chi earthquake at its epicenter, following a few to several pulse-like movements (Figures 18a, 19b–d; Table 4, column 9). The permanent displacements accumulated before the onset of the sliding were 0.62 ~ 1.09. P and S waves arrived in 20.0 s and 25.2 s, respectively, to the landslide site after the earthquake, so that the landslide occurred after 12.8 s or 13.7 s after the S-wave arrival in our Newmark analysis. A survivor witnessed to us that the Tsaoling landslide had started in about 10 s after he felt strong shaking. He did not measure this time interval with a watch or a clock, and his memory in such a chaotic situation may not be reliable. But nonetheless, our result for the landslide initiation is consistent with this witness if the seismic wave he felt was the S wave arrival (12.8–13.7 s versus 10 s). The timing of the landslide initiation in our analysis (pulse-like movements between 31.7 s and 38 s, and the onset of landslide at 38.0 or 38.9 s) agrees with the period of high-frequency signals at 32–40 s identified by the power-spectrum analysis by Chen et al. (2014). Finally, the landslide mass reached point C at 82.2–86.8 s after the generation of Chi-Chi earthquake (Table 4, column on the right side) and this timing is close to the timing of high-frequency signals at 75.8 s (Kuo et al., 2009). Friction coefficient μ drops to the steady-state friction coefficient at infinite velocity μss^{∞} in a few to several seconds after the onset of landslide (Fig. 18 and Fig. 19b–d), so that μss^{∞} primarily determined the speed of a landslide. We assigned 0.076 or 0.098 for μss^{∞} based on the experimental results (3 and 4). However, it should be kept in mind that the high-velocity friction experiments have been conducted at velocities up to 1.3 m/s, more than one order of magnitude smaller than the speed of landslide. The values of μss^{∞} are extrapolated values to higher velocities. Moreover, an accurate measurement of μ below 0.1 is not easy because a precise correction of the measured shear stress for the friction between the Teflon sleeve and rotary specimen is required (see Sawai et al., 2012, Appendix). Despite those uncertainties, we consider that our Newmark analysis using experimentally-determined friction parameters successfully captured the initiation and movement processes of the Tsaoling landslide.

Friction constitutive law and importance of water

We now compare different friction models by previous workers and our friction law with different scenarios in this paper (Figure 20). The friction coefficients determined in the back calculations by Tang et al. (2009) with a discrete element method and by Kuo et al. (2009) using a continuum model of hydraulic flow are shown by orange and brown dashed lines, respectively, in Fig. 20a. Those methods are flexible in handling disaggregated particles such as debris flow and are powerful tools in solving the overall dynamic motion of high-speed landslides. However, the Tsaoling landslide mass cannot stay at a slope dipping 14° with the friction coefficients of 0.15 and 0.09 obtained in those studies, and hence the initiation processes of landslides cannot be solved by the methods. But the steady-state friction coefficients at infinite velocity μss^{∞} from this study (0.076 and 0.098) are comparable to the back-calculated basal friction values of 0.15 and 0.09.

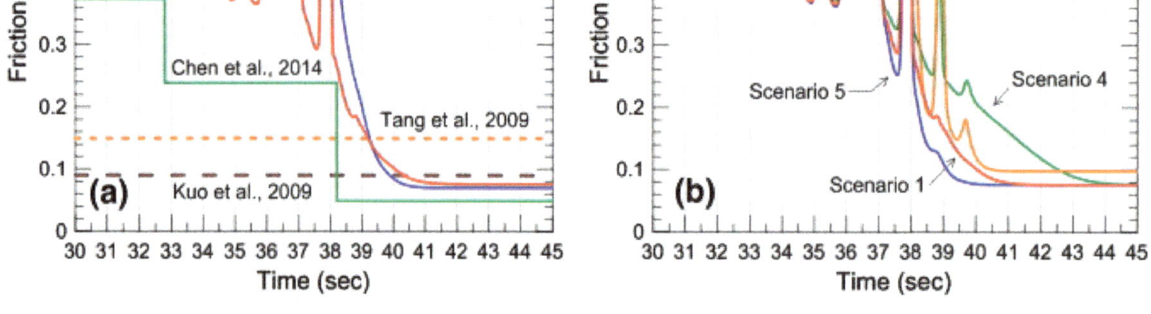

Figure 20: (a) Temporal variation of the friction coefficient proposed by different researchers as shown on each curve, and (b) a comparison of temporal evolution of the friction coefficient for the five scenarios solved in this paper. Time on the horizontal axes gives the time after the generation of the 1999 Chi-Chi earthquake at its epicenter.

All other methods summarized in Fig. 20 are the Newmark analysis with a few different friction laws. Chen et al. (2014) recently attempted to combine results from traditional direct shear tests and from high-velocity friction experiments by Miyamoto et al. (2009), and set up three levels of friction coefficient as shown by the green curve in Fig. 20a. They assumed that the peak friction coefficient (0.37) with cohesion of 152 kPa instantly dropped to the residual friction coefficient (0.24) as soon as sliding began and that the residual friction coefficient dropped instantly to the mobilized friction coefficient (0.05) after a displacement of 0.5 m (they studied the cases for μp of 0.58, 0.47, and 0.27 too). The mobilized friction coefficient corresponds to our steady-state friction coefficient μssand was taken from Miyamoto et al. (2009). Although the Newmark analysis yielded similar results to ours in this study, such two abrupt changes in friction were not observed in any previous experiments. We do not consider that idealized behavior by Chen et al. (2014) is realistic.

Differences among other cases reflect either (1) difference in the friction law or (2) differences in the landslide materials and/or conditions. As for the friction law, Dong et al. (2013) used an exponential slip-weakening law, Eq. (2), with slip-weakening parameters given by Eqs. (3) and (5), for the experimental data by Miyamoto et al. (2009). The only difference between Dong et al. (2013) and the present study is that Dong et al. (2013)used Eq. (3) for the steady-state friction coefficient μss, whereas we used Eq. (4) with an additional parameter μult (an ultimate friction coefficient in the slip-strengthening regime) in addition to the peak friction coefficient μp (Fig. 8, Fig. 14 and Fig. 15). Thus this paper attempted to include the changes in friction with increasing slip rate (Figure 8), but the velocity dependency of friction is variable (e.g., Di Toro et al., 2011) and Eq. (4) is not general. A common feature of Eqs. (3) and (4) is that the steady-state friction coefficientμss is forced back to μp in Dong et al. (2013) and to μult in our friction law as soon as the landslide velocity V drops to zero. Because of this, the friction coefficient μ went back to the high initial values as V decreased during the initial pulse-like movements (see results for Scenarios 1 and 3–5 in Figure 20b). Results by Dong et al. (2013) exhibit similar trends although we do not show their results in the figure.

On the other hand, Togo et al. (2014) idealized the early motion of earthquake-induced landslide as linear accelerating/decelerating slip and conducted friction experiments on dry shale powder from the Tsaoling landslide site with frequencies of 0.3 to 1.2 Hz. This slip history is very similar to the pulse-like early movements in the results from Newmark analysis in Fig. 18 and Fig. 19, except that there are temporary periods of stops between the accelerating/decelerating slip in the landslide motion (e.g., see five pulses of movements between 34 s and 38 s, separated by stops lasting 0.4–0.8 s in Figure 18a). Experimental results by Togo et al. (2014) revealed that the accelerating and decelerating motion caused weakening and strengthening, while undergoing overall slip-weakening. They approximated the overall behavior by an exponential weakening law, Eq. (2), and conducted the Newmark analysis of the Tsaoling landslide with results similar to those of our Scenarios 2 and 3 (Figure 19a, b). The change in μ in Togo et al. (2014) for wet shale powder is shown as blue line in Fig. 20a. This result without strength recovery is more realistic than those with our friction law in view of the results from the oscillatory accelerating/decelerating experiments, and the simple weakening law in Togo et al. (2014) would be better for analyzing the initiation of landslide. However, Togo et al. (2014) did not include changes in slip-weakening parameters with increasing velocities which reduces μ_{ss}^{∞} and Dc. Thus, friction laws in this study and in Dong et al. (2013) are more involved in analyzing the high-speed landslide motion which is dictated by μ_{ss}^{∞}. Empirical approaches are inevitable at this stage until a constitutive law that can describe changes in friction for arbitrary slip histories at high velocities is proposed in the future.

Our results for six scenarios in Table 4 highlight the effects of frictional parameters and landslide materials. Dry shale powder and fault gouge has high μult and μp (Scenario 2 inTable 4) and the landslide motion stopped after a few pulse-like movements (Figure 19a). But wet shale powder has lower values of μult, μp, and Dc (Scenario 3), making it possible to induce a landslide (Figure 19b). Semi-wet fault gouge has even smaller values of μult, and μp than those for the above cases and landslides occurred in Scenarios 1, 4, and 5 with different Dc (Table 4, Fig. 18 and Fig. 19c, d). Because of lowμult and μp, the landslide mass is easy to move and the pulse-like motion started earlier than for wet shale powder cf. Figure 19b, c, and d). A large Dc in Scenario 4 slowed down the reduction in μ with time (cf. green curve with the three curves for Scenarios 1, 3, and 5 in Figure 20b). Our analysis thus demonstrated that the landslide mass began to move at 31.7 s for semi-wet fault gouge and at 34.7 s for wet shale power after the generation of the Chi-Chi earthquake at its epicenter, and that the high-speed landslide started at 38.0 s for semi-wet fault gouge and at 38.9 s for wet shale powder. The calculated early movements are consistent with the high-frequency signals at 32 to 40 s analyzed recently by Chen et al. (2014). The landslide mass is likely to have hit the river bank at 75.8 s as suggested by the seismic record (Kuo et al., 2009). If the leading edge of the landslide mass in Fig. 3b caused the seismic signals, the sliding distance becomes about 1340 m and the average speed of landslide mass becomes about 36 m/s. Whereas if the high-frequency seismic record corresponds to the collision of the centroid of the mass, the sliding distance and the average speed become 1655 m and 44 m/s, respectively. It is hard to imagine that the collision of the frontal part of the landslide mass did not cause any signals, and we consider 35–40 m/s as a proper range of the average speed of the Tsaoling landslide mass during Stages I and II.

It is notable from Fig. 18, Fig. 19 and Fig. 20 that the landslide can be triggered for fairly wide ranges of peak or ultimate frictional strength, μp or μult, and this was possible because of high accelerations to make the acceleration of the landslide mass S, given by Eq. (6), positive. However, another important factor is the amount of slip displacement of the landslide mass to reduce the friction angle below the slope angle to trigger the landslide. Fig. 21a–c exhibits seismic records for the critical time period, 35–40 s, in triggering the Tsaoling landslide, and Fig. 21d and e shows calculated accelerations, Anand Ad, in the normal direction to the sliding surface and in the down-dip direction, respectively. The strike and dip of the sliding surface of the Tsaoling landslide is 136° from N (clockwise) and 14° to SW, respectively, and Eq. (3) in Huang et al. (2001) gives:

$$A_d = -0.70\cdot a_E - 0.67\cdot a_N - 0.24\cdot a_V,\ A_n = -0.17\cdot(a_E + a_N) + 0.97\cdot a_V \quad (7)$$

where aE, aN, and aV are accelerations in the E–W, N–S, and vertical directions, respectively (eastward, northward, and upward positive). Eq. (6) for S indicates that negative An and Ad increase S and promote landslide motion. There are three positiveaN peaks at 37.2 s, 38.1 s, and 39.0 s (Figure 21b), and aE has a broad positive peak at around 37 s and two positive peaks at 38.05 s and 39.0 s (Figure 21c). Those three positive peaks of aN and aE caused three negative Ad peaks and three notable positive peaks of S, at around 37.2 s, 38.1 s, and 39 s (Figure 21f). Negative values of An at 36.9–37.4 s, 38.1–38.4 s, and 38.9–39.2 s also contributed to the three positive Speaks. Large down-dip acceleration S allowed large enough displacement to reduce friction angle below the slope angle, over a slip-weakening distance Dc. Thus, the three S spikes were large enough to trigger the Tsaoling landslide for a fairly wide range of Dc values (up to 8.55 m, Table 4). Apparently, the Tsaoling landslide was triggered by the three strong acceleration pulses in the NE direction at 37–39 s, and unfortunately this was the most favorable orientation to induce the Tsaoling landslide. Forecasting such dangerous seismic ground motion from existing active faults would be the most important task in forecasting earthquake-induced landslides in the future.

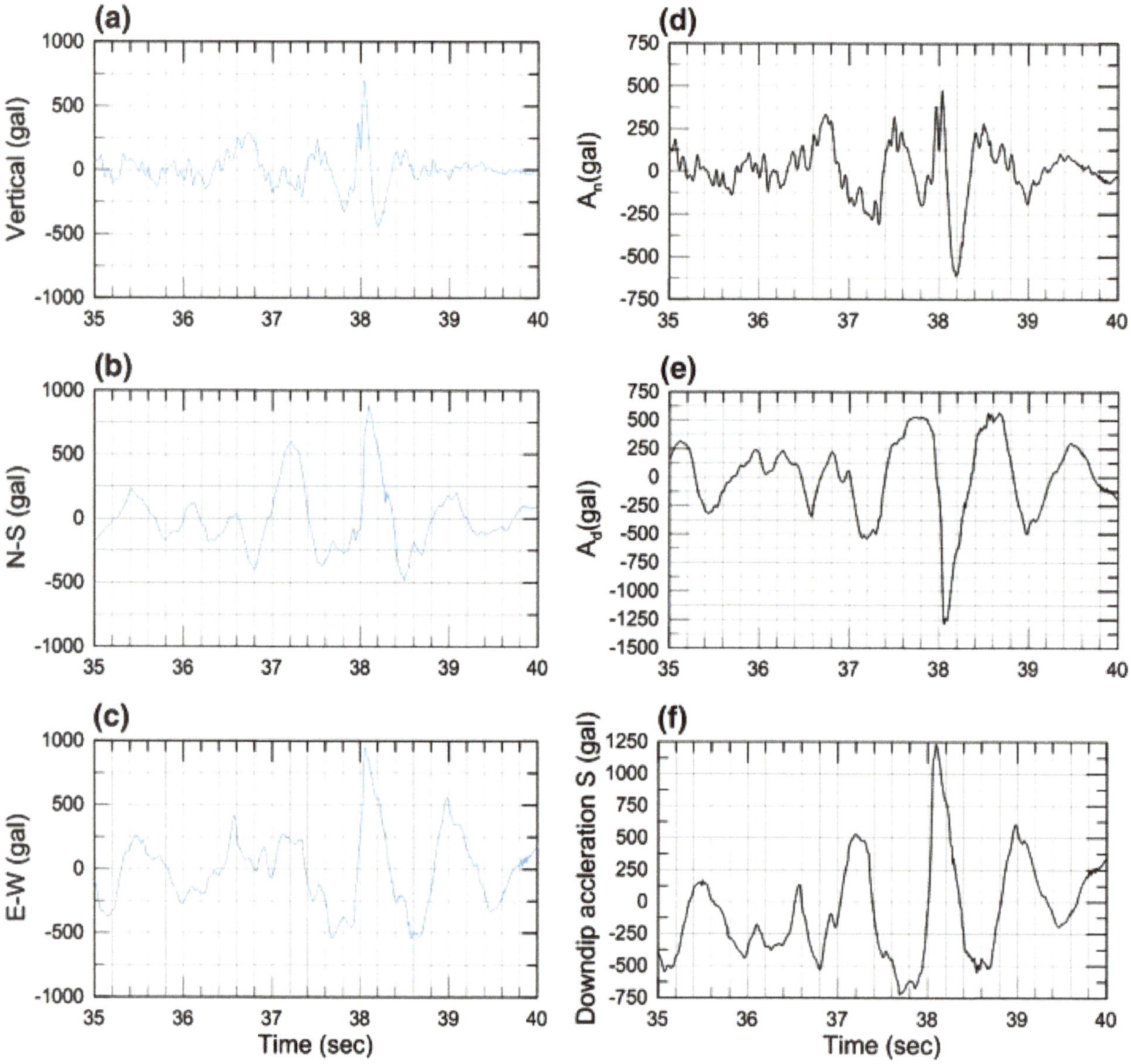

Figure 21: Seismic ground motion during the critical time period of 35 to 40 s since the generation of the Chi-Chi earthquake, in triggering the Tsaoling landslide. Accelerations (a) for the vertical, (b) for the N–S, and (c) for the E–W directions, respectively, recorded at seismic station CHY080. (d) and (e) show calculated accelerations from (a) to (c) in the normal direction to the sliding surface and in the sliding direction, respectively. (f) Calculated acceleration of the landslide mass using the frictional parameters for Scenario 1 (see Table 4).

Finally, it should be kept in mind that the cohesion term is not included in Eq. (6). How intact bedding planes develop into sliding surface with a small or negligible cohesion is an important problem to be studied experimentally. Repeated earthquakes such as the Chi-Chi earthquake might have caused small pulse-like movement to condition the bedding planes to change into easy sliding surfaces during geological time. Moreover, wet fault gouge has a friction angle smaller than the slope angle and cannot support the landslide. We added 25% of water to fault gouge and compacted the gouge under the same axial stress until the axial shortening stopped. Thus, part of water was squeezed out because the specimens were not sealed and this is the reason why we called our fault gouge as "semi-wet" gouge. On the other hand, Miyamoto et al. (2009) conducted friction experiments immediately after setting up the gouge with water, and this is probably the main reason for having higher values of μp and μult in our experiments than in their results. We consider that our case is closer to the formations at the Tsaoling landslide site because there is enough time for excess water to escape from the sliding surface in nature. However, landslides were triggered at the Tsaoling landslide site by heavy rains in 1942 and 1979 (Hung et al., 2002). Heavy rain might have provided excess water into the sliding surface to reduce its friction. Effect of water on friction is very subtle and better controlled experiments on the effects of water are needed in the future. The effect of water is discussed further in the next subsection.

Microstructures of deformed gouge

The friction behaviors of samples with different water contents are quite different. Based on the microstructural observations in this study (Section 3.4), the semi-wet fault gouges sheared at high to low velocities contain a large variety of microstructural features from stratified to uniform (Fig. 9, Fig. 10, Fig. 11, Fig. 12 and Fig. 13). In contrast, the microstructures of the slip zone in wet samples do not change significantly by shearing at different velocities (Ferri et al., 2011). The microstructures of the wet samples were identified as insignificant shape-preferred clast orientations within a strongly foliated matrix. The foliation was locally deflected around the clasts, inclined towards the shear direction, and converged towards the shear localization zone that developed along the fault gouge-rock boundary. CCAs were not present in the sheared wet fault gouge samples at short shear displacements (< 3 m) but appeared at large shear displacements, even in wet samples (Han and Hirose, 2012).

The wet clay-rich fault gouge (w = 40%) was compacted, and water was progressively expelled along a continuous water film that formed at the clay-gouge/wall rock boundary (Ferri et al., 2011). Consequently, the peak friction coefficients of the wet samples are lower than those of the dry samples due to the lubrication of the water film. Another explanation is related to the excess pore pressure that is generated during loading. The degree of saturation dominates the generation of the excess pore pressure when a normal stress is applied. Higher initial water contents could give rise to higher excess pore pressures in the fault gouges, and the shear resistance will decrease accordingly. The peak friction coefficient will decrease with increasing water content of the tested sample. Because the effective stress does not change if the test occurs under saturated undrained shear conditions, similar microstructures are reasonable.

Based on rotary shear tests of dry fault gouge samples, Yao et al. (2013a) reported that a fine-grained deformation zone forms in the upper part of the fault gouge layer near the rotational side of the host rock with a slip displacement of 0.13 m at a normal stress of 2.5 MPa and a slip rate of 0.0014 m/s. Single or overlapping stratified structures of the U layer, CCA-bearing layers, and CCA-free layers were observed in air-dried samples and wet samples after large shear displacements (Boutareaud et al., 2010, Ferri et al., 2011,Han and Hirose, 2012, Sawai et al., 2012, Yao et al., 2013a and Yao et al., 2013b). In conclusion, the microstructures of our sheared semi-wet fault gouges are similar to those of the dry samples. This accounts for the similar steady-state friction coefficients of the samples with different water contents (Fig. 14 and Fig. 15). We infer that the water contents of the tested samples decreased to low values similar to those of dry samples after large shear displacements.

An interesting result is that the slip-weakening is closely related to the appearance of the CCAs. Natural CCAs have frequently been observed in BSIs in association with carbonate-rich or smectite-rich fault rocks (Warr and Cox, 2001, Beutner and Gerbi, 2005 and Boullier et al., 2009). Artificial CCAs have been produced from high-velocity rotary shear tests during investigations of fault dynamics (Boutareaud et al., 2010, Ujiie and Tsutsumi, 2010, Han and Hirose, 2012 and Sawai et al., 2012) and studies of catastrophic landslides (Ferri et al., 2011). The presence of CCAs indicates a liquid-vapor phase transition of pore-water and the interplay of complex physical and chemical processes (Boutareaud et al., 2010). CCAs tend to form in long run-out, large landslides where the shear displacement is large enough to raise the temperature to water vaporization or mineral decomposition conditions. Notably, the appearance of CCAs does not guarantee the onset of thermal pressurization (Ferri et al., 2011). Hungr and Evans (2004) summarized several possible mechanisms that may account for the weakening strength of large landslides. Davies and McSaveney (2009) suggested that the apparently instantaneous pressures resulting from grain fragmentation greatly reduces the effective stress, and the strength subsequently decreases. The mechanisms involved in the strength variations during large landslides can be complex (Legros, 2002), and further studies are required.

Energy budget of the Tsaoling landslide

Results from our study provide the interesting information on the energy budget of the Tsaoling landslide. The DTMs before and after the 1999 Chi-Chi earthquake revealed the horizontal travel distance L of 2524 m and the elevation difference H of 524 m during the Tsaoling landslide (Figure 3b), which gives an apparent friction coefficient of 0.21. The potential energy Ep released during the Tsaoling landslide is given by:

$E_p = m \cdot g \cdot H$

- where m is the landslide mass, g is the acceleration of gravity (9.807 m/s^2). The volume of the Tsaoling landslide is estimated at 1.25×10^8 m^3 (Tang et al., 2009) andHung (1982) reports 2430 kg/m^3 for the density of rocks in Cholan formation. Those values give Ep = 1.56 $\times 10^{15}$ J.
- On the other hand, we analyzed the Tsaoling landslide as a rigid block sliding in Stages I and II in Fig. 3b. After the onset of landslide, the friction coefficient μ(V) drops to the steady-state friction coefficient at infinite velocity μss^∞ after sliding of a few to several seconds (red curves in Fig. 18 and Fig. 19). Then the frictional work Ef during the sliding is given approximately by:

$$E_f \approx mg\mu ss^\infty (L_1 \cos\theta_1 + L_2 \cos\theta_2) \quad (8)$$

where L_1 and L_2 are sliding distances along the sliding surface (784 m and 871 m), and θ_1 and θ_2 are dip angles of the slopes (14° and 20°) for Stages I and II, respectively (Figure 3b).

Those values and $\mu ss^{\infty} = 0.076$ gives 3.58×10^{14} J. Thus the frictional work Ef is only 23% of the released potential energy Ep of the Tsaoling landslide, and most of Ep must have been consumed in Stage III such as crushing of the landslide mass and heat dissipation. Analysis of particle size distribution of the Tsaoling landslide deposits is interesting in estimating the energy used for the fragmentation of the landslide mass, but studying the complex landslide processes in Stage III is beyond the scope of this paper.

CONCLUSIONS

We conducted rotary-shear high-velocity friction experiments on fault gouge from bedding-parallel faults immediately below the sliding surface of the Tsaoling landslide induced by the 1999 Chi-Chi earthquake, compiled existing friction data on the Tsaoling landslide materials, proposed a revised friction law that can describe the velocity–displacement dependent frictional behaviors, and performed Newmark analysis of the triggering and high-speed sliding of the Tsaoling landslide using the experimentally-determined frictional properties. The main outcomes from this study are summarized below.

(1)Frictional behavior of the semi-wet fault gouge in our rotary-shear experiments can be described by a few empirical equations using the ultimate friction coefficient in the slip-strengthening regime μult (0.44), peak friction coefficient μp (0.38), steady-state friction coefficient at infinite velocity μss^{∞} (0.076), critical velocity for the onset of dramatic slip-weakening VCritical (0.15 m/s), and by the slip-weakening distance Dc(4.9 m on the average). The same parameters were compiled for reported frictional properties of shale powder and fault gouge from the Tsaoling landslide site under dry and wet conditions, to determine variations of frictional properties. Deformation textures of the semi-wet gouge are similar to those reported for dry fault gouges, but their frictional strength is smaller than that of dry gouges.

(2)We classified the Tsaoling landslide motion into three stages; i.e., Stage I — rigid-block sliding on a flat bedding place dipping 14°, Stage II — nearly rigid-block sliding on a steeper surface dipping 20°, and Stage III — stopping phase after the hitting of the landslide mass against the bank of the Chinshui River. We performed Newmark analysis of landslide motion in Stages I and II, assuming rigid-block sliding and using measured frictional properties with five scenarios set up for shale powder and fault gouge under dry, semi-wet and wet conditions.

(3)Our Newmark analysis showed that the landslide stopped after a few pulse-like motions with frictional properties for dry shale powder and fault gouge. But analysis for the other four scenarios for wet shale powder and fault gouge revealed that the landslide motion started at 38.0 or 38.9 s after the generation of the earthquake at its epicenter, following a few to several pulse-like motions. This is consistent with high-frequency signals at 32–40 s reported by Chen et al. (2014) who suggested that the signals are associated with the initiation of the landslide. The onset of landslide was at 12.8 s or 13.7 s after the S wave arrival to the landslide site, and our results agree with a survivor's witness that the landslide started in about 10 s after he felt a strong earthquake, possibly due to S wave. Our analysis showed that the landslide mass reached the river bank at 82.2–86.8 s after the generation of Chi-Chi earthquake and this is very close to the high-frequency signals at 75.8 s reported by Kuo et al. (2009)who proposed that the signals are due to hitting of the landslide mass against the river bank. Thus, our analysis reproduced the overall features of the Tsaoling landslide remarkably well with measured frictional properties. A typical average speed of the Tsaoling landslide during Stages I and II is 35–40 m/s. It is interesting to note that the evaluated critical slip distance were from 0.62 m to 1.09 m, which is considerably larger than the documented one for shallow slides.

(4)The sliding surface of the Tsaoling landslide has dip direction and dip angle of 226° (from N clockwise) and 14°, respectively. Accelerations in the NE directions are favorable for triggering the landslide. There were three strong acceleration motions in that direction at 37 to 39 s after the generation of Chi-Chi earthquake, and our Newmark analysis revealed that the Tsaoling landslide was triggered by those seismic ground motions. Forecasting dangerous seismic ground motion based on measured frictional properties of landslide materials and on the information on existing active faults is important for mitigating earthquake-induced landslides.

(5)The centroids of the landslide mass and landslide deposits were determined accurately by using a GIS software and DTMs before and after the Chi-Chi earthquake. The vertical and horizontal displacements during the landslide were 524 m and 2524 m, respectively, which give an apparent friction coefficient of 0.21. This elevation difference and the landslide mass give the released potential energyEp of 1.6×10^{15} J. The frictional work Ef during the sliding of the landslide mass in Stages I and II before hitting the river bank was estimated at 3.6×10^{14} J in the Newmark analysis with our Scenario 1 for semi-wet fault gouge data. This corresponds to about 23% of Ep, and the rest of Ep must have been consumed mostly during the stopping phase of the Tsaoling landslide after hitting the river bank in very complex processes such as fragmentation, heat dissipation, and spreading of the landslide deposits over a wide area. Analysis of particle size distribution of the landslide deposits would be of interest in estimating the energy used in the fragmentation.

ACKNOWLEDGMENTS

This research was funded by the National Science Council, Taiwan under the grants ofNSC 102-2116-M-008-005 and NSC 99-2625-M-009-004-MY3. We thank the Editor, Guest Editors of this special issue, Prof. Theo van Asch and two anonymous reviewers for their useful and constructive reviews which greatly improved this paper.

REFERENCES

1. Beutner, E.C., Gerbi, G.P., 2005. Catastrophic emplacement of the Heart Mountain block slide, Wyoming and Montana, USA. Bull. Seismol. Soc. Am. 117, 724–735.
2. Bouchut, F., Westdickenberg, M., 2004. Gravity driven shallow water models for arbitrary topography. Commun. Math. Sci. 2, 359–389.
3. Boullier, A.M., Yeh, E.C., Boutareaud, S., Song, S.R., Tsai, C.H., 2009. Micro-scale anatomy of the 1999 Chi-Chi earthquake principal slip zone. Geochem. Geophys. Geosyst. 10, Q03016.

4. Boutareaud, S., Boullier, A.M., Andre'ani, M., Calugaru, D.G., Beck, P., Song, S.R., Shimamoto, T., 2010. Clay clast aggregates in gouges: new textural evidence for seismic faulting. J. Geophys. Res. 115, B02408.

5. Byerlee, J., 1978. Friction of rocks. Pure Appl. Geophys. 116, 615–626.

6. Campbell, C.S., 1989. Self-lubrication for long runout landslides. J. Geol. 97, 653–665.

7. Campbell, C.S., Cleary, P.W., Hopkins, M., 1995. Large-scale landslide simulations: global deformation, velocities and basal friction. J. Geophys. Res. Solid Earth 100, 8267–8283.

8. Chang, K.J., Taboada, A., Lin, M.L., Chen, R.F., 2005. Analysis of landsliding by earthquake shaking using a block-on-slope thermo-mechanical model: example of Jiufengershan landslide, central Taiwan. Eng. Geol. 80, 151–163.

9. Chang, K.J., Wei, S.K., Chen, R.F., Chan, Y.C., Tsai, P.W., Kuo, C.Y., 2013. Empirical modal decomposition of near field seismic signals of Tsaoling landslide. Earthquake-Induced Landslides. Springer, Berlin Heidelberg, pp. 421–430.

10. Chen, T.C., Lin, M.L., Hung, J.J., 2004. Pseudostatic analysis of Tsao-Ling rockslide caused by Chi-Chi earthquake. Eng. Geol. 71, 31–47.

11. Chen, R.F., Chan, Y.C., Angelier, J., Hu, J.C., Huang, C., Chang, K.J., Shih, T.Y., 2005. Large earthquake-triggered landslides and mountain belt erosion: the Tsaoling case, Taiwan. Compt. Rendus Geosci. 337, 1164–1172.

12. Chen, R.F., Chang, K.J., Angelier, J., Chan, Y.C., Deffontaines, B., Lee, C.T., Lin, M.L., 2006. Topographical changes revealed by high-resolution airborne LiDAR data: the 1999 Tsaoling landslide induced by the Chi-Chi earthquake. Eng. Geol. 88, 160–172.

13. Chen, T.C., Lin, M.L., Wang, K.L., 2014. Landslide seismic signal recognition and mobility for an earthquake-induced rockslide in Tsaoling, Taiwan. Eng. Geol. 171, 31–44.

14. Chigira, M., Wang, W.N., Furuya, T., Kamai, T., 2003. Geological causes and geomorphological precursors of the Tsaoling landslide triggered by the 1999 Chi-Chi earthquake, Taiwan. Eng. Geol. 68, 259–273.

15. Cleary, P.W., Campbell, C.S., 1993. Self-lubrication for long runout landslides: examination by computer simulation. J. Geophys. Res. Solid Earth 98, 21911–21924.

16. Crosta, G.B., Imposimato, S., Roddeman, D.G., 2003. Numerical modelling of large landslides stability and runout. Nat. Hazards Earth Syst. Sci. 3, 523–538.

17. Davies, T.R., McSaveney, M.J., 2009. The role of rock fragmentation in the motion of large landslides. Eng. Geol. 109 (1), 67–79.

18. Di Toro, G., Hirose, T., Nielsen, S., Pennacchioni, G., Shimamoto, T., 2006. Natural and experimental evidence of melt lubrication of faults during earthquakes. Science 311, 647–649.

19. Di Toro, G., Han, R., Hirose, T., De Paola, N., Nielsen, S., Mizoguchi, K., Ferri, F., Cocco, M., Shimamoto, T., 2011. Fault lubrication during earthquakes. Nature 471, 494–498.

20. Donath, F.A., Parker, R.B., 1964. Folds and folding. Geol. Soc. Am. Bull. 75, 45–62.

21. Dong, J.J., Lee, W.R., Lin, M.L., Huang, A.B., Lee, Y.L., 2009. Effects of seismic anisotropy and geological characteristics on the kinematics of the neighboring Jiufengershan and

22. Hungtsaiping landslides during Chi-Chi earthquake. Tectonophysics 466, 438–457.

23. Dong, J.J., Yang, C.M., Yu, W.L., Lee, C.T., Miyamoto, Y., Shimamoto, T., 2013. Velocity– displacement dependent friction coefficient and the kinematics of giant landslide. In: Ugai, K., Yagi, H., Wakai, A. (Eds.), Earthquake-Induced Landslides. Springer, pp. 395–401.

24. Ferri, F., Di Toro, G., Hirose, T., Shimamoto, T., 2010. Evidence of thermal pressurization in high-velocity friction experiments on smectite-rich gouges. Terra Nova 22, 347–353.

25. Ferri, F., Di Toro, G., Hirose, T., Han, R., Noda, H., Shimamoto, T., Quaresimin, M., de Rossi, N., 2011. Low-to high-velocity frictional properties of the clay-rich gouges from the slipping zone of the 1963 Vaiont slide, northern Italy. J. Geophys. Res. Solid Earth 116, B09208.

26. Han, R., Hirose, T., 2012. Clay-clast aggregates in fault gouge: an unequivocal indicator of seismic faulting at shallow depths? J. Struct. Geol. 43, 92–99.

27. Han, R., Shimamoto, T., Hirose, T., Ree, J.H., Ando, J., 2007. Ultra-low friction of carbonate faults caused by thermal decomposition. Science 316, 878–881.

28. Hart, M.W., 2000. Bedding-parallel shear zones as landslide mechanisms in horizontal sedimental rocks. Environ. Eng. Geosci. 6 (5), 95–113.

29. Heim, A., 1932. Bergsturz und Menschenleben. Fretz und Wasmuth, Zürich.

30. Hirose, T., Shimamoto, T., 2005. Growth of molten zone as a mechanism of slip weakening of simulated faults in gabbro during frictional melting. J. Geophys. Res. 110, B05202.

31. Huang, C.C., Lee, Y.H., Liu, H.P., Keefer, D.K., Jibson, R.W., 2001. Influence of surface-normal ground acceleration on the initiation of the Jih-Feng-Erh-Shan landslide during the 1999 Chi-Chi, Taiwan, Earthquake. Bull. Seismol. Soc. Am. 91 (5), 953–958.

32. Hung, J.J., 1982. Investigation of engineering geology in the Tsaoling landslide. Report of rock mechanics tests, COA, 71-1.3-12(3), (in Chinese).

33. Hung, J.J., 2000. Chi-Chi earthquake induced landslides in Taiwan. Earthq. Eng. Eng. Seismol. 2, 25–33.

34. Hung, J.J., Lee, C.T., Lin, M.L., 2002. Tsao-ling rockslides, Taiwan. Catastrophic landslides: effects, occurrence, and mechanisms. Geol. Soc. Am. Rev. Eng. Geol. 15, 91–115.

35. Hungr, O., 1995. A model for the runout analysis of rapid flow slides, debris flows, and avalanches. Can. Geotech. J. 32, 610–623.
36. Hungr, O., Evans, S.G., 1996. Rock avalanche run out prediction using a dynamic model. Proc., 7th Int. Symp. Landslides 1, Trondheim, Norway, pp. 233–238.
37. Hungr, O., Evans, S.G., 2004. Entrainment of debris in rock avalanches: an analysis of a long run-out mechanism. Geol. Soc. Am. Bull. 116, 1240–1252.
38. Hungr, O., Evans, S.G., Bovis, M.J., Hutchinson, J.N., 2001. A review of the classification of landslides of the flow type. Environ. Eng. Geosci. 7, 221–238.
39. Jibson, R.W., Keefer, D.K., 1993. Analysis of the seismic origin of landslides: examples from the New Madrid seismic zone. Geol. Soc. Am. Bull. 105, 521–536.
40. Keefer, D.K., Wilson, R.C., 1989. Predicting earthquake-induced landslides with emphasis on arid and semi-arid environments. In: Sadler, P.M., Morton, D.M. (Eds.), Landslides in a Semi-arid Environment: Riverside. California. Inland Geological Society. 2, pp. 118–149.
41. Kuo, C.Y., Tai, Y.C., Bouchut, F., Mangeney, A., Pelanti, M., Chen, R.F., Chang, K.J., 2009. Simulation of Tsaoling landslide, Taiwan, based on Saint Venant equations over general topography. Eng. Geol. 104, 181–189.
42. Kuo, C.Y., Wei, S.K., Tsai, P.W., 2013. Ensemble empirical mode decomposition with supervised cluster analysis. Adv. Adapt. Data Anal. 5, 1350005.
43. Lee, C.N., 2001. Preliminary Study on the Tsaoling Landslide Area Under Earthquake. (Master's thesis) Institute of Civil Engineering, National Taiwan University.
44. Lee, C.T., 2011. Geomorphological evolution and geology of the Tsaoling rockslide. J. Chin. Soil Water Conserv. 42, 63–73.
45. Lee, C.T., 2012. Engineering Challenges and Countermeasures of the Difficult Geological Conditions in Taiwan. 2012 Taiwan Rock Engineering Symposium, pp. 29–36.
46. Lee, W.H.K., Shin, T.C., Kuo, K.W., Chen, K.C., Wu, C.F., 2001. CWB free-field strong-motion data from the 21 September Chi-Chi, Taiwan, earthquake. Bull. Seismol. Soc. Am. 91, 1370–1376.
47. Legros, F., 2002. The mobility of long-runout landslides. Eng. Geol. 63, 301–331.
48. Liao, C.J., Lee, D.H., Wu, J.H., Lai, C.Z., 2011. A new ring-shear device for testing rocks under high normal stress and dynamic conditions. Eng. Geol. 122, 93–105.
49. Miyamoto, Y., Shimamoto, T., Togo, T., Dong, J.J., Lee, C.T., 2009. Dynamic weakening of bedding-parallel fault gouge as a possible mechanism for catastrophic Tsaoling landslide induced by 1999 Chi-Chi earthquake. The Next Generation of Research on Earthquake-induced Landslides: An International Conference in Commemoration of 10th Anniversary of the Chi-Chi Earthquake, pp. 398–401.
50. Mizoguchi, K., Hirose, T., Shimamoto, T., Fukuyama, E., 2007. Reconstruction of seismic faulting by high-velocity friction experiments: an example of the 1995 Kobe earthquake. Geophys. Res. Lett. 34, L01308.
51. Newmark, N.M., 1965. Effects of earthquakes on dams and embankments. Geotechnique 15, 139–159.
52. Pirulli, M., Mangeney, A., 2008. Results of back-analysis of the propagation of rock avalanches as a function of the assumed rheology. Rock Mech. Rock. Eng. 41 (1), 59–84.
53. Romeo, R., 2000. Seismically induced landslide displacements: a predictive model. Eng. Geol. 58, 337–351.
54. Sawai, M., Shimamoto, T., Togo, T., 2012. Reduction in BET surface area of Nojima fault gouge with seismic slip and its implication for the fracture energy of earthquakes. J. Struct. Geol. 38, 117–138.
55. Scheidegger, A.E., 1975. Physical Aspects of Natural Catastrophes. Elsevier, (289 pp.). Sepúlveda, S.A., Murphy, W., Jibson, R.W., Petley, D.N., 2005. Seismically induced rock slope failures resulting from topographic amplification of strong ground motions: the case of Pacoima Canyon, California. Eng. Geol. 80 (3–4), 336–348.
56. Shimamoto, T., Tsutsumi, A., 1994. A new rotary-shear high-speed frictional testingmachine: its basic design and scope of research. J. Struct. Geol. 39, 65–78.
57. Sosio, R., Crosta, G.B., Hungr, O., 2008. Complete dynamic modeling calibration for the Thurwieser rock avalanche (Italian Central Alps). Eng. Geol. 100, 11–26.
58. Tang, C.L., Hu, J.C., Lin, M.L., Angelier, J., Lu, C.Y., Chan, Y.C., Chu, H.T., 2009. The Tsaoling landslide triggered by the Chi-Chi earthquake, Taiwan: insights from a discrete element simulation. Eng. Geol. 106, 1–19.
59. Tanner, P.W.G., 1989. The flexural-slip mechanism. J. Struct. Geol. 11, 635–655.
60. Tanner, P.W.G., 1992. Morphology and geometry of duplexes formed during flexural-slip folding. J. Struct. Geol. 14 (10), 1173–1192.
61. Tika, T.E., Hutchinson, J.N., 1999. Ring shear tests on soil from the Vaiont landslide slip surface. Geotechnique 49, 59–74.
62. Togo, T., Shimamoto, T., 2012. Energy partition for grain crushing in quartz gouge during subseismic to seismic fault motion: an experimental study. J. Struct. Geol. 38, 139–155.
63. Togo, T., Shimamoto, T., Ma, S., Hirose, T., 2011. High-velocity frictional behavior of Longmenshan fault gouge from Hongkou outcrop, Sichuan, China and its implications for dynamic weakening of fault during the 2008 Wenchuan earthquake. Earthq. Sci. 24, 267–281.
64. Togo, T., Shimamoto, T., Dong, J.J., Lee, C.T., Yang, C.M., 2014. Triggering and runaway processes of catastrophic Tsaoling landslide induced by the 1999 Taiwan Chi-Chi earthquake, as revealed by high-velocity friction experiments. Geophys. Res. Lett. 41 (6), 1907–1915.
65. Towhata, I., Yamazaki, H., Kanatani, M., Lin, C.E., Oyama, T., 2001. Laboratory shear tests of rock

specimens collected from site of Tsao-ling earthquake-induced landslide. Tamkang J. Sci. Eng. 4, 209–219.

66. Tsutsumi, A., Shimamoto, T., 1996. Frictional properties of monzodiorite and gabbro during seismogenic fault motion. Geol. Soc. Jpn. 102, 240–248.

67. Tsutsumi, A., Shimamoto, T., 1997. High-velocity frictional properties of gabbro. Geophys. Res. Lett. 24, 699–702.

68. Ujiie, K., Tsutsumi, A., 2010. High-velocity frictional properties of clay-rich fault gouge in a megasplay fault zone, Nankai subduction zone. Geophys. Res. Lett. 37, L24310.

69. Voight, B., Faust, C., 1982. Frictional heat and strength loss in some rapid landslides. Geotechnique 32, 43–54.

70. Wang, W.N., Wu, H.L., Nakamura, H., Wu, S.C., Ouyang, S., Yu, M.F., 2003. Mass movements caused by recent tectonic activity: the 1999 Chi-chi earthquake in central Taiwan. Island Arc 12, 325–334.

71. Warr, L.N., Cox, S.J., 2001. Clay mineral transformations and weakening mechanisms along the Alpine Fault, New Zealand. In: Holdsworth, et al. (Eds.), The Nature and Significance of Fault Zone WeakeningGeol. Soc. Lond., Spec. Publ. 186, 85–101.

72. Weissel, J.K., Stark, C.P., 2001. Landslides triggered by the 1999 Mw7.6 Chi Chi earthquake in Taiwan and their relationship to topography. Geoscience and Remote Sensing Symposium, 2001. IGARSS'01. IEEE 2001 International. 2, pp. 759–761.

73. Wieczorek, G.F., Wilson, R.C., Harp, E.L., 1985. Map showing slope stability during earthquake in San Mateo County, California. Miscellaneous Investigation Maps I-1257-E, United States Geological Survey.

74. Wilson, R.C., Keefer, D.K., 1983. Dynamic analysis of a slope failure from the 6 August 1979 Coyote Lake, California, earthquake. Bull. Seismol. Soc. Am. 73, 863–877.

75. WRA, 2000. The Report: Remediation of Tsaoling Landslide. Water Resources Agency, Ministry of Economic Affairs, Taiwan, (in Chinese).

76. Wu, J.H., Chen, C.H., 2009. Back calculating the seismic shear strengths of the Tsaoling landslide associated with accelerograph and GPS data. Iran. J. Sci. Technol. Trans. B Eng. 33, 301–311.

77. Wu, J.H., Chen, C.H., 2011. Application of DDA to simulate characteristics of the Tsaoling landslide. Comput. Geotech. 38, 741–750.

78. Wu, J.H., Tsai, P.H., 2011. New dynamic procedure for back-calculating the shear strength parameters of large landslides. Eng. Geol. 123 (1–2), 129–147.

79. Yano, K., Shimamoto, T., Oohashi, K., Dong, J.J., Lee, C.T., 2009. Ultra-low friction of shale and clayey fault gouge at high velocities: implication for Jiufengershan landslide induced by 1999 Chi-Chi earthquake. The Next Generation of Research on Earthquake-induced Landslides: An International Conference in Commemoration of 10th Anniversary of the Chi-Chi Earthquake, pp. 402–406.

80. Yao, L., Ma, S., Shimamoto, T., Togo, T., 2013a. Structures and high-velocity frictional properties of the Pingxi fault zone in the Longmenshan fault system, Sichuan, China, activated during the 2008 Wenchuan earthquake. Tectonophysics 599, 135–156.

81. Yao, L., Shimamoto, T., Ma, S., Han, R., Mizoguchi, K., 2013b. Rapid postseismic strength recovery of Pingxi fault gouge from the Longmenshan fault system: experiments and implications for the mechanisms of high-velocity weakening of faults. J. Geophys. Res. Solid Earth 118, 1–17.

CHAPTER 5

CORRELATIONS BETWEEN FORMATION PROPERTIES AND INDUCED SEISMICITY DURING HIGH PRESSURE INJECTION INTO GRANITIC ROCK

Mark W. McClure [a], Roland N. Horne [b]

[a] Department of Petroleum Geosystems Engineering, University of Texas at Austin, 200 E. Dean Keeton, C0300, Austin, TX 78704, United States

[b] Energy Resources Engineering, Stanford University, 367 Panama Street, Green Earth Sciences 065, Stanford, CA 94305-4007, United States

ABSTRACT

We reviewed published results from six projects where hydraulic stimulation was performed in granitic rock. At each project, fractures in the formation were well-oriented to slip at the injection pressures used during stimulation. In all but one case, thousands of cubic meters of water were injected, and in every case, flow rates on the order of tens of liters per second were used. Despite these similarities, there was a large variation in the severity of induced seismicity that occurred in response to injection. At the three projects where induced seismicity was significant, observations at the wellbore showed evidence of well-developed brittle fault zones. At the three projects where induced seismicity was less significant, observations at the wellbore indicated only crack-like features and did not suggest significant fault development. These results suggest that assessments of the degree of fault development at the wellbore may be useful for predicting induced seismicity hazard. We cannot rule out that the differences were caused by variations in frictional properties that were unrelated to the degree of fault development (and it is possible that there is a relationship between these two parameters). The projects with more significant seismicity tended to be deeper, and if this is a meaningful correlation, it is unclear whether depth influenced seismic hazard through the degree of fault development, frictional properties, or some other variable. The results of this paper are not conclusive, but they suggest that there may be significant opportunity for future research on identifying geological conditions that increase induced seismicity hazard.

INTRODUCTION

Induced seismicity is an issue of growing importance for the exploitation of geothermal energy (Majer et al., 2007, Majer et al., 2011, Cladouhos et al., 2010 and Evans et al., 2012), wastewater disposal (Frohlich et al., 2011, Kim, 2013 and Zhang et al., 2013), CO_2 sequestration (Zoback and Gorelick, 2012), hydrocarbon production (Suckale, 2009), and other activities (McGarr et al., 2002 and Hitzman, 2012). Appropriate management of induced seismicity requires estimation of induced seismic hazard.

Some methodologies for estimating induced seismic hazard are purely statistical (Bommer et al., 2006 and Bachmann et al., 2011) or hybrid statistical/fluid flow models (Shapiro et al., 2007 and Gischig and Wiemer, 2013). However, these methods are site-specific and must be conditioned by performing the activity for which seismic hazard needs to be assessed. Low natural seismicity does not necessarily indicate that induced

Citation: Mark W. McClure, Roland N. Horne, Correlations between formation properties and induced seismicity during high pressure injection into granitic rock, Engineering Geology, Volume 175, 10 June 2014, Pages 74-80, ISSN 0013-7952, http://dx.doi.org/10.1016/j.enggeo.2014.03.015.

seismic hazard will be low, though there is some correlation (Evans et al., 2012).

In addition to purely statistical methods, numerical simulation has been used for induced seismic hazard analysis. This may involve kinematic modeling of deformation to estimate induced stress on neighboring faults (Segall, 1989, Hunt and Morelli, 2006 and Vörös and Baisch, 2009) or dynamic modeling that couples fluid flow, stresses induced by deformation, and friction evolution (Baisch et al., 2010 and McClure and Horne, 2011). However, model results are dependent on assumptions and input parameters that may be challenging to estimate. For example, change in stress on a fault may be estimated, but it is unclear how this should be related quantitatively to increased hazard (due to uncertainties such as the fault stress state and frictional properties).

All approaches to induced seismicity hazard analysis, whether purely statistical, physically based, or a hybrid of both, could benefit from methodologies that relate geophysical and geological observations to hazard (e.g., Davis and Frohlich, 1993).McGarr (1976) predicted that induced seismic moment release should be proportional to the volume of fluid injected, and this has been borne out by subsequent experience (Rutledge et al., 2004, Bommer et al., 2006, Hunt and Morelli, 2006 and Baisch and Vörös, 2009). However, the constant of proportionality between injection volume and moment release varies over orders of magnitude between different locations. For example, there are over 35,000 wells in the United States that have been hydraulically fractured in unconventional shale resources, and there are only a handful of confirmed instances of an induced event felt at the surface (page 76 of Hitzman, 2012). Induced seismicity does appear to be associated with oil and gas activities in many cases, but this is apparently due to long term fluid injection (Frohlich et al., 2011) or extraction (van Eijs et al., 2006). On the other hand, hydraulic stimulation for exploitation of geothermal resources in crystalline rock has routinely induced seismic events felt at the surface (Majer et al., 2007, Majer et al., 2011, Cladouhos et al., 2010 and Evans et al., 2012). Even among geothermal hydraulic fracturing projects, a huge diversity in induced seismicity hazard has been observed (Table 1; Kaieda et al., 2010). More work is needed to explain how geological conditions cause these large variations in induced seismicity.

Table 1: Summary of experiences with induced seismicity at six EGS projects. Supporting references are given in the text below. The assessments of "volume of fluid injected" are not necessarily complete because in some projects a variety of different injection operations were carried out over many years. We have not made an extensive effort to document all of the injections performed at these projects.

	Depth range	Maximum magnitude	Temperature	Degree of fault development	Volume of fluid injected during stimulation
Basel	4.6–5.0 km	3.4	190 °C at 5.0 km	High	11,570 m^3
Cooper Basin (Habanero 1)	4.1–4.4 km	3.7	250 °C at 4.4 km	High	20,000 m^3 in 2003, 25,000 m^3in 2005
Fjällbacka	0.5 km	− 0.2	16 °C at 0.5 km	Low	400 m^3 in Fjb1 and 36 m^3 in Fjb3
Ogachi (OGC-1)	0.99–1.0 km	− 1.0 except a 2.0 outlier	230 °C at 1.0 km	Low	10,140 t (approximately 9200 m^3)
Rosemanowes	1.7–2.65 km	0.16	100 °C at 2.6 km	Low	100,000 m^3 over two months in RH11 and RH12 (1982) and 5700 m^3 in RH15 (1985)
Soultz (shallow)	2.8–3.4 km	1.9	150 °C at 3.4 km	High	Two stimulations of 20,000 m^3each
Soultz (deep)	4.5–5 km	2.9	200 °C at 5.0 km	High	Three wells stimulated at volumes between 20,000 and 35,000 m^3

It is universally accepted that hydraulic fracturing in an oil and gas settings causes the initiation and propagation of new fractures (Economides and Nolte, 2000). However, during injection, fluid could leakoff into existing faults and cause slip (and potentially seismicity). In EGS, it is typically believed that injection predominantly causes induced slip on preexisting fractures (Cladouhos et al., 2011), though some authors have argued that there is probably more new fracture propagation than is commonly believed (McClure, 2012, Jung, 2013 and McClure and Horne, submitted for publication).

For significant induced seismicity to occur: (1) faults must be oriented properly with respect to the prevailing stress field so that they slip in response to imposed changes in stress and/or fluid pressure, (2) faults must have appropriate frictional properties so that slip occurs rapidly enough to generate seismicity, and (3) faults must be large enough to host significant events.

The first requirement, appropriately oriented faults, can be understood in the context of Coulomb theory (Chapter 2 of Hitzman, 2012). It has been argued that, in general, the crust is in a state of failure equilibrium (Townend and Zoback, 2000), suggesting that nearly everywhere in the subsurface, faults are present that will slip in response to an increase in fluid pressure.

The second requirement, rapid slip, can be explained in the context of results from laboratory friction experiments. Experiments have shown that the tendency for faults to slip seismically (rapidly) or aseismically (gradually) depends on the frictional properties of the minerals contacting in the fracture walls. Rock type, temperature, and other factors affect the tendency for fractures to slip seismically or aseismically (Dieterich, 2007).

There is a significant untapped opportunity to apply the results from friction experiments to help relate induced seismic hazard to lithology and depth. For example, differences in frictional properties may be the reason why hydraulic fracturing in granite has led to much greater induced seismic hazard than hydraulic fracturing in sedimentary formations. Assessments of induced seismic hazard from fluid injection in sedimentary formations (e.g. CO_2 sequestration or wastewater disposal) would benefit from efforts to identify lithologies where fault frictional properties are most favorable for seismic slip. Zhang et al. (2013) identified an apparent correlation between induced seismicity hazard and injection into basal aquifers.

The third requirement is that faults must be large enough to host significant-sized events. Seismic imaging and stratigraphic study can be used to identify major faults in layered formations. But in nonlayered formations such as crystalline basement rock, these techniques are limited because of the lack of seismic reflectors and discernible stratigraphic offsets. A very thick fault zone is required to generate a visible reflection at significant depth in crystalline rock. Faults in the basement may not extend into the overlying sediments. Even in layered sedimentary formations, hidden faults may be capable of hosting significant induced seismicity.

In this paper, we investigate whether wellbore observations could be used to estimate induced seismic hazard by assessing the degree to which large, brittle faults are present in a formation. Wells only sample the formation locally and may not intersect the most seismically important faults. However, formations that contain large faults are likely to contain abundant faults at all levels of development, and the overall degree of fault development in the formation should be observable at the wellbore. This theory is supported by the general observation that induced seismicity typically follows a Gutenberg–Richter distribution (Baisch et al., 2009, Baisch et al., 2010 and Bachmann et al., 2011), that large faults are surrounded by sizable damage zones, and that fracture size distributions are usually found to obey a power law or exponential distribution (Chapter 3 of Scholz, 2002). The mechanistic reason is that faults develop from accumulated deformation over time, starting with small, isolated cracks, which eventually link up and develop into large, continuous features (Chapter 3 of Segall and Pollard, 1983; Scholz, 2002 and Mutlu and Pollard, 2008). We refer to formations that have significantly developed faults that have linked up and formed larger features as having a high "degree of fault development."

To test whether the degree of fault development may be correlated to induced seismic hazard, we reviewed six projects where hydraulic stimulation (high rate fluid injection) was performed in granitic rock for the exploitation of geothermal energy: the projects at Cooper Basin, Australia; Soultz, France; Ogachi, Japan; Rosemanowes, United Kingdom; Basel, Switzerland; and Fjällbacka, Sweden. Projects using hydraulic fracturing to develop geothermal energy reservoirs are sometimes called "Enhanced Geothermal Systems,", or EGS. Some well-known EGS projects in granite, such as the projects at Fenton Hill, USA and Hijiori, Japan, were not included in this study because we were unable to find references that would permit an assessment of the degree of fault development. The Rosemanowes and Fjällbacka projects were performed for research purposes and targeted lower temperature reservoirs than would be typical for geothermal exploitation.

To control for the possible effect of lithology on the frictional properties of the faults, only projects in granite were included. To minimize the potential effect of fault orientation, we included only hydraulic stimulation tests where large injection pressures were used. The bottomhole fluid pressure likely reached or exceeded the minimum principal stress at every project considered, except possibly Basel, where estimates of the minimum principal stress are not available (McClure and Horne, submitted for publication). At these elevated pressures, faults with a wide variety of orientations are capable of slip (though not all orientations). In the "Fault orientation" section, we review literature that shows that fractures were present that were capable of slipping at the injection pressure at all six projects. A source of uncertainty is that pressure decreases with distance away from the wellbore during injection.

Fracture observations from a linear well are biased because wells are more likely to intersect faults perpendicular to their orientation than parallel to their orientation (Mauldon and Mauldon, 1997). This may particularly be an issue for vertical wells in strike-slip faulting regimes, where faults are expected to be vertical. Of the projects reviewed in this paper, Cooper Basin was in a reverse faulting regime (Baisch et al., 2006), Fjällbacka was in a reverse/strike-slip faulting regime (Wallroth et al., 1999), Soultz was in a normal/strike-slip regime (Valley and Evans, 2007), Basel, Ogachi, and Rosemanowes were in strike-slip faulting regimes (Pine and Batchelor, 1984 and Häring et al., 2008), and different references specify Ogachi as either strike-slip (Kiho et al., 1999) or reverse faulting (Ito, 2003). The wells at all these projects were nearly vertical, except at Rosemanowes, where the wells were deviated around 30° (Pine and Batchelor, 1984). Despite the potential for sampling bias, the approach in this paper assumes that wellbore observations will be adequate to characterize the degree of fault development in the formation. This approach relies on the idea that there was enough natural variability in fault orientation to provide an adequate sample of the formation from a borehole, and also on the idea that a formation containing large faults (which may not be intersected by the well) will also contain numerous smaller features that provide evidence of brittle fault development.

The six projects could be placed in two categories. At three of the projects, cataclasite and evidence of well-developed faults were observed at the wellbore, and induced seismicity was significant. At the other three projects, cataclasite and well-developed faults were not observed at the wellbore, and induced seismicity was much less significant. These results seem to indicate that the degree of fault development is a good indicator for induced seismic hazard, as long as the faults are well-oriented for slip in the ambient stress field and have frictional properties favorable for seismicity.

The results cannot be considered conclusive, especially because only six projects were included in the study. It cannot be ruled out that variation in frictional properties and/or depth accounted for the variation in seismicity, nor can their effects be fully disentangled because they may be interrelated. Even though the results are not conclusive, the results of this study suggest that formation frictional properties and degree of fault development could be useful in seismic hazard assessment and that these concepts merit further study.

METHODOLOGY

Six EGS projects in granitic rock were reviewed. Many projects involved multiple wells and stimulations. Other than at Fjällbacka, thousands or tens of thousands of cubic meters of fresh water were injected into the wells during stimulation. At Fjällbacka, injection volume was on the order of hundreds of cubic meters. In all cases, injection was performed into uncased sections of the wellbore at rates in excess of ten liters per second. Maximum magnitudes at each project are quoted directly from literature sources.

The degree of fault development was assessed from the thickness of observed faults and/or the presence of cataclasite and ultracataclasite. In projects with a "high" degree of fault development, fault zone features with thicknesses of meters were observed, or alternatively, drilling cuttings or core showed the presence of cataclasite and ultracataclasite, consistent with the classical concept of a fault with a damage zone (Faulkner et al., 2010). In projects with a "low" degree of fault development, only crack-like features were observed with thickness no greater than a few millimeters. These categorizations are coarse, but this was intentional due to the lack of precision inherent to assembling results from many different sources.

Maximum magnitudes are reported for the period during stimulation or shortly after. At Rosemanowes and Fjällbacka, larger events occurred during long-term unbalanced circulation (more fluid injected that produced) between injector/producer pairs. Long-term unbalanced circulation is a different process from stimulation, involving the net injection of a greater volume of fluid over a larger period of time and contained in a larger spatial region. Therefore, the induced seismic hazard would reasonably be expected to be greater in these cases, and so for consistency, induced events during long-term unbalanced circulation were not included in the assessments of maximum magnitude that are reported in Table 1.

RESULTS

Table 1 summarizes the results. In the following sections, the specific details of each project are reviewed.

Basel

One well was stimulated in an openhole section from 4629 m to 5000 m with 11,570 m^3 of water. The temperature at the bottom of the hole was around 190 °C. When the well was being drilled, drill cuttings were recovered containing ultracataclasite and cataclasite from five discrete zones (Kaeser et al., 2007 and Häring et al., 2008). The fault thickness could not be directly observed with image logs due to enlarged borehole diameter (Häring et al., 2008), but the presence of cataclasite indicates that well-developed faults were present. Mazurek (1998) described wells that were continuously cored through the crystalline basement of northern Switzerland within distances of 20–100 km from the Basel site. In these wells, thick well-developed faults were abundant. We classified the degree of fault development as high. The maximum magnitude was 3.4 (Evans et al., 2012).

Cooper Basin

The well Habanero 1 was completed openhole in granite from 4135 m to 4421 m and stimulated twice, in 2003 and 2005. In 2003, over 20,000 m^3 was injected (Baisch et al., 2006), and in 2005, around 25,000 m^3 was injected (Baisch et al., 2009). The temperature at the bottom of the hole was around 250 °C (Baisch et al., 2006 and Baisch et al., 2009). In wellbore imaging logs, a major fault zone was identified with a core with thickness of a few meters, surrounded by subsidiary fracturing within ten meters of the core [*Doone Wyborn, personal communication*]. Flow from the wellbore localized at this major fault. A broad, horizontal region of microseismicity spread from the well at the same depth as the fault (the minimum principal stress is vertical), and a well drilled subsequently several hundred meters away intersected a large fault at the same depth as the microseismic cloud (Baisch et al., 2006). We classified the degree of fault development as high. Maximum magnitude was 3.7 (Asanuma et al., 2005).

Fjällbacka

Two wells were stimulated from openhole sections around 500 m depth. Fjb1 was stimulated with 400 m^3 of water (Wallroth et al., 1999), and Fjb3 was stimulated with 36 m^3 (Evans et al., 2012). Temperature at 0.5 km was 16 °C (Wallroth et al., 1999). One of the wells was continuously cored, and Eliasson et al. (1990) performed a detailed analysis of the mineralogy of infilling of the fractures found in the wellbore core. Hydrothermally altered fracture zones were found, but there was no report of cataclasite. We classified the degree of fault development as low. The maximum observed magnitude during or after stimulation was − 0.2. During unbalanced circulation between the wells, when more fluid was injected than produced, there was an event that was felt only at the project site, but no estimate of magnitude is available (Evans et al., 2012).

Ogachi

OGC-1 was drilled and stimulated from an openhole section from 990 to 1000 m with 10,140 t (approximately 9200 m^3) (Kaieda et al., 2010). Hundreds of meters of OGC-1 and an adjacent well, OGC-2, were continuously cored. Temperature was around 230 °C at 1 km (Kitano et al., 2000). Faults with gouge less than two millimeters were abundant, but faults were not observed with thickness greater than 5 mm (Ito, 2003). Hydrothermal breccias were observed, but Ito (2003) specifically noted that these features were associated with fluidization, not shear deformation, because they contained neither cataclasites nor slickensides. A major flowing zone in the well OGC-2 was found to occur along an andesite dike, not a brittle fault structure. We classified the degree of fault development as low.

Suzuki and Kaieda (2000) discussed the results of a seismic reflection survey and a magneto-telluric survey that were performed at the Ogachi project. Their interpretation was that there is a large fault roughly 500 m from the Ogachi wells. For several reasons, we did not include this feature in our assessment of the degree of fault development. First, the fault was not observed directly and was inferred based on assumptions made in interpretation and processing.

The interpretation was especially challenging because the formation is crystalline and not layered. Second, even if a structural offset exists, the seismic interpretation did not give any detail about its inner structure — whether it is a brittle fault structure, a ductile shear zone, or some other feature. Because the inferred fault is unconfirmed by direct measurements and no information is available about its inner structure, we feel this report is not enough to change our assessment, particularly because such a detailed core analysis is available from Ito (2003).

The largest event was magnitude 2.0, but the next largest event was − 1.0 (Kaieda et al., 2010). This seismic behavior was unusual compared to other projects, where Gutenberg–Richter distributions of magnitude versus frequency were observed (Baisch et al., 2009, Baisch et al., 2010 and Bachmann et al., 2011). Other than the outlier event, the magnitude–frequency distribution at Ogachi fell along a Gutenberg–Richter distribution with a b value of 1.1 [*Hideshi Kaieda, personal communication*].

Rosemanowes

Three wells were stimulated in openhole sections at different depths between 1.7 and 2.65 km. Temperature at 2.6 km was around 100 °C (Richards et al., 1994). An analysis of wellbore imaging logs at the Rosemanowes wells discusses only thin fractures, not thicker fault features (Pearson et al., 1989). Fracture mapping from surface and mineshafts for the project discuss only jointing, not thicker fault features (Whittle, 1989 and Randall et al., 1990). Heath (1985) studied wells that were continuously cored to 700 m depth at a separate site around 10 km from the Rosemanowes project and reported faults with displacements of no more than a few millimeters. We classified the degree of fault development as low. The maximum magnitude observed during or immediately after stimulation was 0.16, but magnitude 1.7 and magnitude 2.0 events were reported during long term, unbalanced circulation between an injector and producer (when more fluid was being injected than produced) (Evans et al., 2012). In 1982, 100,000 m^3 of water was injected into RH11 and RH12 over two months. In 1985, 5700 m^3 of water was injected into RH15 (Evans et al., 2012).

Soultz (shallow)

One well was stimulated from an openhole section in granite from 2.8 to 3.4 km depth. Two stimulations were performed with 20,000 m^3 each (Evans et al., 2005). Temperature at 3.4 km was around 150 °C (Genter et al., 2010). GPK1 was extensively logged with wellbore imaging logs, and an adjacent well, EPS1, was continuously cored to 2230 m. Thick fault zones (up to 27.5 m thick) with the pattern of cataclasite and breccia in a core surrounded by intensely fractured and hydrothermally altered surrounding zones were observed in the core (Genter et al., 2000). We classified the degree of fault development as high. Maximum magnitude was 1.9 (Evans et al., 2012).

Soultz (deep)

Three wells were stimulated in openhole sections in granite from roughly 4.5 to 5 km. Temperature at 5 km was around 200 °C (Genter et al., 2010). Consistent with the shallower Soultz reservoir, the degree of fault development was high (Dezayes et al., 2010). Maximum magnitude was 2.9. The wells were each stimulated with volumes from 20,000 m^3 to 37,000 m^3 (Evans et al., 2012).

DISCUSSION

Seismicity

Seismicity was significantly greater at the three projects where evidence of well-developed faults was found, Cooper Basin, Soultz, and Basel, than at the three where it was not, Rosemanowes, Fjällbacka, and Ogachi.

At the less seismically active projects, there were a handful of outlier events. At Ogachi, there was a report of a single magnitude 2.0 even though the next largest event was reported to be magnitude − 1.0. At Rosemanowes and Fjällbacka, the maximum magnitudes during stimulation were very low, 0.16 and − 0.2. But there were a few reports of larger events (magnitude 1.7 and 2.0 events at Rosemanowes and an event felt at the project site at Fjällbacka) during long term unbalanced circulation. Despite these outlier events, the overall intensity of seismicity at these projects was much less than at the more active projects. These outlier events do not fall along a typical Gutenberg–Richter distribution of magnitudes, as was observed at Cooper Basin (Baisch et al., 2009), Soultz (Baisch et al., 2010), and Basel (Bachmann et al., 2011).

Degree of fault development

The correlation between the degree of fault development and seismicity could be explained in terms of slip surface continuity or by considering the overall formation deformation that occurs in response to a unit of fluid injected.

Earthquake ruptures have been observed to terminate when they reach fault stepovers (Wesnousky, 2006). Because it is difficult for ruptures to propagate from one fracture to another, large slip events should be prevented from occurring in formations with low fault development, where large, continuous fractures are less likely to be present.

However, the difference in seismicity between the projects cannot be explained only from slip surface continuity. The overall amount of seismic moment release was orders of magnitude greater for the projects with a high degree of fault development. The projects with much lower maximum magnitudes would have required a huge number of small events (with an unusually high b value) to have matched the seismic moment release from the projects with large maximum magnitudes and typical magnitude–frequency distributions. This was not observed.

Formations with a higher degree of fault development may be expected to experience more overall shear deformation in response to injection. Preexisting flaws deform much more easily than intact rock, which is why stiffness and rock strength decrease with increasing length scale (Heuze et al., 1990). As a simple example, injecting fluid into a fault cut in a granite block loaded with shear stress could lead to significant shear deformation. Injecting fluid into an intact granite block loaded under the same conditions would lead to much less shear deformation (unless the intact rock failed in compression or tension). Analogously, highly faulted formations should experience more shear deformation in response to injection than formation that are sparsely fractured. All other factors held equal, the amount of shear deformation caused by injection should be greater for formations with greater fault development. The scaling of this relationship deserves further study. It is possible that for formations with sufficiently developed faults, further fault development does not increase the amount of deformation that will occur in response to injection. van Eijs et al. (2006) performed a statistical investigation across geologically similar formations in the Netherlands and found a correlation between induced seismicity hazard and "fault density" in the formation as estimated from interpretations of seismic reflection surveys.

Fault orientation

A Coulomb analysis can be used to demonstrate that fault orientation in the ambient stress state strongly affects the tendency for a fault to slip in response to injection (Chapter 2 of Hitzman, 2012). In the EGS projects reviewed in this paper, injection pressure reached the minimum principal stress in almost every case (see review inMcClure and Horne (submitted for publication)). With such high fluid pressure, critical stress analysis may be less useful for assessing hazard because many or most fractures in the formation are likely to be able to slip. Of course, there will always be a fluid pressure gradient away from the well during injection, so fluid pressure elevation during injection will decrease as a function of distance.

Analyses performed at Soultz (Evans, 2005 and Meller et al., 2012), Fjällbacka (Jupe et al., 1992), Rosemanowes (Pine and Batchelor, 1984), Ogachi (Kiho et al., 1999 and Moriya et al., 2000), and Basel (Häring et al., 2008 and Delacou et al., 2009) indicate that abundant fractures were present at these projects that were oriented appropriately to slip at the injection pressure. We do not have access to statistics on fracture orientation from the wellbore imaging logs at Cooper Basin, but it is certain that natural fractures slipped in response to injection at this site because there were large magnitude events and the microseismic cloud was oriented subhorizontally in a reverse faulting stress regime (Baisch et al., 2006).

In different applications, where downhole fluid pressure may be more modestly elevated, fracture orientation and stress analysis may be more useful for estimating induced seismic hazard.

Depth and temperature

In our dataset, there was a correlation between depth and induced magnitude. This was most apparent at the Soultz project, where the maximum magnitude during the stimulation of the reservoir at 3 km was lower than in the reservoir at 5 km.

One explanation could be that because stress scales linearly with depth, stress drop also scales with depth. In fact, there is some evidence that stress drop scales with depth for shallower earthquakes. Shearer et al. (2006) reviewed earthquakes in Southern California and found that the median stress drop increased from around 0.6 MPa near the surface to around 2.2 MPa at 8 km. However, the total seismic moment release and maximum magnitude at the sites reviewed in this paper varied by orders of magnitude, more than be explained solely from stress drop.

Depth might relate to seismicity in other ways. A rate and state analysis suggests that the critical patch size for unstable slip decreases for greater normal stress (Dieterich, 2007). This suggests that aseismic slip is more likely under the lower stress conditions at shallower depth, all other factors held equal.

Temperature may play a role. Blanpied et al. (1995) performed laboratory experiments on wet fractures in granite which showed that stable slip should occur above 350 °C (higher temperature than any project reviewed in this project). There were mixed results in the temperature range of 250 °C–350 °C, mostly stable slip at room temperature, and consistently unstable slip between 100 °C and 250 °C. The Fjällbacka project was within the lower temperature range where stable slip appeared to be favored (somewhere below 100 °C), and the Rosemanowes project was in the neighborhood of the (not clearly delineated) lower transition temperature between stable and unstable sliding. The Basel and Soultz projects were within the temperature range consistent with unstable slip. The Cooper Basin and Ogachi projects were approaching the upper temperature range where there were mixed results indicating the possibility of stable slip. Application of these laboratory experiments should be considered highly imprecise because variations in granite mineralogy, fault gouge, hydrothermal alteration, and other factors could lead to considerable differences between the experiments and in-situ conditions.

A final possibility is that deeper formations are more likely to contain well-developed faults because they are under greater stress (though their depth is unlikely to have been constant over geological time), and perhaps deformation at greater stress is more likely to contribute to rock failure and fault zone development. To fully disentangle this issue, more data points would be needed, either shallow projects with well-developed faults or deep projects in formations without well-developed faults.

Seismic and aseismic slip

Because all projects reviewed were in granitic rock, we hoped to control for variations in the frictional properties of the fractures in the formations. However, we cannot independently verify that the frictional properties of the fractures in these formations were the same. The frictional properties of the faults could be related to the variation in composition of the different granites, different hydrothermal alterations and fracture infilling mineralogies, or different temperatures and depths. It is also possible that well-developed cataclasite fault cores in granite have frictional properties that are especially favorable for seismic slip. These issues could only be addressed with careful studies specifically directed at addressing the frictional properties of fractures in the individual formations. Variance of frictional properties of the formation has received little attention in the induced seismicity literature, though Zoback et al. (2012) discussed this possibility with respect to variations in microseismicity during hydraulic fracturing of shale.

Several authors have argued that aseismic deformation must have taken place at the shallow Soultz reservoir (Cornet et al., 1997 and Evans, 1998). Caliper logs indicated that faults had experienced greater shear offsets greater than could be explained from the total seismic moment release. It seems plausible that the greater maximum magnitude observed in the deeper Soultz reservoirs could be related to the frictional properties of the faults at different depths. This would suggest that even among formations with well-developed faults, variations of frictional properties will strongly affect induced seismicity hazard.

Spatial continuity

If our hypotheses are correct that the degree of fault development can be considered a formation property, it can be estimated from wellbore observations, and it can be related to seismic hazard, it would be useful to understand the spatial continuity of "degree of fault development." During hydraulic fracturing, the region of fluid pressure increase is relatively localized around the wellbore. But for long term injection (e.g., wastewater disposal, CO_2 sequestration, or unbalanced circulation in an EGS injector/producer pair) the region of pressure perturbation could spread significant distances from the wellbore. In these cases, assessments of the degree of fault development from observations made at the wellbore may be less reliable.

Analyses have shown induced seismicity at individual sites following predictable statistical behavior, such as the linear relationship between injection volume and moment release (Rutledge et al., 2004, Bommer et al., 2006, Shapiro et al., 2007 and Bachmann et al., 2011). But since these statistical relationships are highly variable between sites, determining the region where statistical stationarity can be assumed becomes an important issue. If a region of stress and/or pressure perturbation is growing over time, it will become increasingly likely that the pressure perturbation will reach regions of the subsurface that behave statistically differently. These transitions may not necessarily be easy to predict. Variations in lithology could potentially cause breaks in stationarity, as could faults that have formed along preexisting interfaces between lithologies. Outlier events were observed at Rosemanowes and Ogachi, and one possible explanation for these events is that the fluid pressure perturbation reached more distant faults that had statistically different behavior than the faults sampled near the wellbore.

As discussed in Section 3.4, a detailed core analysis at Ogachi indicated a low degree of fault development (Ito, 2003), but a seismic reflection survey was interpreted to indicate that a large fault exists 500 m from the wells. One interpretation could be that the inferred fault caused the outlier event at Ogachi and that it behaves statistically differently than the formation immediately surrounding the wells.

Injection volume

There was no apparent correlation between injection volume and maximum induced magnitude. At a particular site, increasing injection volume would be expected to increase overall seismic moment release and maximum magnitude. However, when comparing between projects, no correlation is apparent, evidently because site-specific factors overwhelmed the effect due to injection volume.

CONCLUSIONS

In this paper, we reviewed six projects where hydraulic stimulation was performed in granitic rock. Even though similar injection volumes were used (except Fjällbacka, where a lower volume was used) and the formations contained fractures well-oriented to slip at the injection pressure, there was striking variability in the magnitude of induced earthquakes. With the limited data, it is impossible to draw firm conclusions. However, there was a correlation between the degree of fault development observed at the well and the induced seismic deformation. We cannot rule out that variations in the frictional properties of the faults in the formations caused the observed variability (or that frictional properties are related to the degree of fault development). Comparison with laboratory experiments suggests that temperature may play a role, though the effect on the projects discussed in this paper is unclear. The three projects with the greatest seismicity and degree of fault zone development were also the deepest projects. From the six projects in this comparison, the relationships between depth, degree of fault development, temperature, and induced seismicity cannot be disentangled.

To our knowledge, prediction of induced seismic hazard by characterizing the degree of fault development based on wellbore observations has not previously been proposed in the literature. At many projects, measurements are not taken that would make such characterizations possible. Wellbore imaging logs, mud logs, and (if practical) wellbore core can be used to characterize the degree of fault development observed at the wellbore. Imaging logs can identify large features, and core and mud logs can be used to identify the presence of cataclasite. The results of this study suggest that this approach deserves further consideration, and operators should attempt to characterize the degree of fault development whenever feasible. Characterizing the degree of fault zone development may have other uses, such as helping understand flow and fracture network connectivity in a reservoir. The variability in seismicity between the sites reviewed in this paper may also be related to frictional behavior, another concept that has rarely been discussed in the induced seismicity literature but which has significant promise to improve induced seismicity hazard assessments.

ACKNOWLEDGMENTS

Part of this work was performed with support from the Precourt Institute for Energy at Stanford University. Several individuals generously helped us track down information about various projects. Thank you very much to Ingrid Stober, Albert Genter, Eva Schill, Steve Hickman, Doone Wyborn, and Günter Zimmermann. We discussed several issues with Nick Davatzes and thank you to him for his very stimulating comments. Thank you very much to Art McGarr and four other anonymous reviewers for their very helpful comments, which significantly improved the paper.

REFERENCES

1. Asanuma, H., Soma, N., Kaieda, H., Kumano, Y., Izumi, T., Tezuka, K., Niitsuma, H., Wyborn, D., 2005. Microseismic monitoring of hydraulic stimulation at the Australian HDR project in Cooper Basin, Paper Presented at the World Geothermal Congress, Antalya, Turkey (http://www.geothermal-energy.org/pdf/IGAstandard/WGC/2005/1615.pdf).
2. Bachmann, C.E., Wiemer, S., Woessner, J., Hainzl, S., 2011. Statistical analysis of the induced Basel

2006 earthquake sequence: introducing a probability-based monitoring approach for Enhanced Geothermal Systems. Geophys. J. Int. 186 (2), 793–807. http://dx.doi.org/10.1111/j.1365-246X.2011.05068.x.

3. Baisch, S., Vörös, R., 2009. Deep heat mining Basel — seismic risk analysis: AP 3000, induced seismicity. Q-Con Report (http://www.wsu.bs.ch/geothermie.htm).

4. Baisch, S., Weidler, R., Vörös, R., Wyborn, D., de Graaf, L., 2006. Induced seismicity during the stimulation of a geothermal HFR reservoir in the Cooper Basin, Australia. Bull. Seismol. Soc. Am. 96 (6), 2242–2256. http://dx.doi.org/10.1785/0120050255.

5. Baisch, S., Vörös, R., Weidler, R., Wyborn, D., 2009. Investigation of fault mechanisms during geothermal reservoir stimulation experiments in the Cooper Basin, Australia. Bull. Seismol. Soc. Am. 99 (1), 148–158. http://dx.doi.org/10.1785/0120080055.

6. Baisch, S., Vörös, R., Rothert, E., Stang, H., Jung, R., Schellschmidt, R., 2010. A numerical model for fluid injection induced seismicity at Soultz-sous-Forêts. Int. J. Rock Mech. Min. Sci. 47 (3), 405–413. http://dx.doi.org/10.1016/j.ijrmms.2009.10.001.

7. Blanpied, M.L., Lockner, D.A., Byerlee, J.D., 1995. Frictional slip of granite at hydrothermal conditions. J. Geophys. Res. 100 (B7), 13045–13064. http://dx.doi.org/10.1029/ 95JB00862.

8. Bommer, J.J., Oates, S., Cepeda, J.M., Lindholm, C., Bird, J., Torres, R., Marroquín, G., Rivas, J., 2006. Control of hazard due to seismicity induced by a Hot Fractured Rock geothermal project. Eng. Geol. 83 (4), 287–306. http://dx.doi.org/10.1016/j.enggeo.2005.11.002.

9. Cladouhos, T., Petty, S., Foulger, G., Julian, B., Fehler, M., 2010. Injection induced seismicity and geothermal energy. Trans. Geotherm. Resour. Counc. 34, 1213–1220.

10. Cladouhos, T.T., Clyne, M., Nichols, M., Petty, S., Osborn, W.L., Nofziger, L., 2011. Newberry volcano EGS demonstration stimulation modeling. Trans. Geotherm. Resour. Counc. 35, 317–322.

11. Cornet, F.H., Helm, J., Poitrenaud, H., Echecopar, A., 1997. Seismic and aseismic slips induced by large-scale fluid injections. Pure Appl. Geophys. 150 (3), 563–583. http:// dx.doi.org/10.1007/s000240050093.

12. Davis, S.D., Frohlich, C., 1993. Did (or will) fluid injection cause earthquakes? — criteria for a rational assessment. Seismol. Res. Lett. 64 (3–4), 207–224.

13. Delacou, B., Sartori, M., Philippossian, F., Savary, J., Grosjean, G., 2009. AP2000 Report — Structural Model of Potential Seismogenic Faults, Deep Heat Mining Basel — Seismic Risk Analysis, SERIANEX.

14. Dezayes, C., Genter, A., Valley, B., 2010. Structure of the low permeable naturally fractured geothermal reservoir at Soultz. Compt. Rendus Geosci. 342 (7–8), 517–530. http://dx. doi.org/10.1016/j.crte.2009.10.002.

15. Dieterich, J.H., 2007. 4.04 — Applications of Rate- and State-Dependent Friction to Models of Fault Slip and Earthquake Occurrence. In: Schubert, G. (Ed.), Treatise on Geophysics. Elsevier, Amsterdam, pp. 107–129.

16. Economides, M.J., Nolte, K.G., 2000. Reservoir Stimulation, 3rd ed. John Wiley, Chichester, England, New York.

17. Eliasson, T.M., Tullborg, E.-L., Landström, O., 1990. Fracture filling mineralogy and geochemistry at the Swedish HDR research site, Paper Presented at the Hot Dry Rock: Proceedings Camborne School of Mines International Hot Dry Rock Conference. Robertson Scientific Publications.

18. Evans, K., 1998. Does Significant Aseismic Slip Occur on Fractures in HDR Systems Under Stimulation Conditions, Paper Presented at the 4th International HDR Forum, Strasbourg, Germany.

19. Evans, K.F., 2005. Permeability creation and damage due to massive fluid injections into granite at 3.5 km at Soultz: 2. Critical stress and fracture strength. J. Geophys. Res. 110 (B4). http://dx.doi.org/10.1029/2004JB003169.

20. Evans, K.F., Moriya, H., Niitsuma, H., Jones, R.H., Phillips, W.S., Genter, A., Sausse, J., Jung, R., Baria, R., 2005. Microseismicity and permeability enhancement of hydrogeologic structures during massive fluid injections into granite at 3 km depth at the Soultz HDR site. Geophys. J. Int. 160 (1), 389–412. http://dx.doi.org/10.1111/j.1365-246X. 2004.02474.x.

21. Evans, K.F., Zappone, A., Kraft, T., Deichmann, N., Moia, F., 2012. A survey of the induced seismic responses to fluid injection in geothermal and CO2 reservoirs in Europe. Geothermics 41, 30–54. http://dx.doi.org/10.1016/j.geothermics.2011.08.002.

22. Faulkner, D.R., Jackson, C.A.L., Lunn, R.J., Schlische, R.W., Shipton, Z.K., Wibberley, C.A.J., Withjack, M.O., 2010. A review of recent developments concerning the structure, mechanics and fluid flow properties of fault zones. J. Struct. Geol. 32 (11), 1557–1575. http://dx.doi.org/10.1016/j.jsg.2010.06.009.

23. Frohlich, C., Hayward, C., Stump, B., Potter, E., 2011. The Dallas–Fort Worth earthquake sequence: october 2008 through may 2009. Bull. Seismol. Soc. Am. 101 (1), 327–340. http://dx.doi.org/10.1785/0120100131.

24. Genter, A., Traineau, H., Ledésert, B., Bourgine, B., Gentier, S., 2000. Over 10 years of geological investigations within the HDR Soultz project, France, Paper Presented at the World Geothermal Congress, Kyushu–Tohoku, Japan (http://www.geothermal-energy.org/pdf/IGAstandard/WGC/2000/R0710.PDF).

25. Genter, A., Evans, K., Cuenot, N., Fritsch, D., Sanjuan, B., 2010. Contribution of the exploration of deep crystalline fractured reservoir of Soultz to the knowledge of enhanced geothermal systems (EGS). Compt. Rendus Geosci. 342 (7–8), 502–516. http://dx. doi.org/10.1016/j.crte.2010.01.006.

26. Gischig, V.S., Wiemer, S., 2013. A stochastic model for induced seismicity based on nonlinear pressure diffusion and irreversible permeability enhancement. Geophys. J. Int. 194 (2), 1229–1249. http://dx.doi.org/10.1093/gji/ggt164.

27. Häring, M.O., Schanz, U., Ladner, F., Dyer, B.C., 2008. Characterisation of the Basel 1 enhanced geothermal

system. Geothermics 37 (5), 469–495. http://dx.doi.org/10. 1016/j.geothermics.2008.06.002.

28. Heath, M.J., 1985. Geological control of fracture permeability in the Carnmenellis granite, Cornwall: implications for radionuclide migration. Mineral. Mag. 49 (351), 233–244.

29. Heuze, F.E., Shaffer, R.J., Ingraffea, A.R., Nilson, R.H., 1990. Propagation of fluid—driven fractures in jointed rock. Part 1—development and validation of methods of analysis. Int. J. Rock Mech. Min. Sci. Geomech. Abstr. 27 (4), 243–254. http://dx.doi.org/10. 1016/0148-9062(90)90527-9.

30. Hitzman, M.W., 2012. Induced Seismicity Potential in Energy Technologies. National Research Council of the National Academies.

31. Hunt, S.P., Morelli, C., 2006. Cooper Basin HDR Hazard Evaluation: Predictive Modeling of Local Stress Changes due to HFR Geothermal Energy Operations in South Australia. South Australia Department of Primary, Industries and Resources.

32. Ito, H., 2003. Inferred role of natural fractures, veins, and breccias in development of the artificial geothermal reservoir at the Ogachi Hot Dry Rock site, Japan. J. Geophys. Res. 108 (B9). http://dx.doi.org/10.1029/2001JB001671.

33. Jung, R., 2013. EGS — goodbye or back to the future, in effective and sustainable hydraulic fracturing. In: Bunger, A.P., McLennan, J., Jeffrey, R. (Eds.), InTech, pp. 95–121. http:// dx.doi.org/10.5772/45724.

34. Jupe, A.J., Green, A.S.P., Wallroth, T., 1992. Induced microseismicity and reservoir growth at the Fjällbacka Hot Dry Rocks project, Sweden. Int. J. Rock Mech. Min. Sci. Geomech. Abstr. 29 (4), 343–354. http://dx.doi.org/10.1016/0148-9062(92)90511-W.

35. Kaeser, B., Kalt, A., Borel, J., 2007. The Crystalline Basement Drilled at the Basel-1 Geothermal Site, a Preliminary Petrological–Geochemical Study. Institut de Géologie et d'Hydrogéologie. Université de Neuchâtel, Switzerland.

36. Kaieda, H., Sasaki, S., Wyborn, D., 2010. Comparison of characteristics of microearthquakes observed during hydraulic stimulation operations in Ogachi, Hijiori, and Cooper Basin HDR projects, Paper Presented at the World Geothermal Congress, Bali, Indonesia (http://www.geothermal-energy.org/pdf/IGAstandard/WGC/2010/ 3157.pdf).

37. Kiho, K., Kitano, K., Kaieda, H., Ito, H., Shin, K., Eguchi, Y., Suenaga, H., 1999. Overview of the Ogachi HDR project. Trans. Geotherm. Resour. Counc. 23.

38. Kim, W.-Y., 2013. Induced seismicity associated with fluid injection into a deep well in Youngstown, Ohio. J. Geophys. Res. 118. http://dx.doi.org/10.1002/jgrb.50247.

39. Kitano, K., Hori, Y., Kaieda, H., 2000. Outline of the Ogachi HDR project and character of the reservoirs, Paper Presented at the World Geothermal Congress, Kyushu–Tohoku, Japan (http://www.geothermal-energy.org/pdf/IGAstandard/WGC/2000/R0435. PDF).

40. Majer, E.L., Baria, R., Stark, M., Oates, S., Bommer, J., Smith, B., Asanuma, H., 2007. Induced seismicity associated with Enhanced Geothermal Systems. Geothermics 36 (3), 185–222. http://dx.doi.org/10.1016/j.geothermics.2007.03.003.

41. Majer, E., Nelson, J., Robertson-Tai, A., Savy, J., Wong, I., 2011. Protocol for addressing induced seismicity associated with Enhanced Geothermal Systems. http://www1. eere.energy.gov/geothermal/pdfs/egs-is-protocol-final-draft-20110531.pdf.

42. Mauldon, M., Mauldon, J.G., 1997. Fracture sampling on a cylinder: from scanlines to boreholes and tunnels. Rock Mech. Rock. Eng. 30 (3), 129–144. http://dx.doi.org/10.1007/ BF01047389.

43. Mazurek, M., 1998. Technical Report 93-12: Geology of the Crystalline Basement of Northern Switzerland and Derivation of Geological Input Data for Safety Assessment Models. National Cooperative for the Disposal of Radioactive Waste, Switzerland (http://www.nagra.ch/documents/database/dokumente/%24default/Default%20Folder/ Publikationen/NTBs%201991-1993/e_ntb93-12.pdf).

44. McClure, M.W., 2012. Modeling and Characterization of Hydraulic Stimulation and Induced Seismicity in Geothermal and Shale gas Reservoirs. PhD Thesis Stanford University, Stanford, California.

45. McClure, M.W., Horne, R.N., 2011. Investigation of injection-induced seismicity using a coupled fluid flow and rate/state friction model. Geophysics 76 (6), WC181–WC198. http://dx.doi.org/10.1190/geo2011-0064.1.

46. McClure, M.W., Horne, R.N., 2014. An investigation of stimulation mechanism in Enhanced Geothermal Systems. Int. J. Rock Mech. Min. Sci. (submitted for publication).

47. McGarr, A., 1976. Seismic moments and volume changes. J. Geophys. Res. 81 (8), 1487–1494. http://dx.doi.org/10.1029/JB081i008p01487.

48. McGarr, A., Simpson, D., Seeber, L., 2002. 40. Case histories of induced and triggered seismicity. In: Lee, W.H.K., Kanamori, H. (Eds.), International Geophysics. Academic Press, pp. 647–661.

49. Meller, C., Kohl, T., Gaucher, E., Genter, A., 2012. Approach for Determination of the Failure Probability of Fractures at the Soultz-Sous-Forêts EGS Project, Paper Presented at the Thirty-Seventh Workshop on Geothermal Reservoir Engineering, Stanford, CA.

50. Moriya, H., Niitsuma, H., Kaieda, H., 2000. Reevaluation of Reservoir Structure at Ogachi HDR Field by Precise Source Location of AE Multiplet, Paper Presented at the World Geothermal Congress. Ogachi, Japan.

51. Mutlu, O., Pollard, D.D., 2008. On the patterns of wing cracks along an outcrop scale flaw: a numerical modeling approach using complementarity. J. Geophys. Res. 113 (B6). http://dx.doi.org/10.1029/2007JB005284.

52. Pearson, R.A., Moore, P.W., Harker, C., Lanyon, G.W., 1989. Logging results, section 1 — proprietary logging results. In: Parker, R.H. (Ed.), Hot Dry Rock Geothermal Energy:

53. Phase 2B Final Report of the Camborne School of Mines Project. Pergamon Press, Oxford, New York, pp. 423–558.
54. Pine, R.J., Batchelor, A.S., 1984. Downward migration of shearing in jointed rock during hydraulic injections. Int. J. Rock Mech. Min. Sci. Geomech. Abstr. 21 (5), 249–263. http://dx.doi.org/10.1016/0148-9062(84)92681-0.
55. Randall, M.M., Lanyon, G.W., Nicolls, J., Willis-Richards, J., 1990. Evaluation of jointing in the Carnmenellis granite, Paper Presented at the Hot Dry Rock: Proceedings Camborne School of Mines International Hot Dry Rock Conference. Robertson Scientific Publications.
56. Richards, H.G., Parker, R.H., Green, A.S.P., Jones, R.H., Nicholls, J.D.M., Nicol, D.A.C., Randall, M.M., Richards, S., Stewart, R.C., Willis-Richards, J., 1994. The performance and characteristics of the experimental Hot Dry Rock geothermal reservoir at Rosemanowes, Cornwall (1985–1988). Geothermics 23 (2), 73–109. http://dx.doi.org/10.1016/ 0375-6505(94)90032-9.
57. Rutledge, J.T., Phillips, W.S., Mayerhofer, M.J., 2004. Faulting induced by forced fluid injection and fluid flow forced by faulting: an interpretation of hydraulic-fracture microseismicity, Carthage Cotton Valley Gas Field, Texas. Bull. Seismol. Soc. Am. 94 (5), 1817–1830. http://dx.doi.org/10.1785/012003257.
58. Scholz, C.H., 2002. The Mechanics of Earthquakes and Faulting. Cambridge University Press, Cambridge, UK.
59. Segall, P., 1989. Earthquakes triggered by fluid extraction. Geology 17 (10), 942–946. http://dx.doi.org/10.1130/0091-7613(1989) 017b0942:ETBFEN2.3.CO;2.
60. Segall, P., Pollard, D.D., 1983. Nucleation and growth of strike slip faults in granite. J. Geophys. Res. 88 (B1), 555–568. http://dx.doi.org/10.1029/JB088iB01p00555.
61. Shapiro, S.A., Dinske, C., Kummerow, J., 2007. Probability of a given-magnitude earthquake induced by a fluid injection. Geophys. Res. Lett. 34 (22). http://dx.doi.org/10. 1029/2007GL031615.
62. Shearer, P.M., Prieto, G., Hauksson, E., 2006. Comprehensive analysis of earthquake source spectra in southern California. J. Geophys. Res. 111. http://dx.doi.org/10.1029/ 2005JB003979 B06303.
63. Suckale, J., 2009. Induced seismicity in hydrocarbon fields. Adv. Geophys. 51 (2), 55–106. http://dx.doi.org/10.1016/S0065-2687(09)05107-3.
64. Suzuki, K., Kaieda, H., 2000. Geological structure around the Ogachi Hot Dry Rock test site using seismic reflection and CSAMT surveys, paper presented at the World Geothermal Congress. Kyushu - Tohoku, Japan.
65. Townend, J., Zoback, M.D., 2000. How faulting keeps the crust strong. Geology 28 (5), 399–402. http://dx.doi.org/10.1130/0091-7613(2000)28b399:HFKTCSN2.0.CO;2.
66. Valley, B., Evans, K., 2007. Stress state at Soultz-sous-Forêts to 5 km depth from wellbore failure and hydraulic observations, Paper Presented at the Thirty-Second Workshop on
67. Geothermal Reservoir Engineering. Stanford University (http://www.geothermal-energy.org/pdf/IGAstandard/SGW/2007/valley.pdf).
68. van Eijs, R.M.H.E., Mulders, F.M.M., Nepveu, M., Kenter, C.J., Scheffers, B.C., 2006. Correlation between hydrocarbon reservoir properties and induced seismicity in the Netherlands. Eng. Geol. 84 (3–4). http://dx.doi.org/10.1016/j.enggeo.2006.01.002.
69. Vörös, R., Baisch, S., 2009. Deep Heat Mining Basel — Seismic Risk Analysis: AP 4000, Triggered Seismicity. Q-Con Report (http://www.wsu.bs.ch/geothermie.htm).
70. Wallroth, T., Eliasson, T., Sundquist, U., 1999. Hot Dry Rock research experiments at Fjällbacka, Sweden. Geothermics 28 (4–5), 617–625. http://dx.doi.org/10.1016/ S0375-6505(99)00032-2.
71. Wesnousky, S.G., 2006. Predicting the endpoints of earthquake ruptures. Nature 444 (7117), 358–360. http://dx.doi.org/10.1038/nature05275.
72. Whittle, R., 1989. The granites of South West England, section 2 — jointing in the Carnmenellis granite. In: Parker, R.H. (Ed.), Hot Dry Rock Geothermal Energy: Phase 2B Final Report of the Camborne School of Mines Project. Pergamon Press, Oxford, New York, pp. 119–182.
73. Zhang, Y., Person, M., Rupp, J., Ellett, K., Celia, M.A., Gable, C.W., Bowen, B., Evans, J., Bandilla, K., Mozley, P., Dewers, T., Elliot, T., 2013. Hydrogeologic controls on induced seismicity in crystalline basement rocks due to fluid injection into basal reservoirs. Groundwater 51 (4), 525–538. http://dx.doi.org/10.1111/gwat.12071.
74. Zoback, M.D., Gorelick, S.M., 2012. Earthquake triggering and large-scale geologic storage of carbon dioxide. Proc. Natl. Acad. Sci. 109 (26), 10164–10168. http://dx.doi.org/10. 1073/pnas.1202473109.
75. Zoback, M., Kohli, A., Das, I., McClure, M., 2012. The importance of slow slip on faults during hydraulic fracturing stimulation of shale gas reservoirs. SPE 155476, Paper Presented at the SPE Americas Unconventional Resources Conference, Pittsburgh, Pennsylvania USA. http://dx.doi.org/10.2118/155476-MS.

CHAPTER 6

SIMULATING THE INFLUENCES OF GROUNDWATER ON REGIONAL GEOMORPHOLOGY USING A DISTRIBUTED, DYNAMIC, LANDSCAPE EVOLUTION MODELLING PLATFORM

Andrew Barkwith [a], Martin D. Hurst [a], Christopher R. Jackson [a], Lei Wang [a], Michael A. Ellis [a], Tom J. Coulthard [b]

[a] British Geological Survey, Environmental Science Centre, Keyworth, Nottingham, NG12 5GG, UK

[b] Department of Geography, Environment and Earth Science, University of Hull, Cottingham Road, Hull, HU6 7RX, UK

ABSTRACT

A dynamic landscape evolution modelling platform (CLiDE) is presented that allows a variety of Earth system interactions to be explored under differing environmental forcing factors. Representation of distributed surface and subsurface hydrology within CLiDE is suited to simulation at sub-annual to centennial time-scales. In this study the hydrological components of CLiDE are evaluated against analytical solutions and recorded datasets. The impact of differing groundwater regimes on sediment discharge is examined for a simple, idealised catchment, Sediment discharge is found to be a function of the evolving catchment morphology. Application of CLiDE to the upper Eden Valley catchment, UK, suggests the addition of baseflow-return from groundwater into the fluvial system modifies the total catchment sediment discharge and the spatio-temporal distribution of sediment fluxes during storm events. The occurrence of a storm following a period of appreciable antecedent rainfall is found to increase simulated sediment fluxes.

INTRODUCTION

Cellular automata

Simulation of distributed dynamic environmental systems is often based on the solution of differential equations that can be difficult to solve without simplification (Toffoli, 1984) or large computational resources. Cellular automata (CA), first developed by von Neumann, 1951 and von Neumann, 1966 to investigate self replication, provided an alternative to solving the governing equations, allowing a fast, exact numerical simulation of a physical system. Since their inception CA have been applied to a host of physical research areas focussed around fluid dynamics and natural systems (see for example,Margolus et al., 1986, Somers, 1993, Chen and Doolen, 1998 and Chopard and Masselot, 1999). CA discretises space into regular two- or three-dimensional cells, each of which contain physical property information for that region of space. Each cell is able to pass and gather information about neighbouring cells and subsequently modify its contents based on a transfer function. Interaction with neighbouring cells for two-dimensional cases usually consists of either a Moore-type method, where all surrounding cells interact with the central node, or a von Neumann (Manhattan) neighbour-type method where interaction is solely with adjacent cells (Fig. 1). Initial cell states are user defined, and with each time step states are simultaneously updated based on the transfer function and states of neighbouring cells.

Citation: Andrew Barkwith, Martin D. Hurst, Christopher R. Jackson, Lei Wang, Michael A. Ellis, Tom J. Coulthard, Simulating the influences of groundwater on regional geomorphology using a distributed, dynamic, landscape evolution modelling platform, Environmental Modelling & Software, Volume 74, December 2015, Pages 1-20, ISSN 1364-8152, http://dx.doi.org/10.1016/j.envsoft.2015.09.001.

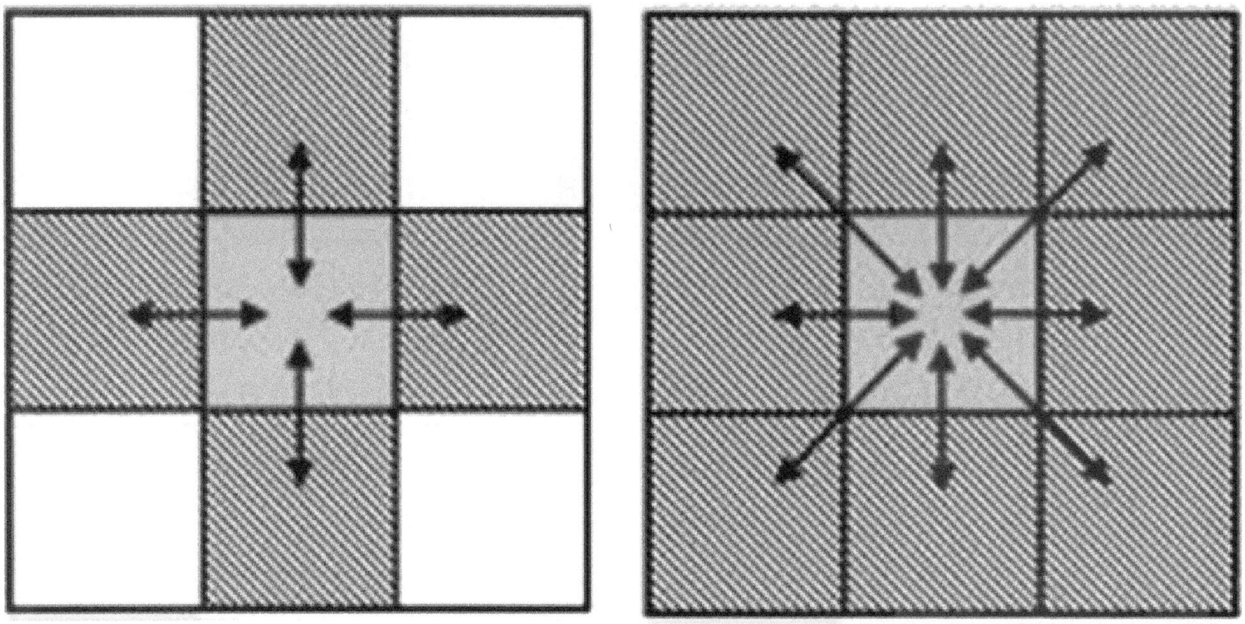

Figure 1: CA neighbourhoods; the von Neumann neighbourhood (left) considers cells in the cardinal directions and the Moore neighbourhood (right) allows interactions with all immediate cells. Arrows represent cell neighbourhood interaction with a central node.

Groundwater and geomorphology

The impacts of surface hydrological processes on geomorphology, and thus topography, have been studied using numerical methods for a number of years. With advances in computer processing power available to researchers, the development of landscape evolution models (LEMs) has progressed from simulating water routing effects on nodal elevation (Ahnert, 1976), to representing detailed fluvial and hillslope processes (for a full review see Tucker and Hancock, 2010). Recent advances in this field have seen increases in domain sizes and resolution, irregular gridding techniques, and the move from hypothetical to real catchments becoming common place (e.g., Pazzaglia, 2003 and Coulthard et al., 2012). Currently, LEMs based on regular, square, gridded digital elevation models (DEM) are popular as slope attribution does not need to be determined manually. These DEM-based models also have the advantage of allowing run-time landscape adjustment in response to runoff, erosion and deposition (Hancock et al., 2011). Current models contain representations of surface hydrology but they all either use highly simplified representations of groundwater influx to rivers (as a constant baseflow) or ignore groundwater processes entirely (Huang and Niemann, 2006). An exception to this is Keesstra et al. (2014), who capture the transient nature of baseflow in an LEM environment suitable for simulating relatively small catchments.

There are a number of mechanisms by which groundwater can have an impact on erosional processes. Within cohesive soils or sediments, in a process similar to overland flow, the formation of subsurface channels from groundwater scour can cause near-surface fractures or cracks in the overlying surface (Dunne, 1988). These pipes can collapse, leading to the formation of extensive gully networks. This process tends to occur in shale-dominated, arid or semi-arid regions (Howard and McLane, 1988). The seepage of groundwater to the surface can locally enhance the erodibility of cohesive soils or sediments by increasing their susceptibility to a range of physical and chemical weathering processes (Laity and Malin, 1985). Chemical weathering at depth, which subsequently leads to enhanced erosion when a unit is exposed, can also be attributed to groundwater processes (Kelly, 2012). In less cohesive sediments the seepage of groundwater to the surface can provide buoyancy to particles, reducing the flow velocities required to entrain them and increasing the development of channel features (Fox et al., 2007). Groundwater flow operates on longer temporal scales than direct surface runoff and, in its simplest form, introduces a mechanism to lag surface water flow and impart a smoothing effect on surface fluvial systems (Changnon et al., 1988). Through this process, the fluvial flow velocities that drive sediment flux can be modified at multiple spatio-temporal scales.

Due to the limited extent of tunnel-scour erosion and long time-scales associated with physical and chemical weathering, these processes are not considered in this study, which focusses on sub-annual to centennial geomorphological evolution. Although the groundwater seepage buoyancy process is known to be important locally, the impact on differing sediment classes and under differing groundwater regimes is not well constrained. Due to its conceptual simplicity and potential applications, it is the representation of baseflow-return to the fluvial system and the impacts that this can have on catchment morphology and sediment transport that forms the basis of this study.

Baseflow

The time taken for water to enter the groundwater system through recharge, until its contribution to a surface water body as baseflow, is influenced significantly by topography, geology, vegetation, land use and climate factors. Aquifer recharge through channel infiltration during storm events can be large enough to modify the commonly observed power law relationship between discharge and basin area (Goodrich et al., 1997), although many LEMs assume this relationship to be linear. If the return of this water to the surface as baseflow coincides with future rainfall events, the increase in storm flow can be significant (Sklash and Favolden, 1979, Buttle, 1994 and

Kirchner et al., 2000). Spatial variability of baseflow is generally highly heterogeneous across a catchment, often increasing surface river flows in the lower reaches. The combination of these spatial and temporal groundwater influences on surface water dynamics can create a highly complex system. Modelling surface water flow dynamics accurately under these complex hydrological conditions without a groundwater component is often not possible. The characteristics of catchments with greatest susceptibility to complex hydrological interactions are: those with large ranges between topographic highs and lows, creating a strong subsurface hydraulic flow gradient; those with highly heterogeneous hydraulic conductivity and specific yield, formed for example by changes in geology; those where groundwater exhibits a strong seasonal signal and hence a variable baseflow-return to rivers, for example sandstone aquifers; and those that are unconfined, allowing the return of water to the surface.

Application

Depending on the baseflow contribution to surface flow components, groundwater process could have major impacts on the development of terrestrial landscapes at a range of scales. If the focus of a particular study is on representing catchment discharges at time-scales beyond millennia, average values could justifiably be used within LEMs to represent the baseflow components in river flow for a simple groundwater dominated catchment. However, there are situations where the spatial and temporal addition of baseflow-return could be too complex to represent using catchment wide, temporal averages, as they are likely to be highly variable at the sub-annual scale and potentially nonlinear in nature. These occur: at shorter time-scales (centennial and below), where variability of storm events becomes important; in complex groundwater situations; or where the spatial distribution of morphological change is of critical importance to a study.

In this study we describe the development of distributed surface and subsurface hydrological components for a LEM designed to operate at sub-annual to the centennial time-scales. The hydrological components of the resultant modelling platform are verified against analytical solutions. A sensitivity analysis of morphological change under differing groundwater regimes in a small, uniform idealised catchment is subsequently undertaken using the platform to assess the groundwater impacts on fluvial and sediment discharge. Finally the platform is applied to a heterogeneous catchment in north-west England where we examine the role that baseflow plays in controlling sub-annual to decadal scale fluvial and sediment transport properties that can shape catchment scale geomorphology.

CLIDE MODELLING PLATFORM

The Dynamic Environmental Sensitivity to Change (DESC) project at the British Geological Survey, on which this paper is based, couples CA modelling from various backgrounds and has produced the CAESAR-Lisflood-DESC (CLiDE) platform; a geomorphological simulator that allows a variety of Earth system interactions to be explored. This platform is based upon the extension of the well established Cellular Automaton Evolutionary Slope and River (CAESAR) model (Coulthard and Van De Wiel, 2006), CAESAR-Lisflood (Coulthard et al., 2013), which incorporates the Lisflood-FP hydrodynamic model (Bates et al., 2010) to simulate channel and overbank flow. The two-dimensional modular design allows great versatility in the range of simulated spatio-temporal scales to which it can be applied. CAESAR has been used to investigate a variety of sediment transport, erosional and depositional processes under differing climatic and land use pressures in river reaches and catchments (e.g., Hancock et al., 2011). The addition of Lisflood-FP to the code (see Coulthard et al., 2013) improves the representation of surface water flow within the model by incorporating momentum. This allows a non-steady flow solution compared to the steady-state flow models of nearly all other LEMs. As with other LEMs, CAESAR-Lisflood simplifies the representation of some of the hydrological processes and their interactions. Specifically, it does not simulate groundwater flow and groundwater discharge to rivers. To address these limitations, the non-Lisflood-FP controlled surface hydrological processes within the CLiDE platform are replaced with a distributed hydrological model that includes a groundwater model. This groundwater model is coupled to the surface model through the exchange of water as groundwater recharge and baseflow-return to rivers.

The coupling of modules in the CLiDE platform, and their distributed nature, allow inter-disciplinary environmental system feedbacks to be explored at a range of scales. A flow chart outlining platform processes and their interactions is shown in Fig. 2. Initialisation of the CLiDE platform requires a number of input items in the form of either gridded or list based ascii files. Spatially discretised initial model inputs are entered as Cartesian grids, with header information containing cell size, domain dimensions and geo-referencing details. Distributed DEM, land use, soil hydrological characteristics, daily rainfall and evapotranspiration datasets are used to initialise and force the surface hydrology model. The sediment transport components of the platform use the same DEM as the hydrological components but also require bedrock elevation referenced to the same datum to determine the thickness of the sediment store layer. Grain size distributions, as determined by field measurement or through a calibration process, are required by CLiDE. A detailed characterisation of the sediment transport, vegetation and slope process components used by the CAESAR algorithms in the CLiDE platform is provided by Van De Wiel et al. (2007).

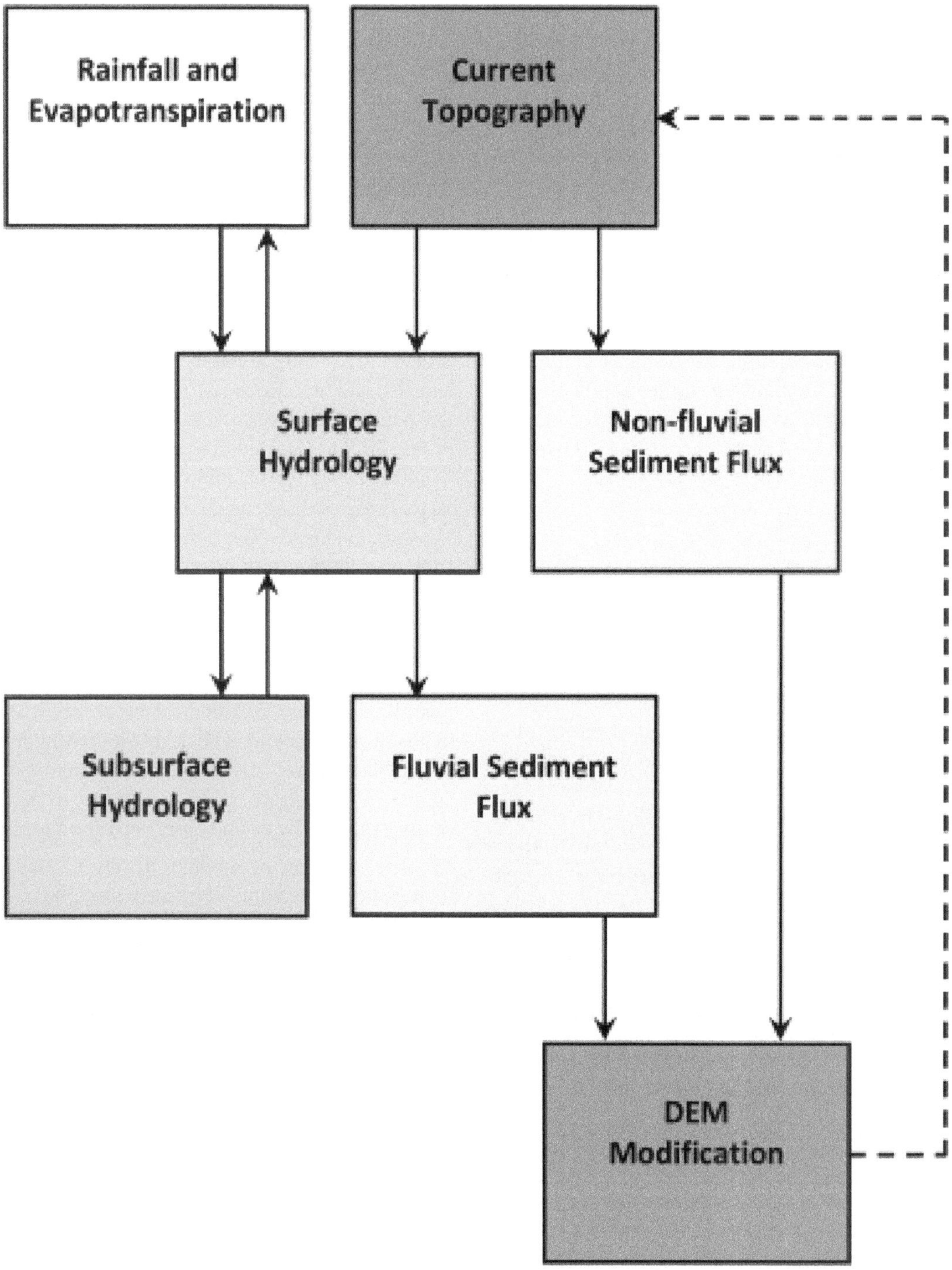

Figure 2: An overview of the coupled processes in the CLiDE platform. The dashed arrow represents the creation of a new topography at the following time step, when the process is repeated.

Hydrological component

Surface and subsurface hydrological processes that ultimately control sediment transport in CLiDE are represented by a distributed coupled CA model. A flow chart of the coupled hydrological components within the platform is presented in Fig. 3. To partition rainfall between evaporatranspiration, surface water runoff and groundwater recharge we implement a soil water balance model (SLiM) based on the conceptualisation of the soil as a single store. Spatially distributed soil types, vegetation types, near surface soil moisture, evapotranspiration and distributed rainfall are considered in calculating the soil water balance and in determining surface water depths and groundwater recharge. Surface water is routed using the Lisflood-FP model (Bates et al., 2010), which takes topographical gradient and terrain frictional proprieties into account. Subsurface flow is modelled using a two-layer, forward-difference or explicit (Wang and Anderson, 1982), finite difference model, which can be considered as a CA approach (Ravazzani et al., 2011). The upper subsurface layer represents water flow along the bedrock-soil interface (saturated soil in Fig. 3) and the lower layer water within the bedrock (groundwater inFig. 3). Laterally heterogeneous hydraulic conductivity and specific yield allow versatility in the range of hydrogeological environments that can be represented. The coupled surface-subsurface hydrological component of CLiDE is driven by climate data and constrained by terrain and subsurface properties.

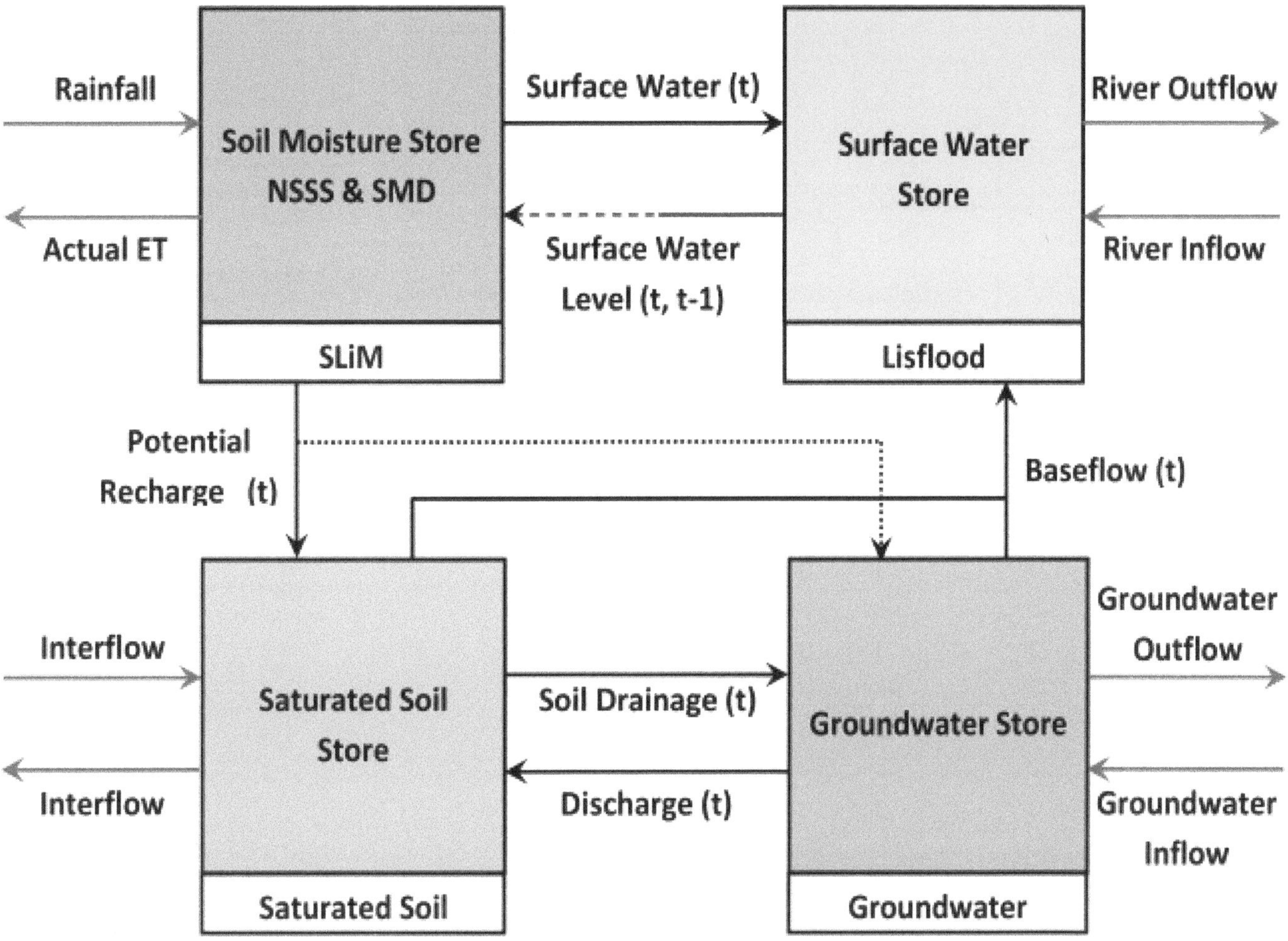

Figure 3: Diagram representing the coupled surface-subsurface CLiDE hydrological processes. With the exception of rainfall and actual evapotranspiration (ET), arrows not linking processes represent flow between nodes (for example, flow between groundwater nodes). Rainfall inputs into SLiM are defined at model initialisation and actual ET is lost to the atmosphere. The dashed surface water line represents an input from the previous time step and the dotted potential recharge line is shown to avoid confusion with the intersecting baseflow-return line.

For each of the platform components, space is discretised into uniform, square cells. The soil water balance and groundwater flow model components employ a daily time step, while surface water routing is implemented on an adaptable sub-daily time step. The spatial resolution of the rainfall and potential evapotranspiration datasets for water partitioning should be at the same or coarser resolution than the model resolution, as the code can factor spatial resolutions (from coarser to finer resolution) automatically. If spatial factoring is required, the finer model cells must be coincident (in terms of alignment and the factor used) with the coarser grid, allowing data to be imprinted directly from the lower to higher resolution grid without interpolation.

Soil water balance

We implemented a technique that represents potential groundwater recharge (soil drainage) and runoff processes based on spatio-temporally distributed soil moisture conditions. Soil moisture is influenced by rainfall, potential evapotranspiration, topography, soil type and vegetation (see Fig. 4). Our approach dynamically captures the impact of these processes on soil water properties by responding to variable soil water storage properties (see Rushton, 2003) and vegetation growth stages (see Allen et al., 1998). Within our scheme, rainfall that does not evaporate or contribute directly to runoff can be either intercepted by plants or reach the ground surface and infiltrate into the soil. The latter has the effect of increasing near-surface soil storage and reducing the soil moisture deficit, SMD [L]. Soil water can be extracted by plant roots for transpiration or drawn to the bare soil surface for evaporation. When soil moisture reaches field capacity, and the soil is unable to store additional water, water drains freely in the saturated soil.

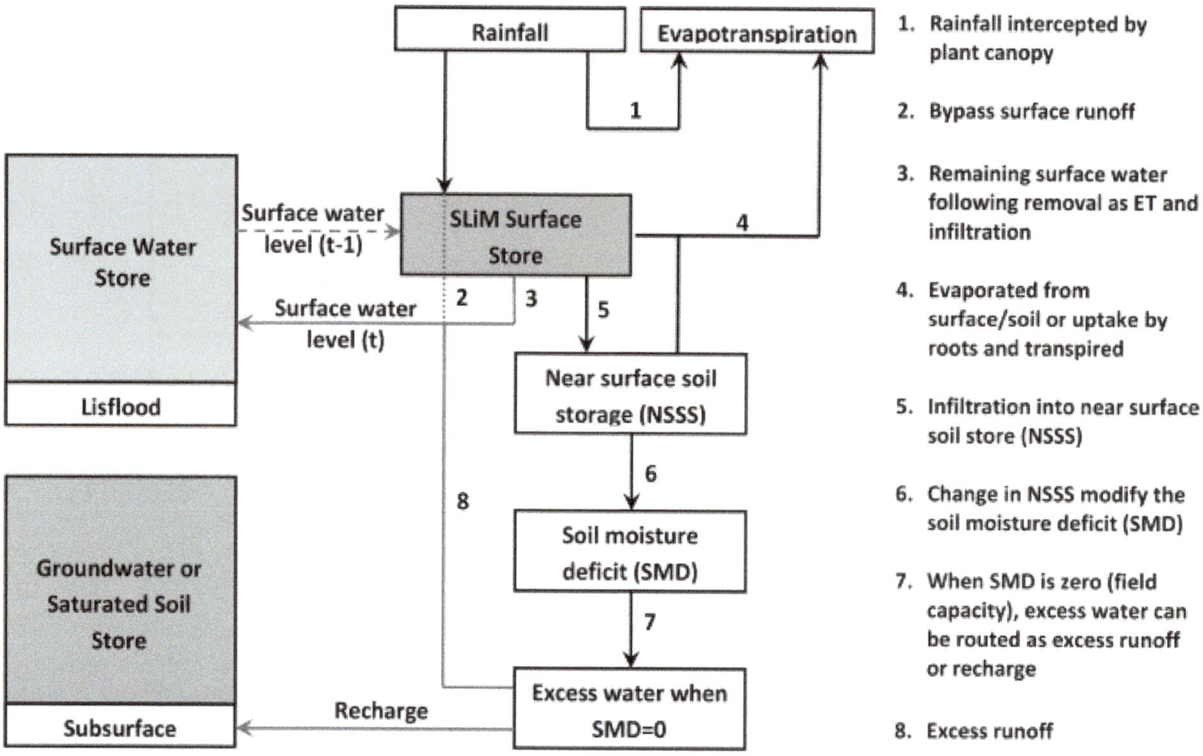

Figure 4: Flow chart for partitioning of water at the soil surface using the SLiM soil water balance method. Connections with the other water components are included.

Additional water inputs to the soil, when field capacity is exceeded, result in the potential for recharge (soil drainage) and lateral surface runoff (routed by Lisflood-FP) if a gradient exists towards adjacent locations. This water, which is not accounted for in soil storage, evapotranspiration or uptake by vegetation is termed excess water, Ew [L^3]. Excess water is divided between runoff and recharge to groundwater based on a baseflow index parameter, BFI [-], except when the rainfall intensity is greater than the infiltration capacity of soil, where all precipitation is routed as surface runoff (bypass surface runoff in Fig. 4). Baseflow index is an average surface to subsurface water partitioning ratio reflecting the permeable nature of the catchment in addition to other catchment characteristics. The baseflow index parameter may be derived through a calibration process or more commonly is estimated from data by performing a baseflow separation on a river flow time-series (see Eckhardt, 2008). Surface runoff, Ro [L^3], and recharge to the groundwater model, Re [L^3], are calculated using Eqs. (1) and (2)).

$$R_e = E_w \cdot BFI \quad (1)$$

$$R_o = E_w \cdot (1 - BFI) \quad (2)$$

In general, greater runoff and reduced recharge is observed in areas with steeper slopes (see for example, Lange et al., 2003 and Haan et al., 2006). The baseflow index represents a long-term average ratio reflecting, in part, the characteristics of average catchment slope. Runoff and recharge can therefore be understood in terms of how local slope at a particular location compares with a catchment average slope. Consequently Eqs. (1) and (2)) can be modified to allow average ($\bar{S}$) and nodal (S) terrain slope [°] to be factored into a calculation of recharge and runoff (Eqs. (3), (4) and (5)). Using this novel technique, runoff on below average slopes (Eq. (3)) is calculated directly as a function of local to average slope ratio, and runoff on greater than average slopes (Eq. (4)) is a product of catchment runoff for an average slope and that additional amount provided by the local slope angle exceeding this.

$$R_o = \frac{E_w \cdot (1 - BFI) \cdot S}{\bar{S}} \quad (\text{if } S \leq \bar{S}) \quad (3)$$

$$R_o = \frac{(S - \bar{S}) \cdot E_w \cdot BFI}{(90 - \bar{S})} + E_w \cdot (1 - BFI) \quad (\text{if } S > \bar{S}) \quad (4)$$

$$R_e = E_w - R_o \quad (5)$$

Surface water routing

Routing of surface runoff and channel flow is controlled by a stripped-down version of the Lisflood-FP model (as used by Coulthard et al., 2013). Lisflood-FP is a one-dimensional inertial model that is applied in the x and y directions to simulate flow in two dimensions over a raster grid. To calculate the flow between cells (Qsw [L^3T^{-1}]), Eq. (6) (Bates et al., 2010) is used:

$$Q_{SW} = \frac{q_{SW} - gh_{flow}\Delta t \frac{\Delta(h+z)}{\Delta x}}{\left(1 + gh_{flow}\Delta t n^2 |q_{SW}| \Big/ h_{flow}^{10/3}\right)} \Delta x \quad (6)$$

Here qsw [L^3T^{-1}]is the surface water flux between cells from the previous time step, g [LT^{-2}] is acceleration due to gravity, n [-] is Manning's roughness coefficient (see Arcement and Schneider, 1989), h [L] is nodal water depth, z [L] is elevation, hflow [L] is the water depth difference between adjacent cells, x [L] is the grid cell width and t [T] is time. Having established the discharge across all four boundaries of a cell, the cell water depth (h) is updated using Eq. (7), where i and j are cell co-ordinates.

$$\frac{\Delta h^{i,j}}{\Delta t} = \frac{Q_{swx}^{i-1,j} - Q_{swx}^{i,j} + Q_{swy}^{i,j-1} - Q_{swy}^{i,j}}{\Delta x^2} \quad (7)$$

To maintain stability, the Lisflood component calculates the maximum allowable length of the next time step (Δtmax) for which a converging solution is highly likely to be maintained. The calculation is based on the shallow water Courant-Friedrich-Levy (CFL) condition, in which:

$$\Delta t_{\max} = \alpha \frac{\Delta x}{\sqrt{gh}} \quad (8)$$

The coefficient α is typically defined as being between 0.3 and 0.7 (Bates et al., 2010), and is used to enhance the model›s robustness, as the CFL condition is a necessary but not sufficient condition for nonlinear systems. As Eq. (8) demonstrates, this is strongly influenced by the grid cell size and the water depth. A typical Δtmax for a 50 m grid spacing, an intermediate α and with surface water present is in the order of a minute.

Lisflood-FP has been extensively tested and benchmarked by Bates et al. (2010) andNeal et al. (2011), demonstrating that the model is capable of simulating flow depths and velocities within 10% of codes that solve the full two-dimensional shallow water equations.

Subsurface flow

Subsurface flow is simulated at the near-surface (saturated soil in Fig. 3) and deeper subsurface (groundwater in Fig. 3) using a two-dimensional lattice of square cells interacting according to the von Neumann-type neighbourhood in a technique similar to that utilised by Rothman (1988). Both the near-surface and deeper subsurface components are similar conceptually and can pass water within and between themselves. The near-surface component represents an unconfined fully-saturated flow within the soil along the bedrock-soil interface, with the top of the bedrock defining the flow base. The deeper groundwater component is again unconfined, but flow is within the bedrock itself. Here the lower surface is at a predefined datum. The flow properties of the near- and deep-subsurface can be defined separately within the platform and both use the same algorithms to simulate flow. To ease model validation and sensitivity analysis (see Section 3), the near-surface component is not enabled in this study (potential recharge and baseflow are passed between the surface and deeper groundwater). However, its description is included for reference and it uses the same equations as the groundwater component described in this section.

Nodes representing groundwater consist of distributed, lateral, hydraulic conductivity, K[LT^{-1}], specific yield, Sy [–], and aquifer head (the unconfined potentiometric surface), h[L]. The difference in aquifer head between two adjacent cells drives inter-cell water flow. The governing equation for steady-state groundwater flow is defined by Bear (1979) as:

$$\nabla T \nabla h + W = 0, \quad (9)$$

where transmissivity, T [L^2T^{-1}], can be approximated by multiplying hydraulic conductivity by groundwater depth, and W [L^2T^{-1}] represents additional source and sink terms. From this, the one-dimensional, steady-state flow, Qgw [L^2T^{-1}], may be defined as:

$$Q_{gw} = -T_s \frac{dh}{ds} \quad (10)$$

where, s is the coordinate aligned with flow direction. Integrating between two different groundwater heads (hi and hj) separated by a distance (we use Δx as it corresponds to the cell width in the flow direction x) yields:

$$Q_{gw} = \overline{T} \frac{h_i - h_j}{\Delta x} \quad (11)$$

Under a heterogeneous transmissivity, where abrupt changes may cause instability in the model, a flux limiting harmonic mean of the transmissivity for the two points i and j is used to calculate an averaged transmissivity, $\overline{T}$:

$$\overline{T} = \frac{2T_i T_j}{T_i + T_j} \quad (12)$$

The total flux to a central cell, Q*gw [L^3T^{-1}], is a combination of fluxes from the neighbouring cells (using Eq. (11) for the four adjacent cells as per the von Neumann scheme, see Fig. 1) and the additional source and sink term. Recharge from the surface hydrological component provides the source term and baseflow-return to the surface acts as a sink. Total fluxes are used to update groundwater heads at each point in the aquifer domain simultaneously at each time step (Δt) using the discrete mass balance equation:

$$h_{t+\Delta t} = h_t + \frac{1}{Sy} \frac{Q^*_{gw}}{\Delta x^2} \Delta t \quad (13)$$

Two user defined lateral groundwater boundary condition types have been implemented into the CLiDE platform. Specified (Dirichlet) boundary conditions fix aquifer head at the boundary and a no-flow (Neumann B) condition sets flux across the boundary to zero. The base of the aquifer is bound with a specified boundary allowing leakage, which is included in the flux algorithm as a secondary source or sink term. The surface boundary allows a flux of water to be returned to the surface component as baseflow where aquifer head is greater than surface water, r [L], for a particular node. The volume of water returned to rivers, qbf [L^3T^{-1}], is dependent on the specific yield of the groundwater node and the thickness (RBb [L^3]) and hydraulic conductivity (RBk [LT^{-1}]) of the river bed using a modified version of the equation derived by Haitjema (1995):

$$q_{bf} = Sy\left(\frac{h - r}{c} \Delta x^2\right) \quad (14)$$

$$c = \frac{RB_b}{RB_k} \quad (15)$$

To ensure model stability when passing water laterally, the cell Reynolds number, Rn [-], is calculated at each node for each time step and a flag is raised if it exceeds unity. Reduction of the Reynolds number is possible by reducing the time step (undertaken automatically by the platform for the entire domain if a flag is raised for any particular cell) or increasing the cell size (this can be undertaken manually by adjusting the resolution of the input datasets). Roache (1976) defines the Reynolds number as:

$$Rn = 4\frac{T}{Sy}\frac{\Delta t}{(\Delta x)^2} \tag{16}$$

Sediment transport component

There are several sediment transport models that could be used to evaluate landscape evolution at the decadal to centennial time-scales required for the CLiDE platform.Coulthard (2001) and Merrit et al. (2003) review these, and an overview of the theory behind modelling landscape evolution is presented by Tucker and Hancock (2010). The ability of current models to employ distributed fluvial and erosional processes is an advantage over previous approaches (see Kinnell, 2010). Current models are based on DEMs, include distributed soil depths, allow surface runoff, and can automatically determine inter-cell slope gradient and transport of sediment. For this study, CAESAR-Lisflood was selected to provide the sediment transport and Graphical User Interface (GUI) functionality for the CLiDE platform, as it has a modular, versatile design and has been verified in a range of environmental settings.

The ability of CAESAR/CAESAR-Lisflood to predict soil erosion has been tested against experimental field data for a small catchment (see Coulthard et al., 2012), where predicted suspended sediment and bedload yields showed a close correspondence in both volume and timing of field measured data. Due to its adaptable nature, CAESAR has been used for many applications including: modelling the response of river systems to environment change (Van De Wiel et al., 2011); human impact on fluvial regimes and fluxes of sediment (Hoffmann et al., 2010); influences of vegetation on river development (Tooth et al., 2008); catchment response to environmental change (Coulthard et al., 2005); and, more recently, determining catchment soil loss rates (Hancock et al., 2011). The remainder of this section describes the major features of CAESAR that have been built upon by CLiDE. See Coulthard and Van De Wiel (2006) and Coulthard et al. (2013) for further detail regarding the equations that drive sediment transport within CAESAR.

Using the water fluxes between cells from the surface hydraulic model (Eq. (6)), velocities in the x and y directions are calculated. The resultant velocities are then used to calculate a cell shear stress that provides an input into a choice of two well established sediment transport laws: Wilcock and Crowe (Wilcock and Crowe, 2003) or Einstein-Brown (Einstein, 1950). These formulae are selected as they both calculate total sediment transport from the sum of individual grain size fractions, which is important for grain the sorting processes within the platform.

The Wilcock and Crowe method can transport sediment within a catchment as both a suspended load and a bedload. Both depend on the volume of sediment transported per time step, where the sediment transported from a central cell to a neighbouring cell is calculated by equating a coefficient to either slope, for bedload transport, or velocity, for suspended load transport. The calculation of suspended load transport is simplified, as it assumes an equal distribution of sediment throughout the water column. Sediment deposition also differs between the two transport types, with bedload sediment deposited (and subsequently re-entrained) at every time step, and suspended sediment deposition based on fall velocities. Under a quasi-uniform sediment grain-size regime, the computationally less expensive Einstein-Brown method for transport may be employed. The formulation evaluates the forces moving and restraining a particle. A bedload transport rate can be calculated from these forces and used to approximate the flux of sediment between cells. The amount of sediment that can be transported is calculated and moved from an active layer, which integrates sediment transport across the multiple grain-sizes (see Van De Wiel et al., 2007).

The incorporation of multiple grain-sizes, selective erosion, transport and deposition of the different size fractions is important and results in spatially variable sediment size distributions. Since this variability is expressed not only horizontally but also vertically, it requires a method of storing subsurface sediment data. This is carried out by using a system of layers comprising: the active layer representing the stream bed; multiple buried layers (strata); a base layer; and, if required, a non-erodible bedrock layer (Van De Wiel et al., 2007). It is worth noting that the size of an individual grain is not allowed to evolve during the simulation.

Non-fluvial sediment transport occurs through instantaneous debris flow and a slower creep process. Debris flow is computed using a two-dimensional sand-pile algorithm (Metha and Barker, 1994) based on a critical slope angle. Where this angle is exceeded, an iterative process, contained within a single time step, moves material down slope to lower cells until the critical angle is no longer exceeded. Soil creep is calculated using a gradient in elevation between cells on a monthly basis through a diffusion equation (Culling, 1960).

Vegetation cover in the model is implemented by specifying areas where vegetation covers the surface layer, which has the effect of binding sediment and reducing erodibility (see Murray and Paola, 2003). A simple alluvial vegetation growth model (seeCoulthard and Van De Wiel, 2006) is included, allowing linear growth of vegetation over time if it is not submerged. When vegetation has grown beyond a specified limit it restricts fluvial erosion to a percentage of that for uncovered sediment (default value is 10%). If flow shear stress exceeds a threshold the vegetation can be removed, thus exposing the sediment below to unhindered erosion. If vegetation becomes covered by sediment it can re-grow through the sediment layer, re-stabilising it.

VERIFICATION AND SENSITIVITY ANALYSIS

Verification of the groundwater component

Bench-marking of the groundwater component was undertaken through comparison to steady-state and transient analytical solutions. To ensure correct functioning of the groundwater model it was subjected to tests similar to those described by Ravazzani et al. (2011). These tests were undertaken on a 2 km wide, square domain with 10 m cell spacing and groundwater parameters representing an unconfined aquifer with a hydraulic conductivity of 2.5 m d^{-1} and a specific yield of 0.1. Two tests allowed an assessment of steady-state and transient flow behaviour.

In the first test a steady-state solution of the subsurface model was compared to an analytical result by simulating the response of groundwater level to constant uniform recharge. Northern and southern boundaries were set to the no-flow condition and eastern and western boundaries to Dirichlet conditions, which represented fixed stage (constant elevation) streams at 22 m (west) and 16 m (east). Recharge was constant in time and space at 0.5 mm d^{-1} and an initial groundwater head of 16 m applied to all regions. The simulation ran until a steady-state solution was reached and a cross section (west to east) of groundwater levels compared to the analytical solution (see Harr, 1962). The simulated and analytical solution show excellent agreement (a root mean square error (RMSE) of below 0.01 m) after a couple of hundred iterations (Fig. 5).

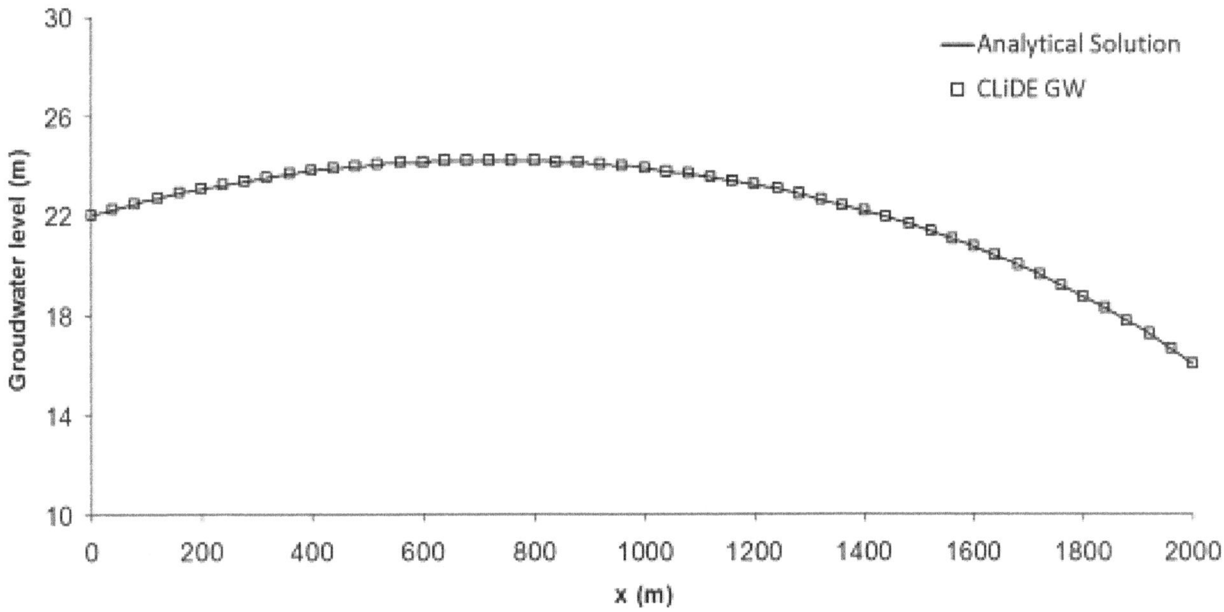

Figure 5: Steady-state comparison of the groundwater component to an analytical solution for water table elevation between fixed streams under uniform recharge conditions.

The second stage of testing involved the simulation of groundwater head in response to a constant pumping rate and comparison to an analytical solution. A modified approximation of the Theis solution, valid for unconfined aquifers (Barry et al., 2000), was used with a small extraction depth to aquifer thickness ratio. Initial head was set at 50 m for all nodes, with a constant pumping rate of 150 m^3 d^{-1} from the central node and a time step of 360 s. Dirichlet boundary conditions were set to match the initial head along all lateral boundaries and groundwater levels monitored at four sites set at 50 m intervals between 100 m and 250 m from the well. The match between transient simulation and analytical solution is good, with close agreement at all well distances (Fig. 6).

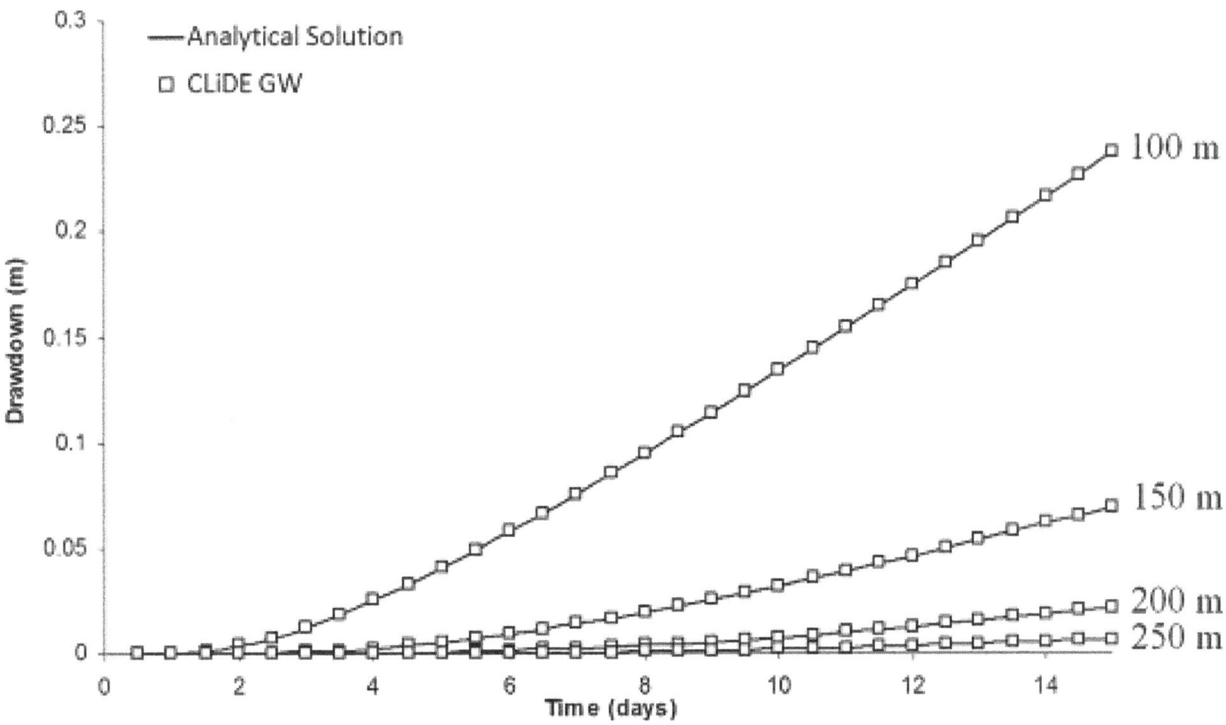

Figure 6: Comparison of drawdown calculated by the modified Theis solution and the groundwater component of CLiDE. Distance of each location from the pumping site is given at the right of the graph.

Sensitivity analysis

By modifying the specific yield and hydraulic conductivity of the subsurface, the groundwater residence times and groundwater storage volume are changed, and subsequently the baseflow-returns to surface flow are perturbed. A numerical experiment was devised to test the sensitivity of baseflow-return and sediment transport within an idealised catchment to differing groundwater regimes with identical boundary conditions (rainfall, potential evapotranspiration and initial topography). Analysis of the sensitivity allows impacts of the differing groundwater regimes to be assessed for a simple catchment.

Platform setup

To keep topographic effects to a minimum a catchment with simple topology of a gentle V shaped valley, as per Coulthard and Van De Wiel (2006), was created (Fig. 7). A cell width of 10 m was selected, allowing a reasonable simulation time (approximately 3 days per simulation for a standard desktop with an octa-core processor) whilst maintaining stability in the groundwater model: a higher spatial resolution would require shorter time stepping (see Eq. (16)).

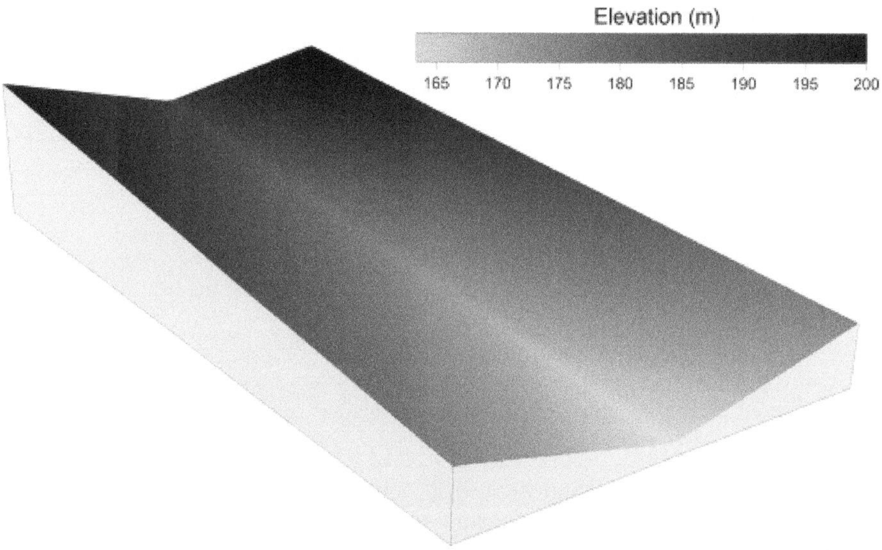

Figure 7: Simple catchment used for the sensitivity simulations. The catchment is 125 by 250 nodes at 10 m grid spacing.

A total of 0.1 m rainfall was applied uniformly over the catchment for the first hour of each day during each of the 3600 days of simulation. Potential evapotranspiration was set to 0.01 m d^{-1} for the duration of each simulation. Land use and soil type, which determine the partitioning of soil water in the SLIM module (see Section 2.1.1.), were fixed and representative of arable land sitting on an unconsolidated, macroporous soil with little by-pass flow (see Boorman et al., 1995). Sediment grainsize was 4 mm across the catchment and all sediment transport treated as bedload.

To assess the changes in water and sediment fluxes, seven scenarios with differing hydrological characteristics were created (see Table 1). Scenario 0 allows both recharge and baseflow-return with specific yield and hydraulic conductivity set at intermediate values. Scenario 1 simulates recharge from the surface to groundwater, but has no baseflow-return component. This is representative of a deep, unconfined water table. Scenario 2 is representative of an impermeable surface, not allowing recharge or baseflow-return. The remainder of simulations (scenarios 3 to 6) allow both recharge and baseflow-return with differing specific yield and hydraulic conductivity. For all scenarios groundwater boundaries were set to the no-flow condition. Groundwater levels and baseflow-return were monitored across the catchment, while sediment yield and fluvial flow were recorded at the outflow point at the base of the catchment.

Table 1: The groundwater conditions for each of the 7 scenarios simulated for the test catchment.

Run number (scenario)	Groundwater recharge from surface?	Baseflow-return to surface?	Specific yield (–)	Hydraulic conductivity (m d^{-1})
0	Yes	Yes	0.1	1
1	Yes	No	–	–
2	No	No	–	–
3	Yes	Yes	0.1	100
4	Yes	Yes	0.1	0.001
5	Yes	Yes	0.5	1
6	Yes	Yes	0.005	1

Hydrological characteristics

The groundwater properties in scenario 0 represent an unconsolidated silt, which produces a relatively stable daily groundwater head under the simulated fluctuating surface hydrological conditions. The result is a permanent river flow at the base of the catchment. Over the length (temporal) of the simulation river flow increases and groundwater heads decrease. The lower areas of the catchment are provided with water for longer periods than they would receive solely through runoff during rainfall events. Scenario 1 represents a deep groundwater table with no baseflow-return to the surface. Here, water can still be stored at the surface as soil moisture, but no water is transported back from the subsurface, creating an ephemeral river, which disappears shortly after the daily rainfall episode ceases. This scenario produces the lowest daily river discharges as water is not returned to the surface as baseflow. Scenario 2 is representative of an impermeable surface that does not allow the transfer of water between the surface and subsurface. Under these conditions all water is either transported across the surface, stored at the surface as soil moisture (with no lateral or horizontal flow), or removed through evapotranspiration.

Scenario 3 uses a high hydraulic conductivity, resulting in a much quicker flow of water within the subsurface compared to the baseline (scenario 0). There is an initial flux of water from the ground to the surface, as groundwater heads adjust themselves to a new equilibrium position from the initial conditions (see Fig. 8). As incision of the central channel progresses, further water is released from the subsurface store to the river as the groundwater levels seek a new equilibrium position within the altered landscape. The low hydraulic conductivity in scenario 4 reduces groundwater flow (when compared to scenario 0) and, as there is sufficient rainfall, creates groundwater levels that are highly dependent on local surface elevation. This is apparent in Fig. 8, where groundwater volumes across the catchment increase from their initial starting point up to the local surface elevation over the first 200 days of simulation. Although a central channel is created during the simulation, groundwater levels at any particular point reflect local surface contours. Therefore, only a minimal amount of water is released from the groundwater store, resulting in a gradual loss of water from 300 days onwards in the simulation (Fig. 8).

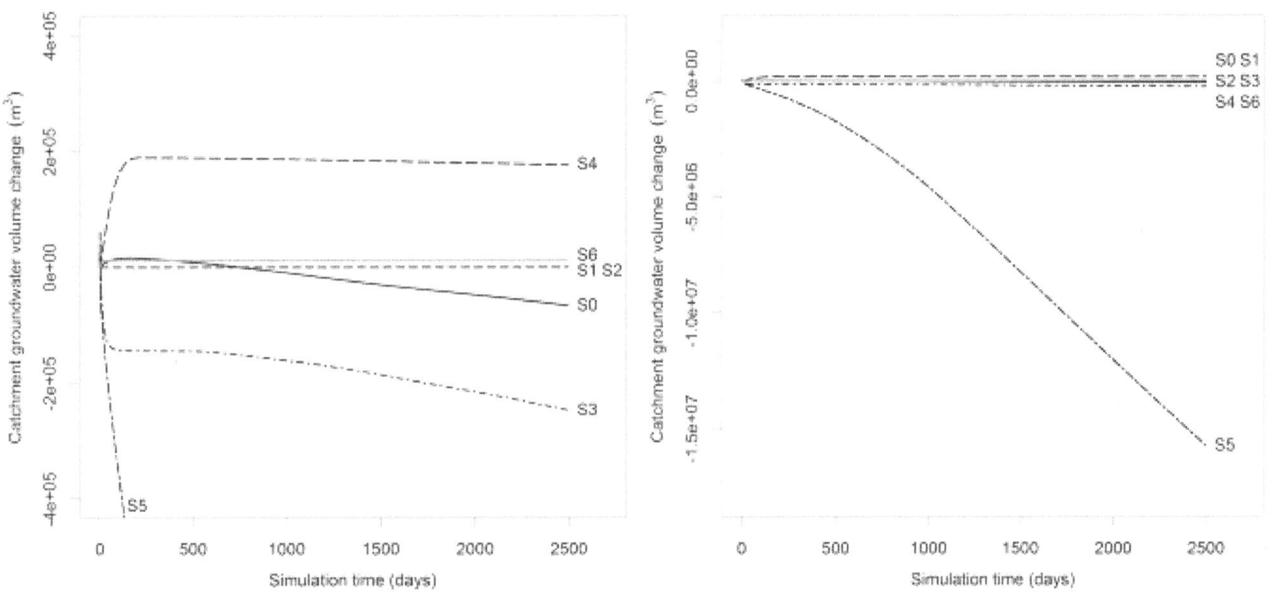

Figure 8: Simulated change in catchment groundwater volumes following initialisation. Negative gradients represent a loss of groundwater to the surface as baseflow-return. The left and right plots show the same data, with differing vertical (groundwater) scales.

The comparatively high specific yields in scenario 5 create a stable groundwater level that is highly influenced by the lowest elevation in the catchment. Recharge is not sufficient to maintain groundwater heads as subsurface water is lost to the river as baseflow. As there is a relatively large volume of water held in the aquifer, channel incision has the potential to release a large amount of water from the subsurface to the surface. This is due to incision along the channel length lowering the water table and introducing water from the groundwater store to the surface, leading to an increase in river discharge (see Fig. 9) and a large drop in groundwater volume (see Fig. 8). Low specific yields create highly variable groundwater levels with a dependency on recharge rates. Assuming no flow in or out of a node, the specific yield used in scenario 6 produces a 1 m rise in groundwater levels for every 5 mm of recharge. As there is adequate rainfall, groundwater levels sit at the ground surface across the catchment during rainfall events and excess water is routed across the surface. During the intervening periods, hydraulic conductivity is not sufficient for groundwater levels to recede. As the overall volume of water contained in the aquifer is low, the carving of the central channel does not release a significant amount of water from the ground for contribution as river baseflow.

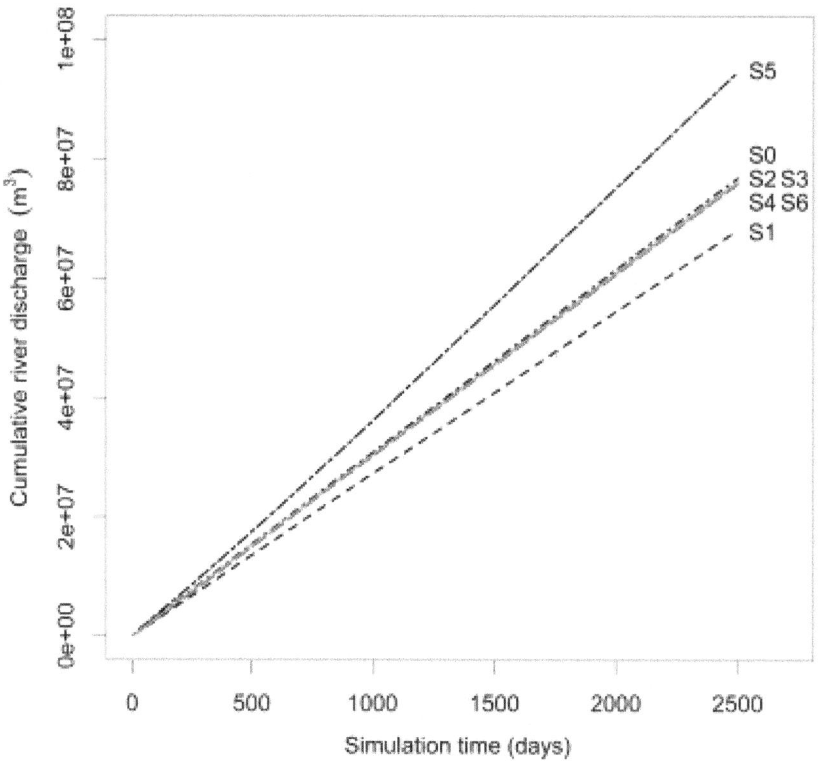

Figure 9: Cumulative river discharge at the catchment outlet.

Sediment transport

Cumulative sediment discharge is presented in Fig. 10. Initially scenario 5 (high baseflow contribution) creates the highest rate of sediment discharge, however after 500 days of simulation scenarios 3 (flashy baseflow) and 6 (low baseflow) create the highest rate of daily sediment discharge. Scenario 1 (no baseflow) initially has the lowest sediment discharge, however a switch in behaviour after 1250 days increases sediment discharge rates similar to those exhibited by scenarios 3 and 6. Catchment sediment discharge for the intermediate groundwater, impermeable surface and low hydraulic conductivity scenarios (0, 2 and 4 respectively) produce quasi-linear rates of catchment sediment discharge without experiencing the same marked change in rates experienced during the other scenarios.

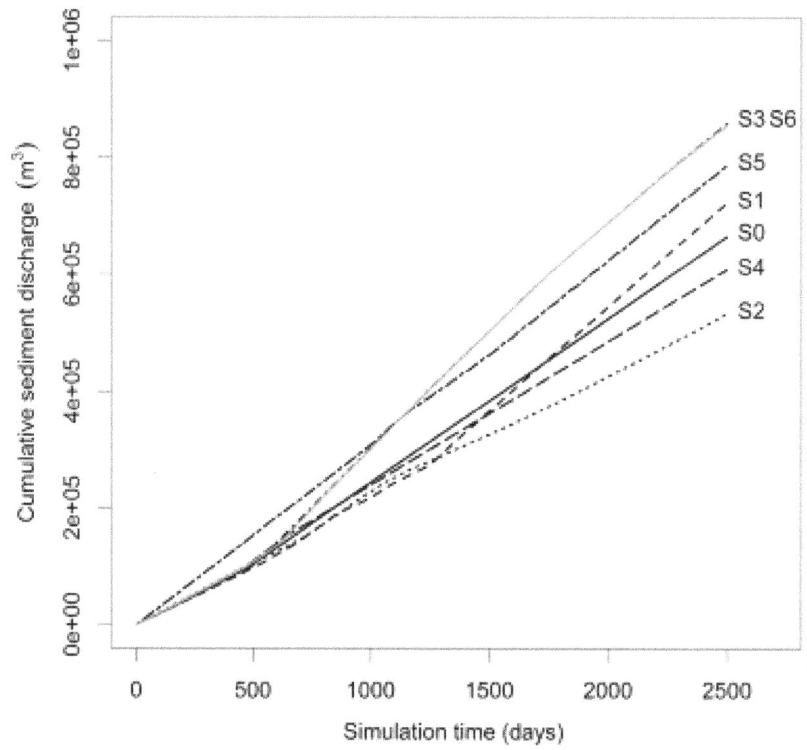

Figure 10: Cumulative sediment yield at the catchment outlet.

There do not appear to be any clear relationships between daily river discharge and sediment discharge when the scenarios are assessed cumulatively. When river and sediment discharge are plotted against each other through time (Fig. 11 and Fig. 12) some clear trends begin to emerge. For each scenario the changes in sediment discharge rate are not constant with respect to time, but instead are held for a number of years before abruptly transitioning to the next discharge rate. The small range in river discharge values throughout each simulation highlights these horizons, by producing a banding of the data with interspersed points in Fig. 11 and Fig 12. A similar behaviour is exhibited in all scenarios (excluding scenario 5), albeit with differences in the sediment discharge rates and the time held at a particular value.

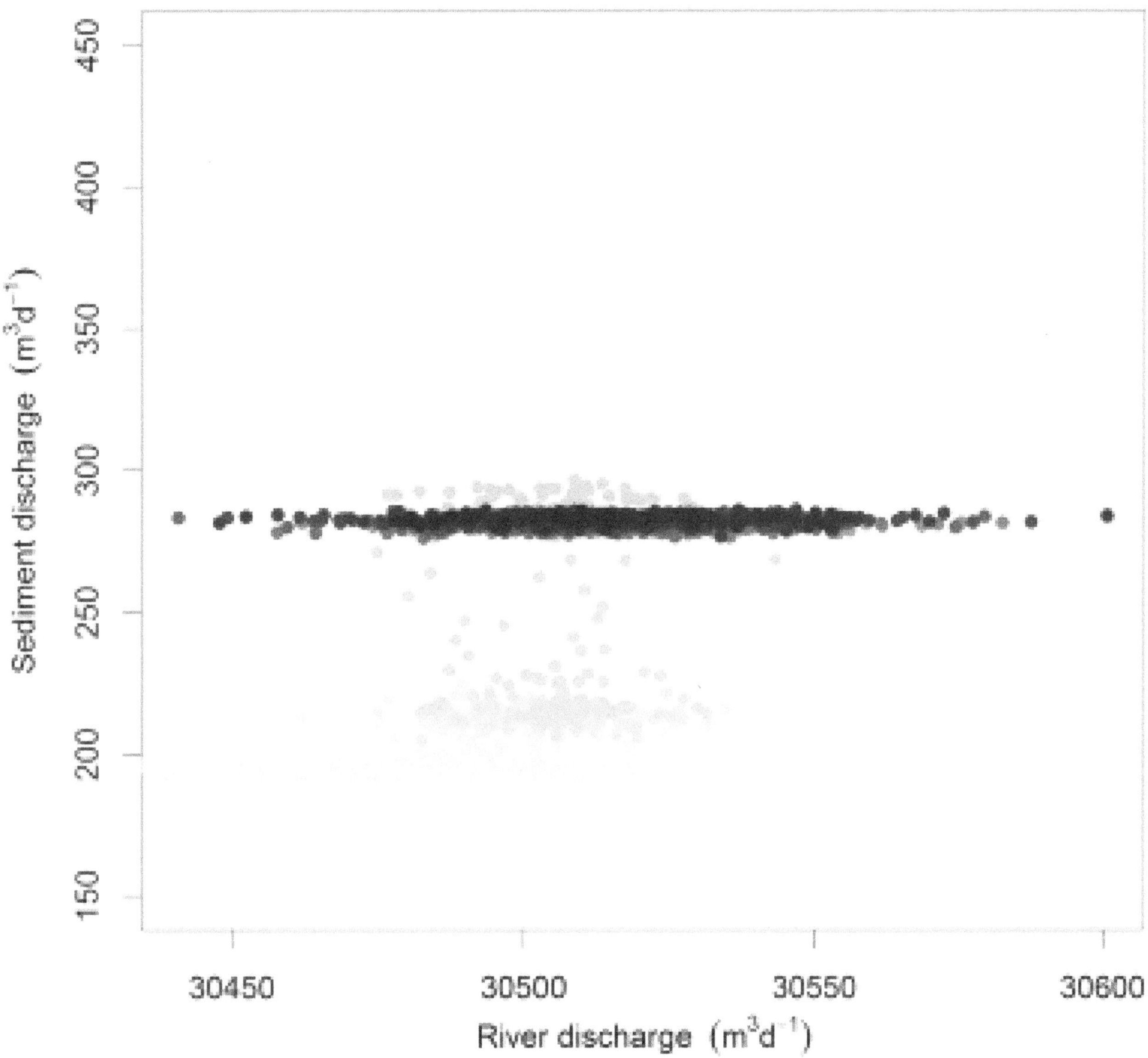

Figure 11: Catchment daily sediment discharge plotted against daily river discharge for the intermediate groundwater scenario (scenario 0). The greyscale of points represents the position in time within the simulation (black at the end of the simulation and light grey at the start).

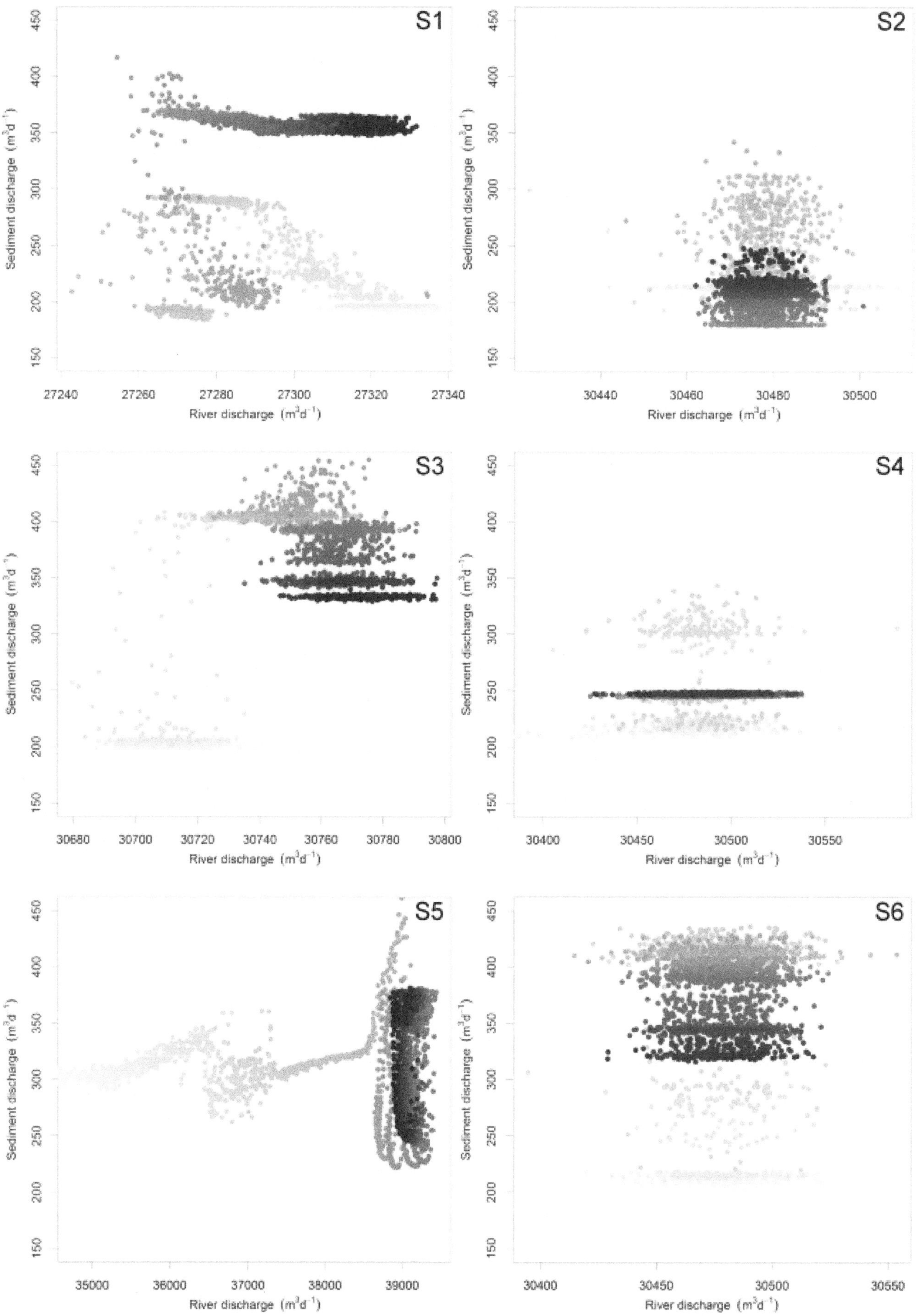

Figure 12: Catchment daily sediment discharge plotted against daily river discharge for each groundwater scenario (numbered in top right of each plot). The greyscale of points represents the position in time within the simulation (black at the end of the simulation and light grey at the start). The same scale is used for sediment discharge, however the river discharge scale is variable.

The intermediate groundwater scenario (scenario 0, Fig. 11) initially outputs 200 m^3 d^{-1}sediment discharge for around 500 days, and then quickly establishes a new catchment sediment discharge rate of 280 m^3 d^{-1}, which it maintains until the end of the simulation. This behaviour is similar to that exhibited by scenario 4 (Fig. 12). The scenarios producing the highest rates of sediment discharge (3 and 6) have the same temporal sediment-river discharge signature. As these simulations progress, sediment discharge increases from 200 m^3 d^{-1} up to peak of around 425–450 m^3 d^{-1} at around 500 days into the simulation. Sediment discharge then drops slowly, to around 325 m^3 d^{-1} where it remains until the end of the run. Scenario 5 (high baseflow) is the only setup where a significant change in daily river discharge occurs, increasing by 5000 m^3 d^{-1} over the simulation. Despite both not having a groundwater input, scenarios 1 and 2 exhibit markedly differing temporal sediment-river discharge signatures. Scenario 2 (no losses to recharge) has a relatively uniform daily river discharge, with sediment discharge initially starting at 200 m^3 d^{-1}, rising for a short period to 300 m^3 d^{-1} before falling back to its initial rate. Scenario 1 (losses to recharge) has a river discharge that changes as the simulation progresses, producing horizons of sediment transport at differing points in time and river discharge. The end of this scenario exhibits the highest sustained rate of discharge from the catchment, despite the lowest river discharge rates.

Comparison of the temporal sediment-river discharge signatures (Fig. 11 and Fig. 12) confirms that there is no overarching relationship that can be defined for river and sediment discharge at the catchment outlet. They highlight the finer details of the sediment discharge rates through time, in particular their propensity to remain at a steady-state sediment discharge for a period of time before abruptly moving to a differing steady-state discharge rate. These changes represent differing internal morphological dynamics occurring within the catchment itself. Differences between the scenarios are formed by the feedbacks that individual hydrological characteristics create with their surrounding morphology. Morphology is initially the same for all scenarios. However, following initialisation each hydrological scenario creates unique distribution of fluvial flow velocities along the river reach, each producing a different value for sediment transport and hence erosion characteristics. Once a morphological feature is well-enough defined for a particular cell, it will start to have an impact on the surrounding cell fluvial characteristics. This feedback can in some cases create rapid changes in morphology between steady-states. For example, a depression forming in the channel at the base of the catchment can accelerate flow velocities immediately upstream, leading to their erosion to the same depth as the initial depression. This process can rapidly cascade upstream until there is insufficient water available to continue the process (the river reach does not extend to the top of the catchment in these scenarios). In the majority of the sensitivity analyses scenarios this process leads to rapid incision of the main channel from the base of the catchment after 500 days of simulation, creating the inflection in sediment discharge responses (see Fig. 10). It is also responsible for the move away from the initial steady-state (lightest grey horizons in Fig. 11 and Fig. 12) to subsequent steady-states in the sediment-river discharge plots. Exceptions to this were scenario 1, where the main channel incised after 1,250 days, and scenario 5, where incision occurs immediately after initialisation.

CONCLUSIONS

Modification of the system hydrology, by means of changing the surface or subsurface properties, creates a range of river discharge responses at the catchment outlet. Sediment discharge at the outlet is not a function of daily river discharge, but transitions through a series of steady-state values independently of the daily river discharge value. This behaviour is controlled by the internal morphological dynamics generated from the feedbacks between localised peak fluvial velocities and the evolving channel morphology.

At the catchment scale, evolving morphology can have an impact on the hydrological processes. This is particularly evident when the channel incises and the drop in river stage reduces surrounding groundwater heads (see Fig. 8, scenario 5) and increases river discharge (see Fig. 9, scenario 5). These effects would be strongest for a scenario with high specific yields and hydraulic conductivities, which would make a large volume of water available for baseflow return as the channel incises and groundwater heads fall over a large area without much time-lag.

Given that the morphology of a simple catchment can evolve in a variety ways based on local hydrological conditions, an accurate representation of baseflow-return influences on regional scale landscape evolution for a groundwater dominated catchment may be essential. To test this hypothesis we assess the impact of baseflow-return on shaping the morphology of a test catchment, the upper Eden Valley, UK.

SEDIMENT TRANSPORT IN THE UPPER EDEN VALLEY

The role of groundwater in shaping catchment-scale surface geomorphology in the upper Eden Valley, a region where baseflow is a major contributor to surface water, is examined. Two scenarios were simulated: with- and without-groundwater recharge and baseflow-return. Aside from this, both have identical initial conditions and external forcings. For the without-groundwater scenario the groundwater module is not enabled. Therefore, baseflow-return is non-existent and all excess water (see Fig. 4) is available at the surface for routing as runoff (equivalent to scenario 2 in Section 3.2). As baseflow is no longer contributing to the surface hydrology, there will be less fluvial flow between rainfall events. As recharge is no longer removed during rainfall events, there is a potential for a higher fluvial flow rate. However, there is also no baseflow-return during these events so the difference in surface peak flow between the two scenarios should be dependent on the previous rainfall history and the hydrogeological characteristics. For both scenarios, the basin sediment yield and DEMs showing morphological changes are recorded. This allows an evaluation of the impact of groundwater on the geomorphology of the basin.

Case study

The Eden catchment, located in north-west England, is an upland area of 2288 km^2 (seeFig. 13). The Eden Valley sits between the Lake District National Park to the west, and the

Pennines, a range of hills which divide North West and North East England, to the east. The most steeply elevated sections of the catchment are located on the eastern side of the valley, whilst the western side is flatter. The land mainly consists of pasture and is considered an important study area due to its water resources (Fowler et al., 2007), significant biodiversity and recent history of large-scale flooding events (Malby et al., 2007). Rainfall in the Eden Valley is above the UK annual average, with approximately 1000 mm y^{-1} on low-lying areas and 1500 mm y^{-1} on the higher ground (Daily et al., 2006). The major river within the region is the River Eden, which runs from the Pennines to the Solway Firth estuary, at the border between Scotland and England.

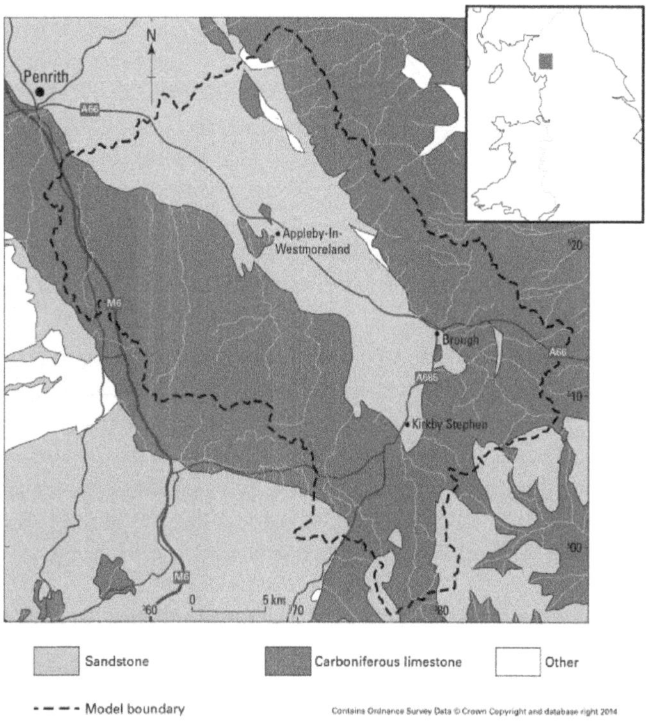

Figure 13: Map showing the location of the upper Eden Valley study site within the north of England (insert) and a geological map of the catchment and model domain (main).

The study area is set in the upper part of the Eden catchment, above Temple Sowerby, an area covering 615 km^2. The Eden tributary has three gauging stations along its reach. Due to their positions relative to the model boundary, only the Temple Sowerby station allows calibration to gauged river flow data.

Catchment geology is composed of sedimentary units deposited between 450 and 220 million years ago, with central valley sections consisting of Permo-Triassic sandstones, and the surrounding upland of Carboniferous limestone (Fig. 13). Two major aquifers are contained in the Permo-Triassic sandstone formations, separated by the Eden Shale Formation. The Penrith Sandstone Formation (early Permian), which supplies large quantities of groundwater for public use through a series of boreholes, is confined to the east by the shale formation and to the west by drift cover. Parts of the Penrith Sandstone Formation are silicified, having a significant effect on the subsurface flow characteristics, with mean hydraulic conductivity of 0.8 m d^{-1} (Allen et al., 1997). Upper sections of the St Bees Sandstone Formation (early mid-Triassic) are also considered to be of hydrological importance, with mean hydraulic conductivity of 0.24 m d^{-1} (Allen et al., 1997). The surrounding Carboniferous Limestone Formation is also significant in this study area as it provides baseflow-return to streams in the sub-catchments of the upper Eden Valley and generates springs along the western margins (Mayes et al., 2006). The sandstones that run down the centre of the Eden catchment were eroded into a valley before the last glaciation. Subsequently, glacial processes formed drumlins and, in some areas, scoured hollows into small basin features.

Due to the shallow depth of the water table in lowland areas, groundwater flow in the upper Eden Valley is dominated by discharge to the River Eden (Butcher et al., 2006 and Bennett et al., 2003), with contributions from the subsurface along the majority of its length. Groundwater levels in upland regions of the study area may be up to 100 m the below surface. However, groundwater recharge in these regions drives subsurface flow and discharge rates in the valleys below. For a detailed review of the baseflow-return mechanisms, the reader is directed to Payn et al. (2012), whose study provides and assesses the spatio-temporal relationships during a seasonal baseflow recession.

Implementation

Initial setup of the upper Eden Valley model requires a calibration of the input parameters through a matching of simulated and observed data. Here we provide a comparison of the calibrated model to observed data. A detailed description of the calibration process, which involves modifying the baseflow index and rooting depth characteristics for surface flows and the hydraulic conductivity and specific yield for groundwater flows, is provided by Barkwith and Coulthard (2014).

The model calculates on a daily time step: the terrain elevation (metres above sea level (masl)), groundwater level (masl), baseflow (m^3), near surface storage (m^3), groundwater recharge (m^3), soil moisture deficit (m) and surface water flow (m^3). To assess the errors associated with differing water flow routes, and to simplify the process, time-series of simulated groundwater and surface flow are calibrated to observed data separately. Comparison of the surface hydrological flows to a baseflow separated record from the Temple Sowerby gauging station is shown in Fig. 14, with the differences between measured and simulated data overlaid. It is important to note that although the RMSE (calculated as 0.03 Ml d^{-1}) of the modelled and measured data suggests a good agreement, the peak differences do not take time-lag errors into account and therefore small temporal errors can create seemingly large differences between simulated and measured datasets.

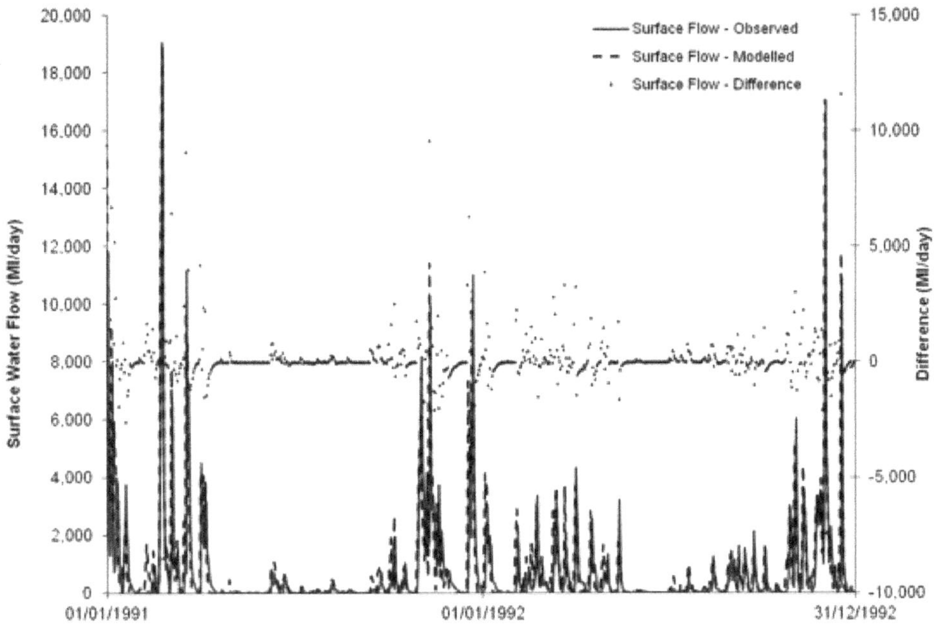

Figure 14: Modelled and observed river hydrograph for Temple Sowerby gauging station, Eden Valley catchment (1991–1992).

Following calibration of the surface routed water, based on reducing RMSE (taking into account the full length of the simulation) and assessment through an observed fitting process, groundwater levels are calibrated using a two-step iterative approach. The model is run with a temporally averaged recharge value until steady-state is achieved, followed by a two-year (1991–1992) transient simulation. Simulated groundwater levels are compared to observed groundwater levels and, where needed, modifications made to the hydraulic conductivity and specific yield, and the process repeated. The post calibration simulated and observed groundwater levels for the Great Musgrave borehole are shown in Fig. 15, with an associated RMSE of 0.27 m. Although the range of measured groundwater levels and recession rates are not well replicated by the model, the position of events and minimum groundwater levels exhibit good correspondence to the observed data. The disparity between simulated and observed groundwater levels arises from the simplistic nature of the conceptual model that underlies the groundwater simulation process. In reality the subsurface hydrology of the upper Eden Valley is a complex, heterogeneous system that is difficult to replicate in time and space using a simplistic approach. However, by including even a relatively poor representation of groundwater, the CLiDE platform achieves a closer conceptual match to the environmental system we are simulating. Using the current setup, and through further calibration and refinement of specific yield and hydraulic conductivity distribution from field data, a reduction in simulated groundwater level errors is expected to be possible. For present purposes, however, the disparity between measured and modelled results during the calibration phase is deemed to be acceptable.

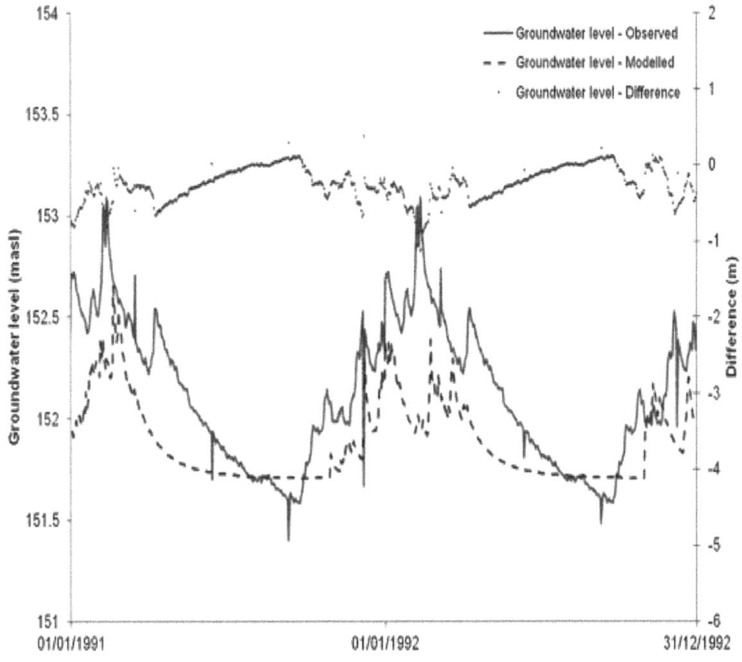

Figure 15: Simulated and observed groundwater levels for the Great Musgrave borehole, Eden Valley catchment (1991–1992).

With calibration complete, the model is setup to run over a two-year period (1991–1992) using daily rainfall and potential evapotranspiration. A 50 m grid spacing was used for the model spatial discretisation, and time stepping limited to between 1 and 3600 s for the sediment and surface fluvial transport processes.

Output and analysis

Time-series of sediment discharges from the catchment are presented for both with- and without-groundwater scenarios in Fig. 16. Initially sediment discharge in both scenarios increases rapidly as the sediment transport component progresses through the spin-up stage (first 5 months, where the model is allowed to reach a dynamic hydrological equilibrium). Significant sediment discharge events only occur during the larger rainfall events. However, as with the sensitivity analysis (Section 3.2), the amount of sediment transported from the system is not proportional to the amount of rainfall. For the majority of the simulation period the volume of sediment discharged is similar for the two scenarios, although they exhibit a differing temporal pattern of sediment flux, which is highlighted later in Fig. 18.

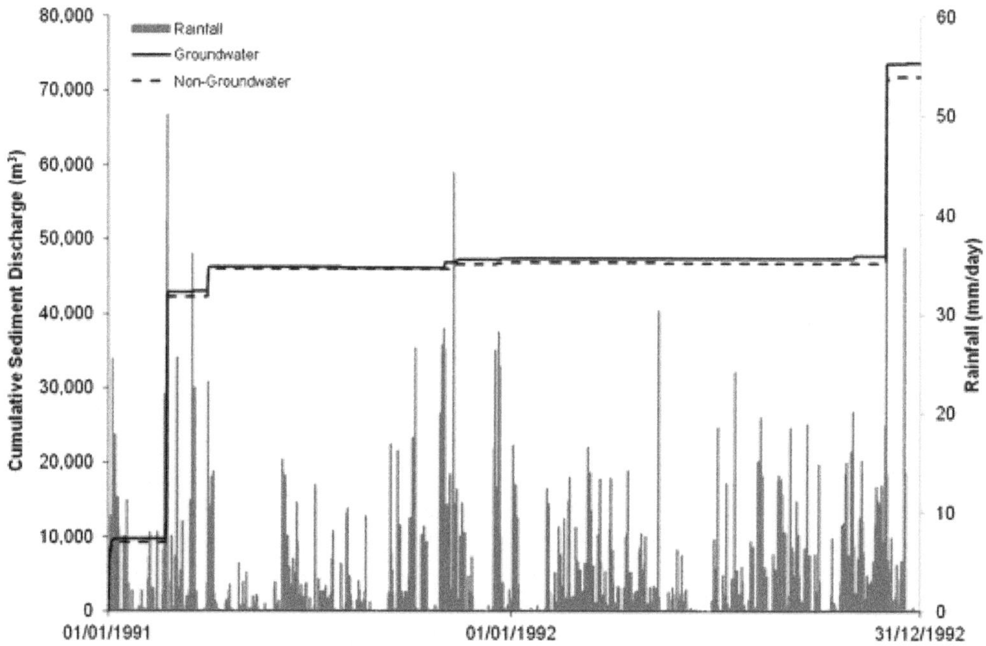

Figure 16: Cumulative sediment discharged from the catchment for the with- and without-groundwater scenarios, and the catchment average rainfall over a two-year period.

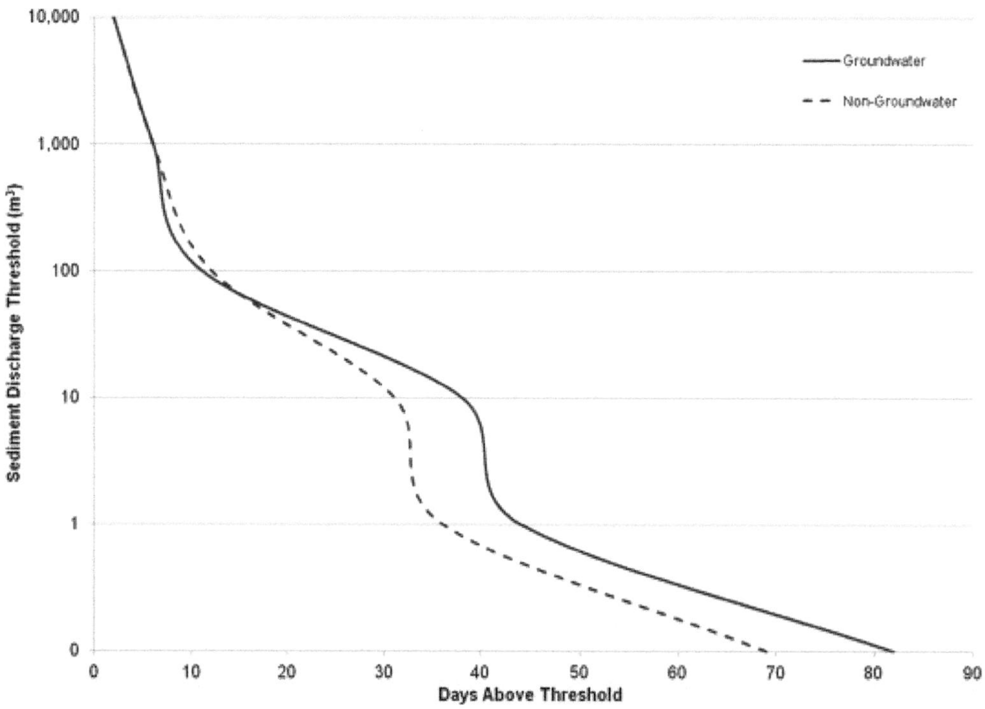

Figure 17: The total number of days when daily sediment flux from the catchment exceeds a particular threshold value. As expected there are fewer days where over 1000 m³ are discharged from the catchment compared to days where over 1 m³ is discharged.

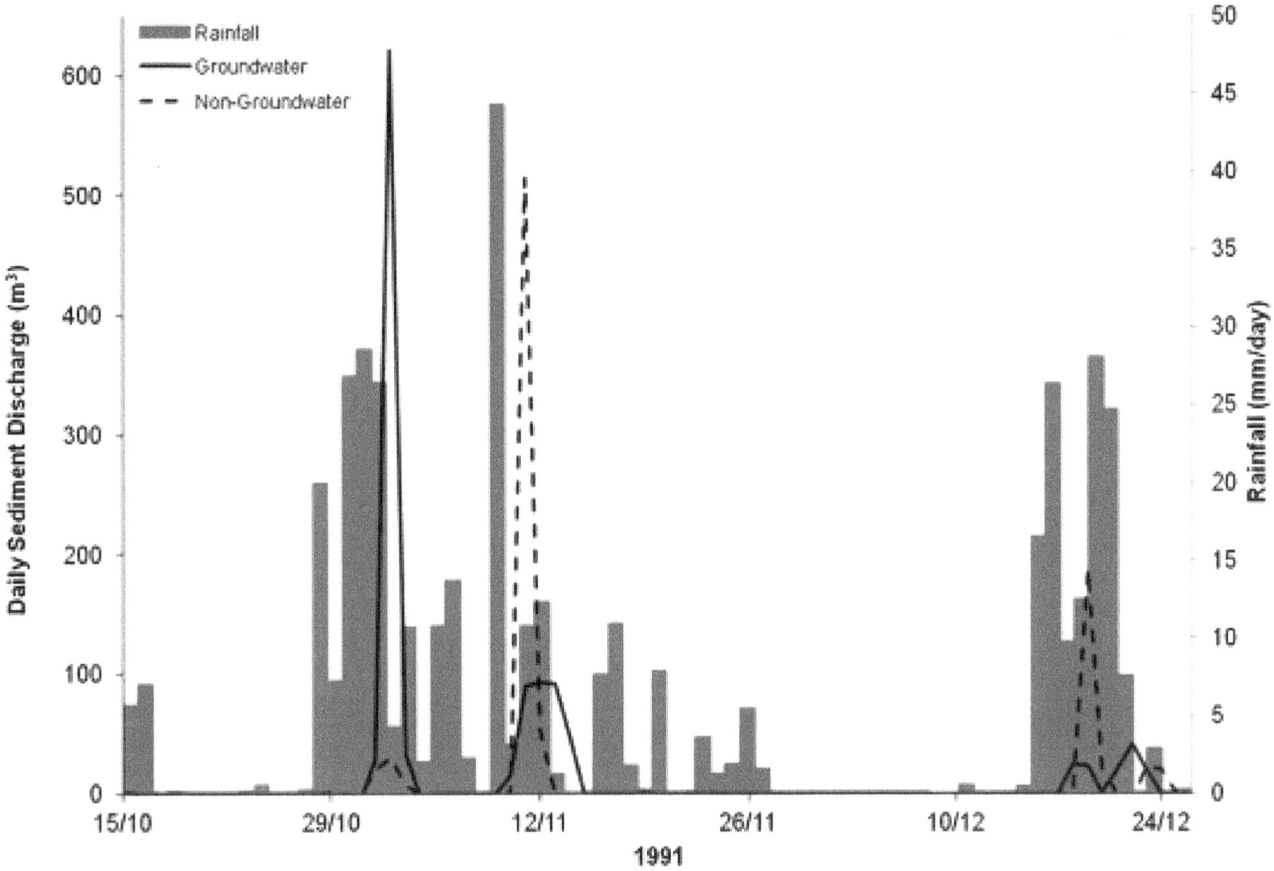

Figure 18: Daily sediment discharge for the with- and without-groundwater scenarios overlaying average catchment rainfall between mid-October and the end of December 1991.

From the end of the spin-up period to the end of the simulation (a period of 19 months), the scenario including groundwater discharges a total of ~1800 m^3 more sediment from the catchment. During this period the greatest sediment transport event for both scenarios occurs on December 1st 1992, where a high rainfall event (47 mm) occurs on the previous day. Sediment discharge over this day increased from near zero in both cases, up to 25,803 m^3 for the groundwater scenario and 24,537 m^3 for the runoff only simulation. It is apparent from the cumulative plot that there are large periods of near zero sediment discharge throughout both simulations, where surface water velocity is not sufficient to transport sediment from the catchment. The total number of days when sediment flux from the catchment exceeds a particular threshold value is plotted for both scenarios (Fig. 17). Both scenarios produce the same number of days with sediment discharge above 1000 m^3. The scenario without groundwater has a few more days with an intermediate sediment discharge (between 100 and 1000 m^3). There are significantly more days where the groundwater inclusive scenario has a sediment discharge of between 0 and 100 m^3. These differences suggest that for this catchment groundwater is responsible for removing low, daily volumes of sediment over long periods of time.

Focussing on a few rainfall events over a short period allows a better descriptive analysis of the difference in the two scenarios. Fig. 18 includes daily sediment discharge for both scenarios and total catchment rainfall between mid-October and the end of December 1991. Sediment discharge events are initiated at similar times, but the groundwater inclusive scenario produces sediment transport over longer periods. During short, intense rainfall events, the without-groundwater scenario often produces a higher sediment discharge value than the subsurface flow inclusive scenario. The difference arises because a certain amount of water is routed to the subsurface store in the groundwater scenario, reducing overland flow and lowering the capacity to move sediment. However, this pattern can be reversed when a series of subsequent rainfall events, or persistent low intensity rainfall events, occur. Under these conditions, the system with baseflow-return transports increased volumes of sediment. The increase in sediment flux occurs where high baseflow levels from previous rainfall events increases total river flow above the threshold required to transport sediment. With groundwater included, this allows rainfall events that are not in themselves large enough to move sediment, to transport sediment with the added water provided by baseflow.

Fluvial channel, cell-averaged sediment flux (post spin-up) is plotted against altitude to assess the overall sediment transport dynamics within the catchment (Fig. 19). Small-scale oscillations (with a wavelength <50 m altitude) represent sediment that has been transported only short distances, creating erosion in the local upper nodes and accretion in the lower nodes. The short transport distances can be attributed to the movement of larger grain sizes that need sustained, high fluvial-velocities to travel long distances. Over the two-year period there is insufficient time for transport of these larger grains further than 50 m.

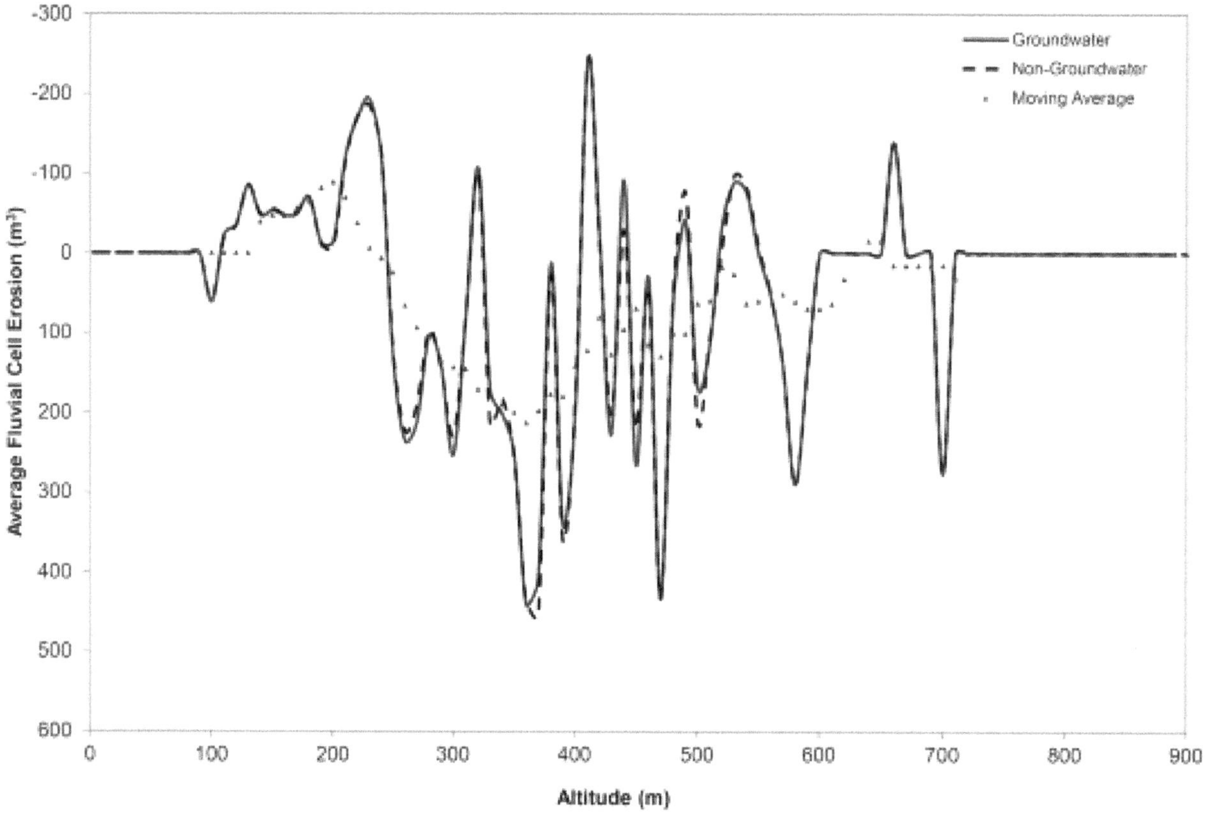

Figure 19: Simulated erosion (or accretion) for fluvial cells averaged across bins of 10 m altitude. The plot highlights differences in erosion characteristics for the with- and without-groundwater scenarios and the larger scale pattern of erosion and accretion within the catchment after two-years of simulation. The moving average is shown for the groundwater inclusive scenario to highlight the larger scale trends.

The difference between the with- and without-groundwater scenarios is not apparent at higher altitudes as the groundwater levels are well below the terrain surface and baseflow does not discharge into the rivers. Between 550 m and 450 m the effects of groundwater become apparent, increasing the peaks of erosion and deposition as surface flow velocities and sediment fluxes are increased. At lower catchment elevations the difference in the amount of sediment transported is diminished due to an overall reduction in slope inclination and river flow velocities.

Using a central moving average over several tens of metres in the up- and down-slope directions, the small-scale oscillations can be removed (example in Fig. 19 for the groundwater inclusive scenario only), revealing a large-scale pattern of increasing erosion from 650 m down to 350 m followed by a reduction in the erosion and deposition at lower altitudes. This large-scale change in the erosional response with altitude is likely to be topographically controlled for upper Eden Valley, as mid to high elevations are associated with increased topographic gradient in comparison to the lower reaches. Above 650 m, surface flow is not concentrated enough to produce sufficient volumes of water to transport sediment. At between 650 m and 350 m surface flow has begun to concentrate into streams and slope is sufficiently steep to favour surface runoff, with velocities great enough to entrain and transport sediment. At between 350 m and 200 m water is further concentrated and steam volumes are increased. However, topographic gradient is no longer sufficient to entrain sediment at the same rates as at higher altitudes (with steeper slopes) and the volume of eroded sediment is reduced with decreasing elevation. Below 200 m, surface flow velocities are diminished to the point where the volume of sediment entrained is smaller than that deposited, and accretion occurs.

Significant sediment discharge events only occur during periods of above average rainfall. As shown by Coulthard et al. (2012) and the sensitivity analysis (Section 3.2), the amount of sediment transported from the system is not necessarily proportional to the amount of rainfall. The same is true for this catchment. Although small, the difference that arises in the total sediment transported under the with- and without-groundwater scenarios is a result of the baseflow sustaining river water fluxes above the sediment pick-up threshold for greater time periods after high rainfall events. The timing of sediment transport events is not equivalent in the two scenarios. The runoff-only simulation exhibits a short-term flashy sediment discharge in response to intense rainfall events, whilst the inclusion of groundwater smooths the sediment discharge and introduces a 'memory' of previous rainfall events into the system. The large-scale trend of increased erosion at mid-elevations is due to the increased fluvial velocities that exist in these regions, and is largely controlled by topographic gradient. These trends are also likely to be, in part, due to the spatially differing distributed catchment surface water inputs.

DISCUSSION

The following discussion covers an assessment of the modelling platform and calibration process, followed by the findings of the sensitivity test and an outline of the impact that groundwater has on the upper Eden Valley geomorphology, with a focus on the traits that may be applicable to other catchments. The final section discusses caveats and uncertainties associated with the platform and modelling process.

Platform assessment

A modelling platform has been created that has the potential to simulate a range of environmental systems. The CLiDE platform was constructed using existing landscape evolution algorithms for the transport of sediment and surface hydrodynamics. These were driven by a new method for the partitioning of water into evapotranspiration, surface, soil and groundwater stores, which was coupled to groundwater flow algorithms via recharge and baseflow-return. Assessment of the new platform was undertaken using comparison of the hydrology to analytical solutions and gauged data, which it was found to reproduce with reasonable accuracy.

The impacts of a range of groundwater scenarios on the fluvial and sediment flux discharges from an idealised catchment were tested. Despite the highly simplified nature of the topography and the identical boundary conditions for each simulation, a range of discharge profiles was created. Sediment discharges were found to switch between steady-states under quasi-steady river discharges as the catchments evolved through time. The initial steady-state attained for the majority of scenarios represents no defined river channel. The subsequent move to the next steady-state represents the rapid carving of the main channel. Under certain subsurface conditions (in particular a high specific yield) a nonlinear catchment discharge response is observed despite the uniform forcing factors (precipitation and potential evapotranspiration). Representation of baseflow-return to fluvial flow using uniform baseflow return rates (common in LEMs) is therefore likely to be inaccurate, particularly in catchments with a highly-heterogeneous subsurface. Analysis of the sensitivity scenarios reveals the ability of catchment sediment transport processes to impact surface and subsurface hydrology. Where adequate connectivity exists between surface and subsurface water, the deepening of channels and resultant drop in average river stage reduces surrounding groundwater heads. The sensitivity analysis suggests a high specific yield could potentially return a large volume of water to the surface when the channel is incised. Hydraulic conductivity places a limit on this return by controlling the spatial extent of the impact on groundwater heads; high hydraulic conductivity will create a large area over which groundwater heads are reduced as the river stage drops and low hydraulic conductivity a local reduction.

Analysis of the platform using observational data was limited to the hydrological modules, as sediment flux data was limited at the catchment scale. The calibration technique of matching simulated flow data to gauged data is applicable to any region where continuous records of river flow are present. Further accuracy in the calibration is possible where multiple, distributed, gauging locations are present across a catchment and with longer flow records that capture a variety of extreme conditions (for example, flooding or drought). For the upper Eden Valley test catchment, calibration was only possible for a single gauging station and although the simulated surface-only flows were well matched to the baseflow separated gauged data, groundwater recession rates were not a good match to the observational data. Obtaining a closer match between observational and simulated groundwater in the upper Eden Valley would be possible through further changes to the distributed hydraulic conductivity and specific yield. However, the calibration process is time consuming and the resources were not available for this study.

Groundwater and the upper Eden Valley geomorphology

Analysis of the upper Eden Valley sediment flux rates suggests that after two years of simulation there are several differences in regional geomorphology that arise when groundwater is introduced into the system. These can be divided into spatial and temporal differences. Spatial differences in sediment erosion and accretion occur throughout the catchment. When averaged these differences are most apparent at elevations of between 550 and 450 m and can be attributed to increased flow velocities from the addition of baseflow. Temporal differences manifest themselves following large rainfall events. At the event-scale (see Fig. 18), the scenario without groundwater predicts both higher and lower daily sediment discharges in comparison to the groundwater scenario, suggesting that antecedent conditions are highly influential. A time-averaged study may predict similar amounts of erosion, but the inclusion of groundwater is critical for accurate representation of the timing and location of sediment flux events. The averaging effect manifests itself in the similar total catchment sediment flux is observed at the end of the simulation. In general, the groundwater response at the event-scale tends to be more protracted. This is highlighted in Fig. 17, where there are nine extra days in the groundwater simulation where sediment discharge events are at 10 m^3 or below. At the annual scale, the difference in total catchment sediment flux is small, but over decades it could make a significant contribution to the total eroded volume. For studies attempting to recreate the evolution of catchments with high baseflow inputs, the representation of baseflow-return within the fluvial processes that drive sediment discharge rates is important, even if they are represented in an averaged sense. It is also important to note that the predicted increase in extreme rainfall events under future climate scenarios (Bates et al., 2008 and Arnell, 2011) should be taken into consideration for studies interested in simulating future landscape morphology.

It is anticipated that application of the CLiDE platform to highly groundwater-dependent alluvial systems with large sediment yields (for example, the Platte River, USA (Knox, 1972); the river Ganges, India (Narayana and Babu, 1983); or the Guadalquivir River, Spain (Lobo et al., 2004)) would show a significant difference in sediment transport with the groundwater module enabled. This would allow an improved assessment of geomorphological evolution over traditional LEM techniques that do not take groundwater into account.

Caveats and uncertainty

There are assumptions inherent in all environmental modelling systems, as the system being represented needs to be simplified in order to be simulated. Process-based system representation carries fewer assumptions than other techniques, and is central to the modules that make up the CLiDE platform. By sticking to the CA principles and avoiding empirical representation through statistical analysis, the platform is suitable for use in a variety of catchments.

As CA discretises the domain into discrete parts and there is assumed to be no heterogeneity at the sub-cell scale. Any sub-cell scale variation is assumed to be integrated into the cell average value for a particular variable. Time is also divided into sections and therefore sub-time step processes are also assumed to be averaged.

The uncertainty associated with input parameters is often unknown, due to the variety of methods available for producing the parameters and interpolating the data into a fixed grid. As environmental systems are often nonlinear, small changes in the initial conditions may have a large effect on the output of the simulation at advanced time steps. By undertaking a calibration, where model attributes are compared to observed values, these issues will have less of an impact, providing a suitable number of comparisons are made over a significant period of time. For the upper Eden Valley case study, simulated hydrological attributes were calibrated to observed values over a two-year period. There was, however, no attempt made to calibrate the simulated sediment transport as it was assumed that the links between hydrology and sediment transport are adequate, as supported by previous comparison of the sediment transport component to observed data (see Coulthard et al., 2012). If records of sediment transport exist for a catchment the platform may be calibrated to these following the hydrological calibration. To better quantify the impact of the initial parameters on the simulation, an ensemble analysis technique may be employed.

The use of a spin-up period in the initial phase of the modelling process allows the modelled domain to attain a dynamic steady-state. During this period non-forcing hydrological and sediment transport parameters reach a relaxed state, producing quasi-uniform responses to similar meteorological events. The hydrological processes reach this state more rapidly than the sediment transport processes as the early phase represents the building of fluvial systems and the latter phase the distribution of sediment. This allows the spin-up period to compensate for any inaccuracies in the initialisation parameters.

One of the major influences on dynamic morphology that was not captured during this modelling case study is the formation of soil at the bedrock-soil interface. This limits the amount of sediment available for transport within the catchment, and in reality would be controlled by the underlying geology as well as the soil hydrology. As the study covers only two-years of simulation, this process was not deemed important. If a centennial scale study was undertaken this process would need to be captured in the model, or the bedrock layer (non-erodible) would be exposed along the river reach and sediment transport would be constrained by the lateral evolution of the channel only. Currently the only influence that geology exerts on the topography within the model (post initialisation) is through the transport of groundwater and the influence this has on surface hydrology and subsequently sediment transport.

CLiDE is aimed at addressing sub-annual to centennial catchment-scale research. Suitability of the platform as a research tool for millennial scale (or longer) studies is not limited by the numerical representation of the underlying processes, but by available processing power and scalability of the platform under increased parallelisation.

SUMMARY AND CONCLUSIONS

A two-dimensional cellular automata platform for exploring the sensitivity of environmental systems over a variety of spatio-temporal scales has been presented to assess the impact of groundwater on regional geomorphological evolution. Physically-based equations were implemented to simulate the flow of surface and subsurface water, which in turn drives sediment transport between adjacent cells. Surface and subsurface hydrology was coupled through the exchange of recharge to groundwater, and baseflow discharge to rivers. Hydrology and sediment transport, both bedload and suspended, were integrated through river flow velocity and DEM modification. The runtime calculation of several soil hydrological properties makes the resulting platform versatile and easily adaptable for simulating a variety of natural systems under differing forcing scenarios (climatic, geologic or anthropogenic).

Testing of the platform's hydrological components, which have been added to CAESAR-Lisflood to form the CLiDE platform, was underpinned by comparison to analytical solutions, which show the model is stable and convergent. The accuracy of the subsurface hydrological components was found to be good when evaluated against analytical solutions under both transient and steady-state conditions. Testing of the platform in a simple, idealised catchment revealed that the impacts of baseflow-return on catchment sediment discharge manifest themselves in a variety of ways and are therefore difficult to represent in LEMs using uniform rates. The sensitivity analysis also reveals some of the feedbacks that can arise between the hydrological components and catchment morphology, and how these can evolve through a series of transitions between steady-states.

To set some of the model parameters for simulating sediment transport in the upper Eden Valley, the CLiDE platform was calibrated using baseflow separated gauged river data and groundwater level time-series at several locations over a two year period. The parameters derived from calibration appear to be sensible, and it was possible to obtain a good match to the surface river flow data. The matching of the groundwater level time-series was not as good as that achieved in the surface component. However, this was expected as we are representing a complex heterogeneous environmental system using a simplified model.

The input of baseflow into rivers creates a less flashy sediment flux responses to extreme rainfall events. During these events, the transport of sediment is maintained over extended periods, as baseflow keeps river velocities above the minimum sediment transport threshold for longer periods. The groundwater component provides the hydrological system with antecedent conditions, allowing previous events to influence future sediment transport rates. At the event-scale, antecedent rainfall lagged in the system as groundwater can result in a substantial difference in predicted volumes of sediment transport for any particular time. Including a mechanism for capturing the spatial and temporal influence of antecedent rainfall conditions on fluvial flow is therefore essential for shorter-term landscape evolution modelling. Assigning a universal transition point (i.e., defining just how short shorter-term is) where process-based or average baseflow-returns are more appropriate is not possible as it is dependent on the focus of any particular study and the individual catchment characteristics.

ACKNOWLEDGEMENTS

The authors are grateful to David Macdonald of the British Geological Survey (BGS) for reviewing this paper and returning feedback that led to its improvement. They would also like to acknowledge Dr Jorge Ramirez (currently at Florida Atlantic University) for the preparation of input data, and Henry Holbrook of the BGS for preparation of Fig. 13. This paper is published with the permission of the Executive Director of the British Geological Survey (NERC), and was supported by the Climate and Landscape Change, and the Environmental Modelling research programmes at the BGS.

REFERENCES

1. Ahnert, F., 1976. Brief description of a comprehensive three-dimensional processresponse model for landform development. Z. Geomorphol. 25, 29e49.
2. Allen, D.J., Brewerton, L.J., Coleby, L.M., Gibbs, B.R., Lewis, M.A., MacDonald, A.M., Wagstaff, S.J., Williams, A.T., 1997. The physical properties of major aquifers in England and Wales. Br. Geol. Surv. Rep. WD/97/034 (333 pages).
3. Allen, R., Pereira, L.A., Raes, D., Smith, M., 1998. Crop Evapotranspiration: Guidelines for Computing Crop Water Requirements. FAO, Irrigation and Drainage Paper No. 56. FAO, Rome, Italy.
4. Arcement, G.J., Schneider, V.R., 1989. Guide for selecting manning's roughness coefficients for natural channels and floodplains. U. S. Geol. Surv. Water-Supply Pap. 2339, 38.
5. Arnell, N.W., 2011. Uncertainty in the relationship between climate forcing and hydrological response in UK catchments. Hydrol. Earth Syst. Sci. 15, 897e912.
6. Barkwith, A., Coulthard, T.J., 2014. CLiDE V1.0 User Guide: British Geological Survey Open Report OR/14/011. British Geological Survey, Nottingham, UK, p. 54.
7. Barry, D.A., Parlange, J.-Y., Li, L., 2000. Approximations for the exponential integral (Theis well function). J. Hydrol. 227, 287e291.
8. Bates, B.C., Kundzewicz, Z., Wu, S., Palutikof, J., 2008. Climate Change and Water. Technical Paper of the Intergovernmental Panel on Climate Change, Geneva, p. 210.
9. Bates, P.D., Horritt, M.S., Fewtrell, T.J., 2010. A simple inertial formulation of the shallow water equations for efficient two-dimensional flood inundation modelling. J. Hydrol. 387, 33e45.
10. Bear, J., 1979. Hydraulics of Groundwater. McGraw-Hill, New York, p. 567.
11. Bennett, S., Dumble, J.P., Ruxton, C., 2003. The Monitoring of "Poor Aquifers" to Meet the Requirements of the EU Water Framework Directive: Phase 1-review and Scoping Report.
12. Boorman, D.B., Hollis, J.M., Lilly, A., 1995. Hydrology of Soil Types: a Hydrologicallybased Classification of the Soils of the United Kingdom. Report No. 126. Institute of Hydrology, Wallingford, UK.
13. Butcher, A.S., Lawrence, A.R., Jackson, C., Cunningham, J., Cullis, E., Hasan, K., Ingram, J., 2006. Investigating rising nitrate concentrations in groundwater in the Permo-Triassic sandstone aquifer in the UK. In: Barker, R.D., Tellam, J.H. (Eds.), Fluid Flow and Solute Movement in Sandstones: the Onshore UK Permotriassic Red Bed Sequence. Geological Society, London, pp. 285e296. Special Publications, 263.
14. Buttle, J.M., 1994. Isotope hydrograph separations and rapid delivery of pre-event water from drainage basins. Prog. Phys. Geogr. 18 (1), 16e41.
15. Changnon, S.A., Huff, F.A., Hsu, C.-F., 1988. Relations between precipitation and shallow groundwater in Illinois. J. Clim. 1 (12), 1239e1250.
16. Chen, S., Doolen, G.D., 1998. Lattice Boltzmann method for fluid flows. Annu. Rev. Fluid Mech. 30 (1), 329e364.
17. Chopard, B., Masselot, A., 1999. The lattice Boltzmann method: a new approach to computational fluid dynamics and particle transport. Future Gener. Comp. Syst. 16, 249e256.
18. Coulthard, T.J., 2001. Landscape evolution models: a software review. Hydrol. Process. 15 (1), 165e173.
19. Coulthard, T.J., Lewin, J., Macklin, M.G., 2005. Modelling differential and complex catchment response to environmental change. Geomorphology 69 (1e4), 224e241.
20. Coulthard, T.J., Neal, J.C., Bates, P.D., Ramirez, J., de Almeida, G.A.M., Hancock, G.R., 2013. Integrating the LISFLOOD-FP 2D hydrodynamic model with the CAESAR model: implications for modelling landscape evolution. Earth Surf. Process. Land. 38, 1897e1906.
21. Coulthard, T.J., Ramirez, J., Fowler, H.J., Glenis, V., 2012. Using the UKCP09 probabilistic scenarios to model the amplified impact of climate change on drainage basin sediment yield. Hydrol. Earth Syst. Sci. 16 (11), 4401e4416.
22. Coulthard, T.J., Van De Wiel, M.J., 2006. A cellular model of river meandering. Earth Surf. Proc. Land. 31, 123e132.
23. Culling, W.E.H., 1960. Analytical theory of erosion. J. Geol. 68 (3), 336e344.
24. Daily, P., Ingram, J., Johnson, R., 2006. Groundwater Quality Review: Eden Valley and Carlisle Basin. Environment Agency Report GWQR004 and ESI Report 6438A R1.
25. Dunne, T., 1988. Hydrology and mechanics of erosion by subsurface flow. In: Back, W., Rosenshein, J.S., Seaber, P.R. (Eds.), Hydrogeology. Geological Society of America, Boulder CO, USA, p. 988.
26. Eckhardt, K., 2008. A comparison of baseflow indices, which were calculated with seven different baseflow separation methods. J. Hydrol. 352, 168e173.
27. Einstein, H.A., 1950. The Bed-load Function for Sediment Transportation in Open Channel Flows. Technical Bulletin 1026. USDA, Soil Conservation Science, pp. 1e77.
28. Fowler, H.J., Tebaldi, C., Blenkinsop, S., 2007. Probabilistic estimates of climate change impacts on

flows in the River Eden, Cumbria. In: 10th National Hydrology Symposium: Sustainable Hydrology for the 21st Century. British Hydrological Society, Exeter, UK.

29. Fox, G.A., Wilson, G.V., Simon, A., Langendoen, E., Akay, O., Fuchs, J.W., 2007. Measuring streambank erosion due to ground water seepage: correlation to bank pore water pressure, precipitation, and stream stage. Earth Surf. Proc. Land. 32 (10), 1558e1573.

30. Goodrich, D.C., Lane, L.J., Shillito, R.M., Miller, S.N., Syed, K.H., Woolhiser, D.A., 1997. Linearity of basin response as a function of scale in a semi-arid watershed. Water Resour. Res. 33 (12), 2951e2965.

31. Haan, M.M., Russell, J.R., Powers, W.J., Kovar, J.L., Benning, J.L., 2006. Grazing management effects on sediment and phosphorus in surface runoff. Rangel. Ecol. Manage. 59, 607e615.

32. Haitjema, H.M., 1995. Analytic Element Modeling of Groundwater Flow. Academic Press, p. 394.

33. Hancock, G.R., Coulthard, T.J., Martinez, C., Kalma, J.D., 2011. An evaluation of landscape evolution models to simulate decadal and centennial scale soil erosion in grassland catchments. J. Hydrol. 308, 171e183.

34. Harr, M.E., 1962. Ground Water and Seepage. McGraw-Hill Book Co., New York.

35. Hoffmann, T., Thorndycraft, V.R., Brown, A.G., Coulthard, T.J., Damnati, B., Kale, V.S., Middelkoop, H., Notebaert, B., Walling, D., 2010. Human impact on fluvial regimes and sediment flux during the Holocene: review and future research agenda. Glob. Planet. Change 72 (3), 87e98.

36. Howard, A., McLane, C., 1988. Erosion of cohesionless sediment by groundwater seepage. Water Resour. Res. 24 (10), 1659e1674.

37. Huang, X., Niemann, J.D., 2006. Modelling the potential impacts of groundwater hydrology on long-term drainage basin evolution. Earth Surf. Proc. Land. 31 (14), 1802e1823.

38. Keesstra, S.D., Temme, A.J.A.M., Schoorl, J.M., Visser, S.M., 2014. Evaluating the hydrological component of the new catchment-scale sediment delivery model LAPSUS-D. Geomorphology 212, 97e107.

39. Kelly, P.J., 2012. Subsurface Evolution: Characterizing the Physical and Geochemical Changes in Weathered Bedrock of Lower Gordon Gulch, Boulder Creek Critical Zone Observatory. Master's thesis. University of Colorado.

40. Kinnell, P.I.A., 2010. Event soil loss, runoff and the Universal soil loss equation family of models: a review. J. Hydrol. 385 (1e4), 384e397.

41. Kirchner, J.W., Feng, X., Neal, C., 2000. Fractal stream chemistry and its implications for contaminant transport in catchments. Nature 403 (6769), 524e527.

42. Knox, J., 1972. Valley alleviation in southwestern Wisconsin. Ann. Assoc. Am. Geogr. 62 (3), 401e410.

43. Laity, J.E., Malin, M.C., 1985. Sapping processes and the development of theaterheaded valley networks on the Colorado Plateau. Geol. Soc. Am. Bull. 96, 203e217.

44. Lange, J., Greenbaum, N., Husary, S., Ghanem, M., Leibundgut, C., Schick, A.P., 2003. Runoff generation from successive simulated rainfalls on a rocky, semi-arid, Mediterranean hillslope. Hydrol. Process 17, 279e296.

45. Lobo, F.J., Sanchez, R., Gonzalez, R., Dias, J.M.A., Hernandez-Molina, F.J., FernandezSalas, L.M., Diaz del Rio, V., Mendes, I., 2004. Contrasting styles of the Holocene highstand sedimentation and sediment dispersal systems in the northern shelf of the Gulf of Cadiz. Cont. Shelf Res. 24, 461e482.

46. Malby, A.R., Whyatt, J.D., Timmis, R.J., Orr, H.G., 2007. Long-term variations in orographic rainfall: analysis and implications for upland catchments. Hydrolog. Sci. J. 52 (2), 276e291.

47. Margolus, N., Toffoli, T., Vichniac, G., 1986. Cellular-automata supercomputers for fluid dynamics modeling. Phys. Rev. Lett. 15 (16), 1694e1696.

48. Mayes, W.M., Walsh, C.L., Bathurst, J.C., Kilsby, C.G., Quinn, P.F., Wilkinson, M.E., Daugherty, A.J., O'Connell, P.E., 2006. Monitoring a flood event in a densely instrumented catchment, the upper Eden, Cumbria, UK. Water Environ. 20 (4), 217e226.

49. Merritt, W.S., Letcher, R.A., Kakeman, A.J., 2003. A review of erosion and sediment transport models. Environ. Modell. Softw. 18, 761e799.

50. Metha, A., Barker, G.C., 1994. The dynamics of sand. Rep. Prog. Phys. 57 (4), 385e416. Murray, A.B., Paola, C., 2003. Modelling the effect of vegetation on channel pattern in bedload rivers. Earth Surf. Proc. Land. 28, 131e143.

51. Narayana, D.V.V., Babu, R., 1983. Estimation of soil erosion in India. J. Irrig. Drain. Eng. 109 (4), 419e434.

52. Neal, J., Schumann, G., Fewtrell, T., Budimir, M., Bates, P., Mason, D., 2011. Evaluating a new Lisflood-FP formulation with data from the summer 2007 floods in Tewkesbury, UK. J. Flood Risk Man. 4, 88e95.

53. Pazzaglia, F.J., 2003. Landscape evolution models. Dev. Quatern. Sci. 1, 247e273.

54. Payn, R.A., Gooseff, M.N., McGlynn, B.L., Bencala, K.E., Wondzell, S.M., 2012. Exploring changes in the spatial distribution of stream baseflow generation during a seasonal recession. Water Resour. Res. 48, W04519.

55. Ravazzani, G., Rametta, D., Mancini, M., 2011. Macroscopic cellular automata for groundwater modelling: a first approach. Environ. Modell. Softw. 26 (5), 634e643.

56. Roache, P.J., 1976. Computational Fluid Dynamics. Hermosa, Albuquerque, NM. Rev. 1976.

57. Rothman, D.H., 1988. Cellular-automaton fluids: a model for flow in porous media. Geophys. 53 (4), 509e518.

58. Rushton, K.R., 2003. Groundwater Hydrology: Conceptual and Computational Models. John Wiley and Sons Ltd, West Sussex, England.

59. Sklash, M.G., Favolden, F.N., 1979. The role of groundwater in storm runoff. J. Hydrol. 45, 45e65.

60. Somers, J.A., 1993. Direct simulation of fluid flow with cellular automata and the lattice-Boltzmann equation. Appl. Sci. Res. 51 (1e2), 127e133.

61. Toffoli, T., 1984. Cellular automata as an alternative to (rather than an approximation of) differential equations in modeling physics. Phys. Nonlinear Phenom. 10 (1e2), 117e127.

62. Tooth, S., Jansen, J.D., Nanson, G., Coulthard, T.J., Pietsch, T., 2008. Riparian vegetation and the late Holocene development of an anabranching river: Magela Creek, northern Australia. GSA Bull. 120 (7e8), 1021e1035.

63. Tucker, G.E., Hancock, G.R., 2010. Modelling landscape evolution. Earth Surf. Proc. Land. 35 (1), 28e50.

64. Van De Wiel, M.J., Coulthard, T.J., Macklin, M.G., Lewin, J., 2007. Embedding reachscale fluvial dynamics within the CAESAR cellular automaton landscape evolution model. Geomorphology 90, 283e301.

65. Van DeWiel,M.J., Coulthard, T.J.,Macklin,M.G., Lewin, J., 2011. Modelling the response of river systems to environmental change: progress, problems and prospects for palaeo-environmental reconstructions. Earth Sci. Rev. 104 (1e3), 167e185.

66. von Neumann, J., 1951. The general and logical theory of automata. Cereb. Mech. Behav. 1e41.

67. von Neumann, J., 1966. Theory of Self-reproducing Automata. University of Illinois Press, Urbana, IL.

68. Wang, J.F., Anderson, M.P., 1982. Introduction to Groundwater Modelling. Freeman, San Francisco, CA, USA.

69. Wilcock, P., Crowe, J., 2003. Surface-based transport modelling for mixed-size sediment. J. Hydraul. Eng. 129 (2), 120e128.

CHAPTER 7

TRACING ORGANIC MATTER COMPOSITION AND DISTRIBUTION AND ITS ROLE ON ARSENIC RELEASE IN SHALLOW CAMBODIAN GROUNDWATERS

Michael Lawson [a], David A. Polya [a], Adrian J. Boyce [b], Charlotte Bryant [c], Christopher J. Ballentine [a,d]

[a] School of Earth, Atmospheric and Environmental Sciences and Williamson Research Centre for Molecular Environmental Science, University of Manchester, Oxford Road, Manchester M13 9PL, UK

[b] Scottish Universities Environmental Research Centre, East Kilbride G75 0QF, UK

[c] NERC Radiocarbon Facility, East Kilbride G75 OQF, UK

[d] Department of Earth Sciences, University of Oxford, South Parks Road, OX1 3AN Oxford, UK

ABSTRACT

Biogeochemical processes that utilize dissolved organic carbon are widely thought to be responsible for the liberation of arsenic from sediments to shallow groundwater in south and southeast Asia. The accumulation of this known carcinogen to hazardously high concentrations has occurred in the primary source of drinking water in large parts of densely populated countries in this region. Both surface and sedimentary sources of organic matter have been suggested to contribute dissolved organic carbon in these aquifers. However, identification of the source of organic carbon responsible for driving arsenic release remains enigmatic and even controversial. Here, we provide the most extensive interrogation to date of the isotopic signature of ground and surface waters at a known arsenic hotspot in Cambodia. We present tritium and radiocarbon data that demonstrates that recharge through ponds and/or clay windows can transport young, surface derived organic matter into groundwater to depths of 44 m under natural flow conditions. Young organic matter dominates the dissolved organic carbon pool in groundwater that is in close proximity to these surface water sources and we suggest this is likely a regional relationship. In locations distal to surface water contact, dissolved organic carbon represents a mixture of both young surface and older sedimentary derived organic matter. Ground–surface water interaction therefore strongly influences the average dissolved organic carbon age and how this is distributed spatially across the field site. Arsenic mobilization rates appear to be controlled by the age of dissolved organic matter present in these groundwaters. Arsenic concentrations in shallow groundwaters (<20 m) increase by 1 μg/l for every year increase in dissolved organic carbon age compared to only 0.25 μg/l for every year increase in dissolved organic carbon age in deeper (>20 m) groundwaters. We suggest that, while the rate of arsenic release is greatest in shallow aquifer sediments, arsenic release also occurs in deeper aquifer sediments and as such remains an important process in controlling the spatial distribution of arsenic in the groundwaters of SE Asia. Our findings suggest that any anthropogenic activities that alter the source of groundwater recharge or the timescales over which recharge takes place may also drive changes in the natural composition of dissolved organic carbon in these groundwaters. Such changes have the potential to influence both the spatial and temporal evolution of the current groundwater arsenic hazard in this region.

Citation: Michael Lawson, David A. Polya, Adrian J. Boyce, Charlotte Bryant, Christopher J. Ballentine, Tracing organic matter composition and distribution and its role on arsenic release in shallow Cambodian groundwaters, Geochimica et Cosmochimica Acta, Volume 178, 1 April 2016, Pages 160-177, ISSN 0016-7037, http://dx.doi.org/10.1016/j.gca.2016.01.010.

INTRODUCTION

The extensive contamination of groundwaters by geogenic arsenic (As) utilized by millions of people as a primary source of drinking water in South and South East Asia has long been recognised as a public health catastrophe (Chakraborty and Saha, 1987,Balga and Kaiser, 1996, Dhar et al., 1997, Smith et al., 2000, Nickson et al., 1998a,Winkel et al., 2008 and Ravenscroft et al., 2009). The consumption of groundwater containing As above 10 μg/l has already resulted in deleterious health effects in millions of people (Yu et al., 2003), with as many as 20% of deaths in arsenic impacted areas of Bangladesh attributable to arsenic (Argos et al., 2010). However, despite more than a decade of active research, several fundamental gaps remain in our understanding of the As release process that limit our ability to better predict locations and anthropogenic activities that risk As contamination. Such information is crucial for governments and those responsible for mitigating the effects of the current crisis and for predicting the potential for secular changes in the As hazard in the future.

The primary source of As in the groundwaters of this region is the rapid weathering of As-bearing rocks in upper Himalayan catchments. This As is then transported by the large rivers that drain these catchments, and deposited in the low-lying young alluvial floodplains of the region. Arsenic is initially sorbed to secondary iron (oxy)hydroxide mineral phases that coat sedimentary grains under aerobic conditions, but is released to solution following the reductive dissolution of these As-bearing phases (Bhattacharya et al., 1997, Nickson et al., 1998b, McArthur et al., 2001, Dowling et al., 2002, Stuben et al., 2003, Horneman et al., 2004, Zheng et al., 2004 and Charlet and Polya, 2006). Iron-reducing bacteria catalyse this release of As to solution, with organic matter (OM) critically implicated as an electron donor in this process (Islam et al., 2004 and van Geen et al., 2004). However, whilst the biogeochemical processes that are responsible for liberating As to groundwater are now widely understood, identification of the source of OM that drives these processes remains unresolved, elusive and even controversial (Fendorf et al., 2010, Lawson et al., 2013 and Neumann et al., 2014).

Several studies have suggested that a sedimentary origin of OM is responsible for driving microbially mediated As release processes; with buried peat (McArthur et al., 2001), aerated root zones and animal burrows (Meharg et al., 2006), or thermally mature hydrocarbons that have migrated from depth (Rowland et al., 2006, Rowland et al., 2007,Rowland et al., 2008, van Dongen et al., 2008 and Al Lawati et al., 2012) amongst suggested potential sources. In contrast, surface derived OM may be utilized in the biogeochemical processes that liberate As, with recharge from ponds, wetlands and through geomorphic features such as abandoned floodplains, meander channel deposits, and scroll bars introducing young dissolved organic carbon (DOC) to shallow groundwaters (Kocar et al., 2008, Papacostas et al., 2008, Polizzotto et al., 2008,Neumann et al., 2010 and Lawson et al., 2013). The introduction of this surface derived OM has also been suggested to be exacerbated by the extensive pumping of groundwaters in this region, resulting in the draw down of increased volumes of organic rich recharge waters (Harvey et al., 2002). This particular model is intensely debated (Aggarwal et al., 2003, Harvey et al., 2003, van Geen et al., 2003, Lawson et al., 2008,Lawson et al., 2013, Sengupta et al., 2008, McArthur et al., 2011, Neumann et al., 2010 and Mailloux et al., 2013), not least because of the implications it holds for future agricultural production in these densely populated countries reliant on rice as a staple food source.

Given the considerable debate surrounding the origin of OM that is responsible for driving As release, the identification and quantification of the different sources of OM present within these aquifer systems is of great importance. This is of particular significance when one considers that it is not necessarily the amount, but the type and reactivity of OM that are the key controls on As release to groundwater (Gault et al., 2005,Rowland et al., 2007, van Dongen et al., 2008 and Mladenov et al., 2010). Indeed, it has been suggested that utilisation of different sources of OM will give rise to different rates of As release to groundwaters (Fendorf et al., 2010). Older sources of OM, such as particulate organic carbon (POC) present in buried peat deposits, are likely to have experienced a long history of involvement and consumption in metabolic processes that began immediately following deposition. It is assumed that such metabolic processing may give rise to a less reactive residual OM that may support a slow rate of As release into the groundwater. In contrast, young surface-derived OM that is introduced into the subsurface has had less time to be utilized in metabolic processes, and as such may provide a higher reactivity source of OM that may support a more rapid rate of As release. Distinguishing between the sources of OM present in any groundwater system, and assessing the vulnerability of the system to changes with respect to its dominant OM source, is therefore of critical importance in predicting future spatial and temporal changes in the current As hazard.

In this study we interrogate the isotopic signature of ground and surface waters at a known As hotspot in Cambodia to (i) determine the residence times of groundwater prior to the onset of extensive groundwater pumping, and (ii) characterize and quantify contributions from different sources of dissolved organic carbon present in this aquifer. The chemical composition of the groundwaters at this site has been well characterized in previous studies (Polizzotto et al., 2008 and Lawson et al., 2013) and as such will not be discussed in detail here. This study focuses on the radiogenic and stable isotopes of hydrogen (tritium) and carbon ($\delta^{13}C$, ^{14}C) to determine groundwater residence times and provenance OM present in the aqueous phase. In addition, the stable isotopes of oxygen ($\delta^{18}O$) and hydrogen (δD) provide insights into the mode and extent of interaction between ground and surface waters at different locations within the field area. This characterization of pre-development conditions allow for a robust evaluation of the relative contributions of in-aquifer and surface derived OM where minimal groundwater-pumping-driven drawdown of surface waters prevail. We consider bulk DOC age as a proxy for the residual reactivity of organic matter because it provides an opportunity to account for the time-integrated impact of the reaction history and consumption it may have experienced. We suggest it is this history that likely controls the remaining potential of residual organic matter for utilization in metabolic processes that could drive As release. This in turn provides an insight into the likely controls on the rate and extent of As release under natural conditions. From this baseline control it will be possible to assess the vulnerability of these aquifer systems to anthropogenic disturbance. In particular, it will provide an opportunity to determine the impact of the extensive pumping of groundwater that is so prevalent in the Bengal basin on groundwater DOC composition and its distribution and ultimately how this influences As release rates to groundwater.

MATERIALS AND METHODS

The field area investigated in this study is situated in the Kandal province of Cambodia and covers an area of approximately 50 km^2 between the Mekong and Bassac rivers (Fig. 1). The site has been both geologically, hydrogeochemically and hydrogeologically well characterised by numerous previous studies (e.g. Polya et al., 2003, Benner et al., 2008, Kocar et al., 2008, Polizzotto et al., 2008, Lawson et al., 2013 and Sovann and Polya, 2014). Groundwater As concentrations have been reported to exceed 1000 μg/l (Polya et al., 2005, Kocar et al., 2008, Polizzotto et al., 2008 and Lawson et al., 2013), with As release shown to occur from near surface river-derived sediments underlying permanently saturated wetlands and abandoned river channels (Berg et al., 2007,Buschmann et al., 2007, Benner et al., 2008, Kocar et al., 2008, Kocar et al., 2014,Papacostas et al., 2008, Polizzotto et al., 2008, Rowland et al., 2008 and Stuckey et al., 2015a). The site exhibits a relatively simple hydrology; groundwater is recharged in the internal wetland basin and is transported on centennial timescales to the Mekong River (Benner et al., 2008 and Kocar et al., 2008). In the monsoon season groundwater flows from the rivers towards the wetlands, with groundwater flowing in the opposite direction for the remaining 8 months of the year (Benner et al., 2008). The simple hydrology at this site reflects natural conditions that would prevail in other As contaminated deltas in the absence of extensive groundwater pumping that otherwise results in the development of complex transient flow paths that make provenancing groundwater difficult (Polya et al., 2005 and Polya and Charlet, 2009).

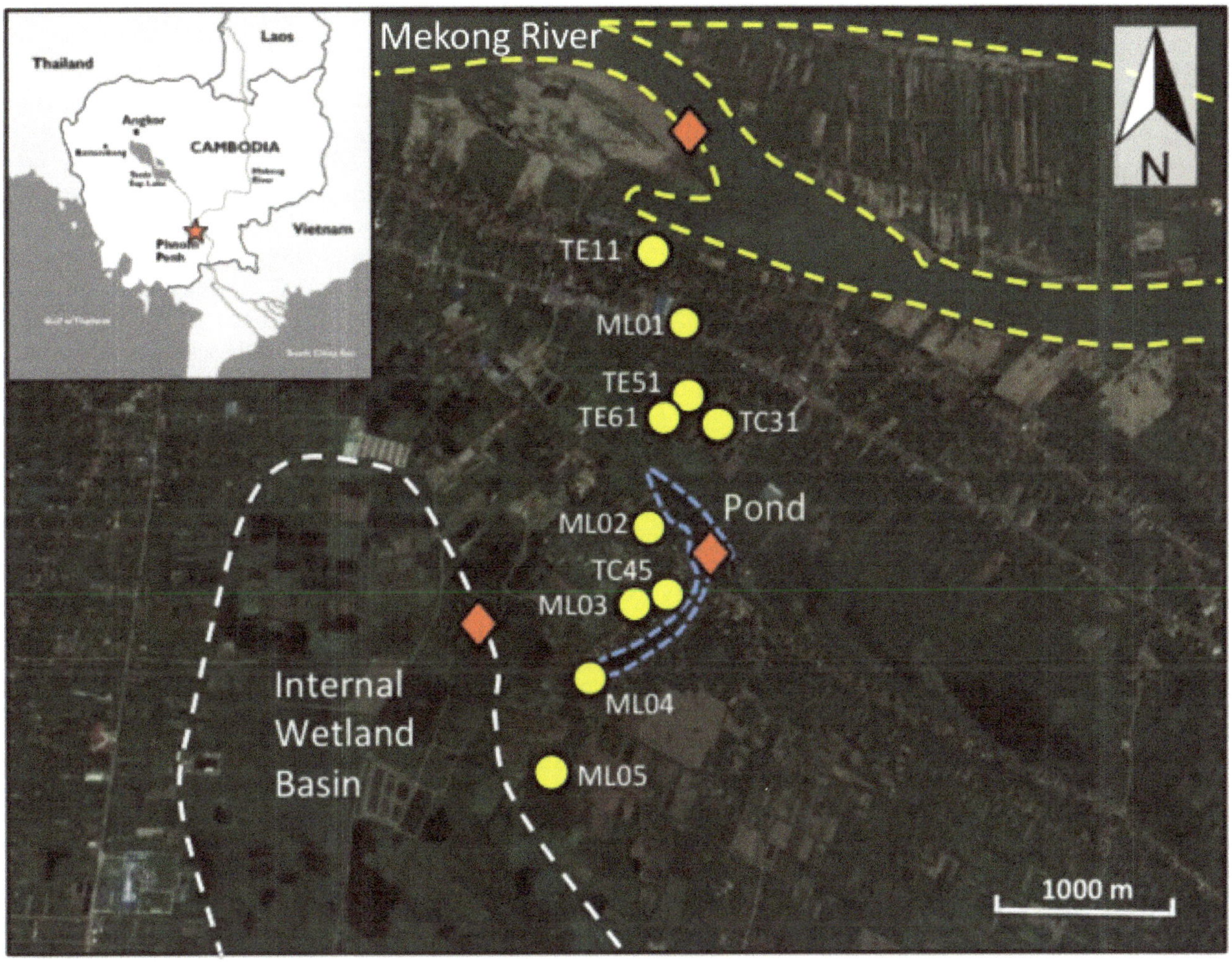

Figure 1: Aerial image of the field area located approximately 20 km SE of Phnom Penh, Cambodia. The tubewells sampled in this study (yellow circles) lie between the Mekong River (dashed yellow line) in the north and the internal wetland basin (white dashed line) in the south. Also shown are sampled surface waters (red diamonds) and the large pond in the center of the study (blue dashed line). (For interpretation of the references to colour in this figure legend, the reader is referred to the web version of this article.)

The site geology is characteristic of the floodplains of the lower Mekong basin, with elevated levees along the banks of the Mekong and Bassac rivers retreating to a seasonally saturated internal wetland basin that lies between the two rivers (Kocar et al., 2008). The aquifer sands that were targeted for sampling extend to depths in excess of 60 m, and comprise of young Holocene levee/overbank deposits of orange/grey clay containing laminations of peat and silt which overlay fine grey aquifer sands (Tamura et al., 2007 and Kocar et al., 2008). Solid phase As concentrations in these sediments vary considerably, with concentrations of 12 mg/kg associated with the surficial clay deposits decreasing to 2 mg/kg in the deeper aquifer sands (Kocar et al., 2008). Clay deposits range in thickness from 3 to 20 m but more commonly are between 10 and 14 m thick

(Kocar et al., 2008). The peat here is approximately 5 m thick and typically extends from a depth of approximately 6 m to 11 m (Tamura et al., 2009). This peat has been shown to host a concentrated zone of As enrichment relative to overlying sediments, with arsenic present in the form of arsenian pyrite (Stuckey et al., 2015b). The timing of this peat formation has been shown to be fairly consistent and comparable to that in the Bengal delta at 8.4–6.3 ka (McArthur et al., 2004 and Tamura et al., 2009). Older Pleistocene deposits, identified by the orange color of the sands and silts, occasionally underlie the Holocene sediments, and have been reported at depths as shallow as 20 m at this site (Tamura et al., 2009).

A set of twenty groundwater samples were collected from ten well nests (locations where multiple tubewells are present that produce from different depths) in March–April 2009 using standard protocols as described in Lawson et al. (2013). Samples cover a 2.6 km transect perpendicular to the Mekong River with a depth range of 7.3–60 m (Fig. 1). These nests and wells were selected to provide a comprehensive depth coverage and a lateral spatial resolution of no less than one groundwater sample per 0.5 km along the flow path from the wetlands to the Mekong River. Wells are screened over a 3ft interval and drilled as described in Lawson et al. (2013). Given the small producing interval, groundwater is sampled over a small and discrete depth interval and as such minimizes mixing of groundwater over multiple flow paths. Work by Richards et al. (2015) suggests residual contamination from drilling fluids is likely to be less than 3%. A subset of tubewells lying along this transect were also sampled on a monthly basis for stable isotope ($\delta^{18}O$ and δD) analysis to capture the seasonal variability in the hydrological cycle. This work is presented in Lawson et al. (2013) and demonstrated that while surface waters experience significant temporal variability in their isotopic composition, groundwater displays very limited seasonal variability at this study site. However, the chemical composition (including As concentrations) of groundwater has previously been shown to exhibit a high degree of temporal variability at this site (see supplementary information of Polizzotto et al., 2008). As yet there has been no systematic study of the controls on this observed temporal variability. While we recognize that the geochemical composition of groundwater may vary seasonally and that this is likely associated with variable recharge conditions driven by the monsoon and dry seasons, understanding the controls on this variability is beyond the scope of this manuscript and we do not consider this further in this study. A further four surface water samples were collected from the Mekong River and two surface waters locations (the central pond and the internal wetlands delineated in white on Fig. 1) over the course of this transect.

A detailed description of the protocols for geochemical and isotopic analysis is provided in Lawson et al. (2013). Briefly, As and other trace elements were measured with inductively coupled plasma–mass spectrometry and major element concentrations were determined by inductively coupled plasma–atomic emission spectroscopy. Anion analysis of water samples for Cl^-, Br^-, NO_2^-, NO_3^-, PO_4^{3-} and SO_4^{2-} was conducted using a Metrohm 761 compact Ion Chromatograph with a Dionex AS9 high capacity column and AG9 high capacity guard column. Stable isotope analysis was performed at the Isotope Community Support Facility, East Kilbride, Scotland using standard techniques. Data are reported as permil variations from the V-SMOW standard and calibrated against both SUERC internal standards (DSW(2), DW2(2) and EKS(2)) and international standards (GISP, V-SMOW and SLAP). $\delta^{13}C$ analysis was performed on a Prism3 (Fisons Instruments) mass spectrometer and calibrated with marble standards (MAB2, medium marble 85/160, $\delta^{13}C$ = 2.48‰), internal sodium bicarbonate ($\delta^{13}C$ −4.67‰) and calcium carbonate ($\delta^{13}C$ −24.23‰). Determination of the tritium content of groundwater samples was achieved using the ^{3}He-ingrowth technique developed byClarke et al. (1976), with measurements of ^{3}He and ^{4}He made at the University of Manchester on a MAP 215 Noble Gas Mass Spectrometer (Mass Analyser Products). Groundwater samples of acidified and sparged $DO^{14}C$, combusted to CO_2, and $DI^{14}C$ acid-hydrolysed to CO_2 were prepared to graphite (Slota et al., 1987). ^{14}C analysis was carried out at the SUERC AMS Laboratory, East Kilbride using either a 5MV tandem accelerator mass spectrometer (Xu et al., 2004) or 250 kV single stage accelerator mass spectrometer (Freeman et al., 2010), both machines by National Electrostatics Corporation, NEC, Wisconsin, US. The ^{14}C enrichment of a sample is measured as a percentage (or fraction) of the ^{14}C activity relative to a modern standard (oxalic acid provided by the US National Bureau of Standards), where 100% modern is defined as the value in AD 1950, in the absence of any anthropogenic influences. Overall analytical precision is quoted at 1σ.

RESULTS

Ground and surface water chemistry

Groundwaters are Ca–Mg–HCO_3 dominated (Table 1 and Table 2) typical of most As-bearing groundwaters in this region (Berg et al., 2001 and Rowland et al., 2008). Concentrations of dissolved As vary significantly over both lateral and vertical profiles at the study site, with a range from 2 to 1100 μg/l and a mean of 209 μg/l. Spatial and lateral variability in As and dissolved redox sensitive cations and anions is discussed in detail in previous studies at this site (Kocar et al., 2008, Polizzotto et al., 2008 and Lawson et al., 2013) and as such is not described in detail here. Nonetheless, we highlight here a region of high As concentrations over the whole depth profile within 1 km of the Mekong River and near the internal wetlands, 2.6 km from the Mekong River. Arsenic concentrations are lower (maximum 35 μg/l) in the centre of the study area and associated with the presence of a large pond (Kocar et al., 2008, Polizzotto et al., 2008 and Lawson et al., 2013).

Table 1: Physical and anion composition of groundwaters and surface waters collected from the study site in April 2009.

Sample	Depth	pH	Eh	Temp	Cl^-	NO_3^-	NH_4^+	ΣPO_4^{3-}	ΣSO_4^{2-}	Br^-	HCO_3^-
	(m)		(mV)	(°C)	mg/l	mg/l	mg/l	mg/l	mg/l	mg/l	mg/l
TE11-12	12	6.4	222	30.0	71	72	ND	ND	37	ND	500
TE11-20	20	6.8	−29	29.6	114	8.8	ND	ND	63	0.2	609
TE11-55	55	6.4	−73	29.8	3.1	1.1	30	0.4	0.2	0.1	511
ML01-9	9	6.1	−44	28.8	64	0.1	12	0.9	7.4	0.2	563
ML01-29	28.6	6.8	−44	30.2	7	ND	15	0.5	0.4	ND	410
ML01-37	37	7.0	−27	30.1	13	ND	15	0.4	0.5	ND	363
TE51-12	12	6.6	−30	34.0	14	0.5	57	0.2	0.6	0.2	465
TE51-20	20	6.9	−116	30.7	5.5	0.1	22	1.2	<0.05	0.3	404
TC31	51	6.8	−76	30.1	5.2	7	50	0.5	0.1	1.1	659
TE61	48	6.3	−100	30.2	4.8	ND	80	ND	ND	0.7	757
ML02	8.5	6.5	−163	29.9	61	ND	0.1	0.5	12	ND	444
TC45-12	12	6.3	−117	31.1	325	0.4	40	3.4	26	0.5	817
TC45-20	20	6.4	135	31.1	479	56	0.5	ND	146	0.7	631
ML03-7	7.3	6.4	−47	29.1	45	0.4	ND	0.3	26	ND	292
ML03-24	23.8	6.1	−70	29.2	30	57	0.3	2.4	0.7	ND	123
ML03-42	42	6.7	ND	30.0	1520	0.3	7.0	1.7	3.3	<0.05	304
ML04-8	8.2	6.4	14	30.4	56	11	ND	<0.05	25	0.2	689
ML04-30	29.1	7.1	60	31.4	21	2.4	ND	ND	2.8	<0.05	187
ML04-44	44	6.9	179	31.2	22	7	ND	ND	4.6	ND	169
ML05	60	6.5	−36	33.2	3.7	20	ND	0.9	0.2	ND	244
Wetlands			ND	ND	26	0.1	1.8	0.4	3	ND	170
Pond			ND	ND	20	0.5	ND	ND	0.8	0.05	176
River Mekong			ND	ND	5.4	1.1	ND	0.2	5.1	ND	62

ND = not determined.

Table 2: Major and minor cation analysis of groundwater samples collected from study site during April 2009.

Sample	DOC	As	Fe	Mn	Ca	K	Mg	Na
	mg/l	ug/l	mg/l	µg/l	mg/l	mg/l	mg/l	mg/l
TE11-12	0.8	6	0.0	109	127.4	1.7	41.6	45.8
TE11-20	1.1	346	14.6	295	135.2	2.2	43.9	100.1
TE11-55	5.1	829	6.7	82	61.2	2.8	16.8	67.7
ML01-9	1.9	212	17.3	915	95.1	2.4	29.7	49.3
ML01-29	6.8	217	2.2	799	45.4	4.6	9.5	27.6
ML01-37	6.5	258	6.2	12,760	40.6	3.6	8.4	19.4
TE51-12	3.8	418	13.1	677	95.5	3.2	26.0	21.4
TE51-20	10.9	1100	2.2	10,187	44.2	7.2	10.7	35.6
TC31	15.3	46	0.0	469	86.1	9.3	20.4	55.9
TE61	11.6	44	0.2	106	76.6	7.3	28.4	29.9
ML02	6.8	4	1.2	474	38.0	1.5	10.7	75.5
TC45-12	4.9	3	3.0	262	194.6	2.0	68.5	138.2
TC45-20	2.5	2	0.4	189	240.8	0.6	92.1	165.4
ML03-7	3.7	10	0.9	11,581	34.7	1.2	15.0	56.6
ML03-24	6.5	35	15.3	477	35.7	2.4	10.1	35.8
ML03-42	63.4	32	0.0	511	52.4	3.0	14.1	58.3
ML04-8	2.2	9	7.8	533	135.4	0.7	42.3	198.3
ML04-30	6.2	13	0.2	88	37.5	5.1	10.2	21.3
ML04-44	6.4	30	15.0	367	42.6	5.0	12.0	21.0
ML05	3.1	778	3.5	323	42.3	1.7	14.9	30.0
Wetlands	6.7	10.2	0.2	207	32.38	5.84	10.14	21.01
Pond	6.1	3.3	0.2	150	31.92	4.38	10.20	19.47
Mekong	2.4	1.2	0.1	0.8	7.3	1.5	2.4	5.7

Stable isotopic composition of ground and surface waters

The connectivity and degree of interaction between groundwater and surface waters was investigated by comparing the isotopic signature (δD and $\delta^{18}O$) of groundwaters from the study site with that of Mekong River, wetland and pond waters located at different points along the transect. A cumulative monthly sample of precipitation was also analysed for the development of a working local meteoric water line (LMWL) from which deviations in the isotopic signature of ground and surface water can be determined. Fig. 2 shows the variation in the isotopic composition of local rainwater, groundwater, Mekong river and surface waters. The LMWL has a gradient of 8.3, which is similar to that of the global meteoric water line (GMWL) of 8.17 (Rozanski et al., 1993) and within the range of gradients (7.95–9.11) determined by Kabeya et al. (2007) for the nearby Kampong Thom province – as such we consider it a reasonable working LMWL for this site. The D excess of the local precipitation of between 8.1 and 17.6 (given here relative to a gradient of 8) is indicative of a contribution of re-evaporated local surface waters to the precipitation, which is typical for such a humid climate (Clark and Fritz, 2000).

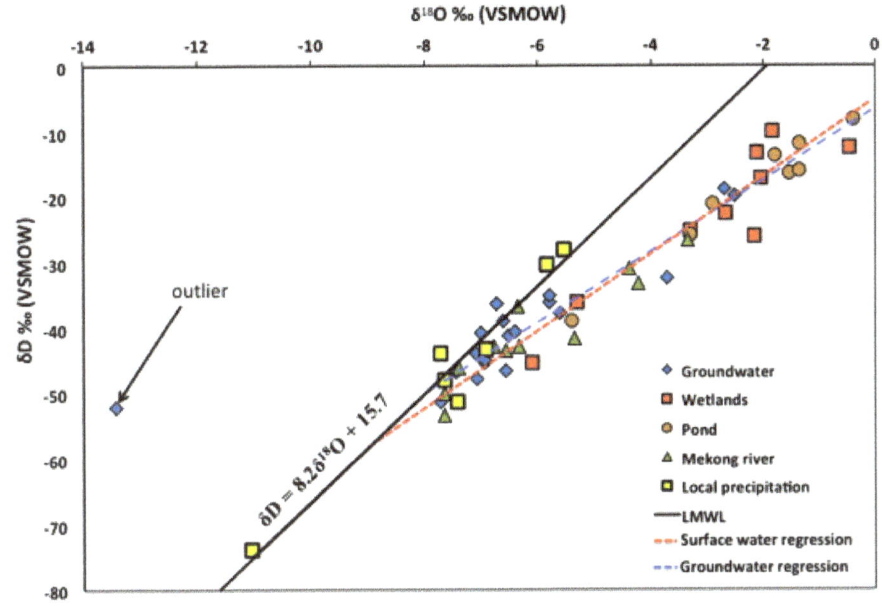

Figure 2: δD vs $\delta^{18}O$ plot of groundwaters, surface waters and precipitation at the study site in the Kien Svay region, Kandal Province, Cambodia. Groundwater samples were collected in April 2009, with surface waters and precipitation (n = 7) collected approximately monthly over the year. Regression lines are given for surface waters (slope of 6.1) and groundwaters (slope of 5.4). The groundwater with measured $\delta^{18}O <$ −13‰ is an outlier (Dixon's Q test; 99% confidence interval) and excluded from the derivation of the groundwater regression line.

Groundwater samples typically fall on or below the LMWL, with a trend to more isotopically enriched values characteristic of surface waters signatures and indicative of contributions from an evaporated source of recharge to the groundwater. The similarity in the gradients of the regression lines between groundwaters (gradient = 5.35) and the wetland and ponds (gradient = 6.1) suggests that surface waters are the most likely source of isotopically enriched recharge to groundwater. These regression lines are calculated by considering the highly variable monthly isotopic composition as measured in Lawson et al. (2013). These gradients are also characteristic of evaporative trends for theoretical evaporation at 70–85% humidity, in line with the range of average monthly humidities for this location.

Tritium concentrations and residence times of ground and surface waters

The tritium content in precipitation from the nearest IAEA GNIP station at Bangkok from 1953 to 2009 is presented in Fig. 3 below. Here it can be seen that the present day tritium content is approximately 2–4 tritium units (TU), and has remained below 6 TU since 1984. We discuss here tritium concentrations previously measured in the study ofLawson et al. (2013). The tritium content of the Mekong River of 3.5 ± 0.2 TU presented inTable 3 is thus consistent with current levels in precipitation and provides a constraint on the initial tritium content of recharging waters.

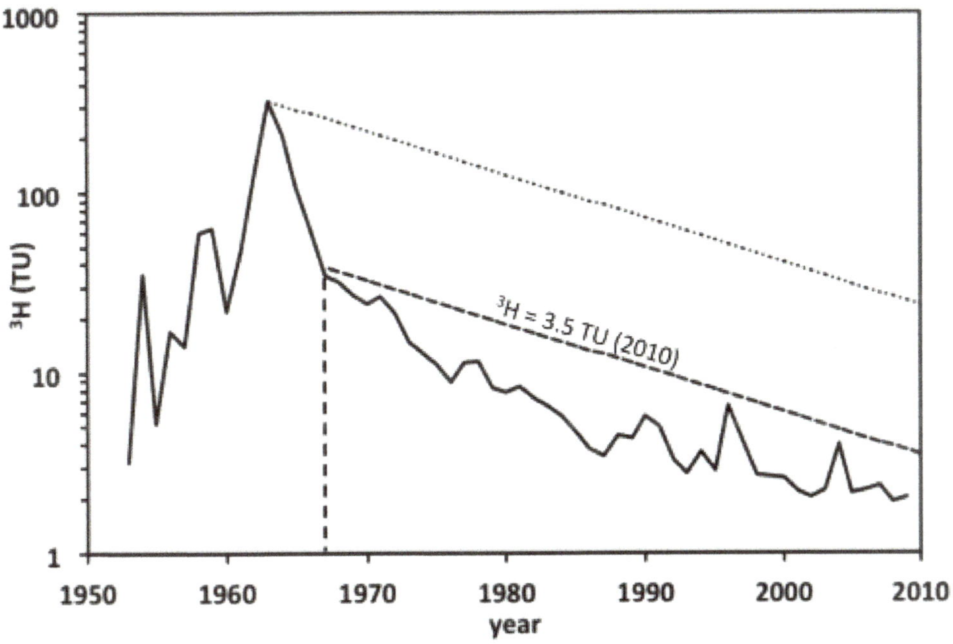

Figure 3: ^{3}H concentration in precipitation. Data for the period 1969–2009 taken from the IAEA GNIP database for the nearest monitoring station at Bangkok (IAEA/WMO, 2006). Data for the period 1953–1968 was calculated from ^{3}H concentration data presented in Stute (2001),

using a correction factor of 1.84. The dashed line represents a compositional trajectory for samples containing 3.5 TU in 2010, and intercepts the tritium in precipitation profile in 1967. Also shown for reference is the decay of bomb peak tritium. Should this be present this would manifest as 2010 tritium concentrations of 23.6 TU, exceeding measured concentrations by an order of magnitude.

Table 3: Summary of ^{3}H concentrations for groundwaters and surface waters and uncorrected and modeled radiocarbon of DIC based mean groundwater residence times. Errors are given as 1σ of the tritium concentration. Tritium ages are given for two tritium input functions. * assumes a 3.5 TU (present day) and # assumes a 38 TU (maximum historic tritium concentration in precipitation observed in 1967) input function. Modeled values could not be determined for locations TE61 and ML03-40 because key input parameters were shown to lie more than 2σ from mean values, and as such considered to be outliers.

Sample	$\delta^{13}C$ (‰)	^{14}C (pmC)	±σ	^{3}H (TU)	±σ	^{3}H age* (years)	^{3}H age# (years)	Uncorrected DIC Age	Modeled Age
TE11-12	−13.1	100.54	0.46	2.53	0.14	5.8	>43	>0	>0
TE11-20	−15.2	102.62	0.41	2.01	0.11	9.9	>43	>0	>0
TE11-55	−15.8	60.71	0.28	0.00		>60	>43	4126	3053
ML01-10	−16.0	101.49	0.41	1.35	0.08	17	>43	>0	>0
ML01-30	−14.1	74.06	0.31	0.00		>60	>43	2483	457
ML01-40	−12.3	72.23	0.34	0.00		>60	>43	2689	>0
TE51-12	−8.7	99.71	0.42	2.22	0.12	8.2	>43	24	>0
TE51-20	−11.3	78.26	0.39	0.00		>60	>43	2026	>0
TC31-51	−17.6	85.7	0.41	0.00		>60	>43	1275	1078
TE61-48	1.1	89.03	0.39	0.00		>60	>43	961	DNC
ML02-8	−17.8	61.03	0.33	1.55	0.08	14.6	>43	4082	3966
TC45-12	−14.9	84.95	0.31	1.91	0.11	10.9	>43	1348	>0
TC45-20	−15.0	96.78	0.41	1.89	0.10	11	>43	271	>0
ML03-10	−14.5	100.77	0.42	1.78	0.09	12.1	>43	>0	>0
ML03-20	−7.7	66.93	0.38	1.63	0.09	13.7	>43	3319	>0
ML03-40	−3.4	41.51	0.37	2.01	0.10	10	>43	7268	DNC
ML04-8	−16.3	81.59	0.38	2.06	0.11	9.5	>43	1682	839
ML04-30	−19.6	93.16	0.49	2.65	0.14	5	>43	586	1259
ML04-44	−13.3	94.73	0.46	2.24	0.12	8	>43	448	>0
ML05-60	−14.4	74.63	0.28	0.00		>60	>43	2419	581
Wetlands				1.91	0.11				
Pond				2.03	0.11				
Mekong River				3.50	0.18				

Tritium concentrations in the groundwater of less than 3 TU suggest that infiltration took place after the cessation of atmospheric bomb testing, which would otherwise exceed 10 TU in low latitude regions when residual "bomb" tritium is present (Clark and Fritz, 2000). This is because the natural decay of tritium in groundwater is significantly slower than the decrease in tritium concentrations in precipitation due to the preferential retention of the isotopically heavier tritium-bearing isotopologue of H_2O in the oceans. If we assume no hydrodynamic dispersion or mixing, which is reasonable given that groundwater flow at this site has been minimally disturbed by groundwater pumping, then estimates of groundwater mean residence times (MRT) can be made from measured tritium concentrations and assumed tritium input functions through the following equation:

$$t = -17.93 \ln \frac{a_t{}^3H}{a_o{}^3H} \quad (1)$$

where t is the mean residence time, $a_t{}^3H$ is the residual tritium activity of the sample, $a_o{}^3H$ is the tritium input function, and −17.93 is the unitless constant to account for the decay of tritium to ^3He (Clark and Fritz, 2000). We discuss our data in the context of mean residence times, and not groundwater ages, because groundwaters very rarely comprise water of a single age. Instead, they more frequently comprise mixtures of waters along flow paths of different ages. Table 3 presents tritium concentrations for each of the wells and surface waters sampled, and provides MRTs based on two tritium input functions that allow us to bracket the possible MRTs at each location. A tritium input function of 3.5 TU was chosen to represent the tritium concentration of present day precipitation at the study site based on concentrations of tritium in the Mekong River. A second tritium input function of 38.5 TU was selected to represent the largest possible input function for groundwaters that recharged after the cessation of thermonuclear bomb testing and is based upon the tritium concentration in precipitation which fell in 1967 (Fig. 3). This was determined by back tracking along a decay line from the present day concentration of 3.5 TU in Mekong River water to the two points at which the decay line intercepts the tritium concentration in precipitation profile. The upper profile Fig. 3 represents the temporal evolution in groundwater tritium concentrations should the bomb peak be present in our samples. The predicted tritium concentration of 23.4 TU in 2010 is over an order of magnitude greater than almost all measured concentrations, which we consider as evidence for the absence of bomb tritium in our samples. Given the absence of peak bomb tritium in our samples after accounting for natural decay, we do not consider the second point at which the decay curve intercepts the tritium concentration in precipitation profile in 1958 as a likely maximum age for tritium-active samples. As such, the maximum possible MRT of any tritium-active sample in this study site cannot exceed 43 years (assuming recharge in 1967 and sample analysis in 2010). However, given that independent sampling of ^3He and noble gases was not possible for this study site, we are not able to determine more precise ages of these groundwaters.

Groundwater was shown to contain an active contribution of modern (<43 years) recharge in 13 of the 20 wells sampled at this study site, and reaches depths of up to 44 m at some locations. The tritium concentrations for tritium-active groundwaters range from 2.65 ± 0.14 TU at site ML04 at a depth of 30 m to 1.35 ± 0.08 TU at site ML01 at a depth of 9 m. This gives a range in groundwater MRT of 5 ± 0.27 years to 17 ± 0.94 years when the 3.5 TU tritium input function we define here is applied and represents the minimum likely MRT of these groundwaters. All tritium-active groundwaters give a MRT greater than 43 years when the 38.5 TU tritium input function is applied, placing a consistent upper limit for tritium-active samples. Groundwater at depths less than 12 m displayed contributions of modern recharge over the entire site, with a range in concentrations of 1.35 ± 0.08 TU at site ML01 to 2.53 ± 0.14 TU at site TE11. Of the 6 intermediate depth (20–30 m) groundwaters sampled, 4 were found to be tritium-active with tritium concentrations of 1.63 ± 0.09 TU–2.65 ± 0.14 TU, representing minimum MRTs of 13.7 ± 0.74 years and 5 ± 0.27 years respectively. In contrast, only 2 of the 7 deep (>30 m) groundwater samples were found to be tritium-active, with concentrations of 2.01 ± 0.10 and 2.24 ± 0.12 TU reported for sites ML03 and ML04 at depths of 42 and 44 m respectively. Tritium dead groundwaters (below the minimum detectable limit of 0.2 TU) were found from depths of 20 m at site TE51 to the maximum depth of sampling at 60 m at site ML05.

The variability in MRTs with depth gives rise to a spatial distribution that allows identification of the locations of rapid recharge through high permeability sediments to underlying groundwater at this site. The presence of tritium-active groundwater suggests an active component of surface water recharge along the entire 2.6 km profile. Groundwater was shown to be tritium-active in both the shallow and intermediate depths at site TE11. This is consistent with stable isotope data and previous hydraulic head measurements and monitoring of Mekong River water levels that suggest that Mekong River water is ingressing into the aquifer at this location (Benner et al., 2008 and Lawson et al., 2013). Three well nests in the centre of the study area report tritium-active groundwater over the entire depth profile (TC45, ML03 and ML04). These groundwaters are all located more than 1.4 km from the Mekong River and as such are outside of the influence of ingressing Mekong River water. Stable isotope data at these locations suggest a possible influence from recharge from the large pond located in close proximity to these well nests (Lawson et al., 2013).

Arsenic concentrations are typically lower in tritium-active groundwaters compared with tritium-dead groundwaters. The concentration of As exceeded 10 μg/l in 7 of the 13 tritium-active groundwaters Fig. 4), with a range of 2–418 μg/l and a mean As concentration of 86.2 μg/l. In contrast, all of the tritium-dead groundwaters contain concentrations of As that exceed 10 μg/l, ranging from 44 to 1100 μg/l and with a mean As concentration of 467 μg/l. Interestingly, three groundwater samples were found to be tritium-active and have As concentrations that exceed 100 μg/l. Sample TE11-20, which has an As concentration of 346 μg/l, appears to be located within the sphere of influence of ingressing Mekong River water during the rainy season. Groundwater at this location has been shown to flow from the internal wetlands to the Mekong River over centennial timescales. The modern tritium signature in groundwaters at this location may arise as a result of mixing between older, high-As groundwater that has migrated from the wetland basin with tritium-active Mekong River water. Similarly, locations ML01-10 and TE51-

12 are located in the zone of high As concentrations within 1 km of the Mekong River but are also tritium-active. This may either reflect hydrologic transport of As from upstream into groundwater that has an active component of modern recharge from above or a rapid removal of oxidants from the system during recharge thus developing the reducing conditions that are required for As mobilisation.

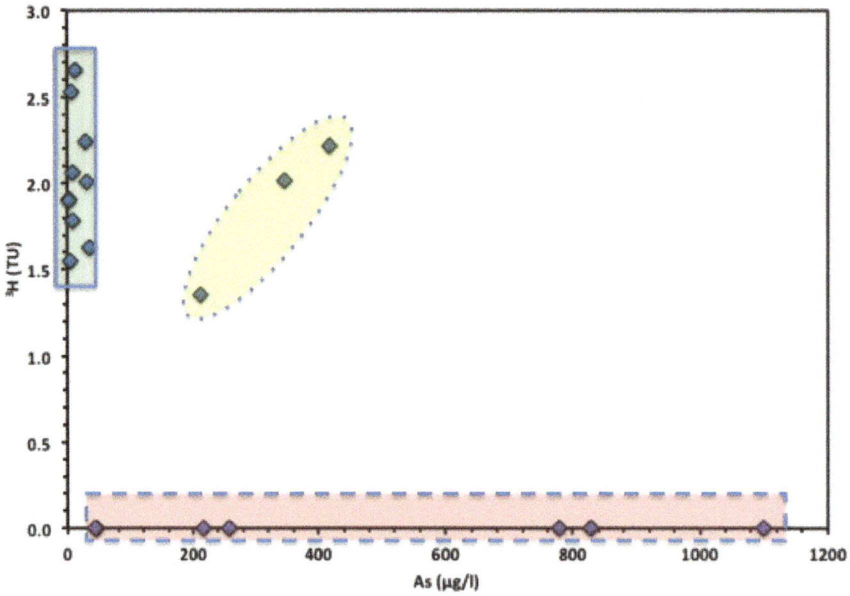

Figure 4: Distinct groups of groundwaters discriminated by ^3H and dissolved As. (i) tritium-active groundwaters with As = 2–35 μg/l (highlighted by green rectangle demarked by solid lines), (ii) tritium-active groundwaters with As > 200 μg/l (highlighted by yellow oval and demarked by dotted lines), and (iii) tritium-dead groundwaters with highly variable (44–1100 μg/l) As (highlighted by red rectangle and demarked by dashed lines). (For interpretation of the references to colour in this figure legend, the reader is referred to the web version of this article.)

Dissolved inorganic carbon radiocarbon concentrations and groundwater residence times

We present here uncorrected radiocarbon ages that are calculated from the measured concentration of ^{14}C of dissolved inorganic carbon (DIC) assuming exponential decay of^{14}C in DIC as a function of its half-life. A detailed description of radiocarbon dating is outside of the scope of this study, and we refer the reader to Clark and Fritz (2000) for details. It is widely accepted that uncorrected ^{14}C groundwater ages often appear to over predict the true age of the sample. This is because contributions of ^{14}C-dead inorganic carbon associated with processes such as carbonate dissolution result in the dilution of^{14}C activity in the sample. In this study, we applied the correction model of Clark and Fritz (2000) to correct measured radiocarbon ages and determine MRTs at this study site for groundwaters older than the 60 years covered by the tritium based technique discussed above. This model utilizes isotopic measurements to calculate the dilution from ^{14}C-dead inorganic carbon given that any process that adds, removes or exchanges carbon with the DIC reservoir would likely alter both the ^{14}C activity of the groundwater and the $\delta^{13}C$ signature of DIC. This model accounts for isotope evolution in the saturated zone by dissolution and two component mixing between the constant $\delta^{13}C$ values of soil CO_2 and inorganic carbon from carbonate dissolution. In addition to this, the model accounts for the variable enrichment of the $\delta^{13}C$ signature that arises because of these reactions and equilibration between $CO_2(g)$ and the dissolved carbonate species in the unsaturated zone. In this study, a $\delta^{13}C$ of −22.9‰ is assumed to be representative of the initial soil CO_2 reservoir. This represents the mean value of $\delta^{13}C$ of dissolved organic carbon determined for these groundwaters.

The uncorrected ^{14}C ages of DIC range from modern (values > 100% modern i.e. containing post AD1950 carbon) at sites TE-11, ML01 and ML03 to 7268 ± 24 years BP for groundwater at ML03 at a depth of 50 m (Table 3). In contrast, modelled groundwater residence times range from modern in 11 of the groundwaters sampled to 3966 ± 27 years BP at site ML02 at a depth of 8 m. The ^{14}C activity does not show any consistent trend along the transect but does decrease with depth in four of the six well nests, suggesting an increase in age and/or contributions of ^{14}C-dead DIC with depth.

Radiocarbon ages of dissolved organic carbon

The radiocarbon age of DOC varies significantly in the groundwater at this study site, with a range of ^{14}C DOC ages from 4036 ± 38 years at site TE11 at a depth of 55 m to modern at depths of 30 and 44 m at site ML04 (see Table 4). Radiocarbon ages of DOC at this site have previously been shown to be almost exclusively younger than sedimentary organic matter (SOM) for equivalent depths (Tamura et al., 2009 and Lawson et al., 2013).

Table 4: Carbon isotope signatures ($\delta^{13}C$ and ^{14}C) and derived C-14 ages for dissolved organic carbon in groundwater at the study site. Note that samples MLO4-30 and ML04-44 contain modern organic carbon signatures, indicating the dominance of surface derived organic carbon in DOC at these depths. ND for tubewell ML04-8 denotes not determined – insufficient sample material.

Tubewell	Depth	C-14 enrichment	+/–	C-14 age	+/–	$\delta^{13}C$
	(m)	(% modern)		(yrs BP)		(‰)
TE11-12	12	99.07	0.46	75	37	–5.3
TE11-20	21	87.94	0.41	1032	37	–18.3
TE11-55	55	60.50	0.28	4036	38	–18.5
ML01-9	9	93.83	0.41	512	35	–13.4
ML01-29	29	71.65	0.31	2679	35	–25.0
ML01-37	37	72.57	0.34	2575	37	–26.5
TE51-12	12	91.08	0.42	751	37	–27.3
TE51-20	20	85.47	0.39	1261	37	–25.7
TC31	51	89.99	0.41	847	36	–26.6
TE61	48	89.61	0.39	881	35	–26.6
ML02	9	75.06	0.33	2304	35	–26.2
TC45-12	12	71.03	0.31	2748	35	–19.1
TC45-20	20	93.75	0.41	519	35	–21.6
ML03-7	7	91.45	0.42	718	37	–21.6
ML03-24	24	87.88	0.38	1038	35	–26.9
ML03-42	42	83.52	0.37	1447	35	–24.4
ML04-8	8	ND	ND	ND	ND	ND
ML04-30	29	105.38	0.49	<50	50	–29.3
ML04-44	44	100.71	0.46	<50	50	–26.2
ML05	60	64.80	0.28	3485	35	–26.2

A depth dependent variation in the rate of As release per year increase in the age of groundwater DOC is observed from the comparison of shallow and deep groundwaters that are tritium dead or contain As concentrations above 100 μg/l when tritium is present (Fig. 5). Shallow groundwaters (<20 m) accrue As at a rate of approximately 1 μg/l for every year increase in associated DOC age compared to approximately 0.25 μg/l for every year increase in DOC age in deep groundwater (20–60 m). This data suggests that older DOC in deeper groundwaters is also able to support As release within the aquifer matrix, albeit at a lower rate of accumulation relative to the age of the associated DOC. This observation is consistent with our assumption that the age of DOC may be an appropriate proxy for organic carbon reactivity. These depth categories were chosen to reflect recharge through shallow surficial sediments, potentially dominated by clays that extend to depths of up to 20 m at this site (Kocar et al., 2008), and deeper aquifer sands. We focus here only on high arsenic or tritium dead waters in this assessment becauseLawson et al. (2013) have previously shown that the low-As groundwater underlying the pond in central region of our study area is dominated by contributions of young pond derived recharge over its entire depth profile. While this recharge introduces modern surface derived DOC, it also has been shown to introduce oxic waters, giving rise to groundwater sulfate concentrations in excess of 200 mg/l (Lawson et al., 2013). As a result, we suggest that the relationship between the age of DOC and As concentration in this region of the site is decoupled because of the impact of oxidants that immobilize As through sorption to Fe(III) phases and likely consume the co-transported surface derived DOC. In contrast, where oxic waters are not present anaerobic processes dominate the consumption of DOC. This may ultimately lead to As release from sediments and may support its continued mobility in the genesis of high-As groundwaters.

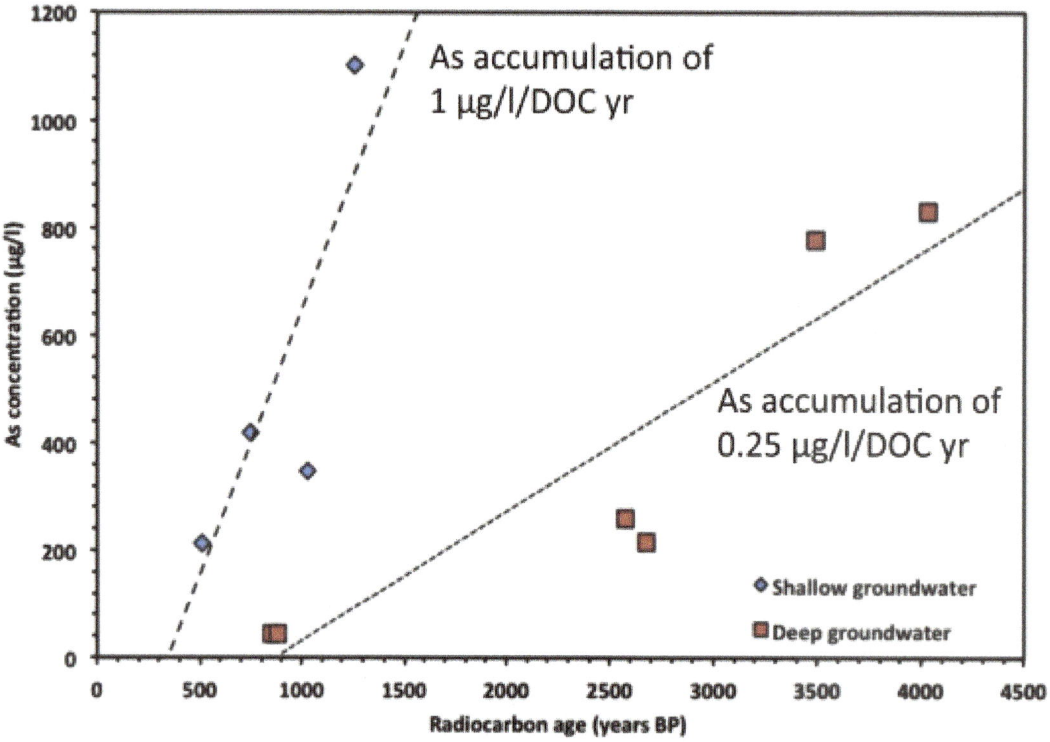

Figure 5: Contrasting release rate (with respect to DOC radiocarbon age) of arsenic in (i) shallow (9–20 m) and (ii) deep (20–60 m) groundwaters, where As concentrations exceeds 10 µg/l and the radiocarbon age is not modern. As concentrations in shallow groundwaters are shown to accumulate at a rate that is four times greater than that in deep groundwaters (~1 µg/l/DOC yr vs ~0.25 µg/l/DOC yr).

DISCUSSION

The unresolved debate as to what source of OM is responsible for driving As release remains one of the few gaps in the understanding of the As release process that limits the ability to predict how, when and where the current As hazard may change in the groundwaters of South and South East Asia. Site-specific variability in the sedimentary profile and the occurrence of high As groundwaters in locations where groundwater pumping is not common practice have led some to suggest that surface (e.g. Sengupta et al., 2008 and McArthur et al., 2011) or sedimentary sources (e.g. Mailloux et al., 2013) of OM cannot be responsible for the generation of high As groundwaters at these sites. However, the lateral transport of OM from subsurface sources and transport of modern surface derived OM could give rise to a pre-development groundwater DOC pool that is made up of contributions from a variety of sources of OM, with both surface and subsurface sources present. If this is the case, the relative contributions of these different sources could control As release rates in these groundwaters (Gault et al., 2005,Rowland et al., 2007, van Dongen et al., 2008, Mladenov et al., 2010 and Fendorf et al., 2010). The potential for groundwater pumping practices to change the relative contributions from these different OM reservoirs (e.g. Harvey et al., 2002) must also be considered.

Hydrological environment

Previous studies at this site have demonstrated significant ground–surface water interaction (Benner et al., 2008 and Lawson et al., 2013). Lawson et al. (2013) show that shallow (<12 m) groundwater is tritium-active over the entire site. This observation conflicts with modelled travel times of 200 years for the migration of recharge waters to depths in excess of 10 m at this site (Benner et al., 2008). This suggests that recharge along natural, rapid flow paths takes place from surface waters or through highly permeable clay windows in direct hydraulic continuity with the groundwater, and that these contributions dominate the shallow groundwater recharge budget at this site. Such conduits have recently been proposed to develop through the burrowing action of crabs providing widespread direct recharge conduits through low permeability surficial sediments to underlying aquifer sands (Stahl et al., 2014).

Of particular importance from previous studies are the observations that there are significant contributions of modern recharge from the large pond to groundwater in the center of the study site 1–2.2 km from the Mekong River (Benner et al., 2008 and Lawson et al., 2013).

Modelled residence times of DIC are internally consistent with residence times based on tritium measurements (i.e. provide modern ^{14}C based residence times where tritium is present and older residence times where tritium is absent) in thirteen of the eighteen groundwaters where a correction for dilution of ^{14}C was possible. Interestingly, four shallow groundwater samples (TE11-12, TE11-20, ML01-10 and ML03-10) containted^{14}C values indicative of recharge during the last 50 years. The observation of modern DIC in groundwater at depths of 12 and 20 m at site TE11 further supports the previous inference that these locations are influenced by ingressing Mekong river water (Lawson et al., 2013). In contrast, the modern DIC at ML01 in shallow groundwater provides additional evidence for the requirement of rapid recharge through surficial clays, with ML01 not located within close proximity to any surface water body.

The groundwater residence time at TE11 at a depth of 55 m of 3053 years in tritium dead groundwater likely represents the upper limit for mean residence times at this study site given that the groundwater at ML02, despite having an older apparent modelled radiocarbon based residence time, is tritium active. The MRT of TE11 at a depth of 55 m is, to first order, consistent with modelling estimates of groundwater age at this location that predict centennial timescales for groundwater flow from the wetland basin to the Mekong River (Benner et al., 2008). Similarly, our modelled MRT of 581 years in tritium dead groundwater at a depth of 60 m at ML05 is in line with age estimates in excess of 280 years for recharge through surficial clay deposits near the wetland basin (Benner et al., 2008).

There are three samples (ML02-8, ML04-8 and ML04-20) where radiocarbon based MRTs exceed constraints provided by co-existing tritium in these groundwaters. We suggest that an alternative source(s) of diluting DIC is made to the groundwater at these sites that is not considered in our age-correction model. Bethke and Johnson (2002)show that mixing of old waters that have been trapped in aquitards with younger groundwater in aquifers occurs at a rate that is controlled only by the volume ratio of fluid in aquitards relative to fluid in aquifers, and is independent of the actual rate of the contribution (a function of permeability). Given that these samples are taken from shallow tubewells that clearly contain significant contributions of young recharge, a large contribution of older water must be sourced from the extensive clay deposits within this vicinity.

This mixing of modern recharge through high permeability sediments and fluids that slowly migrate through clay deposits could have significant implications for the release and spatial distribution of As in the groundwater at this location. Clays have been suggested to be a possible source of arsenic to groundwater at this site and elsewhere regionally (Erban et al., 2013), and as such one may anticipate high As concentrations associated with mixed fluids that contain a component of recharge from these low permeability sediments. However, the groundwater in this region of the transect contains some of the lowest As concentrations measured at this study site over the entire depth profile. These results are more consistent with modelling results from the recent study ofKocar et al. (2014) who suggest that concentrations of As are likely to be low in groundwaters where contributions of high As recharged from low permeability sediment are overwhelmed by recharge through nearby high permeability sediments.

Interestingly, the study of Polizzotto et al. (2008) measured highly variable concentrations of As at site TC45-20, also located within 100 m of this pond, with both relatively high (99 μg/l) and low (9 μg/l) concentrations reported. These are higher than the concentration of As we measured for this location (2 μg/l). Nonetheless, these results broadly support the findings of the recent study by Stuckey et al. (2015a) who suggest that groundwater underlying seasonally saturated water bodies is likely to be characterized by relatively lower concentrations of As because it experiences periods where oxic recharge drives aerobic metabolisms that consume reactive DOC. This process reduces the reactivity of residual DOC for anaerobic processes that may develop deeper in the sediment, rendering it less capable of driving extensive As release.

In contrast to the low As concentrations associated with tritium active groundwater in the center of the field site, we observe high concentrations of As where radiocarbon-based MRT estimates are consistent with the absence of tritium in groundwater (Lawson et al., 2013). The relationship we observe between low As concentrations in tritium-active groundwaters and higher concentrations of As in older groundwaters at similar depths suggests that the mode and timescales of recharge plays a significant role in the genesis of high As groundwaters. Older groundwater at the same depth as young groundwater elsewhere requires large differences in the vertical permeability profile of sediments spatially. Low permeability sediments have previously been shown to act as a barrier to oxidants and lead to the development of reducing conditions (van Geen et al., 2006). This control on As distribution has also been identified in numerous other studies that link high As groundwaters with geomorphologic and surface features such as abandoned or existing channels, swamps and ox bow lakes (Nath et al., 2005 and Acharyya and Shah, 2007). These features have been shown to be areas where the rapid deposition of OM occurs, the decomposition of which facilitates the release of As to groundwaters (Papacostas et al., 2008 and Quicksall et al., 2008).

Organic matter sources to groundwater

The radiocarbon age of DOC, while consistently younger than SOM at similar depths (Lawson et al., 2013), is typically older than co-existing groundwater at this study site. This is consistent with DOC concentration depth profiles that increase with depth and requires a subsurface, sedimentary contribution of DOC to the groundwater DOC pool. This requirement for a contribution from a sedimentary organic carbon pool supports the experimental work of Neumann et al. (2014) who suggest that perturbations to the aquifer matrix could promote the mobilization of bioavailable organic carbon from aquifer sediments. C-14 measurements of the sediment from three drill cores at this site place an upper and lower age limit of 8.4–6.3 kyrs on peat that are present from 6 m to 14 m (Tamura et al., 2009), with deeper sediments shown to contain SOM with C-14 ages of up to 9.0 kyrs (Lawson et al., 2013). Based on these constraints, it has been shown that radiocarbon ages of DOC appear consistent with mixing of 1 mg/l of modern (age corrected) OM with sedimentary derived OM with ages of between 1000 and 6000 years (Lawson et al., 2013).

Given that groundwater DOC ages can be explained by mixing of age corrected modern surface OM and sedimentary OM between 1 and 6 kyrs, a maximum estimate of the relative contributions of surface derived OM and sedimentary derived OM can be calculated (Table 5). From this it can be seen that as the age of the sedimentary OM end member increases, the contributions of surface derived modern OM must also increase to produce the observed DOC age. Similarly, where modern DOC is present, the majority of the DOC must derive from modern surface derived OM. Shallow groundwater is shown to frequently contain a large component of young surface derived OM, with the relative contributions of this young OM typically decreasing with increasing depth. These calculations suggest that sedimentary organic matter can comprise anything from 95% of the total DOC pool at site TC45 at a depth of 12 m to very small relative proportions of DOC with a modern signature at site ML04 at depths of 30 and 44 m.

Table 5: Percentage contributions of sedimentary OM and modern surface derived OM (corrected for decay based on ^{14}C and ^{3}H ages of the groundwater). Scenario A, B and C represents two-component mixing of modern OM with sedimentary OM that is 1000, 3000 and 6000 years old respectively. Where values are not given the age of the sedimentary OM that is mixed is younger than the age of the sample and the model does not converge. ND denotes not determined as a result of insufficient OC in the sample.

Tubewell	C-14 enrichment	% Sedimentary OM			% Modern OM		
	(% modern)	A	B	C	A	B	C
TE11-12	99.07	8	3	2	92	97	98
TE11-20	87.94	–	40	23	–	60	77
TE11-55	60.50	–	–	65.7	–	–	34.3
ML01-9	93.83	54	20	12	46	80	88
ML01-29	71.65	–	93	54	–	7	46
ML01-37	72.57	–	90	53	–	10	48
TE51-12	91.08	78	29	17	22	71	83
TE51-20	85.47	–	47	27	–	54	73
TC31	89.99	87	31	18	13	69	82
TE61	89.61	91	33	19	9	68	81
ML02	75.06	–	82	48	–	18	52
TC45-12	71.03	–	95	56	–	5	44
TC45-20	93.75	55	21	12	45	80	88
ML03-7	91.45	75	28	17	25	72	83
ML03-24	87.88	–	40	24	–	60	77
ML03-42	83.52	–	54	32	–	46	68
ML04-8	ND	ND	ND	ND	ND	ND	ND
ML04-30	105.38	0	0	0	100	100	100
ML04-44	100.71	0	0	0	100	100	100
ML05	64.80	–	–	68	–	–	32

These estimates represent an upper limit on the possible contributions of modern OM to the groundwater DOC pool because they only consider two-component mixing of one modern organic carbon source and one old sedimentary organic carbon source. It is more likely that DOC accumulates from different sources over the course of a flow path during which time a fraction of the DOC will be removed following its involvement in biogeochemical processes. In such a scenario the age of the DOC in the groundwater is actually equivalent to the weighted integrated mean age of the different contributions of DOC along the flow path that remains in the groundwater. As such, the data presented here is heavily model dependent and likely represents an over-simplification of the actual processes associated with the liberation and removal of DOC in groundwater. Nonetheless, this demonstrates that DOC in groundwater must derive from numerous sources and that even in pre-development conditions a significant amount of the total DOC pool derives from a surface source of OM.

The presence of modern bulk DOC in shallow groundwater conflicts with several modeling studies that suggest that modern DOC can only reach depths of a few meters following recharge (Datta et al., 2011, Mailloux et al., 2013 and Kocar et al., 2014). This disparity between modeled and measured transport times for DOC at different sites demonstrates the importance of high-permeability pathways to the transport of surface derived organic matter deep into the subsurface at this site. In this instance, stable and radiogenic isotopes demonstrate that groundwater received a significant contribution of rapidly transported pond derived recharge. We suggest that this observation may also be important regionally, and that the DOC pool of groundwater is likely to be dominated by surface derived DOC where ponds or clay windows act as point source(s) for transport of surface derived DOC through recharge.

In addition to the relationship we report here between groundwater residence time and groundwater As concentrations, we also observe a relationship between the age of organic carbon and As concentrations. These relationships can broadly be separated into three domains and classified according to the recharge environment that governs groundwater residence time and DOC. These domains are presented conceptually inFig. 6. In the northern most domain of our field site within 1 km of the Mekong River, we observe a prevalence of relatively young (<750 years) DOC at depths of less than 15 m in tritium-active groundwater. This DOC likely represents a mixture of DOC transported during rapid recharge through clay windows with older DOC mobilized during slow recharge through surficial clays. The concentration of As in these groundwaters is exclusively greater than 35 μg/l. These higher concentrations are consistent with the proposal of Kocar et al. (2014) that rapid sedimentation and organic matter burial, driven for instance from flooding of the Mekong River during the monsoon season, results in more extensive As release and at greater depths within sediments. DOC ages generally increase with depth during the transition to tritium-dead groundwaters at depths greater than 15 m. Sedimentary organic matter therefore represents an increasingly large contribution of the total groundwater DOC pool in these deeper groundwaters.

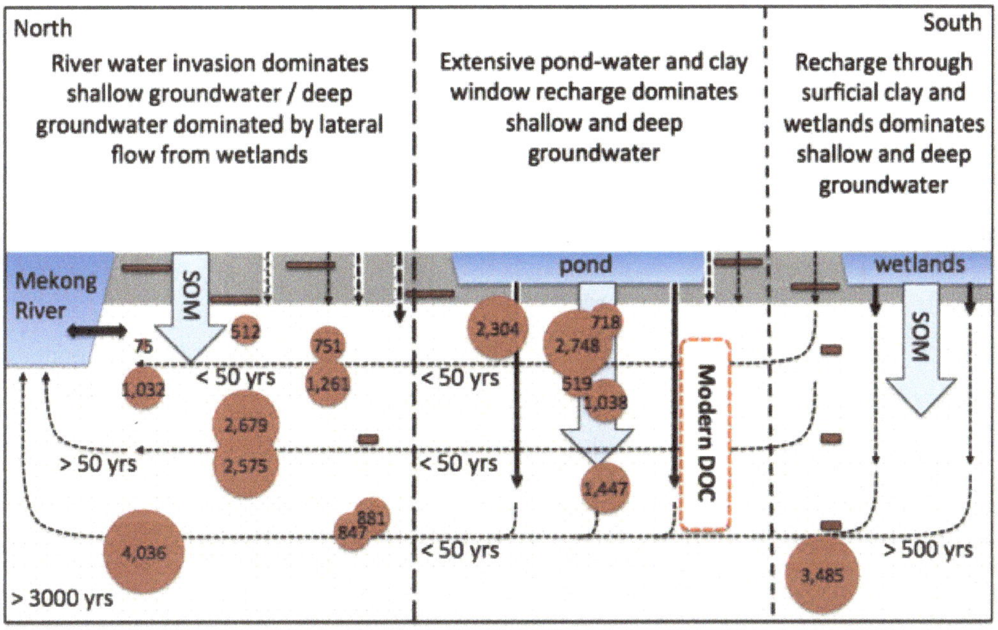

Figure 6: Summary of the three-recharge domains that prevail at this site and the dominant form of organic matter associated with each domain. Circles represent organic matter ages, with larger circles representing older ages. The radiocarbon age of DOC given in the center of each circle. Approximate mean groundwater residence times are provided for the different domains based on integrated tritium and DI^{14}C data assuming flow from the wetland basin to the Mekong River following the study of Benner et al. (2008). Surface derived DOC is transported in recharge through the pond and clay windows in the southern most part of the central domain and dominates the DOC pool in tritium-active shallow and deep groundwater. DOC ages represent mixtures of surface and sedimentary organic matter elsewhere at this site.

In the central domain, rapid tritium-active recharge from ponds and clay windows gives rise to a groundwater DOC pool that is dominated by surface-derived DOC over the entire depth profile in the northern most extent of this domain at site ML04. The DOC in the remainder of the domain represents a mixture of sedimentary and surface derived DOC, despite the fact that the entire depth window is tritium-active. There does not appear to be any relationship with DOC age and depth, which suggests that a complex network of flow paths generates a non-linear age distribution of DOC in this domain. Groundwater As concentrations in this domain are exclusively lower than <35 μg/l. This may reflect the greater number of pore volumes of flushing these sediments have received, resulting in lower concentrations of As in higher permeability sediments compared to lower permeability sediments of similar age. Alternatively, the rapid nature of recharge may allow for the transport of oxidants that are able to consume organic carbon and sequester As from the groundwater. Either way, our results are in agreement with the predictions of Kocar et al. (2014) who suggest that locations where groundwater is dominated by recharge through high permeability pathways will be restricted to As concentrations of less than 60 μg/l.

The final domain represents recharge in the internal wetland basin. We sampled only 1 tubewell in this domain. Nonetheless, our data suggests slow recharge to depths of 60 m through relatively low permeability sediments. The DOC has an age of 3485 years at this depth, which is younger than organic matter in sediments at shallower depths (Lawson et al., 2013). This requires transport of younger OM to this depth. The concentration of As in this domain has previously been shown to exceed 500 μg/l (Polizzotto et al., 2008 and Lawson et al., 2013). The high concentrations of As in this domain are consistent with the study of Stuckey et al. (2015a) who suggest that permanently saturated wetlands at this site experience only anaerobic metabolisms. As such, the consumption of DOC is dedicated to reductive processes that may ultimately result in the extensive release of As to groundwater.

Our results also provide interesting insights on the controls on As release in the aquifer systems of south Asia. We show here that shallow groundwater accrues As at a rate of approximately 1 μg/l/yr increase in DOC ^{14}C age compared to approximately 0.25 μg/l/yr increase in DOC age for deeper groundwaters (see Fig. 6). We recognize that this assessment does not link As concentrations to DOC age evolution along a flow path. Unfortunately, while the previous study of Benner et al. (2008) placed very useful constraints on the overall direction and timing of groundwater flow at the kilometer scale at the site, it still does not provide the appropriate resolution to account for small, local scale (10s–100s meter) fluid flow that is required to perform such a detailed, high resolution assessment. As such, we rely here on the broad classification of shallow and deep groundwater to account for the lack of resolution in our understanding of fluid flow. Ultimately, it is the local scale point source contributions of recharge from ponds and clay windows mixing with the larger scale, deeper fluid flow from the wetlands to the Mekong River that is likely to control the bulk geochemistry of groundwater at different locations at this site. Nonetheless, we believe this assessment provides useful first order constraints on the relationship between DOC age and associated As release. Indeed, this result is consistent with the findings in Vietnam of Postma et al. (2012) who show that As release is inversely proportional to the age of sedimentary organic carbon.

That As release is greater in shallow groundwater compared to deep groundwater at this site suggests that As is not only released in the near surface environment but also in the aquifer sediments, and over the entire length of the flow path. While we show a clear relationship between As concentration and DOC age in most locations, we also show that this relationship breaks down in areas where the DOC pool is dominated by modern DOC. This is because the fluids that transport modern DOC also deliver oxidants that result in the sequestration of As from groundwater to the sediment. This suggests that the rate and extent of arsenic release is therefore likely to be function of the amount of arsenic present in aquifer sediments, the distribution of competing electron acceptors introduced during recharge, and the age and hence residual reactivity of organic carbon that is available for anaerobic processes that release arsenic in these aquifers.

In addition to the mixing of multiple sources of DOC, the consumption of DOC in biogeochemical processes may also influence the radiocarbon age of organic carbon implicated in As release processes in these groundwater systems. The removal of young DOC has been previously reported in the study of Raymond and Bauer (2001) who utilized measurements of $\delta^{13}C$ and changes in the ^{14}C age of DOC and particulate organic matter to demonstrate that bacteria preferentially utilise a ^{14}C enriched (young) source of DOC in estuarine and riverine environments. A more recent study by Mailloux et al. (2013) compared the radiocarbon age of microbial DNA isolated from groundwater with the radiocarbon age of DOC and OM. From this comparison they demonstrate that microbial communities within an aquifer system in Bangladesh where groundwater As concentrations exceed 200 μg/l utilize young OC. The preferential utilization of young DOC over older DOC would therefore result in older apparent radiocarbon ages of residual DOC. In this case, As may accumulate in groundwaters at greater rates than we estimate here because we have characterized only the age of residual DOC, and not that consumed in biogeochemical processes that may ultimately be responsible for driving As mobilization.

CONCLUSIONS

The data we present here demonstrate that recharge in close proximity to ponds has the potential to transport modern DOC to depths of 44 m and dominate DOC in groundwater, even in the absence of groundwater pumping. However, groundwater DOC more typically comprises a mixture of differently sourced OM, with contributions of young surface or near surface derived OM and older, sedimentary sourced OM clearly present. This suggests a potential dual role for both surface and sedimentary organic carbon in controlling the mobilization and distribution of As in these aquifer systems.

We show here a clear association between higher rates of As release in shallow groundwaters containing young DOC, and lower rates of As release associated with older sediments and DOC in deeper groundwater. We suggest that As release not only occurs during recharge through near surface sediments but also during transport through aquifer sands and likely partitions between sediment and groundwater over the entirety of its flow path in the subsurface. The relationship between the radiocarbon age of DOC and groundwater As may also suggest that younger sources of OM are preferentially consumed over older, residual sedimentary sources that have experienced a greater history of consumption in metabolic processes.

This study is the most extensive isotopic characterisation of the hydrological environment at an As hotspot under natural conditions to date. Taken in the appropriate context, this information provides a basis for comparison with aquifer systems where the extensive pumping of groundwater has disrupted subsurface flow and natural recharge conditions, potentially altering the distribution of the OM sources implicated in As release. Such information is of critical importance in developing process oriented models that are capable of accurately predicting how and where the current As hazard will change, and importantly at what rate. The results presented here suggest that the aquifer systems in largely undeveloped countries such as Cambodia are potentially at risk from anthropogenic interference that may influence the groundwater flow regime. In particular, activities such as groundwater pumping and the excavation of surficial clay deposits have the potential to significantly change the sources of groundwater recharge and the timescales over which recharge takes place. Such activities would likely also drive changes in the composition and distribution of groundwater DOC, which may ultimately influence the rate and extent of As release in these groundwaters.

ACKNOWLEDGMENTS

This work was supported by a United Kingdom Natural Environmental Research Council (NERC)-funded CASE (with Thermo Scientific) PhD studentship (NER/S/A/2006/-14038) at The University Manchester. Fieldwork in Cambodia was further supported by the EU-Asia Link CALIBRE project (KH/Asia-Link/04 142966). Stable isotope analysis was conducted at SUERC and was supported by NERC Isotope Geosciences Facilities Access grant IP/1013/110. Radiocarbon analysis was supported by the NERC Radiocarbon Facility NRCF010001 (allocation numbers 1411.0409 and 1029.0403). We thank the late Mickey Sampson, Andrew Shantz and others at Resource Development International (Cambodia) for their assistance and logistical support during field work; Matt Polizzotto, Ben Kocar and Scott Fendorf (Stanford University) for access to wells at the Cambodian study site; and C. Davies, A. Bewsher and P. Lythgoe for analytical support. We also thank three anonymous reviewers who provided valuable feedback and suggestions for improving this manuscript. The views expressed here do not necessarily represent those of any of the funders or those individuals whose support we acknowledge here.

REFERENCES

1. Acharyya S. K. and Shah B. A. (2007) Arsenic-contaminated groundwater from parts of Damodar fan-delta and west of Bhagirathi River, West Bengal, India: influence of fluvial geomorphology and Quaternary morphostratigraphy. Environ. Geol. 52, 489–501.
2. Aggarwal P. K., Basu A. R. and Kulkarni K. M. (2003) Comment on arsenic mobility and groundwater extraction in Bangladesh (I). Science 300, 584b.
3. Al Lawati W. M., Rizoulis A., Eiche E., Boothman C., Polya D. A., Lloyd J. R., Berg M., Vasquez-Aguilar P. and van Dongen B. E. (2012) Characterisation of organic matter and microbial communities in contrasting arsenic-rich Holocene and arsenicpoor Pleistocene aquifers, Red River Delta, Vietnam. Appl. Geochem. 27, 315–325.
4. Argos M., Kalra T., Rathouz P. J., Chen Y., Pierce B., Parvex F., Islam T., Ahmed A., Rakibuz-Zaman M., Hasan R., Sarwar G., Slavkovich V., van Geen A., Graziano J. and Ahsan H. (2010) Arsenic exposure from drinking water, and all-cause and chronic-disease mortalities in Bangladesh (HEALS): a prospective cohort study. The Lancet 376, 252–258.
5. Balga P. and Kaiser J. (1996) India's spreading health crisis draws global arsenic experts. Science 274, 174–175.
6. Benner S. G., Polizzotto M. L., Kocar B. D., Ganguly S., Phan K., Ouch K., Sampson M. and Fendorf S. (2008) Groundwater flow in an arsenic-contaminated aquifer, Mekong Delta, Cambodia. Appl. Geochem. 23, 3072–3087.
7. Berg M., Tran H. C., Nguyen T. C., Pham H. V., Schertenleib R. and Giger W. (2001) Arsenic contamination of groundwater and drinking water in Vietnam: a human health threat. Environ. Sci. Technol. 35, 2621–2626.
8. Berg M., Stengel C., Trang P. T. K., Viet P. H., Sampson M. L., Leng M., Samreth S. and Fredericks D. (2007) Magnitude of arsenic pollution in the Mekong and Red River Deltas – Cambodia and Vietnam. Sci. Total Environ. 372, 413–425.
9. Bethke C. M. and Johnson T. M. (2002) Paradox of groundwater age. Geology 30, 385–388.
10. Bhattacharya P., Chatterjee D. and Jacks G. (1997) Occurence of arsenic contaminated groundwater in alluvial aquifers from delta plains, Eastern India: options for safe drinking water supply. Water Resour. Dev. 13, 79–92.
11. Buschmann J., Berg M., Stengel C. and Sampson M. L. (2007) Arsenic and manganese contamination of drinking water resources in Cambodia: coincidence of risk areas with low relief topography. Environ. Sci. Technol. 41, 2146–2152.
12. Chakraborty A. K. and Saha K. C. (1987) Arsenical dermatosis from tubewell water in West Bengal, India. Indian J. Med. Res. 85, 326–334.
13. Charlet L. and Polya D. A. (2006) Arsenic hazard in shallow reducing groundwaters in southern Asia. Elements 2, 91–96.
14. Clark I. D. and Fritz P. (2000) Environmental Isotopes in Hydrogeology. Lewis Publishers, New York.
15. Clarke W. B., Jenkins W. J. and Top Z. (1976) Determination of tritium by mass spectrometric measurement of tritium. Int. J. Appl. Radiat. Isot. 27, 515–522.
16. Datta S., Neal A. W., Mohajerin T. J., Ocheltree T., Rosenheim B. E., White C. D. and Johannesson K. H. (2011) Perennial ponds are not an important source of water or dissolved organic matter to groundwaters with high arsenic concentrations in West Bengal, India. Geophys. Res. Lett. 38, L20404.
17. Dhar R. K., Biswas B. K., Samanta G., Mandal B. K., Chakraborti D., Roy S., Jafar A., Islam A., Ara G., Kabir S., Khan A. W., Ahmed S. A. and Hadi S. A. (1997) Groundwater arsenic calamity in Bangladesh. Curr. Sci. 73, 48–59.
18. Dowling C. B., Poreda R. J., Basu A. R., Peters S. L. and Aggarwal P. K. (2002) Geochemical study of arsenic release mechanisms in the Bengal Basin groundwater. Water Resour. Res. 38.
19. Erban L. E., Gorelick S. M., Zebker H. A. and Fendorf S. (2013) Release of arsenic to deep groundwater in the Mekong Delta, Vietnam, linked to pumping-induced land subsidence. Proc. Natl. Acad. Sci. U.S.A. 110, 13751–13756.
20. Fendorf S., Michael H. A. and van Geen A. (2010) Spatial and temporal variations of groundwater arsenic in south and southeast Asia. Science 328, 1123–1127.
21. Freeman S. P. T., Cook G. T., Dougans A. B., Naysmith P., Wilcken K. M. and Xu S. (2010) Improved SSAMS performance. Nucl. Instrum. Methods Phys. Res., Sect. B 268, 715–717.
22. Gault A. G., Islam F. S., Polya D. A., Charnock J. M., Boothman C., Chatterjee D. and Lloyd J. R. (2005) Microcosm depth profiles of arsenic release in a shallow aquifer, West Bengal. Mineral. Mag. 69(5), 855–863.
23. Harvey C. F., Swartz C. H., Badruzzaman A. B. M., Keon-Blute N., Yu W., Ashraf Ali M., Jay J., Beckie R., Niedan V., Brabander D., Oates P. M., Ashfaque K. N., Islam S., Hemond H. F. and Ahmed M. F. (2002) Arsenic

mobility and groundwater extraction in Bangladesh. Science 298, 1602–1606.

24. Harvey C. F., Swartz C. H., Badruzzaman A. B. M., Keon-Blute N., Ali M. A., Jay J., Beckie R., Oates P. M., Ashfaque K. N., Islam S., Hemond H. F. and Ahmed M. F. (2003) Response to comments on "Arsenic mobility and groundwater extraction in Bangladesh". Science 300, 584d.

25. Horneman A., van Geen A., Kent D. V., Mathe P. E., Zheng Y., Dhar R. K., O'Connell S., Hoque M. A., Aziz Z., Shamsudduha M., Seddique A. A. and Ahmed K. M. (2004) Decoupling of As and Fe release to Bangladesh groundwater under reducing conditions. Part I: Evidence from sediment profiles. Geochim. Cosmochim. Acta 68, 3459–3473.

26. IAEA/WMO (2006) Global Network of Isotopes in Precipitation. The GNIP database.

27. Islam F. S., Gault A. G., Boothman C., Polya D. A., Charnock J. M., Chatterjee D. and Lloyd J. R. (2004) Role of metalreducing bacteria in arsenic release from Bengal delta sediments. Nature 430, 68–71.

28. Kabeya N., Shimizu A., Chann S., Tsuboyama Y., Nobuhiro T., Keth N. and Tamai K. (2007) Stable isotope studies of rainfall and stream water in forest watersheds in Kampong Thom, Cambodia. In Forest Environments in the Mekong River Basin. Springer, pp. 125–134.

29. Kocar B. D., Polizzotto M. L., Benner S. G., Ying S. C., Ung M., Ouch K., Samreth S., Suy B., Phan K., Sampson M. and Fendorf S. (2008) Integrated biogeochemical and hydrologic processes driving arsenic release from shallow sediments to groundwaters of the Mekong delta. Appl. Geochem. 23, 3059– 3071.

30. Kocar B. D., Benner S. G. and Fendorf S. (2014) Deciphering and predicting spatial and temporal concentrations of arsenic within the Mekong Delta aquifer. Environ. Chem. 11, 579–594.

31. Lawson M., Ballentine C. J., Polya D. A., Boyce A. J., Mondal D., Majumder S. and Biswas A. (2008) The geochemical and isotopic composition of ground waters in West Bengal: tracing ground–surface water interaction and its role in arsenic release. Mineral. Mag. 72, 441–444.

32. Lawson M., Polya D. A., Boyce A. J., Bryant C., Mondal D., Shantz A. and Ballentine C. J. (2013) Pond-derived organic carbon driving changes in arsenic hazard found in Asian groundwaters. Environ. Sci. Technol. 47, 7085–7094.

33. Mailloux B. K., Trembath-Reichert E., Cheung J., Watson M., Stute M., Freyer G. A., Ferguson A. S., Ahmed K. M., Alam M. J., Buchholz B. A., Thomas J., Layton A. C., Zheng Y., Bostick B. C. and van Geen A. (2013) Advection of surfacederived organic carbon fuels microbial reduction in Bangladesh groundwater. Proc. Natl. Acad. Sci. 110, 5331–5335.

34. McArthur J., Ravenscroft P., Safiulla S. and Thirwall M. F. (2001) Arsenic in groundwater: testing pollution mechanisms for sedimentary aquifers in Bangladesh. Water Resour. Res. 37, 109–117.

35. McArthur J. M., Banerjee D. M., Hudson-Edwards K. A., Mishra R., Purohit R., Ravenscroft P., Cronin A., Howarth R. J., Chatterjee A., Talukder T., Lowry D., Houghton S. and Chadha D. K. (2004) Natural organic matter in sedimentary basins and its relation to arsenic in anoxic ground water: the example of West Bengal and its worldwide implications. Appl. Geochem. 19, 1255–1293.

36. McArthur J. M., Ravenscroft P. and Sracek O. (2011) Aquifer arsenic source. Nat. Geosci. 4, 655–656.

37. Meharg A. A., Scrimgeour C., Hossain S. A., Fuller K., Cruickshank K., Williams P. N. and Kinniburgh D. G. (2006) Codeposition of organic carbon and arsenic in Bengal delta aquifers. Environ. Sci. Technol. 40, 4928–4935.

38. Mladenov N., Zheng Y., Miller M. P., Nemergut D. R., Legg T., Simone B., Hageman C., Rahman M. M., Ahmed K. M. and McKnight D. M. (2010) Dissolved organic matter sources and consequences for iron and arsenic mobilization in Bangladesh aquifers. Environ. Sci. Technol. 44, 123–128.

39. Nath B., Berner Z., Mallik S. B., Chatterjee D., Charlet L. and Stuben D. (2005) Characterization of aquifers conducting groundwaters with low and high arsenic concentrations: a comparative case study from West Bengal, India. Mineral. Mag. 69, 841–854.

40. Neumann R. B., Ashfaque K. N., Badruzzaman A. B. M., Ali M. A., Shoemaker J. K. and Harvey C. F. (2010) Anthropogenic influences on groundwater arsenic concentrations in Bangladesh. Nat. Geosci. 3, 46–52.

41. Neumann R. B., Pracht L. A., Polizzotto M. L., Borhon A., Badruzzaman M. and Ahraf Ali M. (2014) Biodegradable organic carbon in sediments of an arsenic-contaminated aquifer in Bangladesh. Environ. Sci. Technol. Lett. 1, 221–225.

42. Nickson R., McArthur J., Burgess W., Ahmed K. M., Ravenscroft P. and Rahman M. (1998a) Arsenic poisoning of Bangladesh groundwater. Nature 395, 338.

43. Nickson R. T., McArthur J. M., Burgess W. G., Ahmed K. M., Ravenscroft P. and Rahman M. (1998b) Response to comments on "Arsenic poisoning of Bangladesh groundwater". Nature 395, 338.

44. Papacostas N. C., Bostick B., Quicksall A. N., Landis J. D. and Sampson M. (2008) Geomorphic controls on groundwater arsenic distribution in the Mekong River Delta, Cambodia. Geology 36, 891–894.

45. Polizzotto M. L., Kocar B. D., Benner S. G., Sampson M. and

46. Fendorf S. (2008) Near-surface wetland sediments as a source of arsenic release to ground water in Asia. Nature 454, 505–508.

47. Polya D. A. and Charlet L. (2009) Rising arsenic risk? Nat. Geosci. 2, 383–384.

48. Polya D. A., Gault A. G., Bourne N. J., Lythgoe P. R. and Cooke D. A. (2003) Coupled HPLC-ICP-MS Analysis Indicates Highly Hazardous Concentrations of Dissolved Arsenic Species Are Present in Cambodian Wellwaters, . Royal Society of Chemistry Special Publication, pp. 127–140.

49. Polya D. A., Gault A. G., Diebe N., Feldman P., Rosenboom J. W., Gilligan E., Fredericks D., Milton A. H., Sampson M., Rowland H. A. L., Lythgoe P. R., Jones J. C., Middleton C. and Cooke D. A. (2005) Arsenic hazard in shallow Cambodian groundwaters. Mineral. Mag. 69, 807–823.

50. Postma D., Larsen F., Thai N. T., Trang P. T. K., Jakobsen R., Nhan P. Q., Long T. V., Viet P. H. and Murray A. S. (2012) Groundwater arsenic concentrations in Vietnam controlled by sediment age. Nat. Geosci. 5, 656–661.

51. Quicksall A. N., Bostick B. C. and Sampson M. L. (2008) Linking organic matter deposition and iron mineral transformations to groundwater arsenic levels in the Mekong delta, Cambodia. Appl. Geochem. 23, 3088–3098.

52. Ravenscroft P., Brammer H. and Richards K. (2009) Arsenic Pollution – A Global Synthesis. Wiley-Blackwell, Chichester.

53. Raymond P. A. and Bauer J. E. (2001) Use of 14C and 13C natural abundances for evaluating riverine, estuarine, and coastal DOC and POC sources and cycling: a review and synthesis. Org. Geochem. 32, 469–485.

54. Richards L. A., Magnone D., van Dongen B. E., Ballentine C. J. and Polya D. A. (2015) Use of lithium tracers to quantify drilling fluid contamination for groundwater monitronig in Southeast Asia. Appl. Geochem. 63, 190–202.

55. Rowland H. A. L., Polya D. A., Lloyd J. R. and Pancost R. D. (2006) Characterisation of organic matter in a shallow, reducing, arsenic-rich aquifer, West Bengal. Org. Geochem. 37, 1101– 1114.

56. Rowland H. A. L., Pederick R. L., Polya D. A., Pancost R. D., Van Dongen B. E., Gault A. G., Vaughan D. J., Bryant C., Anderson B. and Lloyd J. R. (2007) The control of organic matter on microbially mediated iron reduction and arsenic release in shallow alluvial aquifers, Cambodia. Geobiology 5, 281–292.

57. Rowland H. A. L., Gault A. G., Lythgoe P. and Polya D. A. (2008) Geochemistry of aquifer sediments and arsenic-rich groundwaters from Kandal Province, Cambodia. Appl. Geochem. 23, 3029–3046.

58. Rozanski K., Araguas-Araguas L., Gonfiantini R., Swart P. K., Kyger C., McKenzie J. A. and Savin S. (1993) Isotopic patterns in modern global precipitation. In Climate Change in Continental Isotope Records (eds. P. K. Swart, K. C. Lohmann, J. A. McKenzie and S. Savin). AGU American Geophysical Union.

59. Sengupta S., McArthur J., Sarkar A., Leng M. J., Ravenscroft P., Howarth R. J. and Banerjee D. M. (2008) Do ponds cause arsenic-pollution of groundwater in the Bengal Basin? An answer from West Bengal. Environ. Sci. Technol. 42, 5156–5164.

60. Slota P. J., Jull A. J. T., Linick T. W. and Toolin L. J. (1987) Preparation of small samples for 14C accelerator targets by catalytic reduction of CO. Radiocarbon 29, 303–306.

61. Smith A. H., Lingas E. O. and Rahman M. (2000) Contamination of drinking-water by arsenic in Bangladesh: a public health emergency. Bull. World Health Organ. 78, 1093–1103.

62. Sovann C. and Polya D. (2014) Improved groundwater geogenic arsenic hazard map for Cambodia. Environ. Chem. 11, 595–607.

63. Stahl M. O., Tarek M. H., Yeo D. C. J., Badruzzaman A. B. M. and Harvey C. (2014) Crab burrows as conduits for groundwater-surface water exchange in Bangladesh. Geophys. Res. Lett. http://dx.doi.org/10.1002/2014GL061626.

64. Stuben D., Berner Z., Chandrasekharam D. and Karmaker J. (2003) Arsenic enrichment in groundwater of West Bengal, India: geochemical evidence for mobilization of As under reducing conditions. Appl. Geochem. 18, 1417–1434.

65. Stuckey J. W., Schaefer M. V., Kocar B. J., Benner S. G. and Fendorf S. (2015a) Arsenic release metabolically limited to permanently water-saturated soil in Mekong Delta. Nat. Geosci. http://dx.doi.org/10.1038/ngeo2589.

66. Stuckey J. W., Schaefer M. V., Kocar B. D., Dittmar J., Pacheco J. L., Benner S. G. and Fendorf S. (2015b) Peat formation concentrates arsenic within sediment deposits of the Mekong delta. Geochim. Cosmochim. Acta 149, 190–205.

67. Stute M. (2001) Tritium (3 H) in precipitation in Bangladesh. In Groundwater Arsenic Contamination in the Bengal Delta Plain of Bangladesh (eds. G. Jacks, P. Bhattacharya, A. A. Khan). KTH Special Publication, TRITA-AMI REPORT 3084, University of Dhaka, Bangladesh.

68. Tamura T., Saito Y., Sieng S., Ben B., Kong M., Choup S. and Tsukawaki S. (2007) Depositional facies and radiocarbon ages of a drill core from the Mekong River lowland near Phnom Penh, Cambodia: evidence for tidal sedimentation at the time of Holocene maximum flooding. J. Asian Earth Sci. 29, 585–592.

69. Tamura T., Saito Y., Sieng S., Ben B., Kong M., Sim I., Choup S. and Akiba F. (2009) Initiation of the Mekong River delta at 8 ka: evidence from the sedimentary succession in the Cambodian lowland. Quatern. Sci. Rev. 28, 327–344.

70. van Geen A., Zheng Y., Stute M. and Ahmed K. M. (2003) Comment on arsenic mobility and groundwater extraction in Bangladesh (II). Science 300, 584c.

71. van Geen A., Rose J., Thoral S., Garnier J. M., Zheng Y. and Bottero J. Y. (2004) Decoupling of As and Fe release to Bangladesh groundwater under reducing conditions. Part II: evidence from sediment incubations. Geochim. Cosmochim. Acta 68, 3475–3486.

72. van Geen A., Zheng Y., Cheng Z., Aziz Z., Horneman A., Dhar R. K., Mailloux B., Stute M., Weinman B., Goodbred S., Seddique A. A., Hoque M. A. and Ahmed K. M. (2006) A transect of groundwater and sediment properties in Araihazar, Bangladesh: further evidence of decoupling between As and Fe mobilization. Chem. Geol. 228, 85–96.

73. van Dongen B. E., Rowland H. A. L., Gault A. G., Polya D. A., Bryant C. and Pancost R. D. (2008) Hopane,

sterane and nalkane distributions in shallow sediments hosting high arsenic groundwaters in Cambodia. Appl. Geochem. 23, 3047–3058.

74. Winkel L., Berg M., Amini M., Hug S. J. and Johnson C. A. (2008) Predicting groundwater arsenic contamination in Southeast Asia from surface parameters. Nat. Geosci. 1, 536–542.

75. Xu S., Anderson R., Bryant C., Cook G. T., Dougans A., Freeman S., Naysmith P., Schnabel C. and Scott E. M. (2004) Capabilities of the new SUERC 5MV AMS facility for 14C dating. Radiocarbon 46(1), 59–64.

76. Yu W. H., Harvey C. M. and Harvey C. F. (2003) Arsenic in groundwater in Bangladesh: a geostatistical and epidemiological framework for evaluating health effects and potential remedies. Water Resour. Res. 39, 1146–1163.

77. Zheng Y., Stute M., van Geen A., Gavrieli I., Dhar R., Simpson H. J., Schlosser P. and Ahmed K. M. (2004) Redox control of arsenic mobilisation in Bangladesh groundwater. Appl. Geochem. 19, 201–214.

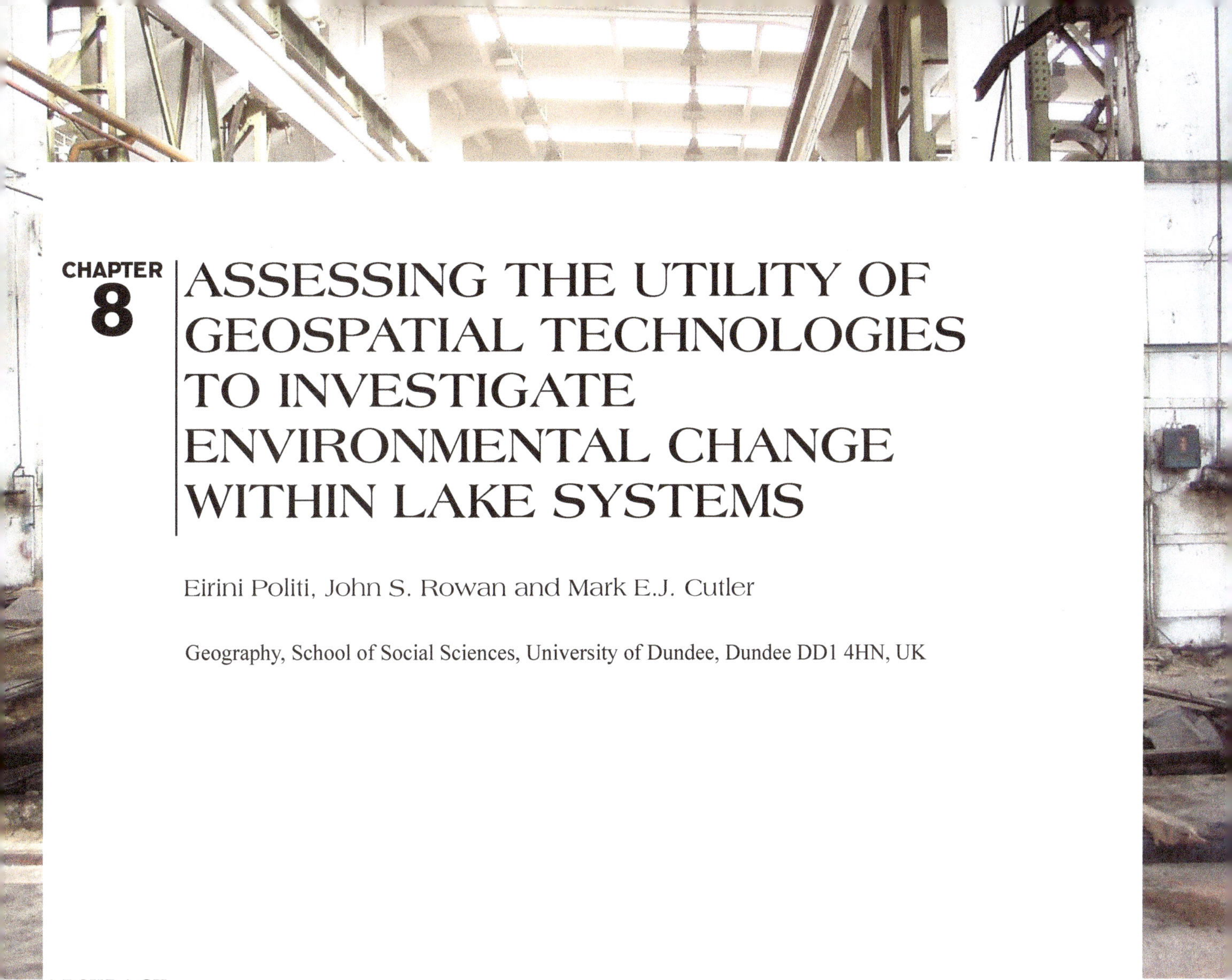

CHAPTER 8

ASSESSING THE UTILITY OF GEOSPATIAL TECHNOLOGIES TO INVESTIGATE ENVIRONMENTAL CHANGE WITHIN LAKE SYSTEMS

Eirini Politi, John S. Rowan and Mark E.J. Cutler

Geography, School of Social Sciences, University of Dundee, Dundee DD1 4HN, UK

ABSTRACT

Over 50% of the world's population live within 3 km of rivers and lakes highlighting the on-going importance of freshwater resources to human health and societal well-being. Whilst covering *c.* 3.5% of the Earth's non-glaciated land mass, trends in the environmental quality of the world's standing waters (natural lakes and reservoirs) are poorly understood, at least in comparison with rivers, and so evaluation of their current condition and sensitivity to change are global priorities. Here it is argued that a geospatial approach harnessing existing global datasets, along with new generation remote sensing products, offers the basis to characterise trajectories of change in lake properties e.g., water quality, physical structure, hydrological regime and ecological behaviour. This approach furthermore provides the evidence base to understand the relative importance of climatic forcing and/or changing catchment processes, e.g. land cover and soil moisture data, which coupled with climate data provide the basis to model regional water balance and runoff estimates over time. Using examples derived primarily from the Danube Basin but also other parts of the World, we demonstrate the power of the approach and its utility to assess the sensitivity of lake systems to environmental change, and hence better manage these key resources in the future.

INTRODUCTION

The world's freshwater ecosystems and the biodiversity they support are vital components of the global biosphere, yet remain fragile and vulnerable to anthropogenic disturbance and climate change (Dokulil, 2013, Schindler and Donahue, 2006,Vörösmarty et al., 2000 and Williamson et al., 2009). Whilst recognition grows of their role in the carbon cycle at regional and continental scales (e.g. Knoll et al., 2013,McDonald et al., 2013, Renwick et al., 2008 and Yang et al., 2008), understanding their global significance to biogeochemical cycling and ecosystem services has been hampered by uncertainties about absolute numbers and the distributions of different types, as well as limited information on their ecological, chemical and physical condition (Downing et al., 2006 and Verpoorter et al., 2012). Verpoorter et al. (2014) enumerated a global inventory of 117 million lakes (> 0.002 km^2), with a combined surface area of about 5×10^6 km^2, equating to 3.7% of the Earth's non-glaciated land mass. In Europe alone there are over 1.5×10^6 lakes with surface areas greater than 0.01 km^2 and at least 500,000 natural lakes larger than 0.1 km^2 (Kristiansen and Hansen, 1994). These systems span a continuum of size, depth, form, altitude, geology, climatic and hydrological regime. Importantly, lakes

Citation: Eirini Politi, John S. Rowan, Mark E.J. Cutler, Assessing the utility of geospatial technologies to investigate environmental change within lake systems, Science of The Total Environment, Volume 543, Part A, 1 February 2016, Pages 791-806, ISSN 0048-9697, http://dx.doi.org/10.1016/j.scitotenv.2015.09.136.

also sit within a wider landscape, catchment and hydrological network that means one needs to look beyond the lake itself to determine the drivers of lake water behaviour and any changes in that. For example, the Danube Basin is some 801,463 km^2 in size and according to the Global Lakes and Wetlands Database (Lehner and Döll, 2004) includes 592 lakes > 0.1 km^2. The large numbers involved and the relative remoteness of many lake districts, especially within large transnational catchments such as the Danube, presents major challenges in terms of managing the standing water resource and emphasises the need to be able to generalise about behaviour based on limited field observations.

The role that aquatic systems play in underpinning society's activities, health and well-being is increasingly being understood through the concept of ecosystem services, which links nature to an economic value to society (TEEB, 2010). Particularly since the launch of the United Nations Millennium Ecosystem Assessment (MEA) in 2005, there has been an increase in the awareness of the ecosystem services provided by aquatic and terrestrial ecosystems amongst policy makers (de Groot et al., 2010), matched by a significant increase in the number of scientific publications on the topic (de Araujo Barbosa et al., 2015, Large and Gilvear, 2014 and Seppelt et al., 2011). Remote sensing has the potential to provide an important source of information for quantifying and mapping dynamic terrestrial and aquatic ecosystem services and hydrological processes, including approaches to hydrologic modelling as reviewed by Xu et al. (2014). In many cases, given the large catchment area of many of the most important river and lake systems in the world (particularly those with large populations dependent upon them and their attendant ecosystem services), the synoptic, wide area coverage and frequent observations provided by satellite-based remote sensing are an important source of standardised information that is not prey to variability in national and regional in situ data collection methodologies and standards. As a result, remote sensing has become an important geospatial technology for deriving information about lakes and their catchments, which alongside advances in in situ sensor systems to capture spatial information in addition to measurements across time (Crawford et al., 2014) and analysis with Geographic Information Systems (GIS), provides an important step forward in our ability to model lake and catchment status and behaviour.

From both passive (e.g. multispectral and hyperspectral) and active (LiDAR and RADAR) remote sensing systems it is possible to retrieve the state of important lake and catchment variables that may vary across space at particular points in time. Recently, either directly or in combination with other data, remote sensing has been successfully used within the Danube Basin to map variations in biomass (Kovàcs, 2007), carbon stocks in riparian forests (Suchenwirth et al., 2014), habitat change (Kollár et al., 2011), historical land cover change (Crăciunescu et al., 2010), evapotranspiration (Rodell et al., 2011) and soil moisture (dall'Amico et al., 2012), to name but a few examples. However, it is important to recognise that whilst some catchment properties may be retrieved directly from remotely sensed data (e.g. canopy height with LiDAR or leaf area index from optical data) some catchment properties, such as land use, may only be inferred from remotely sensed data through contextual associations across, whilst others, such as raw material production, will require remotely sensed data to be coupled with other ancillary datasets, for example in situ measurements and modelled data (de Araujo Barbosa et al., 2015).

Following decades of fundamental research and the launch of several new satellite programmes a report by London Economics to the UK Space Agency suggests that we stand at the dawn of a 'New Space Age' (London Economics, 2015, p. 104), one in which there will be substantial growth in applications of remote sensing data. Indeed the future of European (and global) remote sensing for catchment, ecosystem and lake-based studies looks very promising, particularly with respect to data availability and the new generation Copernicus satellites of the European Space Agency (ESA). The Copernicus satellites build on the success and capabilities of other currently operational (e.g. Landsat series) and non-operational (e.g. Envisat MERIS and AATSR) instruments and include the Sentinel-2 Multispectral Imager (MSI) (launched June 2015), Sentinel-3 Ocean and Land Colour Instrument (OLCI) and Sentinel-3 Sea and Land Surface Temperature Radiometer (SLSTR), the first of which is due for launch late 2015. Sentinel-2 MSI has 13 wavebands in the visible and NIR at 10–60 m spatial resolution and a 5-day revisit time. Sentinel-3 OLCI at a 300 m spatial resolution will employ 21 wavebands in the visible and NIR. Sentinel-3 SLSTR will employ 9 wavebands, with a nominal 500 m spatial resolution in the visible and NIR and 1 km at the TIR. In addition, Sentinel-3 will have a revisit time of 1–2 days. These optical and thermal sensors combine short revisit times, fine spatial resolutions and multiple spectral channels with a free data access policy. Here we signpost some of the new opportunities for catchment and lake studies provided by these new systems by collating experiences of previous research and how these and the plethora of new data sources will contribute to future lake and catchment management.

Several authors have provided comprehensive reviews of the potential and current status of remote sensing to map and quantify a range of ecosystem services (e.g.Andrew et al., 2014 and de Araujo Barbosa et al., 2015). This paper builds upon those comprehensive reviews and others by exploring the utility of remote sensing to capture spatial and temporal patterns of change in key catchment processes and the use of these data to inform policy and management decisions. Particular reference is made to examples from the Danube Basin highlighting information derived from remote sensing data relevant for catchment water balance estimation and modelling, as well as land cover change and other critical ecosystem functions. By focussing on the catchment we evaluate how to derive information on drivers of change in lake behaviour rather than explicitly exploring the use of remote sensing to detect change in the lake water quality itself, which is the subject a separate paper in this Special Issue (Tyler et al., in this volume).

LAKES, LANDSCAPE LIMNOLOGY AND REMOTE SENSING

Soranno et al. (2010) introduced landscape limnology as a new conceptual framework for understanding lake behaviour based on a series of important comparative and regional limnological studies (e.g. Riera et al., 2000, Webster et al., 2000 and Soranno et al., 2009). Landscape limnology is the spatially-explicit study of lakes within freshwater, terrestrial and human landscapes to determine the effects of pattern on ecosystem processes over multiple temporal and spatial scales (Fig. 1). It highlights that landscape settings influence lake functions through the hydromorphological setting mediated by the extent of human modification, superimposed upon which is the effect of climate change.

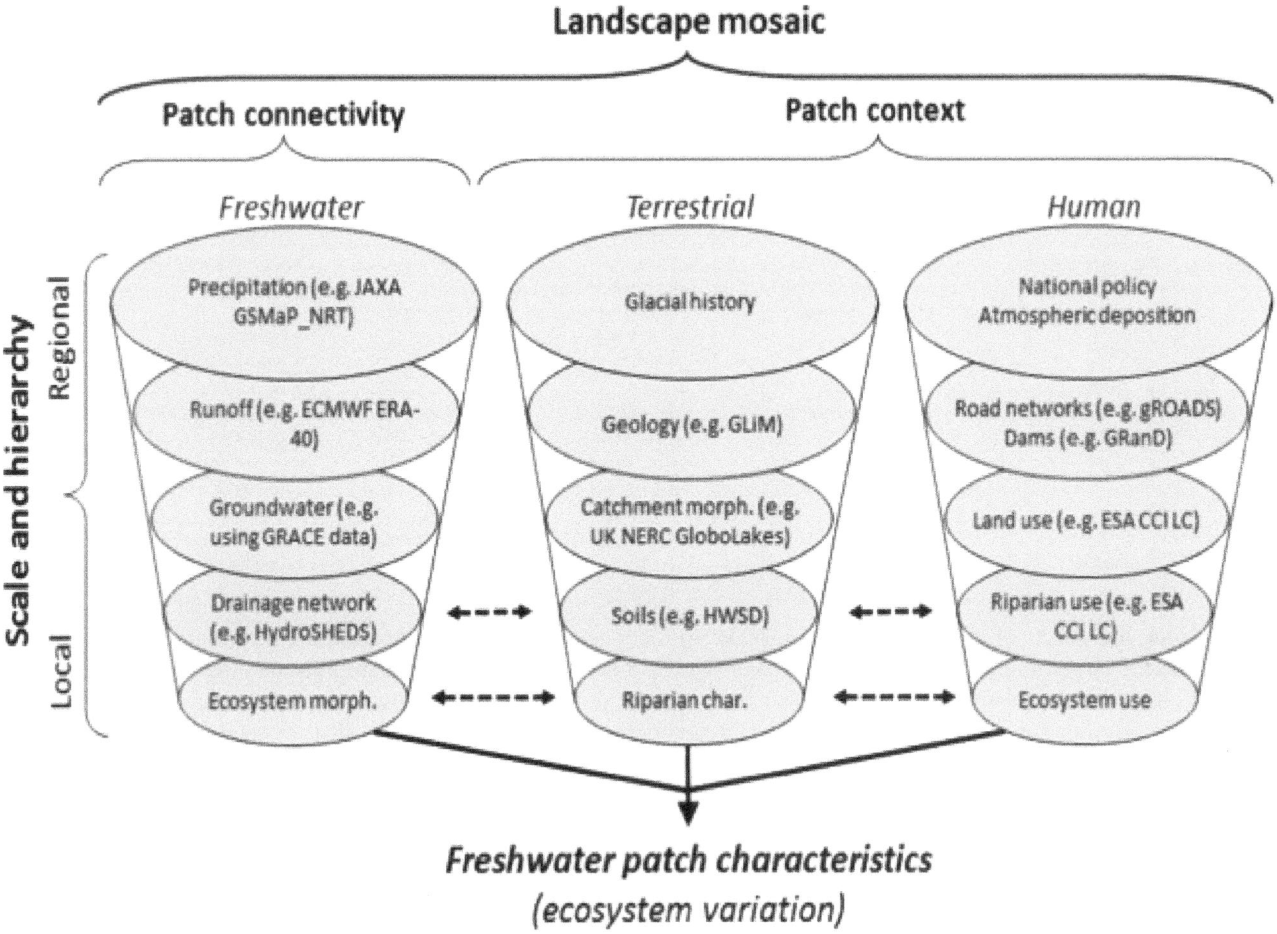

Figure 1: Landscape limnology framework connecting local and regional lake characteristics with examples of globally available datasets and EO products (after Soranno et al., 2009). JAXA GSMaP_NRT (Japan Aerospace Exploration Agency Global Satellite Mapping of Precipitation in Near-Real-Time); ECMWF ERA-40 (European Centre for Medium-Range Weather Forecasts 40-year re-analysis model data); GRACE (NASA/DLR's Gravity Recovery and Climate Experiment); HydroSHEDS (World Wildlife Fund Hydrological data and maps based on SHuttle Elevation Derivatives at multiple Scales); GLiM (Global Lithological Map database); UK NERC GloboLakes (United Kingdom Natural Environment Research Council GloboLakes project); HWSD (Harmonised World Soil Database); gROADS (Global Roads Open Access Data Set); GRanD (Global Reservoir and Dam database); ESA CCI LC (European Space Agency Climate Change Initiative Land Cover).

An extension to this framework is shown in Table 1, which uses four hierarchical structuring elements consistent with the water body classification attributes used in the EU Water Framework Directive (EU WFD, 2000). The first element is the ecoregion which sets out the broadest climatic, physiographic and dominant biomes. Landscape setting establishes the specific characteristics in terms of hydro-climatic factors such as continentality, the elevation of the lake and its catchment (e.g. whether montane, piedmont or lowland) and biodiversity characteristics. The lake-catchment relation captures the relative importance of 'catchment' versus 'lake' processes in relation to water, sediment and nutrient budgets, including the critical role of autochthonous versus allochthonous carbon fluxes (cf. Tranvik et al., 2009 and Kutser et al., 2015). Finally, lake morphometry (size, shape, orientation, water depth, etc.) governs lake processes and characteristics such as mixing regime, residence time, shore zone energetics and riparian habitats (Hutchinson, 1957).

Table 1: WFD compliant eco-geomorphic framework for contextualising lake behaviour.

Scale	Hydro-climatic	Terrestrial	Biotic	Human
Ecoregion	• Latitude/Longitude • Solar radiation regime • Temperature • Precipitation • Wind regime	• Geological structure and lithology • Physiography • Soils	• Regional biomes • Biomass • Phenology	• Airborne pollution • Climate change
Landscape setting	• Continentality • Orographic effects • Environ. lapse rates • Runoff quantity and timing • Ice cover	• Elevation • Topographic variation • Drainage network and density • Glacial history	• Habitat diversity • Species dispersal • Flux of humic substances	• Economy • Demographics • Land cover and management • Water resource management
Catchment-lake relation	• Hydrological regime • Hydrological pathways • Surface:groundwater contribution	• Hydrogeology • Groundwater and solute flux • Sediment budget • Salinity	• Biological continuity • Nutrient cycling • Acidity	• Impoundments • Water transfers • Drainage • Land management • Riparian land cover
Lake hydromorphology (water body scale)	• Size and shape • Stratification and mixing depth • Water residence time • Resuspension and internal loading	• Shoreline complexity • Littoral:pelagic ratio • Riparian habitat structure	• Appropriate species composition • Integrity of trophic function • Alien species	• Water level manipulation • Cumulative pressures and disturbance
Meso-habitat patches	• Microclimate • Exposure • Erosion/sedimentation	• Littoral slope • Substrate • Wave energy	• Local carbon flux • Pollution • Habitat patches	• Local pressures to shore and lake bed

SCALE: ↑ Regional influences — Increasingly site specific ↓

Untangling the effects of multiple pressures on lake structure and function, and determining the environmental pressures that will have the largest effect on a given lake or lake type, is a challenge that requires multi-scale observations of a suite of functionally relevant and reliable indicators of lake ecosystem condition for a larger number of lakes than has been possible with traditional monitoring. Remote sensing provides a plethora of datasets and products that, in combination with ancillary data (e.g. in situ measurements and observations, modelled data), can be used to improve our understanding of lake system response to climate and non-climate pressures within a catchment. For example, land cover/use (and consequently trends in the riparian development of a lakeshore), soil moisture and elevation are, amongst others, standard remote sensing products, whilst water balance modelling techniques have recently assimilated remotely sensed information on climatic variables (e.g. evapotranspiration, precipitation and land surface temperature), soil moisture and vegetation properties. Taking the catchment of Lake Balaton (the largest freshwater body within the Danube Basin) as an example, we hereby present examples of global datasets that are available from various sources at varying temporal frequencies, time periods and spatial resolutions (Fig. 2, Table 2), and which are often used as catchment indicators of lake change. The next sections review the applicability of remote sensing in deriving some of these products in the realm of limnological research at catchment-wide scales across the globe, and specifically in the Danube Basin.

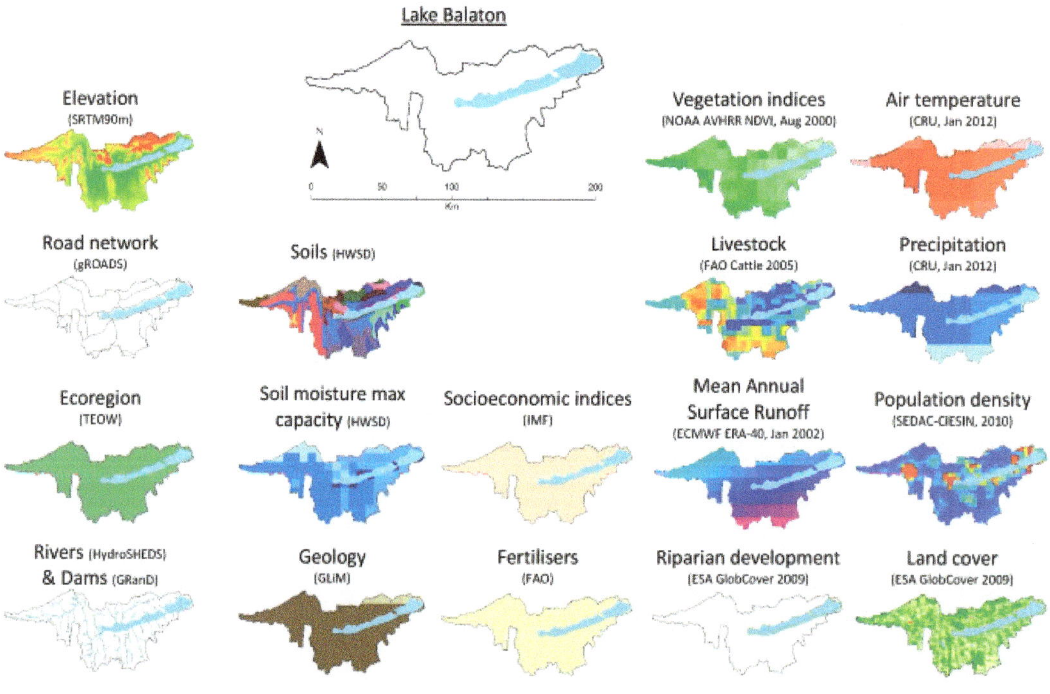

Figure 2: Examples of climate and non-climate datasets depicting lake pressure attributes in the catchment of Lake Balaton. SRTM90m (Shuttle Radar Topography Mission at 90 m spatial resolution); gROADS (Global Roads Open Access Data Set); TEOW (Terrestrial Ecore-

gions of the World); GRanD (Global Reservoir and Dam database); HWSD (Harmonised World Soil Database); GLiM (Global Lithological Map database); IMF (International Monetary Fund); FAO (Food and Agriculture Organisation); NOAA (National Oceanic and Atmospheric Administration); AVHRR (Advanced Very High Resolution Radiometer); NDVI (Normalised Difference Vegetation Index); ECMWF ERA-40 (European Centre for Medium-Range Weather Forecasts 40-year re-analysis model data); ESA GlobCover (European Space Agency Global Land Cover product); CRU (Climatic Research Unit); SEDAC (Socioeconomic Data & Applications Center); CIESIN (Center for International Earth Science Information Network).

Table 2: Examples of global data sources for deriving catchment attributes (period(s) of projected values are shown in square brackets).

Dataset/database	Product parameter(s)	Time period	Format	Units	Spatial resolution
Harmonised World Soil Database (HWSD)	Soil type	2008	Raster	SMU	1 km
Global Lithological Map Database (GLiM)	Geology	2012	Raster	–	62.5 km
Global Reservoir and Dam Database v.1.1 (GRanD)	Location and age of dams; spatial extent of reservoirs	2011	Vector	–	–
Terrestrial Ecoregions of the World (TEOW)	16 ecoregions	2001	Polygons	–	–
HydroSHEDS (USGS)	River network	2008	Vector	–	1 km
Gridded Population of the World v.3 (GPWv3)	Population density	1990, 1995, 2000, [2005], [2010], [2015]	Raster	Persons per km^2	5 km
Shuttle Radar Topography Mission v.4.1 (SRTM90m)	DEM (elevation)	2008	Raster	m	90 m
Climatic Research Unit Time Series v.3.21 (CRU TS 3.21)	Precipitation	1901–2012; monthly	NetCDF	mm	62.5 km
	Potential evapotranspiration			mm	
	Temperature			10 × °C	
European Centre for Medium-Range Weather Forecasts (ECMWF) ERA-40	Runoff	Sep 1957–Aug. 2002; monthly	NetCDF	m/day	~ 15 km
Global Roads Open Access Data Set (gROADSv1)	Road network	1980–2010	Vector	–	–
European Space Agency (ESA) Climate Change Initiative (CCI) Land Cover (LC)	Land cover	1998–2002, 2003–2007, 2008–2012	Raster	–	300 m
Gridded Livestock of the World (FAO, Food and Agriculture Organisation)	Cattle, poultry, buffalo, goat, pig, sheep	2000, 2005	Raster	Animals per km^2	5 km
Fertilisers (FAO)	Nitrogen & phosphate	2002–2010; annual	csv	Tonnes per 1000 ha	–
Irrigated land (actual & potential) (FAO)	Agricultural area irrigated; Total area equipped for irrigation	1961–2011; annual	csv	1000 ha	–
Gross Domestic Product (GDP) per capita, International Monetary Fund (IMF)	GDP per capita	1980–[2018]; annual	csv	$ (USD)	–

CATCHMENT WATER BALANCE

In recent decades the combined effects of economic growth and rapidly increasing global human population has led to increased demand for freshwater, which in turn has resulted in the intensification of water withdrawal from surface freshwater bodies and groundwater pools. Human activities have caused global water scarcity, particularly in arid and semi-arid regions, where overuse of natural water resources has taken place (Wada et al., 2011). At the global scale agriculture is the main water consumer with over 75% of abstractions, far outweighing the amounts used for industrial (20%) or domestic purposes (5%) (FAO, 2014 and UNEP, 2008). However, significant regional variations also exist reflecting local climate and physiographic features (Shiklomanov, 2000), imprinting an often complex picture of change when trying to assess the strong impact that global climate change has on freshwater supplies (WWAP, 2012). Determining the extent and severity of impacts of human activities and climate change on water resources across the globe, particularly in ungauged basins, has hitherto been a major issue. This has often been addressed through water balance modelling that can generate spatial and temporal variations of water availability in lake catchments and river basins, and thus contribute towards the management of water resources from local to global scales (Mitsova, 2014).

In principle, water balance models take into account the amount of water input through precipitation and snowmelt, and the amount of water output through evaporation, transpiration, surface (including river) runoff, groundwater discharge and percolation. There exist numerous water balance models, one of the earlier and more simplistic being the Thornthwaite and Mather (TM) model (Thornthwaite and Mather, 1955 and Thornthwaite and Mather, 1957), a revision of an earlier model by Thornthwaite (1948), which requires precipitation, air temperature and latitude to produce estimates of evapotranspiration, soil moisture storage, snow storage, surplus and runoff (McCabe and Markstrom, 2007). Many studies have since developed new or modified existing water balance models for use in different applications and at various spatio-temporal scales and the reader is referred to Xu and Singh, 1998 and Xu and Singh, 2004 andBoughton (2005) for comprehensive reviews of various water balance modelling techniques.

A key requirement of large hydrological systems, such as the Danube Basin, is that these water balance techniques and models must be applicable to regional or even continental scales, which is potentially where remote sensing can have a part to play. At a continental scale applicable water balance models include FAO AQUASTAT (2001) that was developed to map agricultural water use in Africa, WaterGAP that simulates global water availability and use in large drainage basins (Döll et al., 2003 and Alcamo et al., 2003) and GWAVA that models global water availability with demonstrable applications in Africa, South America and Europe (Meigh et al., 1999). In addition, the European Union Water and Global Change (WATCH) project created a multi-model ensemble bringing together nine large-scale hydrological models to compare their ability to capture mean annual runoff and quantify the uncertainty related to the modelling outputs. It was found that even though the performance of each individual WATCH model differed, their ensemble mean produced reliable estimates of runoff in European catchments (Gudmundsson et al., 2012a and Gudmundsson et al., 2012b).

The application of water balance modelling specifically within the Danube Basin in the past has tended to make use of historic weather station measurements and simulation data as input parameters (e.g. Mauser and Bach, 2009, Petrovič et al., 2010, Klein et al., 2011 and Kling et al., 2012) rather than making direct use of remote sensing observations. However, the GLOWA project (Global Change of the water cycle; funded by the German Ministry of Research and Education (BMBF)) is an example of one such initiative that exemplified the potential of remote sensing to address the impact of global change on regional water resources. Within this initiative is included the GLOWA-Danube project (dedicated to the Upper Danube watershed) that aims to investigate the sustainability of future water use using modelling techniques (Ludwig et al., 2003a). Within the GLOWA-Danube framework satellite-retrieved land cover information was used in a physically-based water balance model to provide spatio-temporal estimates of evapotranspiration, soil moisture, snow cover and runoff (Ludwig et al., 2003b). The study combined multi-temporal coarse spatial resolution NOAA AVHRR images and spectral unmixing techniques to produce sub-pixel land cover information at the fine scales required for the detailed description of various hydrological processes, and showed that the approach produced improved model results compared to the use of conventionally used land cover data (e.g. CORINE maps).

Role of remote sensing in water balance studies

Water balance models require a variety of data ranging from climatic variables to information on soil and vegetation properties. Remote sensing has the potential to produce some of this information at large enough spatial and temporal scales to fulfil the need for continuous water resource monitoring applications at scales varying from local to global (Xu et al., 2014) and it is not uncommon for a combination of remote sensing sensors to be used in hydrological modelling and water balance applications (e.g.Awange et al., 2008 and Milzow et al., 2011). Even though remotely sensed observations complement in situ observations in providing data inputs for water balance modelling techniques, its potential was realised more than four decades ago. For example, evapotranspiration estimations based on remotely sensed canopy temperatures date back to the 1970s (e.g. Stone and Horton, 1974, Blad and Rosenberg, 1976, Heilman et al., 1976 and Jensen and Chery, 1980), as does the estimation of remotely sensed soil water content, also known as soil moisture (Wang and Qu, 2009) and the mapping of snow cover (e.g. Kyle et al., 1978). On the other hand, precipitation can be estimated using remotely sensed data and various global products already exist, but their use in water balance modelling has only recently been realised (e.g. Pereira-Cardenal et al., 2011). Finally, there are no remote sensing products for air temperature, however Deus et al. (2013) propose the use of remotely sensed land surface temperature (LST) as a substitute for air temperature in water balance modelling, and several methods incorporate LST to estimate evapotranspiration (Verstraeten et al., 2008). Whilst Xu et al. (2014) and Karimi and Bastiaanssen (2015) provide excellent reviews on the integration of remotely sensed observations to hydrologic modelling in general, here we briefly review the status of remote sensing as a sources of functional information to understand the role of the catchment in changing lake behaviours.

Remote sensing sensors, scales and products

Evapotranspiration (ET)

Remote sensing is a promising tool for the representation of the spatial distribution of evapotranspiration (ET) across large spatial scales, which is required for distributed water balance modelling (Glenn et al., 2007). Several approaches have been proposed in the literature, commonly using optical and thermal data, and less frequently satellite microwave data. There are two types of methods that are used to estimate ET from remote sensing data; empirical and physical models. Empirical or statistical relationships make use of remotely sensed vegetation indices, such as the normalised difference vegetation index (NDVI) or the soil-adjusted vegetation index (SAVI) (Glenn et al., 2007), which are based on the fraction of the ground area covered or shaded by vegetation (Pereira et al., 2015). However, these methods are insensitive to conditions of water or salinity stress, which is not true for physical models (Pereira et al., 2015). Physical or analytical models solve the surface energy balance (SEB) equation using land surface temperature (LST) estimates based on remote sensing observations. The advantage of SEB models is that they can be used to estimate ET in areas with diverse vegetation types and heterogeneous vegetation cover (e.g. Minacapilli et al., 2009 and Pôças et al., 2013). The reader can refer to, for example, Glenn et al. (2007), Verstraeten et al. (2008),Kalma et al. (2008), Pereira et al. (2015) and Karimi and Bastiaanssen (2015) for detailed reviews of ET estimation methods based on remote sensing data.

The utility of remote sensing in water balance modelling, and particularly in the estimation of ET, depends on the specifications of the sensor and the cloud conditions (Kalma et al. 2008). Medium spatial resolution sensors, such as the Advanced Spaceborne Thermal Emission and Reflection Radiometer (ASTER) on-board the Terra satellite and the Thematic Mapper (TM), Enhanced TM (ETM +) and Operational Land Imager (OLI) on-board the Landsat series have a 16-day revisit capability, which means temporal interpolation techniques are needed to provide the daily inputs often required for hydrological models. However, ASTER has a spatial resolution of 15 m in the visible and NIR wavebands and 90 m in the TIR wavebands, while OLI offers a pixel size of 30 m in the visible and NIR and 100 m in the TIR, which in many cases around the globe and particularly within the Danube Basin, makes them suitable for mapping individual agricultural fields and smaller land parcels (Pereira et al., 2015). On the other hand, the Terra/Aqua Moderate Resolution Imaging Spectrometer (MODIS) and NOAA Advanced Very High Resolution Radiometer (AVHRR) provide daily observations depending on cloud cover at a spatial resolution of 1 km. Both MODIS and NOAA AVHRR can provide LST estimates with an accuracy of 1.0 °C at 1 km or 5 km spatial resolution (Glenn et al., 2007) and less than 1.3 °C at 4.4 km (Pinheiro et al., 2006), respectively. Finally, geostationary satellites at relatively coarse (1–5 km) spatial resolutions in the visible and TIR have very frequent revisit times (15–30 min) and could be used to derive LST, however their constant viewing angle introduces problems at high latitudes which limits their global applicability (Diak et al., 2004).

Overall, whilst the above illustrate the potential of remote sensing to provide reliable information on ET, its practical application in water balance modelling has often been hindered by concerns over accuracy. As Karimi and Bastiaanssen (2015) note, the reliability of remotely-sensed derived ET is often case and location specific, but the ensemble mean ET products currently under development may be a significant step towards addressing these concerns.

Soil moisture

The most common method for the retrieval of soil moisture is the use of microwave remote sensing sensors, both active and passive, and dates back to the 1970s (e.g.Eagleman and Ulaby, 1975, Schmugge et al., 1974, Njoku and Kong, 1977 and Mo et al., 1982). Remotely sensed soil moisture can only be estimated in the uppermost layer of the soil surface (i.e. top few centimetres), but assimilation of these observations within Soil-Vegetation-Atmosphere Transfer (SVAT) models helps retrieve soil moisture in the root zone (Wigneron et al., 2003), which is a key variable in hydrological modelling (e.g.Wilker et al., 2006 and Seneviratne et al., 2010).

There are many different surface soil moisture retrieval techniques that have been proposed in the literature, some used to derive global products, and these can be both empirical and physically-based (de Jeu et al., 2008 and Vereecken et al., 2008). While passive microwave sensors (radiometers) detect microwave radiation that is naturally emitted by the Earth's surface, active microwave sensors (RADAR) transmit an electromagnetic pulse and measure the scattered microwave energy back from the Earth surface. The success of microwave sensors in measuring soil moisture lies in the fact that changes in the soil dielectric constant due to changes in water content are detectable, particularly at low-frequency microwave regions (1–10 GHz) (e.g. Njoku and Kong, 1977, Wagner et al., 2007 and de Jeu et al., 2008). A major advantage of microwave remote sensing is that can it be used day and night regardless of cloud cover. A limitation is that dense vegetation cover and soil roughness introduce noise to soil moisture measurements (e.g. Wagner et al., 2007, de Jeu et al., 2008 and Seneviratne et al., 2010), but low-frequency microwave bands have been found to reduce this effect (Schmugge et al., 2002 and Njoku et al., 2003). In addition, a recent study in the Upper Danube basin demonstrated that soil moisture retrieval errors can be spatially diverse due to different factors (e.g. land surface heterogeneity) that introduce noise in the remote sensing observations, and that the retrieval accuracy strongly depends on the scale of observation (Loew, 2008).

Microwave instruments used in soil moisture retrieval include the non-operational Nimbus 7 Scanning Multichannel Microwave Radiometer (SMMR) (27–148 km spatial resolution), and Aqua Advanced Microwave Scanning Radiometer-EOS (AMSR-E) (38–56 km) for which global soil moisture products exist (e.g. Njoku et al., 2003 and Owe et al., 2008). The F-series Special Sensor Microwave Imager (SSM/I) and Special Sensor Microwave Imager Sounder (SSMIS) lack low-frequency wavebands that are desired for soil moisture retrieval. Nevertheless, soil moisture retrieval methods based on SSM/I data have been proposed in the literature (e.g. Jackson et al., 2002 and Wen et al., 2005). The relatively high-frequency TRMM TMI and WindSat radiometer data have also been used for the retrieval of soil moisture datasets (e.g. Bindlish et al., 2003, Gao et al., 2006 and Parinussa et al., 2012). ESA's Soil Moisture and Oceanic Salinity (SMOS) (30–50 km spatial resolution) was launched in 2009 and employs an L-band, which offers deeper vegetation penetration making it potentially more reliable over densely vegetated regions (Seneviratne et al., 2010). Within the upper Danube region initial results on the validation of soil moisture products suggested that radio-frequency interference can be a problem (dall'Amico et al., 2012) but that the spatial pattern of soil moisture variability was similar to in situ measurements, even if SMOS tended to generally underestimate soil moisture (Schlenz et al., 2012).

Radar scatterometers can also be suitable for the retrieval of soil moisture, depending on the frequency they operate in (Seneviratne et al., 2010). These include ESA's ERS-1/2

Advanced Microwave Instrument Scatterometer (AMI-SCAT) (50 km spatial resolution) and the METOP-A/B/C Advanced Scatterometer (ASCAT) (50 km) (Wagner et al., 2003 and Bartalis et al., 2007). Finally, NASA's Soil Moisture Active and Passive (SMAP) mission carries on-board active and passive L-band sensors and was launched on the 31st January 2015. The mission is designed to collect global observations of soil moisture at a spatial resolution of 9 km, with a revisit time of 2–3 days.

Soil moisture itself is a Global Climate Observing System (GCOS) Essential Climate Variable (ECV) and ESA's Climate Change Initiative (CCI) programme has funded the creation of a more than 30-year long soil moisture dataset by merging active (ERS AMI-SCAT and METOP ASCAT) and passive (SMMR, SSM/I, TMI, and AMSR-E) soil moisture retrievals (Wagner et al., 2012). This product at 0.25° spatial resolution aims to address the discrepancies that have been observed (e.g. Reichle et al., 2007 and Rüdiger et al., 2009) between different products developed for different (or the same) microwave sensors.

Other sensors that have been used in the retrieval of soil moisture include gravity recovery instruments (Seneviratne et al., 2010) and even global positioning system (GPS) receivers (Larson et al., 2008). Seasonal variations in terrestrial water content (TWC), including soil moisture, groundwater, snow and surface water, result in variations of the Earth's gravity field, which can be accurately measured with NASA/DLR's Gravity Recovery and Climate Experiment (GRACE) (e.g. Tapley et al., 2004, Wahr et al., 2004,Andersen et al., 2005, Swenson et al., 2008 and Rodell et al., 2009). However, separation of the soil moisture contribution to the total observed terrestrial water content requires knowledge of the other TWC components, making this technique complex and reliant on additional data being available. In addition, any estimates can only be done at much coarser spatial resolutions (at best 400 km) than those offered by microwave remote sensing. GPS satellite receivers use L-band frequencies, similar to the SMOS and SMAP sensors, and have shown the potential to estimate soil moisture at spatial resolutions of circa 300 m, but the effects of ground roughness and of different vegetation and soil types, requires further investigation (Larson et al., 2008). The reader is referred to Schmugge et al. (2002), Wigneron et al. (2003), de Jeu et al. (2008) and Seneviratne et al. (2010) for excellent reviews on the remote sensing of soil moisture.

Precipitation

Precipitation estimates with remote sensing can be generated using both in situ radar systems and space-borne sensors. Michaelides et al. (2009) and Tapiador et al. (2012)provide excellent reviews of remote sensing techniques for precipitation measurements. Typically, single polarisation weather radar systems have been used to derive rainfall rate, but limitations such as calibration issues and contamination by ground returns amongst others make them less than ideal for precipitation mapping. Recent technological advances have led to the development of more suitable polarimetric radar systems (also referred to as dual-polarisation radars) that overcome some of the limitations attached to single polarisation weather radars (Michaelides et al., 2009). For about three decades, rainfall has been retrieved from satellite instruments employing the visible, infrared and microwave regions of the spectrum. However, most of the satellite observations are discarded due to cloud contamination and issues with the variability of emissivity over land-water areas and especially in coastal areas. Another limitation is the coarse spatial resolution of the satellite sensors that ranges from 10 km^2 over land to 50 km^2 over the ocean, and the common assumption that rainfall is horizontally and vertically homogeneous.

According to Stephens and Kummerow (2007) the optimal approach to retrieve precipitation from space is by combining data from passive and active sensors. A major step towards this direction was made with the launch in 1997 of the space-borne Tropical Rainfall Measuring Mission (TRMM) that carries on-board two primary rainfall sensors; the 13.8 GHz Precipitation Radar (PR) and the TRMM Microwave Imager (TMI). Numerous algorithms for the retrieval of rainfall from TRMM data have since been developed (e.g., Iguchi et al., 2000 and Haddad et al., 1997), leading to the TRMM PR 2A25 and TRMM TMI 2A12 rainfall products. However, these two products exhibit spatial and seasonal differences that have yet to be explained (Wang et al., 2009) and are geographically limited to latitudes between 35°N–35°S. A third sensor on-board TRMM, the Visible and Infrared Radiometer (VIRS) can only provide indirect estimations of rainfall by means of precipitation indices.

Even though most available precipitation retrieval methods focus on rainfall rather than snowfall, high-frequency passive microwave wavebands (e.g. on AMSU) can provide snow detection and frozen precipitation estimations (Kongoli et al., 2004, Vila et al., 2007 and Michaelides et al., 2009). In fact, the Microwave Surface and Precipitation Products System (MSPPS) (now Microwave Integrated Retrieval System (MIRS)) is a suite of operational global products derived from the NOAA Advanced Microwave Sounding Unit (AMSU), including precipitation rate, total precipitable water and snow cover (Ferraro et al., 2005 and Vila et al., 2007).

The need for uniformly calibrated precipitation products that are independent of variation caused by different sensors is to be addressed by the international Global Precipitation Measurement (GPM) mission, which employs a constellation of passive microwave sensors. The GPM aims to provide high resolution satellite precipitation products every 2–3 h for use in various applications, such as hydrology and climatic studies. Building on the capabilities of the TRMM PR and TMI, the GPM-Core Observatory, which was launched in February 2014, covers latitudes between 65°N–65°S and contributes to the GPM constellation of sensors that combined cover the entire globe with datasets (Levels 0–3) freely available online. A global dataset for satellite-derived rainfall is the Global Satellite Mapping of Precipitation in Near-Real-Time (GSMaP_NRT) by the Japan Aerospace Exploration Agency (JAXA) Global Rainfall Watch System (Ver. 3.0), which provides global (60°N–60°S) rainfall rate (mm/h) at hourly intervals by combining various remote sensing data, including the GPM-Core Microwave Imager (GMI), TRMM TMI, DMSP (Defence Meteorological Satellite Programme) SSMIS, NOAA AMSU and METOP AMSU.

Snow cover

Both optical and microwave remote sensing are used for snow cover mapping. Even though optical remote sensing is limited to mapping snow extent, much more information can be derived from microwave data; this includes snow water equivalent (SWE; mass of water per unit area), snow depth (SD) (a relevant measurement because of its relationship to SWE), snow extent, and snow state (wet/dry). The use of optical data is based on the fact that snow reflects radiation strongly in the visible and very poorly in the NIR regions of the EM spectrum. Dozier (1989) mapped snow using the ratio of Landsat TM reflectance in the visible and NIR, a technique that had been previously used in the 1970s (e.g. Kyle et al., 1978) and was later adapted to develop the normalised-difference snow index (NDSI) for MODIS (Hall et al., 2002 and Hall and Riggs, 2007) and Landsat ETM + (Salomonson and Appel, 2004). NOAA AVHRR has also been used to map snow cover (e.g. Gesell, 1989, Voigt et al., 1999, Akyürek and Şorman, 2002, Matikainen et al., 2002 and Hüsler et al., 2012) despite its coarse spatial resolution (1 km), because it has a long data archive and a very frequent revisit time (twice daily). A limitation of using optical data to map snow cover is that it is often difficult to separate clouds from snow, because snow may show a similar spectral response to clouds in the visible and TIR (Akyürek and Şorman, 2002 and Miller et al., 2005). Another challenge is that the reflectance of snow decreases with impurity and age (seasonal and even daily ageing) (König et al., 2001 and Dietz et al., 2012) and vegetation (resulting in surface heterogeneity) might also affect the reflected signal (Nolin, 2010).

Snow attenuates microwave radiation, which is positively correlated with the amount of snow in the snowpack (Chang et al., 1987, König et al., 2001 and Clifford, 2010). In a similar manner to soils, the dielectric constants of ice and water are very different, which enables the use of microwave remote sensing to map snow. The depth of snow that can be mapped depends on the microwave wavelength; ranging from a minimum of around 2 cm (Dietz et al., 2012) to a maximum of 100 times the microwave wavelength (Clifford, 2010). In microwave remote sensing of snow, the size and shape of the snow crystals can affect the estimations (Foster et al., 1999), as well as depth, temperature, snow state and density (Schmugge et al., 2002). In addition, the presence of dense vegetation and liquid water in the snowpack can lead to underestimation of SD and SWE (Dietz et al., 2012), when passive microwave sensors are used. Using early morning satellite overpasses (local time) can minimise the impact of wet snow on snow mapping (Schmugge et al., 2002). Alternatively, active microwave remote sensing is more suitable for mapping wet snow (e.g. Baghdadi et al., 1997 and König et al., 2001), with the added bonus of finer spatial resolutions than passive microwave satellite sensors.

In the 1980s, the US National Weather Service (NWS) developed operational remote sensing products for snow hydrology, using NOAA AVHRR and GOES to produce periodic river basin snow cover extent maps (Schmugge et al., 2002) and since 2002 the Aqua AMSR-E has been used to produce a global SWE Level 3 product. MODIS data have also been used to derive a suite of snow products at various spatial (500 m–30 km) and temporal (daily-monthly) resolutions. In addition, the snowmelt runoff model (SRM) was adapted to use remotely-sensed snow cover information at basin-scales (Martinec et al., 2008). The reader can refer to, for example, König et al. (2001), Schmugge et al. (2002), Clifford (2010), Nolin (2010) and Dietz et al. (2012) for more detailed reviews on the remote sensing of snow cover.

Land surface temperature (LST)

Atmospherically corrected brightness temperature retrieved using remote sensing thermal sensors are used in land surface temperature retrieval. There are three techniques for the retrieval of LST from remote sensing data; single channel, generalised split window (GSW) and dual angle (or dual algorithm) (DA) approach. The single channel method requires a comprehensive radiative transfer model and atmospheric profiles. The split window approach, combines two thermal wavebands and accounts for atmospheric effects based on the differential absorption in adjacent infrared bands. However, because of the large emissivity differences between thermal wavebands over land due to vegetation, topography and soil, this method can be problematic. Land surface emissivity studies have been conducted to overcome this problem and various techniques have been proposed in the literature for NOAA AVHRR and Terra/Aqua MODIS (e.g. Wan and Dozier, 1996, Becker and Li, 1990, Mao et al., 2005 and Pinheiro et al., 2006), (A)ATSR (e.g. Sòria and Sobrino, 2007 and Galve et al., 2009) and MSG Spinning Enhanced Visible and Infrared Imager (SEVIRI) (e.g. Jiang and Li, 2008, Qian et al., 2013 and Freitas et al., 2013). The Terra/Aqua MOD11 LST product uses a split-window approach that utilises thermal bands 31 and 32. The dual angle approach is based on the differential absorption of a single waveband due to different viewing angles. In order to apply this technique two scenes simultaneously acquired by different satellites, e.g. the geostationary MSG and polar-orbiting Television Infrared Observation Satellites (TIROS-series) (Chédin et al., 1982) or a single scene from a satellite with multi-angle acquisition capability, such as the (Advanced) Along-Track Scanning Radiometer ((A)ATSR) on-board Envisat, can be used. The second option is more commonly employed using (A)ATSR data (e.g. Kustas and Norman, 1997, Sobrino et al., 1996, Sobrino et al., 2004 and Sòria and Sobrino, 2007) or GOES and Multifunction Transport Satellite (MTSAT) data (Freitas et al., 2013). Finally, a robust empirical technique has been developed specifically for the Terra Advanced Spaceborne Thermal Emission and Reflection Radiometer (ASTER); the Temperature Emissivity Separation (TES) algorithm (Gillespie et al., 1998) that uses ASTER TIR wavebands to estimate LST. The two main advantages of using Terra ASTER for LST retrievals are its fine spatial resolution (90 m) and more frequent revisit capability (twice daily) compared to other thermal sensors (e.g. NOAA AVHRR, (A) ATSR, MODIS, etc.).

Even though mostly thermal sensors are used in LST-retrievals, there exist a few studies where passive microwave sensors were employed (Tomlinson et al., 2011), which have the advantage of providing measurements even on cloudy days. Despite the coarse spatial resolution of passive microwave sensors (5–70 km depending on frequency), the AMSR-E (Chen et al., 2011) and more commonly the SSM/I (e.g. McFarland et al., 1990,Basist et al., 1998, Peterson et al., 2000, Williams et al., 2000 and Fily et al., 2003) have been used to estimate LST. For examples of in-depth reviews on the remote sensing of land surface temperature, the reader can refer to Prata et al. (1995), Dash et al. (2001),Schmugge et al. (2002), Tomlinson et al. (2011) and Li et al. (2013).

MAPPING CATCHMENT LAND COVER AND OTHER DRIVERS OF LAKE BEHAVIOUR

There are multiple catchment drivers of lake change due to intensification of human activities and climate change. Lakes across the globe can be affected by a combination of different stressors and pressures, based on vicinity to human establishments, type and intensity of human activities within the catchment, type and ratio of natural over altered landscape in the catchment, protection status and geographical location. Human activities such as agriculture, deforestation, urbanisation and industrial activities can be grouped under the umbrella of land cover/land use change (LULC), which may have various direct and indirect effects on lake water quality, for example increased nutrient loading and eutrophication (e.g. Huang et al., 2013 and Keatley et al., 2011), pollution (e.g. IJC, 1980), acidification (e.g. Nghiem et al., 2011), anoxic conditions (e.g. Valero-Garcés et al., 2000). The impact of other human activities such as tourism and commercial fishing on lake ecosystem status and ecosystem services is not possible to map and quantify with remote sensing unless their impact is measured indirectly. For example, by mapping forest area reduction due to tourism-related development (Chaplin and Babyn, 2013) or estimating epipelagic fish distribution and abundance in shallow waters (Churnside et al., 2003).

Despite the potential of remote sensing to provide direct estimation of a wide variety of ecosystem functions and services (cf. those described above) both Andrew et al. (2014) and de Araujo Barbosa et al. (2015) note that the majority of published studies that have attempted to map ecosystem functions and services (and by inference related catchment processes) do so using indirect proxies, such as land cover and land use data, derived mainly from remotely sensed sources (Table 3). It is pertinent, therefore, to reflect briefly on how land cover information is derived, and more importantly, the limitations and considerations to its application for monitoring change in ecosystem functions and services of large areas.

Table 3: Linkage between selected catchment/landscape features, fluvial processes, ecosystems services and their detection with remote sensing. (builds on Andrew et al., 2014 and Large and Gilvear, 2014). *Indicative*means the list is not exhaustive.

Catchment feature, or land cover	Inferred processes and characteristics	Ecosystem functions	Ecosystem services (*indicative*)	RS products (*required*)	Remote sensing source (*indicative*)
Catchment delineation	Catchment area, morphometry and hydrodynamic behaviour	Range of freshwater and terrestrial habitats	Water quantity and quality; seasonality	Topography	DEM derived from varied sources, including LiDAR, RADAR, UAV-based photography and aerial photography
Slope	Low slopes reduce energy gradient for transfer of water, sediment & nutrients, promoting storage and biogeochemical processing; reworking of sediment in active reaches	Hydraulic diversity; channel dynamism; habitat creation; sediment storage; habitat heterogeneity; increased wetted perimeter	Flood mitigation; water quality	Topography	DEM derived from varied sources, including LiDAR, RADAR, UAV-based photography and aerial photography
Catchment water balance and water level fluctuations	Water dynamics, quantity, quality and provenance of water and sediment runoff controlling biogeochemical processes including nutrient flux; snow accumulation; runoff volumes and timings; groundwater recharge and connectivity within alluvial flatlands; seasonal variations in water levels; movement of migratory species; crop phenology	Provisioning; supporting and regulating; cultural climate regulation; removal of pollutants; fish production	Water supply; biodiversity; carbon sequestration; flood mitigation and regulation; migratory species; tourism	Evapotranspiration Precipitation Soil moisture Water, snow/ice extent and duration Water level Groundwater	Thermal RS, VIs, climate data RADAR, PM RADAR (e.g. SMAP) Optical, RADAR, PM RADAR altimetry Gravity surveys, subsidence

Hyporheic zones and groundwater	Links to water flow and water quality in surface waters	Groundwater dependent terrestrial ecosystems including wetlands	Water supply; Flood mitigation and regulation; Biogeochemical filtration; biodiversity	Land cover map Species map, spectral diversity	Multispectral and multitemporal RS, HS. Range or variability of biochemistry, spectral indices or reflectance variation
Hydromorphological alteration	Elimination of flood inundation; loss of channel dynamism; altered supply of sediment & nutrients	Loss of natural land cover; hydrological alteration; habitat change	None	Land cover Feature extraction (e.g. embankment length, presence of dams etc)	Multispectral RS, HS
Riparian/river bank woodland	Shading, allochthonous leaf litter and woody debris input	Habitat creation and hydraulic diversity; cooling of water, food source	Biodiversity; fisheries	Spectral diversity Proxy indicators (e.g. canopy structure, productivity, gap fraction)	Range or variability of biochemistry, spectral indices or reflectance variation Varied e.g. LiDAR, Multispectral, HS, HY, multi-temporal RS
Floodplain physical habitat mosaic	Hydromorphological heterogeneity and channel dynamism; varied land use patterns	Range of freshwater-terrestrial habitats; ecotone creation	Biodiversity; fisheries	Spectral diversity, species map, land cover map, LAI, biomass	Range or variability of biochemistry, spectral indices or reflectance variation, LiDAR, image texture, Multispectral, HS, HY
Catchment disturbance	Potential for changing hydrological regime, channel dynamism, nutrient cycling, water quality, biodiversity	Potential loss/ degradation of natural land cover; hydrological alteration	Changed natural ecosystem services	Change in biomass/plant traits Plant stress Fire detection Drought monitoring	Multi-temporal RS VIs, HY Thermal anomalies, fire products (e.g. MODIS) Water content, surface temperature & ET
Wetlands	Plant and animal succession processes; enhanced nutrient cycling and storage; semi-aquatic habitats	Carbon sequestration, phosphorous uptake and denitrification; habitat heterogeneity; flow attenuation; refugia	Water supply; water quality; biodiversity	Land cover map Species map; spectral diversity	Multispectral and multitemporal RS, HS. Range or variability of biochemistry, spectral indices or reflectance variation
Floodplain forest	Substrate stabilisation; enhanced hydraulic processes	Flow attenuation, enhanced nutrient cycling and storage; habitat heterogeneity	Carbon sequestration; flood mitigation; biodiversity; water quality	Land cover map; biomass	Multispectral and multitemporal RS, HS
Floodplain lakes	Water storage; nutrient cycling	Refugia; habitat heterogeneity	Water supply; carbon sequestration; water quality; fisheries; biodiversity	Land cover map Water colour (phytoplankton, suspended matter, CDOM) and water clarity, LWST, ice cover	Multispectral and multitemporal RS, HS. Spectral analysis, multispectral RS. Thermal RS

Agriculture	Potential for increased runoff; enhanced sediment input; water quality deterioration	Loss of natural land cover; hydrological alteration	Natural ecosystem services reduced with increased crop production	Land cover map, LAI	Multispectral and multitemporal RS, HS, HY, LiDAR
Woodland plantation	Substrate stabilisation; enhanced hydraulic roughness	Flow attenuation; biomass increase	Timber production; flood mitigation	Land cover map, biomass	Multispectral and multitemporal RS, HS, HY, LiDAR
Urban areas	Potential for increased runoff and water quality deterioration	Loss of natural land cover; hydrological alteration	None	Land cover map	Multispectral and multitemporal RS, HS

LiDAR (Light Detection and Ranging); MS (Multispectral, e.g. Landsat Thematic Mapper); HS (High spatial resolution imagery (e.g. WorldView-3, Google Earth); HY (hyperspectral imagery/data); PM (passive microwave); VIs (vegetation indices); ET (evapotranspiration); CDOM (coloured dissolved organic matter); LWST (lake water surface temperature).

Perhaps the most common and one of the most basic forms of land cover analysis using remotely sensed data is land cover classification. This relates spectral and textural information within an image to specific land cover classes (as defined by the user) to produce a land cover map, from which land use may be inferred. There are many methods of image classification, from 'traditional' statistically-based per-pixel methods that assign individual pixels to land cover classes based upon their spectral response, to object-based approaches which segment the landscape into parcels based upon spectral response, shape, texture and other object characteristics. Additionally, a classification may be 'hard' where pixels are assigned to a single class, or 'soft' where pixels are assigned multiple class memberships. The latter is particularly relevant, when information relating to small scale changes in land cover are important and where measuring change requires greater precision. More often it is used when the spatial resolution of the instrument does not match the intrinsic scale of land cover heterogeneity within an area of interest. For example, Probeck et al. (2005) were able to derive subscale land cover information from relatively coarse (1 km) NOAA AVHRR imagery of the upper Danube by using a fuzzy classification approach, from which results were produced that compared favourably with independent datasets. A similar approach was used by Ludwig et al. (2003b). In both cases the fuzzy classification method used was spectral unmixing but alternatives include the use of artificial neural networks (e.g.Aitkenhead et al., 2008) and other types of machine learning.

The methods used to classify remotely sensed data are a field of research in themselves, and whilst helpful reviews of the current state of land cover extraction from remote sensing are available (e.g. Cihlar, 2000, Aplin, 2004 and Li et al., 2014), the myriad of methods and approaches leads to variability in the land cover map outputs derived by individual researchers and groups, limiting their wider applicability. To address this, some studies have employed the use of standardised LULC products (e.g. Table 4). These products tend to be produced from a single satellite sensor, removing variability associated with multiple sensor characteristics. Additionally, products may be single-year or multi-year products, enabling change to be detected using a standardised set of data characteristics and land cover classes. Thus, the use of a LULC product may reduce some of the uncertainty associated with individuals undertaking a classification exercise in isolation. However, the uptake in use of LULC products has been slow, put down partly to a lack of awareness but also a lack of confidence in spatial accuracy (Pfeifer et al., 2012 and Congalton et al., 2014). Most operational products are still restricted to coarse spatial resolution data (Table 4), which whilst providing frequent temporal updates may be a deterrent towards their use for some ecosystem studies.

Table 4: Example global land cover datasets suitable for large area land cover analysis.

Product	Sensor	Satellite	Spatial resolution	Temporal resolution
GLC2000[1]	VGT	SPOT	1 km	Single date: 2000
MCD12Q1[2]	MODIS	Aqua, Terra	500 m	Annual: 2001–2007
ESA CCI-LC[3]	MERIS	ENVISAT	300 m	Three epochs: 2008–2012; 2003–2007; 1998–2002
ESA CCI LC Seasonality (Three products: vegetation, snow cover & burned area)	MERIS	ENVISAT	300 m	Annual

1 European Commission (EC) Global Vegetation Monitoring (GVM) Unit Global Land Cover (GLC) 2000 product.
2 NASA Land Processes Distributed Active Archive Center (LP DAAC) Land Cover Type product.
3 European Space Agency (ESA) Climate Change Initiative (CCI) Land Cover (LC) product.

An additional limitation of LULC analysis is determining a biophysical basis for linking particular LULC classes to ecosystem functions and services. For individuals and groups undertaking their own classification there is often a tension when defining appropriate LULC classes to map between those features and classes that can be extracted from remotely sensed data, i.e. those that are spectrally separable, and those features and classes that are directly correlated with ecosystem function and/or catchment processes. The resulting class definitions are, therefore, often a pragmatic solution using features and LULC classes that can be detected remotely and using these as proxy indicators of the environmental variable of interest (e.g. Table 3). These proxies, however, may have limited relevance, being often the result of hypothesised but largely untested relationships (Andrew et al., 2014 and Seppelt et al., 2011) leading to unreliable outputs when compared to field observations (e.g. Eigenbrod et al., 2010). This potentially becomes even more problematic when using standardised LULC products, where there are often inconsistencies in class definitions between products (Congalton et al., 2014) and where different products are produced at different spatial resolutions, thus limiting their use as proxy indicators.

Also, by assuming that the same ecosystem value can be applied to the whole of a particular class ignores any scale dependency in ecosystem processes and leads to inaccuracies when compared with independently observed data (Eigenbrod et al., 2010). Thus ecosystem services derived from land cover information will be highly dependent upon the efficacy of linking a land cover class to a particular ecosystem service/function, the spatial resolution of the data and the accuracy of image classification (de Araujo Barbosa et al., 2015).

The accuracy of image classification is an important, but often ignored factor in determining the reliability of areal change in land cover class(es) derived from a classified map. Change in key indicator classes, such as woodland or urban areas, are important in understanding changes in lake water quality, habitat and other critical ecosystem services. However, as Olofsson et al. (2013) effectively argue, information relating to the accuracy of land cover outputs and products, including commonly used measures such as overall accuracy and the kappa coefficient, fail to provide enough information to effectively judge the effect that classification error has on estimates of areal change. To address this they suggest mapped areas should be adjusted using a stratified estimator approach (Olofsson et al., 2013). At the very least, the uncertainty and error associated with the mapping of ecosystem properties should be addressed and reported, with Rocchini et al. (2013) providing a useful summary of good practise for this very purpose. Even when using globally available LULC products the stated accuracy is not always acceptable for the purpose of mapping change (Congalton et al., 2014).Karimi and Bastiaanssen (2015) noted that the overall accuracies reported for global land cover maps varied between 69 and 87%, suggesting that such products should be used with caution in water accounting applications.

All of the limitations above suggest that whilst LULC information derived from remotely sensed sources has the potential for quantifying and mapping selected catchment and ecosystem functions and services across large areas, the limitations in terms of applicability (biophysical basis of land cover classes), data quality (accuracy) and data properties (spatial and temporal resolution) should be considered carefully when deciding how such data should be used. Fuller et al. (2003) provide a useful commentary on experiences of mapping the UK with remotely sensed data highlighting the limitations of comparing between products, whilst Seppelt et al. (2011) provide a useful review of critical questions that should be raised when reviewing ecosystem service assessments at a regional scale, which are directly relevant to remote sensing-based studies.

REMOTE SENSING OF LAKE LEVEL AND VOLUME

Finally, whilst this review has thus far focussed on catchment functions and characteristics, it is also pertinent to briefly assess the potential of remote sensing to provide information regarding lakes themselves. The remote sensing techniques and sensors used to map lake water quality are reviewed by Tyler et al., in this Special Issue. Remote sensing can also have a role to play in detecting water bodies (e.g. Verpoorter et al., 2014), and estimating lake water level change and volume.

Both water level and volume are important with respect to lake ecology, being determinants of water residence time, flushing rate and mixing. They are also highly influenced by human activities, with the construction of dams and impoundments, and associated freshwater withdrawal for domestic, agricultural and industrial use responsible for significantly altering water level (and, thus, volume) in some lakes and reservoirs globally (in Lake Chad, for example). However, not all lake water level change is caused by direct human intervention. Climate change and natural climatic fluctuations are also important drivers of water level change/fluctuation in freshwater bodies, particularly in closed lake basins in endorheic regions (Hutchinson, 1957).

Remote sensing of Lake water level

Water level change can be retrieved from remote sensing altimetry data (e.g. Birkett, 1995, Crétaux and Birkett, 2006, Hwang et al., 2011, Mercier et al., 2002, Ponchaut and Cazenave, 1998 and Swenson and Wahr, 2009) with very high accuracy (e.g. 3–33 cm;Birkett and Beckley, 2010; 25–53 cm; Frappart et al., 2006). Altimeters can be radar (microwave) or laser (visible and IR), but the second are sensitive to the presence of clouds. They both transmit pulses and receive the reflected signal, so the theoretical basis for measurements is similar (Duan and Bastiaanssen, 2013). Lake surface water level is estimated by calculating the difference between satellite height in respect to a reference surface (e.g. Earth's centre) and the satellite-to-surface range (calculated by the time taken for the pulse to return to the sensor since transmission), correcting at the same time for instrumental, atmospheric and other effects (Duan and Bastiaanssen, 2013). The disadvantage of altimeters is that they can only return measurements from along their track, which does not cover the globe. As a result, only specific water bodies (that fall into the satellite's track) can be detected. Laser altimeters such as the Geoscience Laser Altimeter System (GLAS) on-board ICESat (Ice, Cloud, and land Elevation Satellite) are more suitable for relatively small water bodies due narrower footprint size (~ 100 m) compared to radar altimeters (several kilometres).

Several radar altimeters were until recently or currently operational, including ERS Radar Altimeter (RA) and Envisat RA-2, the Poseidon sensors on-board TOPEX/Poseidon, Jason-1 and Jason-2 (or Ocean Surface Topography Mission, OSTM), and GeoSat Fellow On (GFO) Radar Altimeter. These sensors were used to create three distinguished databases that include lake water level estimates: (1) ESA's River and Lake project, (2) US Department of Agriculture (USDA) Global Reservoir and Lake Monitoring (GRLM) programme, and (3) Laboratoire d'Etudes en Géophysique et Océanographie Spatiales (LEGOS) and Geodesy, Oceanography and Hydrology from Space (GOHS) Hydroweb. A fourth database is based on ICESat laser altimetry; ICESat-GLAS level 2 Global Land Surface Altimetry data (ICESat-GLAS). For an excellent overview of these databases see Duan and Bastiaanssen (2013).

Remote sensing of lake water quantity (volume)

Lake water quantity (volume) cannot be measured directly from remote sensing. Water level information is combined with bathymetric information of the water body in order to produce estimates of lake volume. Apart from using traditional sonar techniques, bathymetric maps can also be produced using GPS and laser transit survey data (e.g.Wilcox and Los Huertos, 2005), and in fact the notion of a laser bathymeter is not new (Muirhead and Cracknell, 1986). To replace the need for bathymetric predictions, new techniques that make use of visible and IR-based lake surface area estimations have been developed for the retrieval of lake water volume (e.g. Duan and Bastiaanssen, 2013 and Lu et al., 2013). Finally, satellite altimetry-derived water levels of lakes can provide estimates of lake water volume change, when combined with satellite or otherwise-derived surface water area (e.g. Becker et al., 2010) or detailed bathymetric maps (Crétaux and Birkett, 2006).

OBSERVATIONS AND CONCLUSIONS

We began this review by presenting a suite of datasets (Fig. 2), some derived from remote sensing, that have the potential to capture both status and change in lake catchments. This is presented as a precursor to understanding both the causes and drivers of lake change, and which can complement direct observations of lake behaviour by satellite remote sensing (Tyler et al., in this issue). Whilst by no means an exhaustive review of the capabilities of remote sensing and geospatial technologies in this field, this paper points towards some of the key catchment variables that can be estimated, both directly and indirectly, as well as the limitations and challenges currently faced.

From the work that is reviewed it is clear that remote sensing is an important tool for mapping ecosystem functions and aiding our understanding of catchment drivers of lake change, with promising results frequently reported. However, it is also clear that many of the results are highly specific to a particular time and/or place, and thus a manifestation of the 'one time one place' approach identified by Woodcock (2002) as a common cause of uncertainty in the application of remotely sensed data for other applications. Time and again various reviews point to continued issues with respect to multiple scales, sensors, reliability and other methodological constraints that limit the operational application of remote sensing for extracting critical ecosystem functions. Many suggest further work is needed to develop techniques that combine multi-source information to provide disaggregated products for catchment monitoring and modelling.

In part a response to these issues, several reviews refer to increasing efforts to integrate information across a variety of sensors and scales, and the development of many standardised remotely sensed products (Andrew et al., 2014 and Pfeifer et al., 2012). Issues of spatial and biophysical relevancy persist, but with the advent of higher spatial resolution global datasets becoming available (e.g. Hansen and Loveland, 2012) and the introduction of new complementary in situ sensor systems capable of measurements at high temporal and spatial -frequencies (Crawford et al., 2014) at least some of these concerns will continue to be addressed.

As stated in our introduction, the future of remote sensing for catchment, ecosystem and lake-based studies looks very promising. ESA's Copernicus programme will not only include Sentinel's 2 and 3 (which themselves will include multiple versions, e.g., Sentinel 3A, 3B and 3C, thereby providing a constellation of 'observers') but also includes a C-band SAR sensor on-board Sentinel-1 (launched April 2014) to provide observations of ice, oil spills and land surface monitoring. Additionally, future missions such as ESA's BIOMASS (planned for 2020) will provide enhanced information on above-ground biomass dynamics at 50 m spatial resolution, whilst ESA's FLEX (Fluorescence Explorer), Germany's EnMAP (30 m spatial resolution) and NASA's HYSPIRI (60 m) satellite sensors will allow regional to global assessments of plant biophysical and biochemical status, biogeochemical cycling, erosion monitoring and ecosystem fragmentation and disturbance. Thus the range of spaceborne sensors for deriving quantitative estimates of critical ecosystem and catchment variables appears to be in good shape looking forward.

In addition to new optical and SAR sensors described above and elsewhere in this review, there is clearly much progress in developing geospatial technologies which will have the potential to revolutionise the mapping of ecosystem functions. These include the development of UAV platforms for low cost, high resolution data gathering, as well as new hyperspectral sensor technologies and multispectral LiDAR that have the potential for direct estimation of biophysical variables (Woodhouse et al., 2011) and plant traits, removing the need for reliance on simple proxies. Furthermore, the existence of platforms that facilitate the use of extensive datasets and long remote sensing archives, such as the Google Earth Engine (GEE) for environmental data analysis, can and will provide, in part, the necessary infrastructure that the era of 'Big Data' asks for.

To maximise the exploitation of new technologies for ecosystem service assessmentAndrew et al. (2014) note the need for close collaboration between remote sensing scientists, ecologists and social scientists. Clearly, for this application the same is true of remote sensing scientists and limnologists, hydrologists, ecologists and many more to truly understand the drivers of lake change and behaviour across the globe, with the potential for much of this to be informed by remote sensing observations. Multidisciplinary collaboration is a necessity for the integration and assimilation of geospatial data and new technologies in hydrological modelling and traditional limnological studies, and in order to be able to understand and interpret both the uncertainties of the data used and the research outputs. The inclusion of remote sensing scientists in limnological research is therefore considered prudent and often an essential step towards understanding remote sensing data and the specific uncertainties, interpretation and limitations related to remote sensing products before their use. In this respect, projects such as the UK Natural Environment Research Council-funded GloboLakes project (Global Observatory of Lake Responses to Environmental Change) is one of several initiatives to address issues of lake change in just such a way, and which will lead to paradigm shifting developments in the characterisation of the planet's freshwater resources and responses to environmental change.

ACKNOWLEDGEMENTS

This work was supported by NERC Consortium Grant NE/J024279/1 — GloboLakes (Global Observatory of Lake Responses to Environmental Change): www.globolakes.ac.uk The Authors also wish to thank the constructive feedback from the anonymous reviewers.

REFERENCES

1. Aitkenhead, M.J., Flaherty, S., Cutler, M.E.J., 2008. Evaluating neural networks and evidence pooling for land cover mapping. Photogramm. Eng. Remote. Sens. 74, 1019–1032.
2. Akyürek, Z., Şorman, A.Ü., 2002. Monitoring snow-covered areas using NOAA-AVHRR data in the eastern part of Turkey. Hydrol. Sci. J. 47, 243–252. http://dx.doi.org/10. 1080/02626660209492927.
3. Alcamo, J., Döll, P., Henrichs, T., Kaspar, F., Lehner, B., Rösch, T., Siebert, S., 2003. Development and testing of the WaterGAP 2 global model of water use and availability. Hydrol. Sci. J. 48, 317–337. http://dx.doi.org/10.1623/hysj.48.3.317.45290.
4. Andersen, O.B., Seneviratne, S.I., Hinderer, J., Viterbo, P., 2005. GRACE-derived terrestrial water storage depletion associated with the 2003 European heat wave. Geophys. Res. Lett. 32, L18405. http://dx.doi.org/10.1029/2005GL023574.
5. Andrew, M.E., Wulder, M.A., Nelson, T.A., 2014. Potential contributions of remote sensing to ecosystem service assessments. Prog. Phys. Geogr. 38, 328–353.
6. Aplin, P., 2004. Remote sensing: land cover. Prog. Phys. Geogr. 28, 283–293.
7. Awange, J.L., Sharifi, M.A., Ogonda, G., Wickert, J., Grafarend, E.W., Omulo, M.A., 2008. The falling Lake Victoria water level: GRACE, TRIMM and CHAMP satellite analysis of the lake basin. Water Resour. Manag. 22, 775–796.
8. Baghdadi, N., Gauthier, Y., Bernier, M., 1997. Capability of multitemporal ERS-1 SAR data for wet-snow mapping. Remote Sens. Environ. 60, 174–186.
9. Bartalis, Z., Wagner, W., Naeimi, V., Hasenauer, S., Scipal, K., Bonekamp, H., Figa, J., Anderson, C., 2007. Initial soil moisture retrievals from the METOP-A Advanced Scatterometer (ASCAT). Geophys. Res. Lett. 34, L20401. http://dx.doi.org/10.1029/ 2007GL031088.
10. Basist, A., Grody, N.C., Peterson, T.C., Williams, C.N., 1998. Using the Special Sensor Microwave/Imager to monitor land surface temperatures, wetness and snow cover. J. Appl. Meteorol. 37, 888–911.
11. Becker, F., Li, Z.-L., 1990. Towards a local split window method over land surfaces. Int. J. Remote Sens. 11, 369–393. http://dx.doi.org/10.1080/01431169008955028.
12. Becker, M., LLovel, W., Cazenave, A., Güntner, A., Crétaux, J.-F., 2010. Recent hydrological behavior of the East African great lakes region inferred from GRACE, satellite altimetry and rainfall observations. Compt. Rendus Geosci. 342, 223–233.
13. Bindlish, R., Jackson, T.J., Wood, E., Gao, H., Starks, P., Bosch, D., Lakshmi, V., 2003. Soil moisture estimates from TRMM Microwave Imager observations over the Southern United States. Remote Sens. Environ. 85, 507–515.
14. Birkett, C., 1995. The contribution of TOPEX/POSEIDON to the global monitoring of climatically sensitive lakes. J. Geophys. Res. 100, 25179–25204.
15. Birkett, C.M., Beckley, B., 2010. Investigating the performance of the Jason-2/OSTM radar altimeter over lakes and reservoirs. Mar. Geod. 33, 204–238.
16. Blad, B.L., Rosenberg, N.J., 1976. Evaluation of resistance and mass transport evapotranspiration models requiring canopy temperature data. Agron. J. 68, 764–769.
17. Boughton, W., 2005. Catchment water balance modelling in Australia 1960–2004. Agric. Water Manag. 71, 91–116.
18. Chang, A.T.C., Foster, J.L., Hall, D.K., 1987. Nimbus-7 SMMR derived global snow cover parameters. Ann. Glaciol. 9, 39–44.
19. Chaplin, J., Babyn, L., 2013. Using remote sensing and GIS to investigate the impacts of tourism on forest cover in the Annapurna Conservation Area, Nepal. Appl. Geogr. 43, 159–168.
20. Chédin, A., Scott, N.A., Berroir, A., 1982. A single-channel, double-viewing angle method for sea surface temperature determination from coincident METEOSAT and TIROSN radiometric measurements. J. Appl. Meteorol. 21, 613–618.
21.
22. Chen, S., Chen, X., Chen, W., Su, Y., Li, D., 2011. A simple retrieval method of land surface temperature from AMSR-E passive microwave data — a case study over Southern China during the strong snow disaster of 2008. Int. J. Appl. Earth Obs. Geoinf. 13, 140–151.
23. Churnside, J.H., Demer, D.A., Mahmoudi, B., 2003. A comparison of Lidar and echosounder measurements of

fish schools in the Gulf of Mexico. ICES J. Mar. Sci. 60, 147–154. http://dx.doi.org/10.1006/jmsc.2002.1327.
24. Cihlar, J., 2000. Land cover mapping of large areas from satellites: status and research priorities. Int. J. Remote Sens. 21, 1093–1114.
25. Clifford, D., 2010. Global estimates of snow water equivalent from passive microwave instruments: history, challenges and future developments. Int. J. Remote Sens. 31, 3707–3726. http://dx.doi.org/10.1080/01431161.2010.483482.
26. Congalton, R.G., Jianyu, G., Yadav, K., Thenkabail, P., Ozdogan, M., 2014. Global land cover mapping: a review and uncertainty analysis. Remote Sens. 6, 12070–12093.
27. Crăciunescu, V., Flueraru, C., Stăncălie, G., 2010. The usage of the historical cartographic datasets and the remote sensing data for the better understanding and mapping of the 2006 Danube floods in Romania. Acta Geod. Geophys. Hung. 45, 112–119.
28. Crawford, J.T., Loken, L.C., Casson, N.J., Smith, C., Stone, A.G., Winslow, L.A., 2014. Highspeed limnology: using advanced sensors to investigate spatial variability in biogeochemistry and hydrology. Environ. Sci. Technol. 49, 442–450.
29. Crétaux, J.-F., Birkett, C., 2006. Lake studies from satellite radar altimetry. Compt. Rendus Geosci. 338, 1098–1112.
30. dall'Amico, J.T., Schlenz, F., Loew, A., Mauser, W., 2012. First results of SMOS soil moisture validation in the Upper Danube Catchment. IEEE Trans. Geosci. Remote Sens. 50, 1507–1516.
31. Dash, P., Göttsche, F.-M., Olesen, F., Fischer, H., 2001. Retrieval of land surface temperature and emissivity from satellite data: physics, theoretical limitations and current methods. J. Indian Soc. Remote Sens. 29, 23–30.
32. de Araujo Barbosa, C.C., Atkinson, P.M., Dearing, J.A., 2015. Remote sensing of ecosystem services: a systematic review. Ecol. Indic. 52, 430–443. de Groot, R.S., Alkemade, R., Braat, L., Hein, L., Willemen, L., 2010. Challenges in integrating the concept of ecosystem services and values in landscape planning, management and decision making. Ecol. Complex. 7, 260–272.
33. de Jeu, R.A.M., Wagner, W., Holmes, T.R.H., Dolman, A.J., van de Giesen, N.C., Friesen, J., 2008. Global soil moisture patterns observed by space borne microwave radiometers and scatterometers. Surv. Geophys. 29, 399–420. http://dx.doi.org/10.1007/s10712- 008-9044-0.
34. Deus, D., Gloaguen, R., Krause, P., 2013. Water balance modeling in a semi-arid environment with limited in situ data using remote sensing in Lake Manyara, East African Rift, Tanzania. Remote Sens. 5, 1651–1680. http://dx.doi.org/10.3390/rs5041651.
35. Diak, G.R., Mecikalski, J.R., Anderson, M.C., Norman, J.M., Kustas, W.P., Torn, R.D., DeWolf, R.L., 2004. Estimating land surface energy budgets from space; review and current efforts at the University of Wisconsin-Madison and USDA-ARS. Am. Meteorol. Soc. http://dx.doi.org/10.1175/BAMS-85-1-65.
36. Dietz, A.J., Kuenzer, C., Gessner, U., Dech, S., 2012. Remote sensing of snow — a review of available methods. Int. J. Remote Sens. 33, 4094–4134.
37. Dokulil, M.T., 2013. Impact of climate warming on European inland waters. Inland Waters 4, 27–40. http://dx.doi.org/10.5268/IW-4.1.705.
38. Döll, P., Kaspar, F., Lehner, B., 2003. A global hydrological model for deriving water availability indicators: model tuning and validation. J. Hydrol. 270, 105–134.
39. Downing, J.A., Prairie, Y.T., Cole, J.J., Duarte, C.M., Tranvik, L.J., Striegl, R.G., McDowell, W.H., Kortelainen, P., Caraco, N.F., Melack, J.M., Middelburg, J.J., 2006. The global abundance and size distribution of lakes, ponds, and impoundments. Limnol. Oceanogr. 51, 2388–2397.
40. Dozier, J., 1989. Spectral signature of alpine snow cover from the Landsat Thematic Mapper. Remote Sens. Environ. 28, 9–22.
41. Duan, Z., Bastiaanssen, W.G.M., 2013. Estimating water volume variations in lakes and reservoirs from four operational satellite altimetry databases and satellite imagery data. Remote Sens. Environ. 134, 403–416.
42. Eagleman, J.R., Ulaby, F.T., 1975. Remote sensing of soil moisture by Skylab radiometer and scatterometers sensors. J. Astronaut. Sci. 23, 147–159.
43. Economics, L., 2015. The Case for Space. Report to the UK Space Agency Available at http://www.ukspace.org/wp-content/uploads/2015/07/LE-Case-for-Space-2015-Full-Report.pdf.
44. Eigenbrod, F., Armsworth, P.R., Anderson, B.J., Heinemayer, A., Gillings, S., Roy, D.B., Thomas, C.D., Gaston, K.J., 2010. The impact of proxy-based methods on mapping the distribution of ecosystem services. J. Appl. Ecol. 47, 377–385.
45. FAO, 2014. Coping with water scarcity: An action framework for agriculture and food security. FAO Water Report 38. Food and Agriculture Organization of the United Nations, Rome.
46. FAO AQUASTAT, 2001. Available at http://www.fao.org/nr/water/aquastat/main/index.stm.
47. Ferraro, R.R., Weng, F., Grody, N.C., Zhao, L., Meng, H., Kongoli, C., Pellegrino, P., Qiu, S., Dean, C., 2005. NOAA operational hydrological products derived from the Advanced Microwave Sounding Unit. IEEE Trans. Geosci. Remote Sens. 43, 1036–1049.
48. Fily, M., Royer, A., Goïta, K., Prigent, C., 2003. A simple retrieval method for land surface temperature and fraction of water surface determination from satellite microwave brightness temperatures in sub-Arctic areas. Remote Sens. Environ. 85, 328–338.
49. Foster, J.L., Hall, D.K., Chang, A.T.C., Rango, A., Wergin, W., Erbe, E., 1999. Effects of snow crystal shape on the scattering of passive microwave radiation. IEEE Trans. Geosci. Remote Sens. 37, 1165–1168.
50. Frappart, F., Calmant, S., Cauhopé, M., Seyler, F., Cazenave, A., 2006. Preliminary results of ENVISAT

RA-2-derived water levels validation over the Amazon Basin. Remote Sens. Environ. 100, 252–264.

51. Freitas, S.C., Trigo, I.F., Macedo, J., Barroso, C., Silva, R., Perdigão, R., 2013. Land surface temperature from multiple geostationary satellites. Int. J. Remote Sens. 34, 3051–3068. http://dx.doi.org/10.1080/01431161.2012.716925.

52. Fuller, R.M., Smith, G.M., Devereux, B.J., 2003. The characterisation and measurement of land cover change through remote sensing: problems in operational applications? Int. J. Appl. Earth Obs. Geoinf. 4, 243–253.

53. Galve, J.M., Coll, C., Prata, F., 2009. Monthly land surface temperature maps over European zone using advanced along track scanning radiometer data for 2007. Proceedings of the Geoscience and Remote Sensing Symposium. 2009 IEEE international, IGARSS 2009 4, pp. 292–295.

54. Gao, H., Wood, E.F., Jackson, T.J., Drusch, M., Bindlish, R., 2006. Using TRMM/TMI to retrieve surface soil moisture over the Southern United States from 1998 to 2002. J. Hydrometeorol. 7, 23–38.

55. Gesell, G., 1989. An algorithm for snow and ice detection using AVHRR data, an extension to the APOLLO software package. Int. J. Remote Sens. 10, 897–905. http://dx.doi.org/ 10.1080/01431168908903929.

56. Gillespie, A., Rokugawa, S., Matsunaga, T., Cothern, J.S., Hook, S., Kahle, A.B., 1998. A temperature and emissivity separation algorithm for advanced spaceborne thermal emission and reflection radiometer (ASTER) images. IEEE Trans. Geosci. Remote Sens. 36, 1113–1126.

57. Glenn, E.P., Huete, A.R., Nagler, P.L., Hirschboeck, K.K., Brown, P., 2007. Integrating remote sensing and ground methods to estimate evapotranspiration. CRC Crit. Rev. Plant Sci. 26, 139–168.

58. Gudmundsson, L., Tallaksen, L.M., Stahl, K., Clark, D.B., Dumont, E., Hagemann, S., Bertrand, N., Gerten, D., Heinke, J., Hanasaki, N., Voss, F., Koirala, S., 2012a. Comparing large-scale hydrological model simulations to observed runoff percentiles in Europe. J. Hydrometeorol. 13, 604–620.

59. Gudmundsson, L., Wagener, T., Tallaksen, L.M., Engeland, K., 2012b. Evaluation of nine large-scale hydrological models with respect to the seasonal runoff climatology in Europe. Water Resour. Res. 48, W11504. http://dx.doi.org/10.1029/2011WR010911.

60. Haddad, Z., Smith, E.A., Kummerow, C.D., Iguchi, T., Farrar, M., Darden, S., Alves, M., Olson, W., 1997. The TRMM 'Day-1' radar/radiometer combined rain-profiling algorithm. J. Meteorol. Soc. Jpn. 75, 799–809.

61. Hall, D.K., Riggs, G.A., 2007. Accuracy assessment of the MODIS snow-cover products. Hydrol. Process. 21, 1534–1547.

62. Hall, D.K., Riggs, G.A., Salomonson, V.V., DiGirolamo, N., Bayr, K.J., 2002. MODIS snowcover products. Remote Sens. Environ. 83, 181–194.

63. Hansen, M.C., Loveland, T.R., 2012. A review of large area monitoring of land cover change using Landsat data. Remote Sens. Environ. 122, 66–74.

64. Heilman, J.L., Kanemasu, E.T., Rosenberg, N.J., Blad, B.L., 1976. Thermal scanner measurement of canopy temperatures to estimate evapotranspiration. Remote Sens. Environ. 5, 137–145.

65. Huang, J., Zhan, J., Yan, H., Wu, F., Deng, X., 2013. Evaluation of the impacts of land use on water quality: a case study in the Chaohu Lake Basin. Sci. World J. 2013, 329187. http://dx.doi.org/10.1155/2013/329187.

66. Hüsler, F., Jonas, T., Wunderle, S., Albrecht, S., 2012. Validation of a modified snow cover retrieval algorithm from historical 1-km AVHRR data over the European Alps. Remote Sens. Environ. 121, 497–515.

67. Hutchinson, G.E., 1957. A treatise on limnology, Volume I, Geography, physics and chemistry. Wiley, New York.

68. Hwang, C., Kao, Y., Tangdamrongsub, N., 2011. A preliminary analysis of lake level and water storage changes over lakes Baikal and Balkhash from satellite altimetry and gravimetry. Terr. Atmos. Ocean. Sci. 22, 97–108.

69. Iguchi, T., Kozu, T., Meneghini, R., Awaka, J., Okamoto, K., 2000. Rain-profiling algorithm for the TRMM precipitation radar. J. Appl. Meteorol. 39, 2038–2052.

70. IJC, 1980. Pollution in the Great Lakes basin from land use activities. Report, International Joint Commission.

71. Jackson, T.J., Hsu, A.Y., Neill, P.E.O., 2002. Surface soil moisture retrieval and mapping using high-frequency microwave satellite observations in the Southern Great Plains. J. Hydrometeorol. 3, 688–699.

72. Jensen, J.R., Chery Jr., D.L., 1980. Landsat crop identification for watershed water balance determinations. Int. J. Remote Sens. 1, 345–359. http://dx.doi.org/10.1080/ 01431168008948247.

73. Jiang, G.-M., Li, Z.-L., 2008. Split-window algorithm for land surface temperature estimation from MSG1-SEVIRI data. Int. J. Remote Sens. 29, 6067–6074. http://dx.doi.org/10. 1080/01431160802235860.

74. Kalma, J.D., McVicar, T.R., McCabe, M.F., 2008. Estimating land surface evaporation: a review of methods using remotely sensed surface temperature data. Surv. Geophys. 29, 421–469.

75. Karimi, P., Bastiaanssen, W.G.M., 2015. Spatial evapotranspiration, rainfall and land use data in water accounting — part 1: review of the accuracy of the remote sensing data. Hydrol. Earth Syst. Sci. 19, 507–532.

76. Keatley, B.E., Bennett, E.M., MacDonald, G.K., Taranu, Z.E., Gregory-Eaves, I., 2011. LandUse Legacies Are Important Determinants of Lake Eutrophication in the Anthropocene. PLoS ONE 6, e15913. http://dx.doi.org/10.1371/journal.pone.0015913.

77. Klein, B., Krahe, P., Lingemann, I., Nilson, E., Kling, H., Fuchs, M., 2011. Assessing climate change impacts on water balance in the Upper Danube Basin based on a 23 member RCM Ensemble. Proceedings of the XXVth Conference of the Danube Countries, Hungary.

78. Kling, H., Fuchs, M., Paulin, M., 2012. Runoff conditions in the upper Danube basin under an ensemble of climate change scenarios. J. Hydrol. 424–425, 264–277.

79. Knoll, L.B., Vanni, M.J., Renwick, W.H., Dittman, E.K., Gephart, J.A., 2013. Temperate reservoirs are large carbon sinks and small CO2 sources: Results from high-resolution carbon budgets. Glob. Biogeochem. Cycles 27, 52–64. http://dx.doi.org/10.1002/gbc. 20020.
80. Kollár, S., Vekerdy, Z., Márkus, B., 2011. Forest habitat change dynamics in a riparian woodland. Spat. Stat. 7, 371–376.
81. Kongoli, C., Grody, N.C., Ferraro, R.R., 2004. Interpretation of AMSU microwave measurements for the retrievals of snow water equivalent and snow depth. J. Geophys. Res. - Atmos. 109, D24111.
82. König, M., Winther, J.-G., Isaksson, E., 2001. Measuring snow and glacier ice properties from satellite. Rev. Geophys. 39, 1–27.
83. Kovàcs, F., 2007. Assessment of regional variations in biomass production using satellite image analysis between 1992 and 2004. Trans. GIS 11, 911–926.
84. Kristiansen, P., Hansen, H.O. (Eds.), 1994. European rivers and lakes: Assessment of their environmental stateEEA Environmental Monographs 1. Denmark, European Environment Agency.
85. Kustas, W.P., Norman, J.M., 1997. A two-source approach for estimating turbulent fluxes using multiple angle thermal infrared observations. Water Resour. Res. 33, 1495–1508.
86. Kutser, T., Verpoorter, C., Paavel, B., Tranvik, L.J., 2015. Estimating lake carbon fractions from remote sensing data. Remote Sens. Environ. 157, 138–146.
87. Kyle, H.L., Curran, R.J., Barnes, W.L., Escoe, D., 1978. A cloud physics radiometer. Proceedings of Third Conference on Atmospheric Radiation, Davis, CA, pp. 107–109.
88. Large, A.R.G., Gilvear, D.J., 2014. Using Google Earth, a virtual globe-imaging platform, for ecosystem services-based river assessment. River Res. Appl. http://dx.doi.org/10. 1002/rra2798.
89. Larson, K.M., Small, E.E., Gutmann, E.D., Bilich, A.L., Braun, J.J., Zavorotny, V.U., 2008. Use of GPS receivers as a soil moisture network for water cycle studies. Geophys. Res. Lett. 35, L24405. http://dx.doi.org/10.1029/2008GL036013.
90. Lehner, B., Döll, P., 2004. Development and validation of a global database of lakes, reservoirs and wetlands. J. Hydrol. 296, 1–22.
91. Li, Z.-L., Tang, B.-H., Wu, H., Ren, H., Yan, G., Wan, Z., Trigo, I.F., Sobrino, J.A., 2013. Satellitederived land surface temperature: current status and perspectives. Remote Sens. Environ. 131, 14–37.
92. Li, M., Zang, S.Y., Zhang, B., Li, S.S., Wu, C.S., 2014. A review of remote sensing image classification techniques: the role of spatio-contextual information. Eur. J. Remote Sens. 47, 389–411.
93. Loew, A., 2008. Impact of surface heterogeneity on surface soil moisture retrievals from passive microwave data at the regional scale: the Upper Danube case. Remote Sens. Environ. 112, 231–248.
94. Lu, S., Ouyang, N., Wu, B., Wei, Y., Tesemma, Z., 2013. Lake water volume calculation with time series remote-sensing images. Int. J. Remote Sens. 34, 7962–7973. http://dx.doi. org/10.1080/01431161.2013.827814.
95. Ludwig, R., Mauser, W., Niemeyer, S., Colgan, A., Stolz, R., Escher-Vetter, H., Kuhn, M., Reichstein, M., Tenhunen, J., Kraus, A., Ludwig, M., Barth, M., Hennicker, R., 2003a. Web-based modelling of energy, water and matter fluxes to support decision making in mesoscale catchments — the integrative perspective of GLOWA-Danube. Phys. Chem. Earth 28, 621–634.
96. Ludwig, R., Probeck, M., Mauser, W., 2003b. Mesoscale water balance modelling in the Upper Danube watershed using sub-scale land cover information derived from NOAA-AVHRR imagery and GIS-techniques. Phys. Chem. Earth 28, 1351–1364.
97. Mao, K., Qin, Z., Shi, J., Gong, P., 2005. A practical split-window algorithm for retrieving land-surface temperature from MODIS data. Int. J. Remote Sens. 26, 3181–3204. http://dx.doi.org/10.1080/01431160500044713.
98. Martinec, J., Rango, A., Roberts, R., 2008. Snowmelt Runoff Model (SRM) User's Manual. New Mexico State University Press, New Mexico.
99. Matikainen, L., Kuittinen, R., Vepsäläinen, J., 2002. Estimating drainage area-based snowcover percentages from NOAA AVHRR images. Int. J. Remote Sens. 23, 2971–2988. http://dx.doi. org/10.1080/01431160110071923.
100. Mauser, W., Bach, H., 2009. PROMET — large scale distributed hydrological modelling to study the impact of climate change on the water flows of mountain watersheds. J. Hydrol. 376, 362–377.
101. McCabe, G.J., Markstrom, S.L., 2007. A monthly water-balance model driven by a graphical user interface. U.S. Geological Survey Open-File report 2007–1088 (6 pp.).
102. McDonald, C.P., Stets, E.G., Striegl, R.G., Butman, D., 2013. Inorganic carbon loading as a primary driver of dissolved carbon dioxide concentrations in the lakes and reservoirs of the contiguous United States. Glob. Biogeochem. Cycles 27, 1–11.
103. McFarland, M.J., Miller, R.L., Neale, C.M.U., 1990. Land surface temperature derived from the SSM/I passive microwave brightness temperatures. IEEE Trans. Geosci. Remote Sens. 28, 839–845.
104. Meigh, J.R., McKenzie, A.A., Sene, K.J., 1999. A grid-based approach to water scarcity, estimates for Eastern and Southern Africa. Water Resour. Manag. 13, 85–115.
105. Mercier, F., Cazenave, A., Maheu, C., 2002. Interannual lake level fluctuations (1993–1999) in Africa from Topex/Poseidon: connections with ocean–atmosphere interactions over the Indian Ocean. Glob. Planet. Chang. 32, 141–163.
106. Michaelides, S., Levizzani, V., Anagnostou, E., Bauer, P., Kasparis, T., Lane, J.E., 2009. Precipitation: measurement remote sensing, climatology and modelling. Atmos. Res. 94, 512–533.
107. Miller, S.D., Lee, T.F., Fennimore, R.L., 2005. Satellite-based imagery techniques for daytime cloud/snow

delineation from MODIS. J. Appl. Meteorol. 44, 987–997.

108. Milzow, C., Krogh, P.E., Bauer-Gottwein, P., 2011. Combining satellite radar altimetry, SAR surface soil moisture and GRACE total storage changes for hydrological model calibration in a large poorly gauged catchment. Hydrol. Earth Syst. Sci. 15, 1729–1743.

109. Minacapilli, M., Agnese, C., Blanda, F., Cammalleri, C., Ciraolo, G., D'Urso, G., Iovino, M., Pumo, D., Provenzano, G., Rallo, G., 2009. Estimation of actual evapotranspiration of

110. Mediterranean perennial crops by means of remote-sensing based surface energy balance models. Hydrol. Earth Syst. Sci. 13, 1061–1074. http://dx.doi.org/10.5194/ hess-13-1061-2009.

111. Mitsova, D., 2014. Coupling land use change modelling with climate projections to estimate seasonal variability in runoff from an urbanizing catchment near Cincinnati, Ohio. Int. J. Geo-Inf. 3, 1256–1277.

112. Mo, T., Choudhury, B.J., Schmugge, T.J., Wang, J.R., Jackson, T.J., 1982. A model for microwave emission from vegetation-covered fields. J. Geophys. Res. 87, 11,229–11,237. http://dx.doi.org/10.1029/ JC087iC13p11229.

113. Muirhead, K., Cracknell, A.P., 1986. Airborne lidar bathymetry. Int. J. Remote Sens. 7, 597–614. http://dx.doi.org/10.1080/01431168608954714.

114. Nghiem, V.T., Nedjai, R., Messaoud, N.N., 2011. The consequences of changes in forest land cover in the Alpine and Jurassic massifs on the physico-chemical status of deep lake waters. J. Alp. Res. 99–3.

115. Njoku, E.G., Kong, J.A., 1977. Theory for passive microwave remote sensing of near-surface soil moisture. J. Geophys. Res. 82, 3108–3118.

116. Njoku, E.G., Jackson, T.J., Lakshmi, V., Chan, T.K., Nghiem, S.V., 2003. Soil moisture retrieval from AMSR-E. IEEE Trans. Geosci. Remote Sens. 41, 215–229.

117. Nolin, A.W., 2010. Recent advances in remote sensing of seasonal snow. J. Glaciol. 56, 1141–1150.

118. Olofsson, P., Foody, G.M., Stehman, S.V., Woodcock, C.E., 2013. Making better use of accuracy data in land change studies: estimating accuracy and area and quantifying uncertainty using stratified estimation. Remote Sens. Environ. 129, 122–131.

119. Owe, M., de Jeu, R., Holmes, T., 2008. Multisensor historical climatology of satellite-derived global land surface moisture. J. Geophys. Res. 113, F01002. http://dx.doi.org/ 10.1029/2007JF000769.

120. Parinussa, R.M., Holmes, T.R.H., de Jeu, R.A.M., 2012. Soil moisture retrievals from the WindSat spaceborne polarimetric microwave radiometer. IEEE Trans. Geosci. Remote Sens. 50, 2683–2694.

121. Pereira, L.S., Allen, R.G., Smith, M., Raes, D., 2015. Crop evapotranspiration estimation with FAO56: past and future. Agric. Water Manag. 147, 4–20. http://dx.doi.org/10.1016/j. agwat.2014.07.031.

122. Pereira-Cardenal, S.J., Riegels, N.D., Berry, P.A.M., Smith, R.G., Yakovlev, A., Siegfried, T.U., Bauer-Gottwein, P., 2011. Real-time remote sensing driven river basin modelling using radar altimetry. Hydrol. Earth Syst. Sci. 15, 241–254.

123. Peterson, T.C., Basist, A.N., Williams, C.N., Grody, N.C., 2000. A blended satellite-in situ near-global surface temperature dataset. Bull. Am. Meteorol. Soc. 81, 2157–2164.

124. Petrovič, P., Mravcová, K., Holko, L., Kostka, Z., Miklánek, P., 2010. Basin-wide water balance in the Danube River Basin. In: Brilly, M. (Ed.), Hydrological Processes of the Danube River Basin. Springer Science + Business Media http://dx.doi.org/10.1007/978- 90-481-3423-6_7.

125. Pfeifer, M., Disney, M., Quaife, T., Marchant, R., 2012. Terrestrial ecosystems from space: a review of earth observation products for macroecology applications. Glob. Ecol. Biogeogr. 21, 603–624.

126. Pinheiro, A.C.T., Mahoney, R., Privette, J.L., Tucker, C.J., 2006. Development of a daily long term record of NOAA-14 AVHRR land surface temperature over Africa. Remote Sens. Environ. 103, 153–164.

127. Pôças, I., Cunha, M., Pereira, L.S., Allen, R.G., 2013. Using remote sensing energy balance and evapotranspiration to characterize montane landscape vegetation with focus on grass and pasture lands. Int. J Appl. Earth Obs. Geoinf. 21, 159–172.

128. Ponchaut, F., Cazenave, A., 1998. Continental lake level variations from Topex/Poseidon (1993–1996). C. R. Acad. Sci. Ser. IIa 326, 13–20.

129. Prata, A.J., Caselles, V., Coll, C., Sobrino, J.A., Ottlé, C., 1995. Thermal remote sensing of land surface temperature from satellites: current status and future prospects. Remote Sens. Rev. 12, 175–224. http://dx.doi.org/10.1080/02757259509532285.

130. Probeck, M., Ludwig, R., Mauser, W., 2005. Fusion of NOAA AVHRR imagery and geographical information system techniques to derive subscale land cover information for the upper Danube watershed. Hydrol. Process. 19, 2407–2418.

131. Qian, Y.-G., Li, Z.-L., Nerry, F., 2013. Evaluation of land surface temperature and emissivities retrieved from MSG/ SEVIRI data with MODIS land surface temperature and emissivity products. Int. J. Remote Sens. 34, 3140–3152. http://dx.doi.org/10.1080/ 01431161.2012.716538.

132. Reichle, R.H., Koster, R.D., Liu, P., 2007. Comparison and assimilation of global soil moisture retrievals from the Advanced Microwave Scanning Radiometer for the Earth Observing System (AMSR-E) and the Scanning Multichannel Microwave Radiometer (SMMR). J. Geophys. Res. 112, D09108.

133. Renwick, W.H., Vanni, M.J., Zhang, Q., Patton, J., 2008. Water quality trends and changing agricultural practices in a Midwest US watershed, 1994–2006. J. Environ. Qual. 37, 1862–1874.

134. Riera, J.L., Magnuson, J.J., Kratz, T.K., Webster, K.E., 2000. A geomorphic template for the analysis of lake

districts applied to the Northern Highland lake district, Wisconsin, USA. Freshw. Biol. 43, 301–318.

135. Rocchini, D., Foody, G.M., Nagendra, H., Ricotta, C., Anand, M., He, K.S., Amici, V., Kleinschmit, B., Förster, M., Schmidtlein, S., Feilhauser, H., Ghisla, A., Metz, M., Neteler, M., 2013. Uncertainty in ecosystem mapping by remote sensing. Comput. Geosci. 50, 128–135.

136. Rodell, M., Velicogna, I., Famiglietti, J.S., 2009. Satellite-based estimates of groundwater depletion in India. Nature 460, 999–1002. http://dx.doi.org/10.1038/nature08238.

137. Rodell, M., McWilliams, E.B., Famiglietti, J.S., Beaudoing, H.K., Nigro, J., 2011. Estimating evapotranspiration using an observation based terrestrial water budget. Hydrol. Process. 25, 4082–4092.

138. Rowan, J.S., 2010. Developing a lake hydromorphology typology for the UK. SNIFFER Research Report WFD104 (50 pp.).

139. Rüdiger, C., Calvet, J.-C., Gruhier, C., Holmes, T.R.H., de Jeu, R.A.M., Wagner, W., 2009. An intercomparison of ERS-Scat and AMSR-E soil moisture observations with model simulations over France. J. Hydrometeorol. 10, 431–447.

140. Salomonson, V.V., Appel, I., 2004. Estimating fractional snow cover from MODIS using the normalized difference snow index. Remote Sens. Environ. 89, 351–360.

141. Schindler, D.W., Donahue, W.F., 2006. An impending water crisis in Canada's western prairie provinces. Proc. Natl. Acad. Sci. U. S. A. 103, 7210–7216.

142. Schlenz, F., dall'Amico, J.T., Loew, A., Mauser, W., 2012. Uncertainty assessment of the SMOS validation in the Upper Danube Catchment. IEEE Trans. Geosci. Remote Sens. 50, 1517–1529.

143. Schmugge, T., Gloersen, P., Wilheit, T.T., Geiger, F., 1974. Remote sensing of soil moisture with microwave radiometers. J. Geophys. Res. 79, 317–323.

144. Schmugge, T.J., Kustas, W.P., Ritchie, J.C., Jackson, T.J., Rango, A., 2002. Remote sensing in hydrology. Adv. Water Resour. 25, 1367–1385.

145. Seneviratne, S.I., Corti, T., Davin, E.L., Hirschi, M., Jaeger, E.B., Lehner, I., Orlowsky, B., Teuling, A.J., 2010. Investigating soil moisture–climate interactions in a changing climate: a review. Earth Sci. Rev. 99, 125–161.

146. Seppelt, R., Dormann, C.F., Eppink, F.V., Lautenbach, S., Schmidt, S., 2011. A quantitative review of ecosystem service studies: approaches, shortcomings and the road ahead. J. Appl. Ecol. 48, 630–636.

147. Shiklomanov, I.A., 2000. Appraisal and assessment of world water resources. Water Int. 25, 11–32. http://dx.doi.org/10.1080/02508060008686794.

148. Sobrino, J.A., Li, Z.-L., Stoll, M.P., Becker, F., 1996. Multi-channel and multi-angle algorithms for estimating sea and land surface temperature with ATSR data. Int. J. Remote Sens. 17, 2089–2114. http://dx.doi.org/10.1080/01431169608948760.

149. Sobrino, J.A., Sòria, G., Prata, A.J., 2004. Surface temperature retrieval from Along Track Scanning Radiometer 2 data: algorithms and validation. J. Geophys. Res. 109, D11101. http://dx.doi.org/10.1029/2003JD004212.

150. Soranno, P.A., Webster, K.E., Cheruvelil, K.S., Bremigan, M.T., 2009. The lake landscapecontext framework: linking aquatic connections, terrestrial features and human effects at multiple spatial scales. Verh. Int. Ver. Limnol. 30, 695–700.

151. Soranno, P.A., Cheruvelil, K.S., Webster, K.E., Bremigan, M.T., Wagner, T., Stow, C.A., 2010. Using landscape limnology to classify freshwater ecosystems for multi-ecosystem management and conservation. Bioscience 60, 440–454.

152. Sòria, G., Sobrino, J.A., 2007. ENVISAT/AATSR derived land surface temperature over a heterogeneous region. Remote Sens. Environ. 111, 409–422.

153. Stephens, G.L., Kummerow, C.D., 2007. The remote sensing of clouds and precipitation from space: a review. J. Atmos. Sci. 64, 3742–3765.

154. Stone, L.R., Horton, M.L., 1974. Estimating evapotranspiration using canopy temperatures: field evaluation. Agron. J. 66, 450–454.

155. Suchenwirth, L., Stümer, W., Schmidt, T., Förster, M., Kleinschmit, B., 2014. Large-scale mapping of carbon stocks in riparian forests with self-organising maps and the knearest neighbor algorithm. Forests 5, 1635–1652.

156. Swenson, S.C., Wahr, J., 2009. Monitoring the water balance of Lake Victoria, East Africa, from space. J. Hydrol. 370, 163–176.

157. Swenson, S., Famiglietti, J., Basara, J., Wahr, J., 2008. Estimating profile soil moisture and groundwater variations using GRACE and Oklahoma Mesonet soil moisture data. Water Resour. Res. 44, W01413.

158. Tapiador, F.J., Turk, F.J., Petersen, W., Hou, A.Y., García-Ortega, E., Machado, L.A.T., Angelis, C.F., Salio, P., Kidd, C., Huffman, G.J., de Castro, M., 2012. Global precipitation measurement: methods, datasets and applications. Atmos. Res. 104, 70–97.

159. Tapley, B.D., Bettadpur, S., Ries, J.C., Thompson, P.F., Watkins, M.M., 2004. GRACE measurements of mass variability in the earth system. Science 305, 503–505. http://dx. doi.org/10.1126/science.1099192.

160. TEEB, 2010. The economics of ecosystems and biodiversity available at http://www. teebweb.org/resources/ecosystem-services/. Thornthwaite, C.W., 1948. An approach toward a rational classification of climate. Geogr. Rev. 38, 55–94.

161. Thornthwaite, C.W., Mather, R.J., 1955. The water balance. Publications in Climatology 8. DIT, Laboratory of Climatology, Centerton, New Jersey, USA, pp. 1–86.

162. Thornthwaite, C.W., Mather, R.J., 1957. Publ. no. 10. Laboratory of Climatology, Centerton, New Jersey, USA.

163. Tomlinson, C.J., Chapman, L., Thornes, J.E., Baker, C., 2011. Remote sensing land surface temperature for meteorology and climatology: a review. Meteorol. Appl. 18, 296–306. http://dx.doi.org/10.1002/met.287.

164. Tranvik, L.J., Downing, J.A., Cotner, J.B., Loiselle, S.A., Striegl, R.G., Ballatore, T.J., Dillon, P., Finlay, K., Fortino, K., Knoll, L.B., Kortelainen, P.L., Kutser, T., Larsen, S., Laurion, I., Leech, D.M., Leigh McCallister, S., McKnight, D.M., Melack, J.M., Overholt, E., Porter, J.A., Prairie, Y., Renwick, W.H., Roland, F., Sherman, B.S., Schindler, D.W., Sobek, S., Tremblay, A., Vanni, M.J., Verschoor, A.M., Von Wachenfeldt, E., Weyhenmeyer, G.A., 2009. Lakes and reservoirs as regulators of carbon cycling and climate. Limnol. Oceanogr. 54, 2298–2314.

165. Tyler, A.N., Hunter, P.D., Spyrakos, E., Groom, S., Constaninescua, A., Kitchen, J., Paun, A., 2015. Developments in Earth observation for the assessment and monitoring of inland, transitional, coastal and shelf-sea waters. Sci. Total Environ. (in this issue).

166. UNEP, 2008. Vital water graphics — an overview of the state of the world's fresh and marine waters. 2nd edition. UNEP, Nairobi, Kenya 92-807-2236-0.

167. Valero-Garcés, B.L., Navas, A., Machin, J., Stevenson, T., Davis, B., 2000. Responses of a saline lake ecosystem in a semiarid region to irrigation and climate variability. AMBIO: J. Hum. Environ. 29, 344–350.

168. Vereecken, H., Huisman, J.A., Bogena, H., Vanderborght, J., Vrugt, J.A., Hopmans, J.W., 2008. On the value of soil moisture measurements in vadose zone hydrology: a review. Water Resour. Res. 44. http://dx.doi.org/10.1029/2008WR006829.

169. Verpoorter, C., Kutser, T., Tranvik, L., 2012. Automated mapping of water bodies using Landsat multispectral data. Limnol. Oceanogr. Methods 10, 1037–1050. http://dx. doi.org/10.4319/lom.2012.10.1037.

170. Verpoorter, C., Kuster, T., Seekell, D.A., Travnik, L.J., 2014. A global inventory of lakes based on high-resolution satellite imagery. Geophys. Res. Lett. 41. http://dx.doi.org/10. 1002/2014GL060641.

171. Verstraeten, W.W., Veroustraete, F., Feyen, J., 2008. Assessment of evapotranspiration and soil moisture content across different scales of observation. Sensors 8, 70–117.

172. Vila, D., Ferraro, R., Joyce, R., 2007. Evaluation and improvement of AMSU precipitation retrievals. J. Geophys. Res. 112, D20119. http://dx.doi.org/10.1029/2007JD008617.

173. Voigt, S., Koch, M., Baumgartner, M.F., 1999. A multichannel threshold technique for NOAA AVHRR data to monitor the extent of snow cover in the Swiss Alps. Proceedings of IUGG99 Symposium HS2, interactions between the cryosphere, climate and greenhouse gases, Birmingham. IAHS Publ. no. 256.

174. Vörösmarty, C.J., Green, P., Salisbury, J., Lammers, R.B., 2000. Global water resources: vulnerability from climate change and population growth. Science 289, 284–288.

175. Wada, Y., van Beek, L.P.H., Bierkens, M.F.P., 2011. Modelling global water stress of the recent past: on the relative importance of trends in water demand and climate variability. Hydrol. Earth Syst. Sci. 15, 3785–3808.

176. Wagner, W., Scipal, K., Pathe, C., Gerten, D., Lucht, W., Rudolf, B., 2003. Evaluation of the agreement between the first global remotely sensed soil moisture data with model and precipitation data. J. Geophys. Res. 108, 4611. http://dx.doi.org/10.1029/ 2003JD003663.

177. Wagner, W., Blöschl, G., Pampaloni, P., Calvet, J.-C., Bizzarri, B., Wigneron, J.-P., Kerr, Y., 2007. Operational readiness of microwave remote sensing of soil moisture for hydrologic applications. Nord. Hydrol. 38, 1–20.

178. Wagner, W., Dorigo, W., de Jeu, R., Fernandez, D., Benveniste, J., Haas, E., Ertl, M., 2012. Fusion of active and passive microwave observations to create and essential climate variable data record on soil moisture. ISPRS Annals of the Photogrammetry, Remote Sensing and Spatial Information Sciences, I-7, eXXII ISPRS Congress, Melbourne, Australia.

179. Wahr, J., Swenson, S., Zlotnicki, V., Velicogna, I., 2004. Time-variable gravity from GRACE: First results. Geophys. Res. Lett. 31, L11501. http://dx.doi.org/10.1029/ 2004GL019779.

180. Wan, Z., Dozier, J.A., 1996. Generalized split-window algorithm for retrieving land-surface temperature for space. IEEE Trans. Geosci. Remote Sens. 34, 892–905.

181. Wang, L., Qu, J.J., 2009. Satellite remote sensing applications for surface soil moisture monitoring: a review. Front. Earth Sci. China 3, 237–247. http://dx.doi.org/10.1007/ s11707-009-0023-7.

182. Wang, N.-Y., Liu, C., Ferraro, R., Wolff, D., Zipser, E., Kummerow, C., 2009. TRMM 2A12 Land precipitation product — status and future plans. J. Meteorol. Soc. Jpn. 87, 237–253.

183. Webster, K.E., Soranno, P.A., Baines, S.P., Kratz, T.K., Bowser, C.J., Dillon, P.J., Campbell, P., Fee, E.J., Hecky, R.E., 2000. Structuring features of lake districts: landscape controls on lake chemical responses to drought. Freshw. Biol. 43, 499–515.

184. Wen, J., Jackson, T.J., Bindlish, R., Hsu, A.Y., 2005. Retrieval of soil moisture and vegetation water content using SSM/I data over a corn and soybean region. J. Hydrometeorol. Spec. Sect. 6, 854–863.

185. Wigneron, J.-P., Calvet, J.-C., Pellarin, T., Van de Griend, A.A., Berger, M., Ferrazzoli, P., 2003. Retrieving near-surface soil moisture from microwave radiometric observations: current status and future plans. Remote Sens. Environ. 85, 489–506.

186. Wilcox, C., Los Huertos, M., 2005. A simple, rapid method for mapping bathymetry of small wetland basins. J. Hydrol. 301, 29–36.

187. Wilker, H., Drusch, M., Seuffert, G., Simmer, C., 2006. Effects of the near-surface soil moisture profile on the assimilation of L-band microwave brightness temperature. J. Hydrometeorol. 7, 433–442.

188. Williams, C.N., Basist, A., Peterson, T.C., Grody, N.C., 2000. Calibration and validation of land surface temperature anomalies derived from the SSM/I. Bull. Am. Meteorol. Soc. 81, 2141–2156.

189. Williamson, C.E., Saros, J.E., Vincent, W.F., Smol, J.P., 2009. Lakes and reservoirs as sentinels, integrators, and

regulators of climate change. Limnol. Oceanogr. 54, 2273–2282.

190. Woodcock, C.E., 2002. Uncertainty in remote sensing. In: Foody, G.M., Atkinson, P.M. (Eds.), Uncertainty in Remote Sensing and GIS. Wiley, Chichester.

191. Woodhouse, I., Nichol, C., Sinclair, P., Jack, J., Morsdorf, F., Malthus, T.J., Patenaude, G., 2011. A multispectral canopy LiDAR demonstrator project. IEEE Geosci. Remote Sens. Lett. 8, 839–843.

192. WWAP, 2012. The United Nations World Water Development Report 4: Managing Water under Uncertainty and Risk. World Water Assessment Programme, Paris, UNESCO.

193. Xu, C.-Y., Singh, V.P., 1998. A review on monthly water balance models for water resources investigations. Water Resour. Manag. 12, 31–50.

194. Xu, C.-Y., Singh, V.P., 2004. Review on regional water resources assessment models under stationary and changing climate. Water Resour. Manag. 18, 591–612.

195. Xu, X., Li, J., Tolson, B.A., 2014. Progress in integrating remote sensing data and hydrologic modelling. Prog. Phys. Geogr. 38, 464–498.

196. Yang, H., Xing, Y., Xie, P., Ni, L., Rong, K., 2008. Carbon source/sink function of a subtropical, eutrophic lake determined from an overall mass balance and a gas exchange and carbon burial balance. Environ. Pollut. 151, 559–568.

CHAPTER 9

INVESTIGATING THE TRAFFIC-RELATED ENVIRONMENTAL IMPACTS OF HYDRAULIC-FRACTURING (FRACKING) OPERATIONS

Paul S. Goodman, Fabio Galatioto, Neil Thorpe, Anil K. Namdeo, Richard J. Davies and Roger N. Bird

School of Civil Engineering and Geosciences, Newcastle University, Newcastle upon Tyne, Tyne and Wear NE1 7RU, UK

ABSTRACT

Hydraulic fracturing (fracking) has been used extensively in the US and Canada since the 1950s and offers the potential for significant new sources of oil and gas supply. Numerous other countries around the world (including the UK, Germany, China, South Africa, Australia and Argentina) are now giving serious consideration to sanctioning the technique to provide additional security over the future supply of domestic energy. However, relatively high population densities in many countries and the potential negative environmental impacts that may be associated with fracking operations has stimulated controversy and significant public debate regarding if and where fracking should be permitted. Road traffic generated by fracking operations is one possible source of environmental impact whose significance has, until now, been largely neglected in the available literature. This paper therefore presents a scoping-level environmental assessment for individual and groups of fracking sites using a newly-created Traffic Impacts Model (TIM). The model produces estimates of the traffic-related impacts of fracking on greenhouse gas emissions, local air quality emissions, noise and road pavement wear, using a range of hypothetical fracking scenarios to quantify changes in impacts against baseline levels.

Results suggest that the local impacts of a single well pad may be short duration but large magnitude. That is, whilst single digit percentile increases in emissions of CO_2, NO_x and PM are estimated for the period from start of construction to pad completion (potentially several months or years), excess emissions of NO_x on individual days of peak activity can reach 30% over baseline. Likewise, excess noise emissions appear negligible (< 1 dBA) when normalised over the completion period, but may be considerable (+ 3.4 dBA) in particular hours, especially in night-time periods. Larger, regional scale modelling of pad development scenarios over a multi-decade time horizon give modest CO_2 emissions that vary between 2.5 and 160.4 kT, dependent on the number of wells, and individual well fracking water and flowback waste requirements. The TIM model is designed to be adaptable to any geographic area where the required input data are available (such as fleet characteristics, road type and quality), and we suggest could be deployed as a tool to help reach more informed decisions regarding where and how fracking might take place taking into account the likely scale of traffic-related environmental impacts.

Citation: Paul S. Goodman, Fabio Galatioto, Neil Thorpe, Anil K. Namdeo, Richard J. Davies, Roger N. Bird, Investigating the traffic-related environmental impacts of hydraulic-fracturing (fracking) operations, Environment International, Volumes 89–90, April–May 2016, Pages 248-260, ISSN 0160-4120, http://dx.doi.org/10.1016/j.envint.2016.02.002.

INTRODUCTION

Aim of the paper

The exploration and potential production of oil and gas from shale reservoirs using hydraulic fracturing (fracking) technology has raised significant public concerns about a range of possible environmental impacts as diverse as the potential for fracking to cause earthquakes (e.g. Davies et al., 2013) to the contamination of water supplies (Osborn et al., 2011 and Davies et al., 2014). The environmental impacts of road traffic associated with drilling and fracking operations has been an additional concern (King, 2012), but there are few analyses of these potential impacts in the peer-reviewed literature.

Fracking an exploration or production well requires additional, primarily Heavy Duty Vehicle (HDV) traffic associated with the use and disposal of water and chemicals used in the fracking process. Since gas resource estimates, for example in the UK and across Europe, are significant (DECC, 2010 and Andrews, 2013) and there is an active debate around the environmental risks, assessment of the impact of road traffic is both timely and important.

The aim of this paper is therefore to produce a scoping-level environmental assessment for the short-term local impact of an individual site, as well as a longer-term assessment of the temporal impact of a number of sites operating in a particular region, over a timeframe of several decades, using a newly-created Traffic Impacts Model (TIM). This should help inform the current debate around the impact of increased traffic associated with the deployment of this technology, specifically for the UK (on which our hypothetical modelling scenarios are based) and the rest of Europe, but with relevance to other countries where fracking is active, such as the US and Canada, or being considered (e.g. Australia, China and South Africa).

Well drilling and fracking

The drilling of a well requires the building of a well pad, movement of a rig and related equipment and materials to the site, for example casing, cement and the chemicals required to make up drilling mud. The fracking process is undertaken after the well has been drilled. Rather than one fracking operation, fracking is normally carried out along different sections of a horizontal or vertical well in what is called a multi-stage fracking operation. Total volumes of water used per well vary considerably, from 1500 m^3 to 45,000 m^3 (e.g. King, 2012). The process requires water to be pumped down the well along with chemicals and proppant (usually ceramics or sand). After pumping has finished, the fracking fluid (including natural contaminants) returns to the surface. As with initial water demand, flowback volumes are variable. In US operations the volumes returning can be 5–50% of the injected volume (King, 2012). Over a period of a few days to a few weeks flowback decreases and gas production commences. A production well pad could include 12 or more wells, which may be re-fracked several times, once production has declined. It subsequently becomes necessary to remove flowback water from those sites both during and after fracturing. If this transportation is done by road, as has typically been the case in the US and Canada, then considerable volumes of HDV (i.e. tanker) traffic may be generated, albeit for relatively short periods (i.e. weeks) of time.

Traffic impacts

HDVs cause visible disruption and traffic congestion, especially if routed along roads that may be considered inappropriate, such as those carrying light traffic in rural areas, or at inappropriate times of the day. Vehicle movements are associated with production of greenhouse gas (GHG) emissions, primarily in the form of Carbon Dioxide (CO_2). The production of CO_2 from traffic emissions due to shale gas exploitation has been previously examined by Broderick et al. (2011), with on-road emissions estimated at 38 t–59 t CO_2 per well. Broderick et al. (2011) also suggested that total CO_2 emissions associated with extraction of shale gas from a well were small (0.2–2.9%) compared to total emissions from combustion of the gas produced by the well.

HDVs are associated with disproportionate annoyance arising from noise emissions when compared to lighter vehicles (Sandberg, 2001), and their diesel engines are perceived as 'dirty' in terms of local air quality emissions of Oxides of Nitrogen (NO_x) and particulate matter (PM), even though modern vehicles are fitted with a variety of exhaust treatment devices to meet emissions standards (the 'Euro' standards) in effect across the European Union, and adopted globally in other countries, such as China and Australia. At the time of writing, the EURO VI standard (Reg (EC) No. 595/2009) is the highest in effect for HDVs (OJEC, 2009).

Emissions of gases affecting local air quality may have been historically of less importance in the US and Canada, with much drilling taking place away from urban areas. However, this may not be the case in more densely populated areas such as in Europe, where exploration may occur relatively close (i.e. within kilometres) to sub-urban locations resulting in HDVs travelling through inhabited areas. Historically, large urban areas, and areas adjacent to major highways, especially in the UK, have had problems meeting statutory limits on concentrations of Nitrogen Dioxide (NO_2 — a component of NO_x) and PM_{10} (particulate matter with aerodynamic size under 10 μm).

Aside from physical emissions of gases and noise, concern has also been voiced about the damage caused to pavement surfaces and the underlying road structure through additional loading by HDVs. Such vehicles cause a disproportionate amount of loading when compared to lighter vehicles, and may substantially shorten the lifespan and increase deterioration of roads not designed with their presence in mind.

Other traffic-related issues that have been raised in the debate about fracking and require rigorous research are: potential increases in the number of accidents (involving direct injury and damage to property, or accidental spillage of materials or chemicals), issues with vibration, community severance, delay (to pedestrians and other drivers) and disruption of normal activity patterns, increases in traffic violations, which all feed into marginal economic effects where the cost of operation of a site may be borne by the local community, rather than the site operator, if appropriate policy mechanisms to redress imbalances are not provided.

Traffic volume and intensity

Estimation of the potential volume of traffic generated by shale gas exploitation associated with a particular drilling site, or over a given region, is compounded by a number of factors. Aside from uncertainty over total quantities and precise locations of resources to be exploited, a number of other issues and considerations may be highlighted.

First, modern techniques allow for multiple wells to be operated from a single well pad during full production. NYS DEC (2011) reports that 90% of Marcellus Shale development (Pennsylvania, US) is expected to be by 'horizontal wells on multi-well pads'. Examination of literature, online resources and anecdotal evidence suggests 6-, 8-, 10- or 12-well pads are now standard operating practice in the US (DEC, 2008, SEAB, 2011 and DrillingInfo, 2014).

Second, there is a large amount of variability reported in the amount of water and proppants required at each well, which depends on the underlying geology, and the number of potential stages of fracking. The single well to date in the UK at Preese Hall (Lancashire) required 8400 m^3 of water (Cuadrilla, 2014a). A summary of US literature, produced for the European Parliament (EP DGIP, 2011) suggests volumes per well (including initial drilling) in the range of 1500 to 45,000 m^3. Jiang et al. (2013) reported using a normal distribution with mean 15,000 m^3 and overall range from 3500 to 26,000 m^3 to model demand for the Marcellus Shale deposits. Broderick et al. (2011)suggest that each stage of a fracking operation for a single well will require between 1100 and 2200 m^3 of water, leading to a total demand of 9000 to 29000 m^3 per well, or 54,000 to 174,000 m^3 for a six-well pad. If a base requirement of 20,000 m^3 of water and 5% by volume proppant is assumed, this equates to a transportation need for delivery of 1000 m^3 of sand, or 1700 t of sand (assuming dry sand with density 1700 kg/m^3), though fracturing additive requirements in some countries may be lower than typical US values, based on exploratory data (Broderick et al., 2011).

Third, there is large variability reported in the amount of flowback material produced. The US Environmental Protection Agency (US EPA, 2010) suggests that the rate recovery of injected fluids from hydraulic fracturing is variable — ranging between 15 and 80%. NYC DEP (2009) suggests use of a 'worst case' option of 100% for the calculation of tanker demand, whilst NYS DEC (2011) reports 9% to 35% returned water for Marcellus Shale wells in Pennsylvania, and Cuadrilla report values of 20 to 40% (Cuadrilla, 2014b). For modelling purposes Broderick et al. (2011) assumed 50%.

Fourth, during the operational lifetime of the well, which may be a period of 5 to 20 years, there may be the need for further stimulation via periodic re-fracturing (or re-fracking). Each re-fracturing event may entail similar, if not higher, water and waste demands to the initial process (Sumi, 2008). Broderick et al. (2011) assumed a single re-fracturing of 50% of wells for UK-wide shale gas scenarios.

Finally, aside from the actual fracking and waste disposal transport demands, there will also be vehicle movements associated with: construction of access roads and site facilities, excavation and concrete pouring for the pad, transportation of drilling equipment, well casings, water tanks and pump equipment to site, excavation equipment used to dig waste pits, completion and capping material transport, and general movement of workers to and from a site.

However, the vast majority of movements (typically 70% or greater) are associated with the fracking and flowback processes. Table 1 presents data from NYC DEP (2009), cited by Broderick et al. (2011), illustrating total truck movements associated with a fracking operation requiring between approximately 12,500 m^3 (low) and 18,750 m^3 (high) of water and proppants per well, with approximately 50% flowback of fluid waste.

Table 1: Truck visits over the lifetime of a six-well pad.

Purpose	Per well		Per pad	
	Low	High	Low	High
Drill pad and road construction			10	45
Drilling rig			30	30
Drill fluid and materials	25	50	150	300
Drilling equipment (casing, drill pipe etc.)	25	50	150	300
Rig completion			15	15
Completion fluid and materials	10	20	60	120
Completion equipment (pipes, wellheads)	5	5	30	30
Hydraulic fracture equipment (pumps and tanks)			150	200
Hydraulic fracture water	400	600	2400	3600
Hydraulic fracture sand proppant	20	25	120	150
Flowback water removal	200	300	1200	1800
Total (parentheses in the 'Per well' column includes additional truck visits associated with pad construction)	685 (890)	1050 (1340)	4315	6590
...of which associated directly with fracturing and flowback processes			3870	5750
			90%	87%

Re-fracturing activities could add a further 2000–3000 HDV trips to the estimates inTable 1 (Broderick et al., 2011). Limited tanker movements (2–3 per well per annum) to remove produced water during the operational lifetime of a well may also be necessary (NYS DEC, 2011).

The duration of activities at a site also determines the intensity of demand for transportation over a given period. NYC DEP (2009) suggests that total operations for the six well pad outlined in Table 1 would span between 500 and 1500 days. However, the peak of HDV demands occur in the delivery of water and proppant between 30–60 days prior to commencement of fracking, with flowback fluid return occurring for 42–56 days after. Transport demands may be complicated by the overlapping of phases between individual wells on a pad (e.g. water deliveries occurring during horizontal drilling), or by operation of multiple well activities in parallel (though actual drilling of more than two wells simultaneously is considered unlikely). Demand intensity may also be mitigated by the provision of water storage facilities and lined waste pits on-site, acting as buffers to demand.

It cannot be ruled out that water transportation to the well pad during exploration, development or after production has commenced could be via pipeline, as was the case for the UK's first fracked shale exploration well at Preese Hall, Lancashire (Mair, 2012).

METHODS

The model

A Traffic Impacts Model (TIM) has been developed to produce a broad, scoping-level environmental assessment for a small number of well pads operating in a region, over a timeframe of several years. As outlined in the introduction, the primary input drivers of the model are the water and proppant demands of fracking at well pads, whilst the primary outputs are local air quality, GHG and noise emissions, plus estimates of axle loadings in the region.

The TIM consists of four main components:

1. a traffic demand model, which utilises information about the scale and duration of activities in a region where fracking is being undertaken, in order to produce the pattern of traffic associated with those activities;
2. a traffic assignment model, which uses the traffic demand data to calculate temporal and spatial patterns in vehicle flow and speeds on the roads being used to supply the region;
3. an environmental model, which uses the patterns of assigned traffic to calculate impacts for a number of assessment criteria; and
4. a post-processing module that collates and statistically summarises information, to compare the impact of operations with regards to the baseline traffic.

Each component is discussed in more detail in the following sections. The traffic demand, assignment and post-processing components are novel in this work, whilst the environmental model forms part of the pre-existing PITHEM (Platform for Integrated Traffic, Health and Environmental Modelling) tool (Namdeo and Goodman, 2012).

Traffic demand

The purpose of the traffic demand model is to calculate the temporal pattern of vehicles within the study region associated with fracking activities. The model is implemented using a hierarchical series of C ++ programming language classes that reflect real-world components (both physical and procedural) in a fracking operation. Class names are initially shown in single quotes ('…') in the following text.

At the top of the hierarchy is the 'region' class, which represents in abstract form an area of a country being studied. The region contains a number of 'well pads' where fracking activities take place. Each well pad may contain a number of 'wells'. Activities at the well pads and wells, outlined below, drive the generation of traffic demand. Additionally, a region contains a transport 'network', comprised of road 'links' that carry both baseline (i.e. without fracking) and fracking 'traffic'. The amalgamation of fracking demand, plus the baseline activity, give the overall level of spatiotemporal demand. The physical and spatial concepts in the TIM model are shown in Fig. 1.

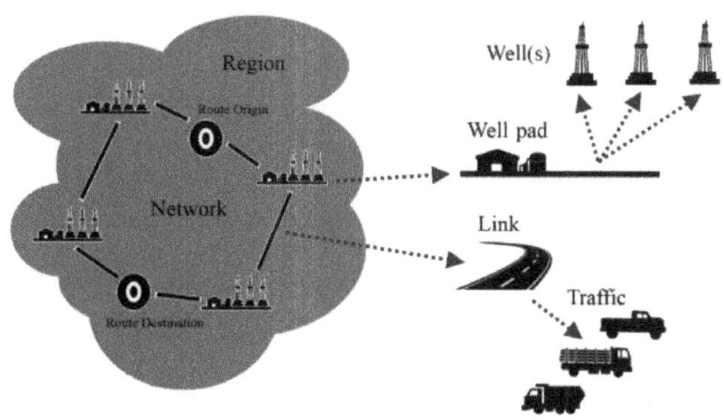

Figure 1: Physical and spatial concepts in the TIM.

As experience with fracking operations has developed, the number of possible wells per pad has increased (Section 1.4). At present the upper limit in the model is set to twelve wells per pad, though this is adjustable. It is presently assumed that pads in a given region are spaced at approximately 1.5 km distant, with a 15 km × 15 km (225 km^2) region therefore having around 100 well pads, though not all would be active at any given time (AEA, 2012).

Both well pads and wells are associated with a number of 'phases' of operation. Each phase possesses a number of attributes and classes:

1. a defined start and end time;
2. the total volume of materials needing to be serviced over the duration of the phase. Note that this volume may be either an input requirement to the phase (e.g. required building materials), or a waste product produced by the phase (e.g. flowback water);
3. links to other phases where required, defined by the volume of materials passed to subsequent phases;
4. a list of 'vehicles' that may be used to service the needs of the phase; and
5. the 'routes' used by the vehicles to and from the well pad.

In the default TIM model, the well pad has two key phases defined:

1. commissioning and construction — encompassing initial survey requirements, the development of access roads to the pad location, pad ground- and earthworks, pad concrete pouring, set-up of shared drilling equipment for wells, excavation of shared waste pits used by wells and delivery of ancillary and site equipment (e.g. site offices); and
2. decommissioning and landscaping — representing the removal of equipment from site and making good via any required landscaping at the end of the pad's operational lifespan. The decommissioning phase may be linked to the commissioning phase so that a certain volume of materials used in construction need to be removed from site.

For each well the following phases are assumed:

1. construction and drilling — representing the arrival of drilling equipment, drill casings and drilling water at the well pad, and the removal of bored material and drilling water from site;
2. hydraulic fracturing — encompassing the delivery of water, proppant materials and chemicals to the well;
3. flowback treatment — representing the removal of waste water from the well. The flowback phase is linked directly to the fracturing phase, so that a certain percentage of waste water needs to be removed from the well; and
4. miscellaneous — encompassing all other operations and activities associated with the well, such as routine movements of site staff.

Phases themselves may overlap in time, both between individual well pads and wells, as well as internally, between activities at a specific pad or well. For example, hydraulic fracturing may be taking place at one well, whilst another well is being drilled, and flowback water is being removed from a third. Likewise flowback water removal may begin whilst a well is still in the process of being fractured. Default durations of phases have been derived from Broderick et al. (2011).

The vehicle class contains information on utilisation rate and capacity, which are used to determine the overall number of vehicles required by the phase. The vehicle class also contains the description of the vehicle, required by the environmental model (Section 2.4), and the passenger-car unit (PCU) scaling factor required by the traffic assignment model (Section 2.3). The vehicle class also contains flow 'profile' information. The profile determines when vehicles servicing the phase are considered active on the road network, and holds data for each day of the week and hour of the day. As an example, a profile may be used to apply all traffic for the phase outside of peak traffic hours, as a form of traffic management policy. Finally, the route class defines a simple static list of network links that vehicles associated with a profile will use to service the site. Fig. 2summarises the phases and classes associated with activities at a well pad.

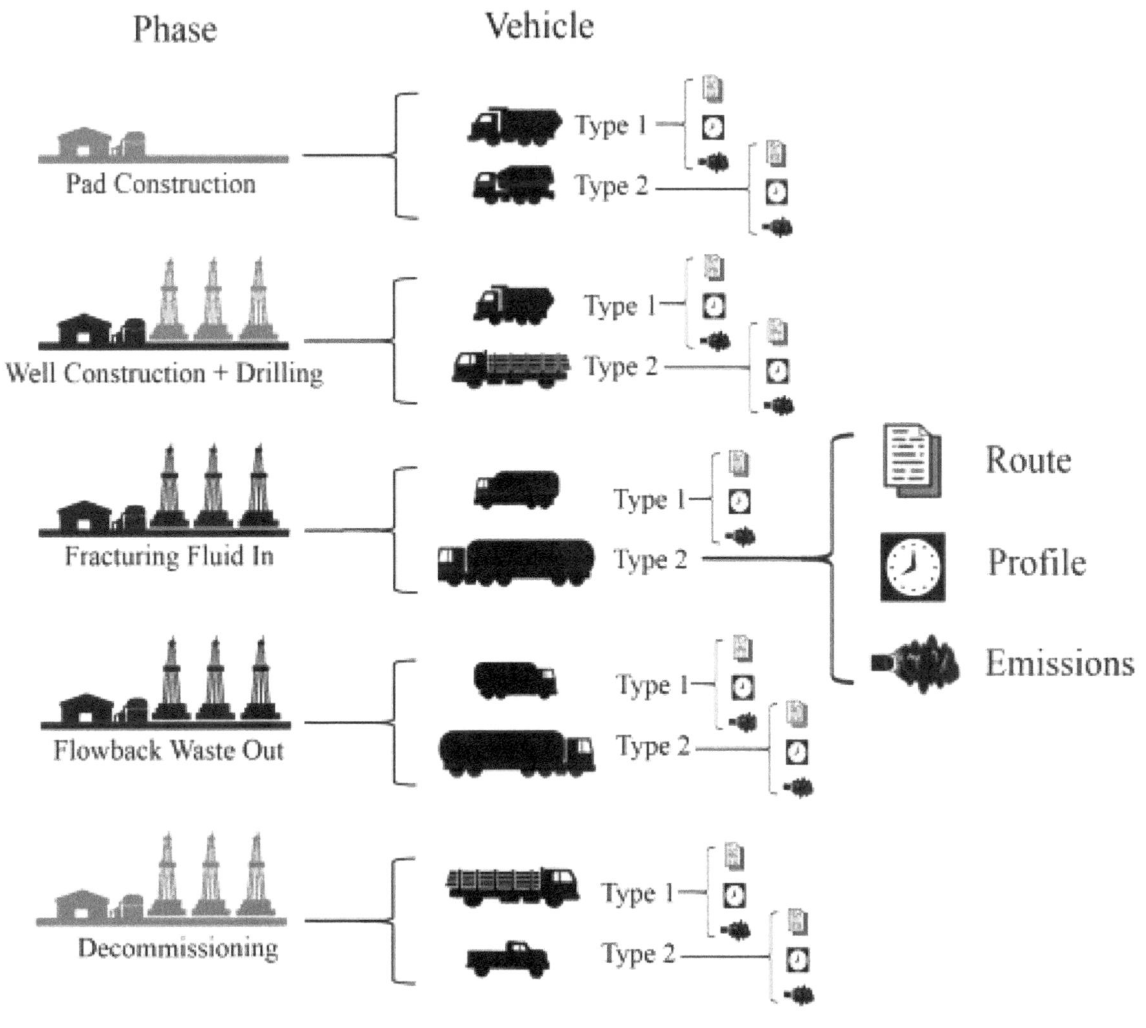

Figure 2: Temporal phases and objects associated with well pad activities.

When run, the traffic demand model produces a temporally-ordered series of time-sliced profiles linked to their parent vehicle classes and route information for its parent region.

Whilst the initial, six phase model (two phases for the well pad and four for the well) is considered an abstract, 'high-level' model, it is considered sufficiently representative of pad activities. Further disaggregate phases, in order to model at a higher level of detail (e.g. for separate treatment of input water from one location, proppant from another, chemicals from a third etc.), or to incorporate additional technological scenarios (e.g. water is brought to the pad site via pipeline, but waste-water is still removed by tanker) are possible.

Re-fracturing of wells (Section 1.4) in order to extend their operational gas production is not explicitly covered, though may feasibly be modelled through the combination of individual runs at differing epochs, using appropriate assumptions for demand and flowback.

Traffic assignment

The traffic assignment model combines information on the baseline status of a transport network in a region, with the temporal profiles, plus vehicle and routing information, output from the traffic demand model, to produce vehicle kilometres travelled (VKT) and average speed information for individual roads, which forms input to the environmental model.

The 'network' class contains a collection of 'links'. Each link contains an identifying name, the type and length of the road, and baseline 'traffic' information. This baseline traffic information consists of a total Annual Average Daily Total (AADT) flow in vehicles per day, coupled with a diurnal 'profile', which is applied to the AADT flow to give hourly variation throughout a week. Open information from the UK Department for Transport (DfT) 'Road Traffic Statistics' datasets (DfT, 2013) was used to produce the baseline diurnal profiles for our hypothetical scenarios. Each type of road has an associated 'speed-flow' curve, which alters the calculated speed on the link, based on the flow in each hour. The road type and calculation year, are used to give the traffic fleet composition on the link. Fleet composition tables are inherited from the data in the parent PITHEM model, and are discussed further in Section 2.4.

Appropriate capacity values for the speed-flow curves are defined for types of link, with reference to UK Design Manual for Roads and Bridges (DMRB) guidelines (Highways Agency, 1999 and Highways Agency, 2002). At present there are parameter values for 30 road types defined in the model, broadly covering categories of rural, sub-urban, central-urban and town and village roads. It is recognised that speed-flow curves are an abstract representation of the real-world behaviour of traffic. A congested situation will be represented as a high-flow, low-average-speed situation, rather than a more realistic low-flow, static or slow-moving queue of vehicles. This has potential implications for the accurate modelling of emissions in congested conditions in that emissions may be under-estimated. Additionally, the parameters for rural links exclude the effects of delay at junctions, which should be separately modelled (Highways Agency, 2002), but are not represented in the current approach.

Environmental assessment

Assessment criteria and fleet composition

As mentioned in Section 2.1, the PITHEM software (Namdeo and Goodman, 2012) has been used to provide the assessment of key environmental criteria based on the output from the traffic assignment model. The environmental criteria used are split into four broad categories:

1. greenhouse gas emissions (tailpipe 'ultimate CO_2' or 'uCO_2' emissions);
2. local air quality emissions (tailpipe NO_x, HC and primary NO_2 emissions, PM_{10}, $PM_{2.5}$ including brake and tyre wear components);
3. noise level at the roadside; and
4. standardised axle loading applied to the road.

Calculation of all of the above criteria depends on appropriate vehicle characteristics, combined with tabulated fleet composition information, to produce correctly weighted additions of contributions from both the vehicles used to describe fracking traffic demand and from the baseline fleet. For our demonstration scenarios in 3.2 and 3.3, fleet data are taken from the UK National Atmospheric Emissions Inventory (NAEI, 2011), supplemented with data from the Emissions Factor Toolkit (EFT), Version 5.1.3 (DEFRA, 2012). These data provide a description of current and projected vehicle types (broken down by chassis type, weight, fuel, engine size and emissions control technology) for a range of years from 2008 to 2035. Separate fleet descriptions are given for broad categories of road: urban roads, rural roads and motorways.

Within the fleet composition tables, there exist temporally evolving splits of Euro emission control measures, driven by underlying vehicle uptake, renewal and scrappage assumptions. Sample 'Euro class' splits for rigid and articulated heavy goods vehicles for the years 2010, 2015 and 2020, from EFTv5.1.3 are given in Table 2 below.

Table 2: Example percentage splits of emissions control technologies (EURO classes) for rigid and articulated HGVs in select years.

Technology	2010		2015		2020	
	Rigid	Artic	Rigid	Artic	Rigid	Artic
Pre-EURO	–	–	–	–	–	–
EURO I	0.49	0.06	–	–	–	–
EURO II	10.41	2.69	0.74	0.07	–	–
EURO III	41.42	30.69	13.74	3.07	0.71	0.06
EURO IV	25.82	31.57	12.58	6.01	2.78	0.33
EURO V (EGR)	5.46	8.75	9.65	10.51	3.08	1.29
EURO V (SCR)	16.39	26.25	28.94	31.52	9.25	3.87
EURO VI	–	–	34.36	48.83	84.18	94.44

NB: Figures may not add up to 100% due to rounding. EGR = Exhaust Gas Recirculation technology, SCR = Selective Catalytic Reduction technology. Both EGR and SCR are 'de-NOx' technologies, seeSection 2.4.3.

As can be seen from Table 2, in 2010 the majority of goods vehicles were assumed to fall within the EURO III and IV emissions control categories, whilst by 2015 the EURO V and VI categories form the largest segments. At 2020 (and beyond) the majority of goods vehicles fall into the cleanest EURO VI control category. Regarding the weight and capacity of vehicles, it has been assumed that a 40 t laden-weight articulated tanker has a capacity of 30,000 l (30 m^3) of water, whilst a 26 t laden-weight rigid body tanker has a capacity of 15,000 l (15 m^3). Further assumptions on the weights and capacities of other vehicles are required for each operational phase. These have primarily been taken from analysis of data in NYS DEC (2011), supplemented with information from the transport section Environmental Statement made for the potential Roseacre Wood, UK site (ARUP, 2014).

Greenhouse gas emissions

Greenhouse gas (GHG) emissions are represented by ultimate CO_2 emissions (total mass of CO_2 after all exhaust components of fuel exhaust have oxidised). The calculation of uCO_2 is based on the speed-emissions curves presented in Boulter et al. (2009), which also form part of the UK Emissions Factor Toolkit, Version 5.1.3c.

The emissions function forms a 'U-shaped' curve, considered valid over a type-specific speed range (typically 5–120 km/h for light vehicles and 10–90 km/h for heavy vehicles). Outside of these ranges, values are clamped to prevent excessive emissions rates. Total emissions for a link are calculated from the summation of individual contributions for all fracking traffic types active in the period, for comparison to the baseline fleet contribution.

Local air quality emissions

Emissions for those pollutants considered detrimental to local air quality (LAQ) are calculated using the same methodology for uCO_2, outlined above, with additional fuel quality and vehicle mileage scaling correction factors from the EFT applied (DEFRA, 2012) after calculation of the base emissions rate. For particulate matter and hydrocarbons, the emissions functions have the same form as that for uCO_2 above, with coefficients from Boulter et al. (2009). For NO_x a variety of functions are used, based on those found in the COPERT4 (COmputer Program for Emissions from Road Transport) version 8.1 methodology (EEA, 2007 and Nael, 2012).

In accordance with the EFT, particulate matter emission rates (PM_{10} and $PM_{2.5}$) are scaled from an initial tailpipe emissions calculation, plus contributions from brake and tyre wear and re-suspended particles. Primary emissions of NO_2 (i.e. those emissions of NO_2 directly from the vehicle tailpipe, prior to the additional generation of NO_2 from photochemistry) are calculated based on the initial NO_x emission, with the percentage conversion factors presented in Boulter et al. (2009) applied.

At present, both the GHG and LAQ emissions functions assume that heavy vehicles are 56% laden over a round trip, based on UK DfT freight transport statistics (DfT, 2013), which is considered equivalent to a tanker arriving on site full, then departing almost empty. Ideally, separate emissions rates would be calculated for each journey leg. The same limitation exists in the standard EFT (DEFRA, 2012) emissions factors.

Noise

Roadside noise levels are calculated using a methodology derived from the CNOSSOS-EU (Common Noise assessment methods in Europe) procedure (Kephalopoulos et al., 2012), developed to fulfil the mapping requirements of the European Noise Directive (OJEC, 2002). The CNOSSOS-EU methodology predicts octave-band noise emissions for five vehicle classes (light vehicles, vans, heavy vehicles, powered two wheelers (PTW) and other), based on separate emissions functions for rolling noise and power-train noise.

Noise emissions are described in terms of an energy-averaged sound power level. Sound power levels are difficult to interpret, as they cannot be readily compared to conventional predictions of noise based on sound pressure levels. Hence, a further conversion to an equivalent roadside noise level (L_{Aeq} level) at 10 m from the road is applied, based on a number of propagation assumptions. This conversion methodology is outlined in Watts et al. (2004).

Axle loading

The final component of the environmental assessment is the calculation of axle loading over the period of fracking operations. Axle loading is assumed to be a proxy for damage to both the road surface and underlying structure caused by traffic (Highways Agency, 2006) and is presented in units of 'Equivalent Standard Axle Loads' (ESAL) for a road over a period of time. This calculation is based on the well-established 'fourth-power law' (ratio of axle load to standard 80 kN load, raised to the fourth power, AASHTO, 1986).

This calculation represents a crude measure of loading damage, which may only be applicable for permanent or semi-permanent flexible (i.e. asphalt) or rigid (i.e. concrete) pavements. Estimation of damage to unsealed, temporary roads or tracks (i.e. construction access roads, farm tracks) is problematic, given the dependency on specific site soil, drainage and weather conditions. Unlike the Highways Agency's DMRB methodology (Highways Agency, 2006), the calculation also includes contributions from light vehicles and vans.

Post-processing

The post-processing procedure collates the individual, period-based pollutant results and calculates absolute and percentage changes in each parameter. The time horizon considered for the calculation of overall impacts in comparison to the baseline may be based on one of three approaches:

1. calculation of the baseline over the overall total duration of fracking activities in all phases across the region, i.e. the time from the start of construction of the first well pad in the region, to decommissioning of the last well pad in the region (i.e. including the operational, gas-producing, lifetimes of the well pads);
2. calculation of the baseline over the activities from the start of construction of the first well pad, to the end of hydraulic fracturing activities (coinciding with the assumed start of operational gas production) at the last well pad; and

3. direct comparison, with the equivalent baseline results only calculated over those periods when any phase activity is occurring in the region.

Adopting the first approach could result in a time horizon of 20 + years, whilst the second or third may give a time horizon of 1–5 years. Comparisons of emissions over a longer baseline reduce the percentage increases reported by the post-processor.

RESULTS

Scenario testing

This section presents results from the TIM in relation to short-term temporal outputs for a single well pad. This is followed by subsequent analysis using the assumptions made for that well pad, applied to multiple well pads in a region, over an extended time horizon.

Impacts of a single well pad

TIM output for a single well pad has been compared to the analysis performed inBroderick et al. (2011). That analysis considered traffic-related uCO_2 emissions from a six-well pad. The assessment has been expanded here to include the additional environmental outputs available from the TIM model.

The analysis uses the 'low' and 'high' water and proppant demands, flowback percentages and ancillary vehicle demands given in Table 1 — summarised as 685 HDV movements per well in the 'low-water use' scenario estimate, and 1050 HDV movements per well in the 'high-water use' scenario estimate. As per Broderick et al. (2011) vehicle routes form a 60 km round trip. Operations at the well pad are assumed to take 85 weeks to complete. The TIM model was run for three initial years, 2010 (as per Broderick et al., 2011), 2015 and 2020, using the vehicle technology assumptions from Table 2. The TIM was run applying vehicle demand to four different road types: a rural village road (baseline flow of 2000 veh/day speed limit 30 mph, 48 km/h); a good quality suburban road (8 veh/day, 40 mph, 64 km/h); a good quality rural road (2500 veh/day, 50 mph, 80 km/h) and a motorway (40,000 veh/day, 70 mph, 113 km/h). Hence, for each entry the range of emissions per well from the four road types is presented. Air pollutants are expressed as mass of additional emission over baseline emissions/well, noise as the increase in roadside L_{Aeq} levels in dB, averaged over the period, whilst ESAL is given as the additional standard axle loading (thousands of axles) per well throughout the period.

At 35 to 59 t per well, the CO_2 emission rates from the TIM model for a start year of 2010 are in line with those reported by Broderick et al. (2011). This is to be expected given that both models share a common source of data in the DEFRA Emissions Factor Toolkit (DEFRA, 2012). Even though Broderick et al. (2011) used a previous version of the EFT, CO_2 emissions factors have remained constant through a number of successive EFT versions. It is also notable that CO_2 emissions factors for the HDV segment show little change across the three periods considered, as it is assumed in this instance that fuel consumption savings through advancements in vehicle technology are counteracted by increased fuel requirements caused by exhaust treatment systems (particle filters, de-NO_x catalysts etc.) in later vehicles. The CO_2 emissions are also relatively invariant with speed, when compared to the LAQ pollutants.

The results for the individual pollutants show large changes with both road type (primarily due to the underlying speed assumptions) and base year selection, though the key driver of emissions reductions with time is the increasing presence of EURO VI vehicles within the fleet composition. Indeed, fleet and technology changes over the decade are predicted to decrease HDV NO_x emissions by around 90%. Progress in reductions of particle matter is substantial, but less noticeable, given that large reductions achieved through the introduction of diesel particulate filters (DPFs) were achieved previously in earlier EURO classes.

Fig. 3 presents the effect of road type, and underlying baseline traffic assumptions on the relative increases in VKT, GHG and LAQ emissions, noise and axle loading over the 85 week period of pad operation. Given the high capacity and baseline flow, the results for the motorway show negligible (< 0.5%) relative change, with the suburban, rural and village road scenarios show increasing relative change over the decreasing baseline flow.

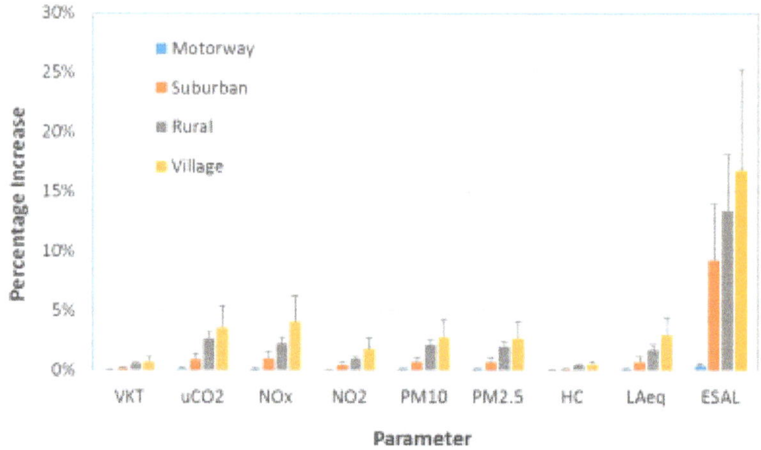

Figure 3: Percentage increases over baseline values for vehicle kilometres travelled, GHG and LAQ pollutants, noise and axle loading, assuming six-well pad operation using differing road types. Vertical bars show the difference between the 'low-water use' and 'high-water use' scenarios.

In all cases, for all emission parameters except hydrocarbons, relative percentage increases are disproportionate compared to the additional VKT travelled, given the higher emissions from HDVs compared to light vehicles. The effect is most noticeable in terms of axle loading over the period, with a 0.9% increase in VKT in the 'village road/low-water consumption' scenario accounting for a 17% increase in the ESAL value. NO_x emissions increase by 6%, uCO_2 by 5% and emissions of other pollutants by 0.7 to 4.5%. In the 'village road/high-water consumption' scenario, a 1.1% increase in VKT leads to a 25% increase in ESAL, compared to a 0.06% increase in VKT and a 0.5% increase in ESAL for the 'motorway/high-water consumption' scenario.

In addition to the comparison of overall operations to baseline, the TIM model has been used to produce outputs for individual phases of operation. Fig. 4 and Fig. 5 examine diurnal-period NO_x emission rates and roadside L_{Aeq} levels for the 'village roads' scenario, at the periods of maximum predicted traffic to the pad, under a variety of different site access policies, being: 24/7 'all-day' access to the site, movements limited to 'daytime' (0600–1900), movements limited to the overnight period, movements limited to the daytime, inter-peak traffic period (1000–1600) and movements restricted to peak traffic hours (0700–1000, 1600–1900). The maximum predicted traffic occurs assuming that two wells are in operation on the pad simultaneously, one undergoing final water and proppant deliveries in the week before initial fracturing, whilst the second is undergoing initial drilling. In the low and high scenarios this situation equates to 71 or 109 vehicle movements to and from site per day. In the 600 day lifecycle to completion of the pad, this situation is assumed to occur three times.

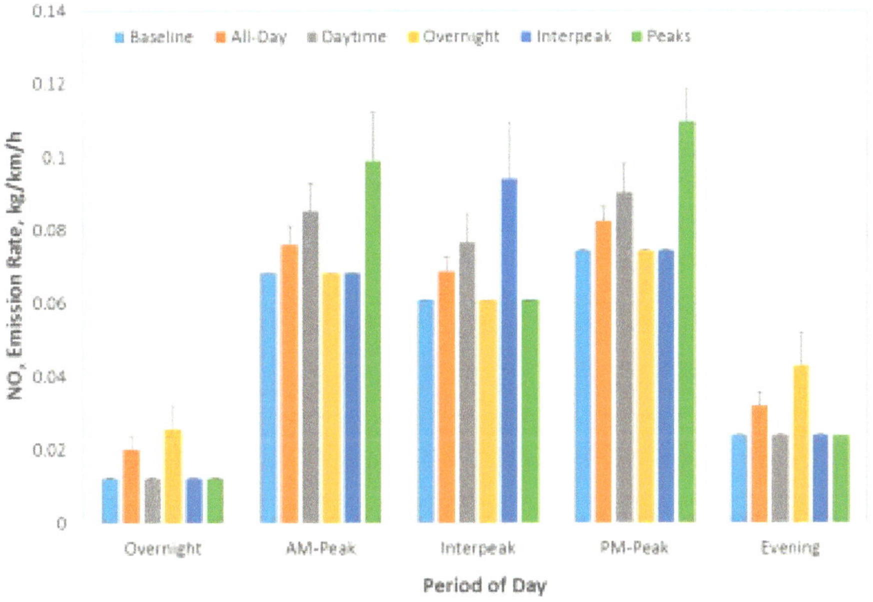

Figure 4: Peak operational NO_x emission rates compared to the baseline profile under differing vehicle arrival and departure policies.

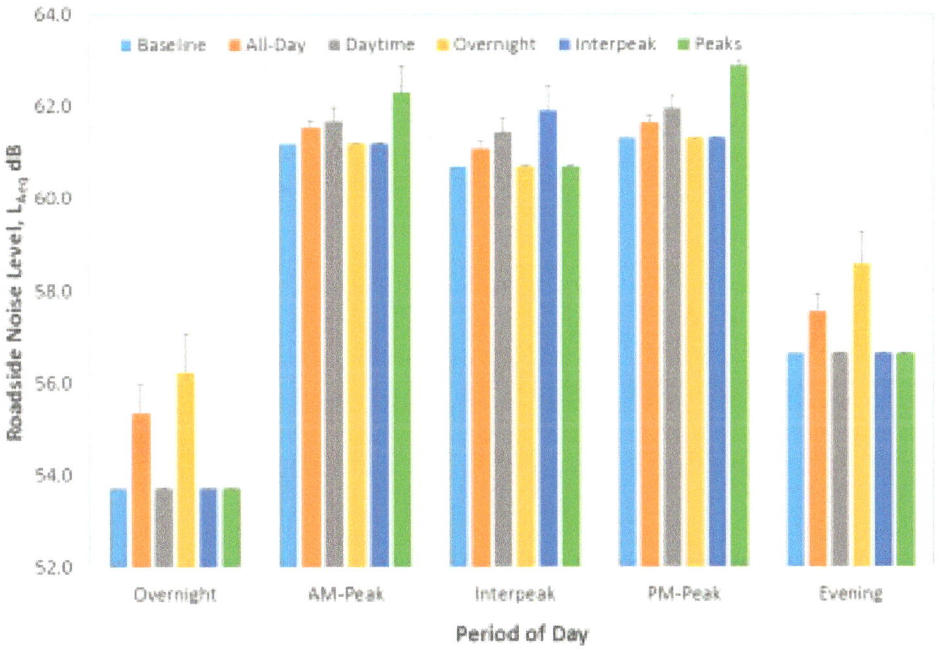

Figure 5: Peak roadside L_{Aeq} noise levels compared to the baseline noise levels under differing vehicle arrival and departure policies.

Total daily NO_x emissions increase by 18%–20% (low scenario), and 27%–30% (high scenario) depending on access policy. In both scenario runs the best policy for diurnal total emissions is overnight access, which creates negligible conflict with baseline traffic, though the relative emissions in the overnight period increase by a factor of 2–2.6. The worst policies for diurnal total emissions are, predictably, allowing access only in peak hours, or during the inter-peak periods, where baseline traffic flows are high, though relative increases are lower, by a factor of 1.4–1.8.

Conversely, the overnight policy is viewed as very detrimental to noise levels, with 2.5 to 3.4 dBA increases predicted to the hourly L_{Aeq} values, effectively doubling the sound pressure at roadside, and being clearly perceptible, even discounting effects to sleep disturbance, due to peak events (L_{Amax} levels) associated with HDV passbys (WHO, 2009). Increases in relation to the peak- or inter-peak policies are more marginal at 1.1–1.8 dBA.

The noise results highlight an issue with the calculation of long-term average L_{Aeq} values, as would be required for noise mapping activities compliant with the European Noise Directive (OJEC, 2002). The actual contribution of fracking operations normalised over the period of operations may be considered miniscule, with increases of under 1 dBA being generally un-perceptible as a change to humans, and well within the expected range of uncertainty in such modelling. However, this interpretation ignores the fact that the change in noise level due to a particular phase of operation may be far greater (seeFig. 5), and people's perception of the whole period may be skewed by such events. Nor does it consider that subjective annoyance to noise may be further exaggerated by frequent peak-events, substantive increases in levels over previously very quiet background levels and noise in evening or night periods. All of these issues are not adequately represented by a single L_{Aeq} value.

Region and UK wide scenarios

In addition to the localised impacts studied above, a number of larger scenarios have been simulated to explore the magnitude and temporal profile of emissions (CO_2, NO_x and PM_{10}) regarding the development of a region, over a number of years. Four initial development scenarios were assumed: low (190 wells), medium (400 wells) high (810 wells) and ultra (2970 wells). The first three well number assumptions were taken from Regeneris (2011), regarding potential exploitation of Bowland Shale resources in Lancashire, UK, the 'ultra' value being taken from the estimated total calculated byBroderick et al. (2011) to meet 10% of UK annual gas consumption (2010 baseline) over a 20 year period.

Four well development timelines were considered over the 2015 to 2050 horizon, to give 16 baseline scenarios:

- an assumed normal distribution of wells developed per year ($\mu = 2025$, $\sigma = 3$ years);
- an 'early' development timeline (Norm dist. $\mu = 2018$, $\sigma = 3$ years, skew = − 3);
- a 'late' development timeline (Norm dist. $\mu = 2032$, $\sigma = 3$ years, skew = + 3); and
- uniform development over the years 2015 to 2035.

Under these assumptions the timelines based on the normal distributions peak at 25 (Low) to 395 (ultra) wells being developed in a single year, whilst the uniform timeline provides between 9.5 (low) and 150 (ultra) wells per year.

Each of the baseline scenarios was subsequently further tested with two additional assumptions:

- in order to sustain production, 50% of the wells coming on-stream in a given year would be re-fracked five years later; and
- that 90% of fracking water demand could be met by pipeline, with flowback and produced water removed by tanker.

The same assumptions for the localised low-water demand well analysis outlined inSection 3.2 were applied for each well pad. Finally, all scenario combinations were tested using four road networks: a motorway network, a primary-rural network, a sub-urban network and a minor rural road network. Baseline traffic growth on the network over the period was assumed from the 'Scenario 2' forecast scenarios outlined in the UK National Transport Model (NTM). All combinations of the above result in a total of 256 scenario runs.

Fig. 6 presents the additional CO_2 emissions results due to fracking operations under 'early' and 'late' timelines, assuming all water demands are met solely by tanker, on the minor rural road network. Total CO_2 emissions over the 20 year period vary from 6.6 kT in the 'late low' scenario to 108 kT in the 'early ultra' scenario.

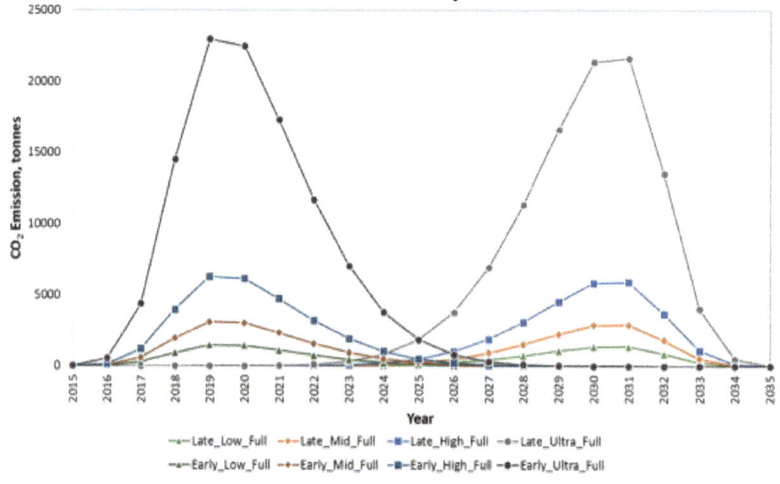

Figure 6: Annual CO_2 emissions for 'early' and 'late' timelines, under 'low', 'medium', 'high' and 'ultra' scenarios, assuming full fracking water demand met by tanker and no re-fracturing.

The lower total values and peaks in the late development timeline reflect the technology-based fuel consumption savings assumed for HDVs, taken from the NTM, amounting to an additional 6% reduction in each of the late adoption scenarios out to 2035.

Fig. 7 and Fig. 8 present the additional annual CO_2 and NO_x emissions associated with the low development scenario (190 wells), for all timelines, both with and without re-fracturing. Total CO_2 emissions vary between 6.6 kT and 10.3 kT, with the 'early' timeline producing the highest emissions. The need for re-fracturing adding an additional 48% to the emissions total in the 'early' and 'normal' timelines and 49% in the 'uniform' and 'late' timelines. For NO_x and PM_{10} the total ranges were 3.40–7.70 t and 0.85–1.29 t respectively, with the 'early' timeline producing 70% more emissions than the 'late' timeline, given the predominance of Euro V (and earlier) vehicle technologies in the initial fleets. The decrease in NO_x production with fleet turnover to Euro VI is especially noticeable in the 'uniform' timeline, which exhibits a rapid decay to almost constant values beyond the mid-2020s. Later adoption makes smaller absolute, but greater relative changes, to NO_x and PM_{10} emissions totals when considering the impact of re-fracturing.

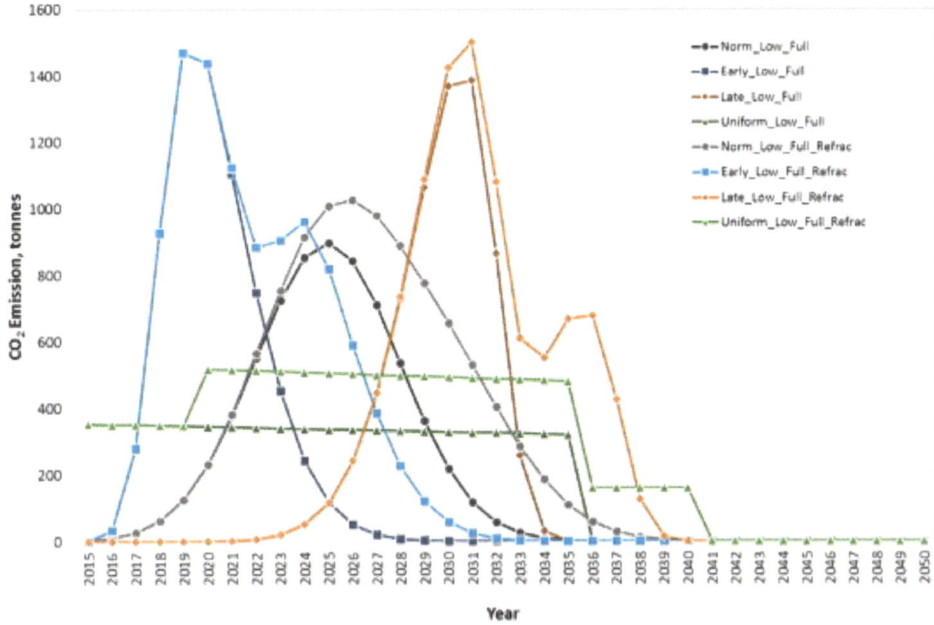

Figure 7: Annual CO_2 emissions for all timelines, under low well development, assuming full fracking water demand met by tanker, with and without re-fracturing.

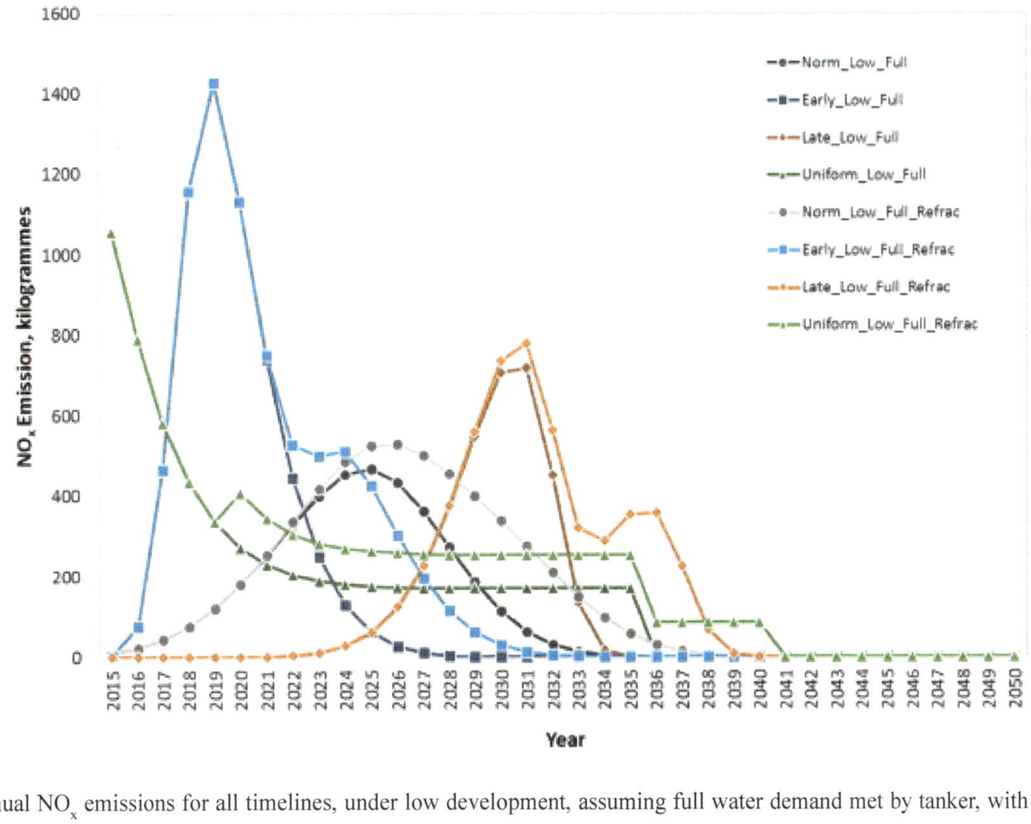

Figure 8: Annual NO_x emissions for all timelines, under low development, assuming full water demand met by tanker, with and without re-fracturing.

Fig. 9 presents the CO_2 emission results for the 'early low development' timeline, with and without re-fracturing and with and without the assumption that 90% of initial demand for drilling and fracturing water can be met by the existing water network. All flowback and produced water are then removed from site by tanker. This reduces the total HDV VKT for a given well to just under 40% of that assuming full demand is met by tanker. Commensurate reductions occur for all pollutants, driven primarily by the VKT changes, with only minor variation potentially attributable to changes in speed with reduction in traffic.

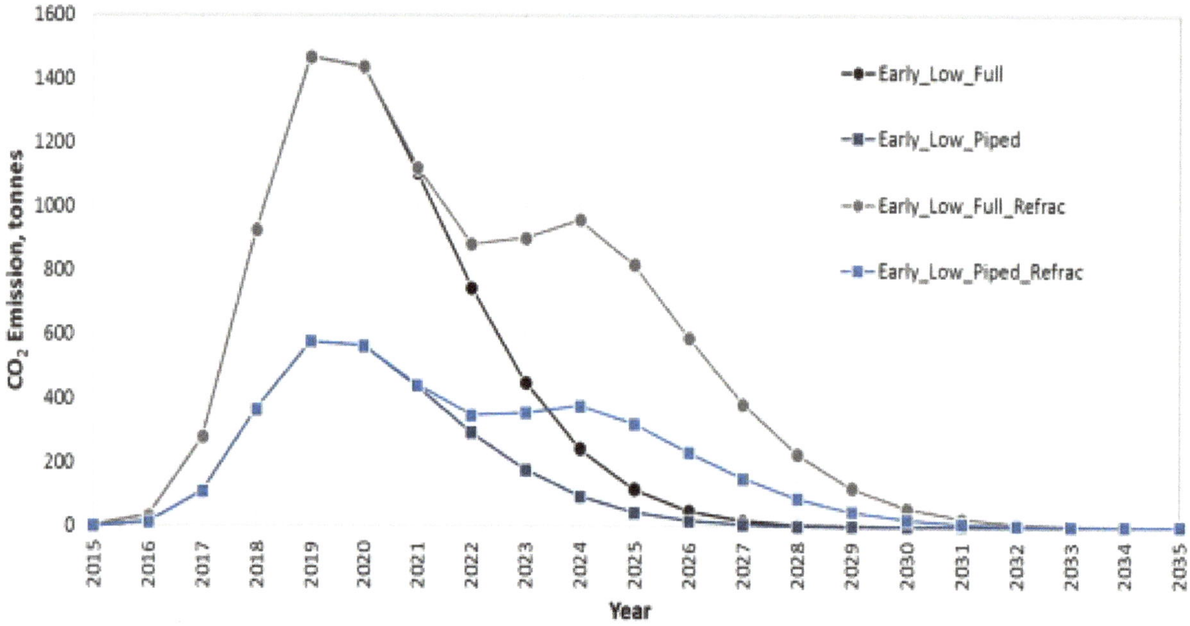

Figure 9: Annual CO_2 emissions for the early uptake timeline, under low development, assuming either full demand for fracking water met by tanker or 90% met by pipeline, with and without re-fracturing.

Further to the results presented in Table 3, the choice of road network, and hence underlying base flow, traffic growth, speed and congestion assumption, made relatively little impact on total additional CO_2 estimates, but could drastically alter the estimates for PM_{10} and especially NO_x, with NO_x estimates for the predominantly motorway-based network being only 40–50% of those values shown in Fig. 8, reflecting the behaviour of the underlying emissions curve to gross changes in speed (i.e. > 20 km/h) between dominant categories of road in the network.

Table 3: Comparison of transport impacts model to Broderick et al. (2011).

Source	Broderick et al. (2011)	TIM	TIM	TIM	Broderick et al. (2011)	TIM	TIM	TIM	
Year	2010	2010	2015	2020	2010	2010	2015	2020	
Scenario	Low	Low	Low	Low	High	High	High	High	Unit
uCO_2	38.0	35.1–38.8	34.9–38.3	34.6–38.2	58.7	53.4–58.9	52.6–58.1	52.5–58.0	t/well
NO_x	–	171–253	54.4–111.2	11.9–21.3	–	261–385	83.3–170	18.3–44.0	kg/well
pNO_2	–	22.7–32.6	6.3–12.2	1.2–2.9	–	34.7–49.6	9.6–18.7	1.9–4.5	kg/well
PM_{10}	–	6.4–8.7	4.1–6.5	3.4–5.9	–	9.8–13.2	6.3–10.0	5.2–9.0	kg/well
$PM_{2.5}$	–	4.9–5.9	2.8–3.9	2.1–3.3	–	7.5–9.0	4.1–5.9	3.2–4.9	kg/well
HC	–	5.6–6.6	1.0–1.2	0.3–0.4	–	8.5–10.1	1.6–1.9	0.4–0.5	kg/well
L_{Aeq}	–	0–0.15	0–0.15	0–0.15	–	0.01–0.20	0.01–0.20	0.01–0.20	dBA
ESAL	–	4.9	4.9	4.9	–	7.4	7.4	7.4	× 1000 axles/well

Table 4 provides a summary of the mass of fracking water moved, total vehicle kilometres travelled and emissions associated with the minor road dominant (i.e. worst case) scenarios.

Table 4: Summary of results from minor road network scenarios.

Scenario	No. of wells	Fracking water supply	Refrack	Fracking + flow-back water, Mt	VKT, million km	CO_2, kT	NO_x, T	PM_{10}, T
Low	190	Tanker	No	3.53	8.20	6.6–6.9	3.8–5.9	0.84–0.87
		Tanker	Yes	5.30	12.30	9.7–10.2	5.1–7.7	1.26–1.29
		90% pipe	No	1.38	3.23	2.5–2.7	1.3–2.4	0.33–0.34
		90% pipe	Yes	2.08	4.86	3.9–4.1	2.0–3.1	0.49–0.51
Medium	400	Tanker	No	7.44	17.26	13.9–14.5	7.1–12.5	1.77–1.82
		Tanker	Yes	11.16	25.89	20.6–21.6	10.7–16.2	2.68–2.71
		90% pipe	No	2.92	6.82	5.4–5.7	2.8–5.0	0.70–0.71
		90% pipe	Yes	4.37	10.23	8.1–8.6	4.2–6.4	1.05–1.07
High	810	Tanker	No	15.07	34.95	28.0–29.4	14.5–25.3	3.59–3.68
		Tanker	Yes	22.59	52.42	41.7–43.8	21.7–32.8	5.39–5.50
		90% pipe	No	5.90	13.81	11.0–11.6	5.7–10.0	1.41–1.45
		90% pipe	Yes	8.85	20.71	16.4–17.2	8.6–13.0	2.13–2.16
Ultra	2970	Tanker	No	55.25	128.15	102.8–107.8	53.2–92.6	13.2–13.5
		Tanker	Yes	82.86	192.22	153.1–160.4	79.7–120.4	19.8–20.1
		90% pipe	No	21.65	50.63	40.5–42.5	21.0–36.8	5.20–5.33
		90% pipe	Yes	32.47	75.95	60.3–63.2	31.5–47.8	7.79–7.94

The values in Table 4 may be placed in context with regional or national data. For example, the 'ultra' scenario, assuming no re-fracturing and no water supply by pipeline, involves lifting a tonnage water over a 20 year period comparable to total production of petroleum products in the UK in a single year (~ 64 Mt produced in 2013, DECC, 2015; ~ 59 Mt lifted in 2013, DfT, 2015a). Likewise, the 'ultra' scenarios produce VKT values over a 20 year period that are comparable to 10–42% of total VKT travelled by bulk liquid transporters in the UK in 2013 (DfT, 2015b). The 'medium' scenario assuming one re-fracturing and supply of water by tanker is equivalent in tonnage to forecourt petrol deliveries to the four major UK supermarket chains in a single year (12.2 Mt in 2013,DECC, 2015). The CO_2 emissions for the 'ultra' scenarios are approximately equivalent to those from transport associated annually with a mid-sized town (e.g. CO_2 emissions from the UK market town of Burnley in Lancashire (pop. 73,000) were estimated at 131 kT in 2012, emissions from all minor roads in the county as a whole were 737 kT, and for all transport sources were 2.57 Mt, DECC, 2014). It is noted that the number of wells involved in the 'ultra' scenarios is two orders of magnitude below the currently active gas wells in the United States (~ 493,000 in 2009 of which 90% have been fracked, Urbina, 2011). Additionally, it should be kept in mind that Table 4 only relates to traffic emissions, and does not include use of diesel fuels for drilling or pumping, nor does it include non-CO_2 GHG emissions (i.e. methane) from well sites.

DISCUSSION

We believe that the TIM model presented in this paper is a simple, yet flexible, approach to deal with spatial and temporal scenarios of fracking operations. It has been designed to be adaptable to different circumstances within any country, given available information on fleet characteristics, road type data, road quality data and haulage length characteristics, though the default emission rate values are based on European assumptions, which can be substituted with another set of emission factors with little effort. Pad operations may be modelled as subject to a timetable of events, with the compound effects on the road network allowed to arise. The phase and period approach also may be adapted to different levels of maturity in the industry, from initial exploratory wells in rural locations, to fully-developed facilities using pipelines for the majority of water transfer.

However, the literature cited in the introduction to the paper, and the subsequent results from our exploratory scenario testing, raise a number of interesting points to be considered. We believe a case can be made for caution in discussing impacts, with clear separation in that discussion between local and broader impacts. Citing total fracking-related emissions, for an individual well, multi-well pad, or even a region may appear negligible compared to those associated with transport in the region as a whole, or those emissions associated with another, established industrial sector. However, this does not negate discussion and analysis of impacts at the local level, where short-duration but large-magnitude events may occur, to the detriment of local ambient air quality and noise.

Comparison of impacts or normalisation of additional emissions over a long temporal baseline can distort the reporting of possible impact, leading to underestimation and undervaluing of potential public exposure and response to those impacts. In this work, we have chosen to report impacts for a six-well pad over the period from initial pad development to completion of all wells. This leads to mostly single-digit percentage increases in emissions, for all road types considered. Reporting over a longer baseline, e.g. the entire operational lifetime of a pad to decommissioning would result in negligible relative increases compared to baseline traffic impacts. However, examination of maximal results for phases with high traffic demand, even though the duration of those phases may be short, show substantial increases over the baseline, potentially leading to pollution exceedance events and breaches of local air quality standards, or increased annoyance and sleep disturbance in relation to noise.

A key, fundamental issue in the current discussion of fracking operations, identified through model development, is the considerable uncertainty in a large number of associated parameters, including, but not limited to, the overall scale of operations within a region, the individual water and waste demands at specific pad locations, and the amount of supporting infrastructure (both in terms of the underlying road transport network, and in the water supply network that could mitigate need for road transport) are all critical. For example, fracking activities in the UK may be more able to follow existing onshore oil and gas exploration practice with water supply by pipeline, thus reducing reliance on road tanker transport. Much of the literature examined giving the possible scope, demands and impacts of fracking in Europe cite experience from the US especially that contained in New York City and state documents (NYC DEP, 2009 and NYS DEC, 2011). Ultimately the number of wells may well be determined by public acceptability, as well as economics and geology.

Whilst the impact of an individual pad or well may relatively be low and generally localised, multiple pads operating in a region can have a non-linear compounding effect on those localised impacts, "particularly in a densely populated nation" (Broderick et al., 2011) such as the UK or other EU member states. The traffic assignment sub-model in TIM, using the 'speed-flow curve' approach is recognised as simplistic, with a potential lack of compounding congestion effects potentially leading to an under-prediction in modelled emissions. A subsequent enhancement will be the linking of the traffic demand sub-model to a microscopic traffic modelling framework for comparison to PITHEM calculated results, as per the methodology of O'Brien et al. (2014).

Current results from the TIM are based on hypothetical assumptions. Additional data is being sought to further enhance capability, with the intention of performing Monte-Carlo analysis to give additional data on the distribution and range of results. Our results do demonstrate that, even for a single, multi-well pad, relative impacts are disproportionate to VKT increases, especially for the case of axle loadings.

The expected technology improvements arising from the Euro class regulations are critical to the emissions calculations. Traffic-related NO_x emission rates per well in the mid-2020s are presently modelled as only 7%–11% of those of an operation commencing a decade earlier, based on the EFTv5.1.3 factors. However, there are potentially systematic issues with NO_x/NO_2 modelling predictions from road transport. Reductions in tailpipe emissions expected through the introduction of the EURO V and VI standards, forming part of the emissions assumptions in the EFT, have not been reflected in measured concentrations. The reasons for these discrepancies are discussed at length elsewhere (e.g. Carslaw et al., 2013) and remain an area of active research for both LDVs and HDVs. It is anticipated and understood that NO_x emissions rates will need to be updated as more knowledge on the real-world performance of EURO V HDVs, but especially EURO VI vehicles come to light. Performance of EURO V HDVs using de-NO_x catalysts (SCR), in conjunction with particle traps, has not lived up to initial promise. Oxley et al. (2012) suggest that the current EURO V/VI situation is "effectively equating to a failure of the EURO standards" in urban areas, though initial EURO VI on-road trials were encouraging (Hagman and Amundsen, 2013), with EURO VI HDV NO_x emissions potentially being lower than those for LDV diesel vehicles. Another future avenue of work is the use of the TIM to examine emissions under alternate HDV fuel scenarios, such as Compressed Natural Gas (CNG), Bio-methane or, ultimately, hydrogen (Oxley et al., 2012 and Pang and Murrells, 2013). Modelling of particulate emissions and resuspension from unpaved roads is another area requiring attention.

An integrated approach of using PITHEM in conjunction with air-quality modelling tools has already been demonstrated (Goodman et al., 2014), and subsequent heath impact or environmental justice analysis (O'Brien et al., 2014) represents an area for future work. The CNOSSOS-EU approach for noise mapping exposure assessment is also an active subject, with the methodology being relatively untested for real-world scenarios. There is also the possibility of embedding the TIM within a larger life-cycle assessment (LCA) framework for a more holistic approach to analysis, exemplified by the US GREET (Argonne, 2014) or Canadian GHGenius (NRC, 2013) models.

The disproportionate increases in ESAL found in the non-motorway cases are a cause for concern. It is likely that such roads were not originally designed with substantial volumes of HDV traffic in mind, and, given current economic conditions, may not fall on relevant authorities priority lists for reconstruction. Further assessment of impacts is problematic, given the diverse range of road ages and construction types possible. In older bituminous roads the main structural element of the roads, the roadbase, may have cured sufficiently to be able to withstand HDV loading, though surface layers may have aged, become brittle and be subject to cracking. Conversely, newer roads that have not reached full structural strength could be induced to failure early in their lifetimes (Nunn, 2000).

CONCLUSIONS

This paper has introduced a Traffic Impact Model (TIM) for studying the environmental impact of hydraulic fracturing operations. Implementation of the model has involved development of a traffic demand model for fracking operations, an abstract traffic assignment model to represent road network effects, and linking both to pre-existing, environmental modelling software that implements a number of standard approaches to assess GHG, local air quality, noise and axle loading impacts on roads.

Exploratory analyses using the model have revealed that the traffic impact of a single well pad can create substantial increases in local air quality pollutants during key activity periods, primarily involving the delivery of water and materials for fracking to the site. Modelling of NO_x emissions showed increases reaching 30% over non-fracking periods and noise levels doubling (+ 3.4 dBA), dependent on access policy implemented to the site, potentially exacerbating existing environmental issues. Normalisation of these values over a longer period, such as the time to completion of all wells on a pad, mitigates the 'raw values' but may present a distorted picture of the actual impact on the local populous.

Somewhat conversely, use of the model to explore hypothetical future technology timelines over a range of well development scenarios covering several decades, show that the overall impact to a region, or a country as a whole, appear somewhat negligible compared to general traffic or industrial activities, though it is recognised that the methodology used may underestimate emissions associated with network congestion.

ACKNOWLEDGEMENTS

This research was carried out as part of the ReFINE (Researching Fracking in Europe) consortium led by Newcastle and Durham Universities and funded from 2013 to 2015 by the Natural Environment Research Council (UK), Shell and Chevron, and now by Ineosand Centrica. We are grateful to the Faculty of Science and Faculty of Science, Agriculture and Engineering ethics committees at Durham and Newcastle Universities respectively, as well as the UK Research Integrity Office for their advice on research ethics. We thank the ReFINE Independent Science Board (http://www.refine.org.uk/how-we-work/independent-scienceboard) for spending time prioritising the research projects undertaken by ReFINE. The results and conclusions are solely those of the authors.

REFERENCES

1. AASHTO, 1986. Guide for the Design of Pavement Structures. American Association of State Highway and Transportation Officials, Washington D.C., US.
2. AEA, 2012. Support to the identification of potential risks for the environment and human health arising from hydrocarbon operations involving hydraulic fracturing in Europe.
3. Report by AEA for the European Commission DG Environment. AEA/R/ED57281 (Didcot, Oxfordshire, UK). http://ec.europa.eu/environment/integration/energy/pdf/ fracking%20study.pdf.
4. Andrews, I.J., 2013. The Carboniferous Bowland Shale gas study: geology and resource estimation. Report by the British Geological Survey on Behalf of the Department for Energy and Climate Change. London, UK (Online resource: https://www.gov.uk/ government/uploads/system/uploads/attachment_data/file/226874/BGS_DECC_
5. BowlandShaleGasReport_MAIN_REPORT.pdf [Accessed: 13/07/2015]).
6. Argonne National Laboratory, 2014. The Greenhouse Gases, Regulated Emissions, and Energy Use in Transportation (GREET) Model. Argonne Transportation Technology R&D Centre, Chicago, Il. USA (Online resource: https://greet.es.anl.gov/[Accessed: 07/07/2015]).
7. ARUP. 2014. Environmental statement: temporary shale gas exploration: Roseacre Wood, Lancashire. Report by ARUP Consultants on Behalf of Cuadrilla Elswick Ltd.
8. Boulter, P.G., Barlow, T.J., McCrae, I.S., 2009. Emissions factors 2009: report 3 — emissions factors for road vehicles in the UK (version 6). TRL Published Project Report PPR356 for the UK Department for Transport (via Gov.uk) (Online resource: https://www. gov.uk/government/uploads/system/uploads/attachment_data/file/4249/report-3. pdf [Accessed: 13/07/15]).
9. Broderick, J., Anderson, K., Wood, R., Gilbert, P., Sharmina, M., Footitt, A., Glynn, S., Nicholls, F., 2011. Shale gas: an updated assessment of environmental and climate change impacts. A Report Commissioned by the Co-operative and Undertaken by Researchers at the Tyndall Centre, University of Manchester. Tyndall Centre for Climate
10. Change Research, University of Manchester (Online resource: http://www.tyndall.ac. uk/sites/default/files/coop_shale_gas_report_update_v3.10.pdf [Accessed: 13/07/ 2015]).
11. Carslaw, D.C., Williams, M.L., Tate, J.E., Beevers, S.D., 2013. The importance of high vehicle power for passenger car emissions. Atmos. Environ. 68, 8–16.
12. Cuadrilla, 2014a. Preese Hall Site, Weeton, Lancashire. Online resource: http://www. cuadrillaresources.com/our-sites/locations/weeton/ ([Accessed: 13/07/15]).
13. Cuadrilla, 2014b. Water disposal. Online resource: http://www.cuadrillaresources.com/ protecting-our-environment/water/water-disposal/ ([Accessed: 13/07/15]).
14. Davies, R.J., Foulger, G., Mathias, S.A., Moss, J., Hustoft, S., Newport, L., 2013. Reply: Davies et al. (2012), Hydraulic fractures: how far can they go? Mar. Pet. Geol. 43, 519–521.
15. Davies, R.J., Almond, S., Ward, R.S., Jackson, R.B., Adams, C., Worrall, F., Herringshaw, L.G., Gluyas, J.G., Whitehead, M.A., 2014. Oil and gas wells and their integrity: implications for shale and unconventional resource exploitation. Mar. Pet. Geol. 56, 239–254.
16. DEC, 2008. Multi-well Pads are the Norm. Devon Energy Corporation (Online resource: http://www.dvn.com/CorpResp/initiatives/Pages/Multi-wellPads.aspx#terms? disclaimer=yes [Accessed: 13/08/2014]).

17. DECC, 2010. The unconventional hydrocarbon resources of Britain's onshore basins — shale gas. Report by the British Geological Survey on Behalf of the Department for Energy and Climate Change (Online resource: https:// www.og.decc.gov.uk/UKpromote/ onshore_paper/UK_ onshore_shalegas.pdf [Accessed: 13/07/15]).

18. DECC, 2014. UK Local Authority and Regional Carbon Dioxide Emissions National Statistics: 2005–2012. Department for Energy and Climate Change (Online resource: https://www.gov.uk/government/statistics/ local-authority-emissions-estimates [Accessed: 13/07/15]).

19. DECC, 2015. Digest of UK Energy Statistics (DUKES) 2014. Department for Energy and Climate Change (Online Resource: https://www.gov.uk/government/ uploads/system/ uploads/attachment_data/file/338750/ DUKES_2014_printed.pdf [Accessed: 13/07/ 15]).

20. DEFRA, 2012. Emission Factors Toolkit for Vehicle Emissions (Version 5.1.3. August 2012 Update). Department for Environment, Food and Rural Affairs (Online resource: http://laqm.defra.gov.uk/review-and-assessment/tools/emissions-factors-toolkit. html [Accessed: 13/07/15]).

21. DfT, 2013. Average Annual Daily Flow and Temporal Traffic Distributions (TRA03). Road Traffic Statistics Tables. Department for Transport (Online resource: https://www. gov.uk/government/collections/road-traffic-statistics [Accessed: 10/11/14]).

22. DfT, 2015a. Goods lifted by commodity: Annual 2004–2013. Road Freight Statistics Tables. Department for Transport (Online resource: https://www.gov.uk/ government/ statistical-data-sets/road-freight-statistical-tables-index [Accessed: 07/07/15]).

23. DfT, 2015b. Goods Lifted, Goods Moved and Loaded Vehicle Kilometres by Mode of Appearance, 2013. Road Freight Statistics Tables. Department for Transport (Online resource: https://www.gov.uk/government/ statistical-data-sets/road-freightstatistical-tables-index [Accessed: 07/07/15]).

24. DrillingInfo, 2014. On the launch pad: the rise of pad drilling. Online resource: http://info. drillinginfo.com/ launch-pad-rise-pad-drilling/ ([Accessed: 13/07/2015]).

25. EEA, 2007. EMEP/EEA air pollutant emissions inventory guidebook 2009: part B: sectorial guidance chapters, 1. Energy, 1.A. Combustion, 1.A.3.b. Road transport. Technical Report 16/2007 (Formerly EMEP/ CORINAIR). European Environment Agency (Online resource: http://www.eea.europa.eu/publications/emep-eea-emission-inventoryguidebook-2009 [Accessed: 13/07/15]).

26. EP DGIP, 2011. Impacts of shale gas on the environment and on human health. European Parliament: Directorate-general for Internal Policies. Policy Department A: Economic and Scientific Policy, Environment, Public Health and Food Safety (IP/A/ENVI/ST/ 2011-07 PE 464.425. Online resource: http://www.europarl. europa.eu/RegData/ etudes/etudes/join/2011/464425/ IPOL-ENVI_ET(2011)464425_EN.pdf [Accessed: 13/07/2015]).

27. Goodman, P.S., Namdeo, A.K., Galatioto, F., Bell, M.C., Foster, E.T., Shield, C., 2014. Investigation of the emission and air quality impacts of low emission zone scenarios in Newcastle and Gateshead, UK. Proceedings of the 16th International Conference on Harmonisation Within Atmospheric Dispersion Modelling for Regulatory Purposes (HARMO 16), Varna, Bulgaria (2014, September 8–11).

28. Hagman, R., Amundsen, A.H., 2013. Exhaust emissions from vehicles with Euro 6/VI technology. English Language Summary of TOI Report 1259/2013. Institute for Transport Economics, Norwegian Centre for Transport Research, Olso, Norway.

29. Highways Agency, 1999. Traffic capacity of urban roads. Design Manual for Roads and Bridges (DMRB)(Assessment and Preparation of Road Schemes), Section 1 (Preparation and Implementation), Part 3 Vol. 5. UK Highways Agency (DMRB 5.1.3 TA 79/99. Online resource: http://www.standardsforhighways.co.uk/dmrb/ vol5/section1/ ta7999.pdf [Accessed: 13/07/2015]).

30. Highways Agency, 2002. Speeds on links. Design Manual for Roads and Bridges (DMRB)(Economic Assessment of Road Schemes), Section 1 (The COBA Manual), Part 5 Vol. 13. UK Highways Agency (DMRB 13.1.5. Online resource: http:// webarchive. nationalarchives.gov.uk/20100304070241/http://www. dft.gov.uk/pgr/ economics/software/coba11usermanual/ part5speedsonlinks.pdf [Accessed: 13/07/ 2015]).

31. Highways Agency, 2006. Traffic assessment. Design Manual for Roads and Bridges (DMRB)(Pavement Design and Maintenance), Section 2 (Pavement Design and Construction), Part 1 Vol. 7. UK Highways Agency, DMRB 7.2.1 HD 24/06 Online resource: http://www. standardsforhighways.co.uk/dmrb/vol7/section2/ hd2406.pdf ([Accessed: 13/07/2015]).

32. Jiang, M., Hendrickson, C., VanBriesen, J., 2013. Life cycle water consumption and wastewater generation impacts of a Marcellus shale gas well. Environ. Sci. Technol. 48 (3), 1911–1920. http://dx.doi.org/10.1021/ es4047654. Online resource (http://www. ncbi.nlm. nih.gov/pmc/articles/PMC3915742/ [Accessed: 13/07/2015]).

33. Kephalopoulos, S., Paviotti, M., Anfosso-Lédée, F., 2012. Common noise assessment methodologies in Europe. To be Used by the EU Member States for Strategic Noise Mapping Following Adoption as Specified in the Environmental Noise Directive 2002/49/ EC. Joint Research Centre (JRC) Institute for Health and Consumer Protection, Ispra, Italy (Report EUR 25379 EN. Online resource: http://publications. jrc.ec.europa. eu/repository/bitstream/JRC72550/cnossos-eu%20 jrc%20reference% 20report_final_on%20line%20 version_10%20august%202012.pdf [Accessed: 13/07/ 2015]).

34. King, G.E., 2012. Hydraulic fracturing 101. What every representative, environmentalist, regulator, reporter, investor, university researcher, neighbor and engineer should know about estimating frac risk and improving frac performance in unconventional gas and oil wells.

Proceedings of the Society of Petroleum Engineers' Hydraulic Fracturing Technology Conference. The Woodlands, Texas, USA (2012, February 6–8. Online resource: http://www.kgs.ku.edu/PRS/Fracturing/Frac_Paper_SPE_152596.pdf [Accessed: 07/07/15]).

35. Mair, R., 2012. Shale Gas Extraction in the UK: A Review of Hydraulic Fracturing. The Royal Society and the Royal Academy of Engineering, London. UK (Online resource: https://royalsociety.org/~/media/Royal_Society_Content/policy/projects/shale-gas/2012-06-28-Shale-gas.pdf [Accessed: 13/07/2015]).

36. NAEI, 2011. Vehicle Fleet Projections (Base 2011). UK National Atmospheric Emissions Inventory (Online Resource: http://naei.defra.gov.uk/data/ef-transport?f_datachunk_ id=328 [Accessed: 13/07/2015]).

37. NaeI, 2012. NOx Speed-related Emissions Functions COPERT4v8.1. UK National Atmospheric Emissions Inventory (Online resource: http://naei.defra.gov.uk/data/eftransport?f_datachunk_id=328 [Accessed: 13/07/2015]).

38. Namdeo, A.K., Goodman, P.S., 2012. Platform for integrated traffic, health and environmental modelling (PITHEM). Proceedings of the Better Air Quality Conference, Hong Kong (2012, December 5–6).

39. NRC, 2013. GHGenius: a model for lifecycle assessment of transportation fuels. Version 4.03. Report by (S&T)2 Consultants on Behalf of Natural Resources Canada. Ontario, Ottowa, Canada (Online resource: http://www.ghgenius.ca/reports.php?tag= GHGenius%204.03 [Accessed: 07/07/15]).

40. Nunn, M.E., 2000. Development of Bituminous Pavement Design in the UK and Related Research Thesis for DPhil by Published Works University of Ulster, Belfast. NI.

41. NYC DEP, 2009. Impact assessment of natural gas production in the New York City water supply watershed. Rapid Impact Assessment Report. Hazen and Sawyer Report on Behalf of the New York City Department of Environmental Protection. New York, NY, US (September 2009. Online resource: http://www.nyc.gov/html/dep/pdf/ natural_gas_drilling/rapid_impact_assessment_091609.pdf [Accessed: 13/07/15]).

42. NYS DEC, 2011. Revised draft supplemental generic environmental impact statement on the oil, gas and solution mining regulatory program. Well Permit Issuance for Horizontal Drilling and High-volume Hydraulic Fracturing to Develop the Marcellus Shale and Other Low-permeability Gas Reservoirs. New York State Department of Environmental Conservation, Albany, NY, US (September 2011. Online resource: http://www.dec.ny.gov/energy/75370.html [Accessed: 13/07/15]).

43. O'Brien, J., Namdeo, A., Bell, M.C., Goodman, P., 2014. Congestion sensitive approach to modelling road network for air quality management. Int. J. Environ. Pollut. 54 (2/3/ 4), 213–221.

44. OJEC, 2002. Directive 2002/49/EC of the European Parliament and of the Council of 25 June 2002 relating to the assessment and management of environmental noise. Official Journal of the European Communities, Brussels, BE (18th July 2002. Online resource: http://eur-lex.europa.eu/legal-content/EN/TXT/PDF/?uri=CELEX:32002L0049&from=EN [Accessed: 10/05/15]).

45. OJEC, 2009. Regulation (EC) No 595/2009 of the European Parliament and of the Council of 19 June 2009 on type-approval of motor vehicles and engines with respect to emissions from heavy duty vehicles (Euro VI). Official Journal of the European Communities, Brussels, BE (18th June 2009. Online resource: http://eur-lex.europa.eu/LexUriServ/ LexUriServ.do?uri=OJ:L:2009:188:0001:0013:EN:PDF [Accessed: 10/05/15]).

46. Osborn, S.G., Vengosh, A., Warner, N.R., Jackson, R.B., 2011. Methane contamination of drinking water accompanying gas-well drilling and hydraulic fracturing. Proc. Natl. Acad. Sci. 108, 8172–8176.

47. Oxley, T., Elshkaki, A., Kwiatkowski, A., Castillo, T., Scarborough, T., ApSimon, H., 2012. Pollution abatement from road transport: cross-sectoral implications, climate cobenefits and behavioural change. Environ. Sci. Policy 19–20, 16–32.

48. Pang, Y., Murrells, T., 2013. Emissions Factors for Alternative Vehicle Technologies. National Atmospheric Emissions Inventory (NAEI Reference3 ED57423001. Online resource: http://naei.defra.gov.uk/resources/NAEI_Emisison_factors_for_alternative_ vehicle_technologies_Final_Feb_13.pdf [Accessed: 10/05/15]).

49. Regeneris, 2011. Economic impact of shale gas exploration & production in Lancashir and the UK. Report Prepared on Behalf of Cuadrilla Resources, Regerneris Consulting Limited (2011, September).

50. Sandberg, U., 2001. Noise emissions of road vehicles: effect of regulations. I-INCE Working Party on Noise Emissions of Road Vehicles. Swedish National Road and Transport Research Institute, Linköping, Sweden.

51. SEAB, 2011. Shale gas production sub-committee: 90 day report. Secretary of Energy Advisory Board Report to the US Department of Energy, 18 August 2011 (Online resource: http://www.shalegas.energy.gov/resources/081811_90_day_report_final.pdf [Accessed: 13/07/2015]).

52. Sumi, L., 2008. Shale gas: focus on Marcellus shale. Report for the Oil & Gas Accountability Project/ Earthworks. May 2008 (Online Resource: http://www.earthworksaction.org/ files/publications/OGAPMarcellusShaleReport-6-12-08.pdf [Accessed: 13/07/2015]).

53. Urbina, I., 2011. Regulation lax as gas wells' tainted water hits rivers. Article in the New York Times, 26th Feb 2011. New York, NY, USA. Online Resource: http://www. nytimes.com/2011/02/27/us/27gas.html?_r=2&pagewanted=all ([Accessed: 07/ 07/2014]). US EPA, 2010. Hydraulic fracturing research study. Online resource: http://www.epa. gov/ogwdw000/uic/pdfs/hfresearchstudyfs.pdf ([Accessed:16/07/2015]).

54. Watts, G., van Maercke, D., van Leeuwen, H., Barelds, R., Beuving, M., Defrance, J., Jonasson, H., Nota, R., Taraldsen, G., Witte, J., 2004. Effects of speed distributions on the Harmonoise model predictions.

Proceedings of the 33rd International Congress and Exposition on Noise Control Engineering (InterNoise 2004), Prague, Czech Republic, 22–25th August 2004 (Online Resource: http://dgmr.nl/uploads/files/Effects%20of%20speed%20distributions%20on%20the%20Harmonoise%20model% 20predictions.pdf [Accessed: 13/07/2015]).

55. WHO, 2009. Night noise guidelines for Europe. World Health Organisation Regional Office for Europe, Copenhagen, Denmark (Online resource: http://www.euro.who.int/__ data/assets/pdf_file/0017/43316/E92845.pdf [Accessed: 13/07/15]).

CHAPTER 10

CHALLENGES IN GROUNDWATER RESOURCE MANAGEMENT IN COASTAL AQUIFERS OF EAST AFRICA: INVESTIGATIONS AND LESSONS LEARNT IN THE COMOROS ISLANDS, KENYA AND TANZANIA

Jean-Christophe Comte[a,c], Rachel Cassidy [b,c,] Joy Obando [d], Nicholas Robins [c,e], Kassim Ibrahim[f] , Simon Melchioly[g], Ibrahimu Mjemah[h], Halimu Shauri[i] , Anli Bourhane[j,k], Ibrahim Mohamed[f] , Christine Noe [g], Beatrice Mwega [d], Mary Makokha [d], Jean-Lambert Join[j] , Olivier Banton[l] , Jeffrey Davies [e,m]

School of Civil Engineering and Geosciences, Newcastle University, Newcastle upon Tyne, Tyne and Wear NE1 7RU, UK

[a] University of Aberdeen, School of Geosciences, Aberdeen, UK
[b] Agri-Food and Biosciences Institute, Belfast, UK
[c] Previously at Queen's University Belfast, Groundwater Research Group, Belfast, UK
[d] Kenyatta University, Department of Geography, Nairobi, Kenya
[e] Previously at British Geological Survey, Wallingford, UK
[f] University of Comoros, Faculty of Sciences, Moroni, Comoros
[g] University of Dar Es Salaam, Department of Geology, Dar Es Salaam, Tanzania
[h] Sokoine University of Agriculture, Department of Physical Sciences, Morogoro, Tanzania
[i] Pwani University, Department of Social Sciences, Kilifi, Kenya
[j] University of Reunion Island, Laboratory Geosciences Reunion, Saint-Denis, France
[k] Water Office Reunion, Saint-Denis, France
[l] University of Avignon, Laboratory of Hydrogeology, Avignon, France
[m] University College London, London, UK

Citation: Jean-Christophe Comte, Rachel Cassidy, Joy Obando, Nicholas Robins, Kassim Ibrahim, Simon Melchioly, Ibrahimu Mjemah, Halimu Shauri, Anli Bourhane, Ibrahim Mohamed, Christine Noe, Beatrice Mwega, Mary Makokha, Jean-Lambert Join, Olivier Banton, Jeffrey Davies, Challenges in groundwater resource management in coastal aquifers of East Africa: Investigations and lessons learnt in the Comoros Islands, Kenya and Tanzania, Journal of Hydrology: Regional Studies, Volume 5, March 2016, Pages 179-199, ISSN 2214-5818, http://dx.doi.org/10.1016/j.ejrh.2015.12.065.

ABSTRACT

Study region

Coastal areas of Kenya (Kilifi County), Tanzania (Kilwa district) and Comoros (Ngazidja island), East Africa.

Study focus

Research aimed to understand the physical and societal drivers of groundwater accessibility and identify critical aspects of groundwater access and knowledge gaps that require further monitoring and research. Interdisciplinary societal, environmental and hydrogeological investigations were consistently undertaken in the three areas considered as exemplars of the diversity of the coastal fringes of the wider region. This paper focuses on the hydrogeological outcomes of the research, framed within the principal socio-environmental issues identified.

New hydrological insights

Results confirm the fundamental importance of coastal groundwater resources for the development of the region and the urgent need to match groundwater development with demographic and economic growth. Hydrogeological knowledge is fragmented, groundwater lacks a long-term monitoring infrastructure and information transfer from stakeholders to users is limited. Current trends in demography, climate, sea-level and land-use are further threatening freshwater availability. Despite possessing high-productivity aquifers, water quality from wells and boreholes is generally impacted by saltwater intrusion. Shallow large-diameter wells, following the traditional model of these areas, consistently prove to be less saline and more durable than deeper small-diameter boreholes. However, promoting the use of large numbers of shallow wells poses a significant challenge for governance, requiring coherent management of the resource at local and national scales and the engagement of local communities.

GROUNDWATER IN COASTAL REGIONS OF EAST AFRICA

Africa is the continent with the fastest growing populations (United Nations, 2011) with coastal regions projected to experience the highest rates of population growth in coming decades (Vafeidis et al., 2011). At the continental level, East Africa has the second highest rate of population growth, after the Central African region (Ashton and Turton, 2009), while having the lowest renewable freshwater resource of Sub-Saharan Africa (Braun and Xu, 2010). In the Mozambique Channel region, the populations of the Comoros Islands, Kenya and Tanzania have quadrupled in the last 50 years (World Bank, 2014a), twice the global average and growing. Most of the population increase is in urban areas, which are already densely populated and where water resources, particularly groundwater which is often the only source of water of acceptable quality, are already under intense pressure (Steyl and Dennis, 2010; MacDonald et al., 2012; Walraevens et al., 2015). Recent papers have shown the importance of groundwater resources that are under growing pressure in developing regions elsewhere, but crucial for economic development (e.g., Mukherjee et al., 2015 and Watto and Mugera, 2015). Adequate provision of water is central to human health and economic development in these regions where water scarcity is a serious impediment to growth and poses a threat to political stability.

In a demographic context within Sub-Saharan Africa, groundwater resources are heavily relied upon and the focus of strategic development (Ashton and Turton, 2009) because of both the relative resilience of aquifers to anticipated climate change and the widespread contamination of surface water resources. However, as exemplified by findings from previous inter-African reviews of regional and national groundwater management frameworks (e.g. Robins et al., 2006, Adelana et al., 2008, Braune and Xu, 2008 and Braune and Xu, 2010; Knüppe, 2011), groundwater resources, except in countries totally dependent on them, (i) still suffer from an under-evaluation of their importance and significance, (ii) are often managed separately from surface water and (iii) management institutions (governments, communities, NGOs, consultants) are largely fragmented and lack a central strategy. Groundwater information services, i.e. databases and systematic long-term monitoring, are non-existent or of inadequate quality and fragmented. The involvement of stakeholders, including communities, in decision-making processes and resource utilisation is insufficient, under-acknowledged by managers and governments and requires urgent capacity building of all parties, from individual to institutional (e.g. BGR, 2007; Foster et al., 2008 and Braune and Xu, 2010).

Efforts towards the integration of groundwater management within holistic water management frameworks, such as river basin organisations have only emerged recently, mainly through national Integrated Water Resource Management (IWRM) frameworks (Global Water Partnership, 2000 and Global Water Partnership, 2002; World Water Council, 2006) that aim to address these deficiencies and contribute to the Millennium Development Goals (MDGs). Nonetheless, groundwater management appears subsumed under broader policy, legal and institutional frameworks dealing with the management of water resources; hence the integration of groundwater into national policy requires development of adequate cross-sector dialogue within government (Mumma et al., 2011 and Foster et al., 2012). The management of coastal groundwater poses a further challenge due to its vulnerability to seawater contamination and the specific physical and socio-economic characteristics of the coastal zone. The review ofSteyl and Dennis (2010) is one of the very few works that provides insights on common issues with regards to groundwater management in coastal aquifers of Africa. There is a notable gap in the literature regarding the inter-disciplinary aspects of the management of coastal groundwater resources in Sub-Saharan Africa and the regionally-specific socio-environmental drivers.

In the East African coastal region, demographic change has led to an overall increase in groundwater abstraction, with increased drilling of deep boreholes with higher abstraction rates than traditional dug wells and shallow boreholes. High abstraction rates and concentrated well fields are incompatible with the nature of coastal aquifers. These aquifers are mainly low-lying with shallow water tables and are susceptible to seawater intrusion if not carefully managed, regardless of aquifer productivity or recharge rates (Robins, 2013 and Werner et al., 2013). The abstraction of inland groundwater, by way of contrast is generally only limited by aquifer productivity and available recharge.

The geology of much of the East African coastal aquifers is young (Mesozoic to present), composed of soft or unstable sediments and volcanic deposits, which need supporting during and after drilling to avoid collapse and bloc fall. Though slower and labour intensive to construct, large diameter traditional wells are more suited to such environments; causing less drawdown than boreholes and giving access for clearance of debris (e.g. Bourhane et al., 2015). A larger number of shallow wells widely distributed across an area is effective in minimising the risk of seawater intrusion. Such strategies, however, pose a management challenge in urban areas (Edmunds, 2012) where high densities of abstraction points may be necessary, initial drilling costs are high and risk of contamination is considerable.

East Africa is identified as one of the regions at greatest risk globally from the impacts of climate change (Hinkel et al., 2012). Most coastal areas are low-lying and will experience significant inundation even with modest rises in sea level. The most recent climate change projections (IPCC, 2014a and Cai et al., 2014) anticipate an accelerated rise in global mean sea level in coming decades (Watson et al., 2015) combined with an increase in rainfall during the wet seasons and higher annual temperatures. These may increase potential evapotranspiration and increase seawater intrusion in low-lying coastal settings and islands (Comte et al., 2014). According to the latest report of the Intergovernmental Panel on Climate Change (IPCC, 2014a and IPCC, 2014b), the available information or evidence on projected impacts of climate change on coastal aquifers is globally limited. More specifically, the impact on aquifer recharge in East Africa remains poorly constrained (Döll, 2009) due to a lack of hydrometric monitoring which could be used to inform forward modelling and proactive development of management approaches (Robins et al., 2006, Adelana, 2009 and Steyl and Dennis, 2010).

The combined effects of demographic, socio-economic, climatic and political changes on coastal groundwater resources negatively impact both the availability and sustainability of potable groundwater. Although an adverse impact is expected, the extent of the problem is not well established because of the large number of driving factors involved and the paucity of data with which to test them. Better understanding requires the implementation of integrated multidisciplinary research and expertise (including hydrogeologists, hydrologists, geographers, and socio-environmentalists) in order to improve the knowledge and understanding both of the vulnerability of coastal groundwater and its long-term response to change. This will permit identification of appropriate management strategies to mitigate negative impact on resources. This constitutes a necessary and major step towards the sustainable development and management of water accessibility for people in coastal East Africa.

AIMS AND OBJECTIVES

A suite of investigations was undertaken at three contrasting coastal areas in East Africa; Grande Comore (Ngazidja) Island (11°40′S; 43°20′E) in the Comoros Archipelago, Kilifi County, North of Mombasa, in Kenya (3°37′S; 39°50′E), and Kilwa, a rural area in southeast Tanzania (8°55′S; 39°30′E). These represent three of the lowest-income countries (LIC) (as classified by the World Bank, 2014b) of the East Africa/South-western Indian Ocean zone (Fig. 1) and are exemplars of the physical, social and cultural diversity of the coastal fringes of the region. An important component of the work was testing whether the concerns raised by Steyl and Dennis (2010), regarding the sustainable management of African coastal groundwater resources, apply consistently to the three study areas.

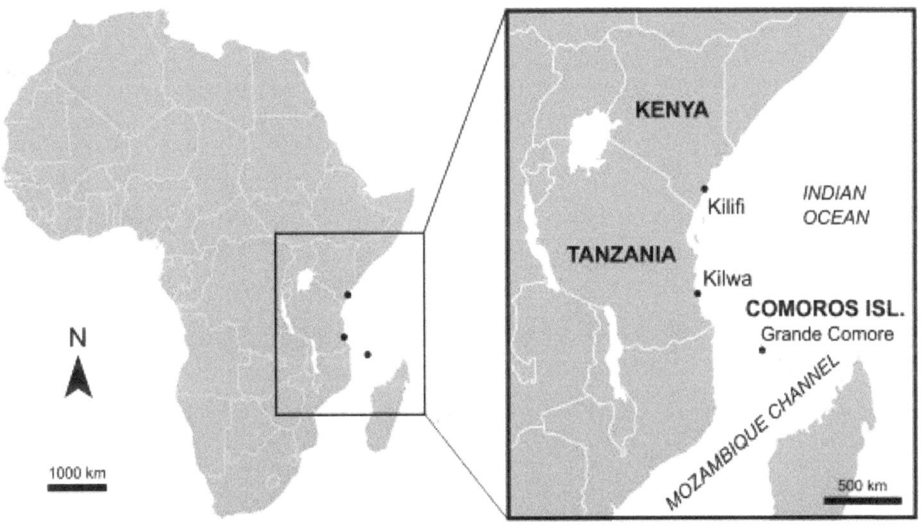

Figure 1: Location of the three study sites in the coastal zones of the Comoros Islands, Kenya and Tanzania.

Three research sites covering the (1) societal, (2) environmental and (3) hydrogeological status of the study areas were established and within each a consistent work programme was applied. The overall research programme aimed to understand the physical and societal aspects of groundwater accessibility, to provide up-to-date regional groundwater databases and identify critical aspects of groundwater access as well as knowledge gaps that will require further longer-term monitoring and research. In this paper we focus on the hydrogeological outcomes of the research, framed within a summary of the principal societal and environmental issues identified in each area. Issues relating groundwater management with governance and institutional frameworks in the study areas would be addressed in detail in another publication (Obando et al., in preparation) focussing specifically on the socio-economic aspects of this research programme.

STUDY AREAS-OVERVIEW

Comoros Islands—Grande Comore (Ngazidja) (see map in Fig. 2 a)

In the Comoros archipelago, the island of Grande Comore (1148 km^2) was selected for the study. As the most populated island in the region, and growing at over 2% per annum (~400,000 inhabitants estimated in 2015), more than half of the population is concentrated within 5 km of the shoreline due to the steep gradients of the elevated and afforested interior of the island. About 65% of the population do not have permanent access to groundwater and primarily harvest rainwater from roofs into tanks, which are only sufficient during the wet season and have issues with bacterial contamination. During the dry season, water use is rationed. For families that can afford it, freshwater is distributed by water trucks delivering from the few fresh abstraction wells on the island. These comprise 54 wells drilled in the volcanic aquifers of the coastal zone which supply 20 localities representing about 35% of the island's population (Mohamed and Othman, 2006). Fewer than 30% of the wells provide groundwater of acceptable quality, i.e. Total Dissolved Solids (TDS) < 1 g L^{-1} (Bourhane et al., 2015) and consequently the local drinking water salinity guideline is usually taken at 3 g L^{-1} instead of 1 g L^{-1} as recommended by the World Health Organisation (WHO, 2003). Wells with higher salinity continue to be used for irrigation, livestock or washing. This situation is not projected to improve in coming decades with anticipated population growth (demand is increasing at about twice the rate of groundwater development) and projected climate change (observed decrease in rainfall; Vincent et al., 2011). The large variations in groundwater salinity are locally responsible for community conflicts with regard to water costs.

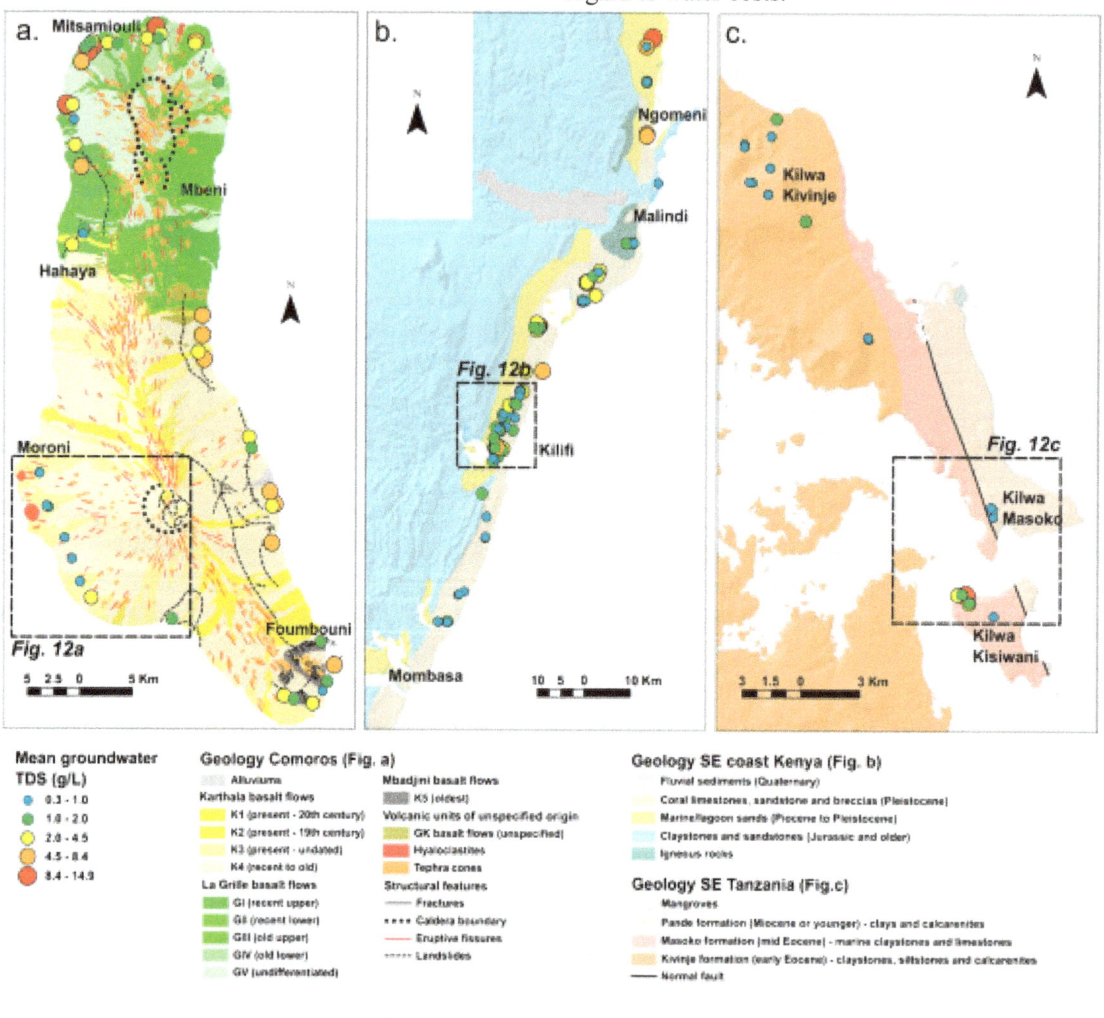

Figure 2: Overview of the study areas, including the main geological units and groundwater salinities of surveyed wells in (a) Grande Comores Island (Comoros), (b) Kilifi region (Kenya) and (c) Kilwa district (Tanzania). See Fig. 1 for the general location on these sites in East Africa. Dashed boxes indicate more detailed areas plotted in Fig. 12.

Kenya–Kilifi County (see map in Fig. 2b)

Kilifi County (13,006 km^2) lies along the Kenyan coast between Malindi to the north and Mombasa to the south, with a total population of 1,109,700 growing at a rate of 3.1% per annum. The area experiences a semi-arid climate, with irregular wet seasons and frequent drought (Onyancha et al., 2010). Although much of the piped water supply in the county is currently supplied from surface water sources, it is expected that reliance on groundwater in the area will increase as the demand for domestic and agricultural water increases. Groundwater yields vary along the coastal zone with poor quality and low yields in areas of Jurassic shales and Pleistocene coral limestones and higher quality and yields in Triassic sandstone and Quaternary sands. Excessive abstraction has already resulted in water contamination and saline intrusion (Musingi et al., 1999 and Munga et al., 2006). Surface water sources in the county are generally inland or to the north where the Athi-sabakia and Tana rivers, which originate in the Kenyan highlands, flow towards the Indian Ocean. The security of these sources is currently threatened by planned large-scale agricultural projects in the Kenyan interior and plans for dams upstream. Most of the water supplied to the densely populated Kilifi township and the linked coastal zone, which is the focus of this study, is piped from these rivers and inland lakes. Daily water demand in the township is estimated to be 200,000 m^{3}against the available 130,000 m^3 (NEMA, 2010) and is leading to pressure and conflict in the area. The contrast between tourist developments and affluent coastal residences and the amenity-poor, densely populated urban zone is pronounced. Many of the existing boreholes yield either saline or brackish water, and the remaining freshwater aquifers are in danger of overexploitation and are susceptible to pollution.

Tanzania—Kilwa district (see map in Fig. 2c)

Kilwa district (19,998 km^2) is a largely rural, low-lying coastal and island environment in south-eastern Tanzania consisting of coral limestone to the east and clays in the west, which dip eastward. Overburden comprises weathered lateritic soils and some Quaternary to Recent deposits of blown sand which are the dominant water bearing formations. The district is predominantly agricultural focussing on crop production such as cashews, simsim and coconut with some livestock rearing. Kilwa Masoko is a growing urban centre, the centre of local administration, with a developing tourism industry. In 2012 it was designated a service port for the gas pipeline installations to the north at Songo Songo island. The population has grown in recent years (currently ~15,000), fuelled by migration from the coastal hinterland. Water demand exceeds supply in the town and across the district. Currently the town is supplied by 4 public supply boreholes at a well field at the centre of the urban area, about 2 km from the coast, supplying approximately 1500 m^3/day. Demand is twice this volume so additional piped water is taken from open springs 15 km to the north at Mpara, with more wells under construction. The islands of Kilwa Kisiwani and Songo Mnara are historic stone-built towns (12thC) now designated a UNESCO world heritage site and a tourist attraction. Only Kilwa Kisiwani (area 17 km^2) is populated, with about 850 people settled in the more elevated (15 m) northwest of the island. The island is composed of coral limestone, overlying marls and clays (Nicholas et al., 2006). Groundwater occurs in shallow lenses, replenished during seasonal rains. Drainage is localised with no rivers or streams. The main employment is subsistence farming and fishing. Water shortages are major issues for the island community. The current sources of water are open shallow wells, the less saline dating from the 12th Century, which were added to by installation of boreholes and hand pumps in 2006 (Marobhe and Songo, 2006). All but one of the boreholes are saline or collapsed and the hand pumps are broken.

METHODOLOGY

Systematic investigations were undertaken in each of the study areas and hydrometric monitoring infrastructure installed at each site. The societal component of the study targeted issues of accessibility facing local communities and stakeholders in each area. An environmental characterisation, covering land use change and examination of existing data sets on sea-level and climate provided insight into the potential factors limiting groundwater availability and sustainability both at present and in the future. The hydrogeological component aimed to establish a quantitative and qualitative background status of coastal groundwater resources.

Societal constraints on water accessibility—perspectives of stakeholders and user groups

Gaining a local perspective on issues affecting access to water was one of the core objectives. Surveys of water users (householders, farmers and industries) and stakeholders (e.g. drilling companies, water managers) provided valuable complementary data to existing demographic observations and census data. Direct interviews with households (a total of 142 households in Kilifi, 154 in Kilwa and 185 in Grande Comore) were undertaken using a stratified random sampling method to select households within local electoral areas. Focused group discussions with community members and health providers were used to identify sources and quality of water supply and public health issues including common waterborne and water-related diseases.

Local stakeholder groups in each target area were established and a Participatory Action Research PAR (Pretty and Vodouhê, 1997) approach used to understand and address the social-economic and environmental factors of water resource management (Van Niekerk and Van Niekerk, 2009 and Mapfumo et al., 2013). This involved working with the selected communities through an inception workshop followed by community workshops and meetings and a final dissemination workshop. The PAR included use of questionnaires, Focused Group Discussion (FGD), in depth interviews and visioning (all in a participatory manner) where the stakeholders (youth groups, women's groups, elders) reflect on the past, look at the current situation and develop a 'vision' of what should be in the future.

The outputs provide an overview of: (a) existing issues with water access in each study area and perceptions of causes; (b) societal issues affecting water resources management; (c) current water resources management strategies and environmental factors affecting the same; (d) past practices and traditional methods of water management; (e) community stakeholder analysis on best practices of water resources management; (f) community perceptions on changing water resources and effects on their livelihoods; and (g) water resources management and environmental health.

Environmental change—land use change and climatic trends

For each study area available secondary data sources on climate, land use and recharge were compiled, and supplemental monitoring implemented where data gaps were identified. An automatic weather station provided rainfall and potential evapotranspiration time-series data at hourly intervals to promote understanding of borehole and well level and electrical conductivity monitoring.

Land cover and use were mapped from available aerial imagery for each area, and validated in the field. Temporal change was evaluated by comparison of the most recent aerial imagery with past imagery of equal resolution. Sources were limited by budget and as such the period varied depending on the availability of imagery for each site. Land use change was mapped and classified within a Geographic Information System (GIS) to provide information on settlement growth/decline and the clearance of land for agriculture, industry and infrastructure.

Long-term trends in temperature and rainfall in the study areas were examined, including the frequency and intensity of rainfall events, and temperature trends. Weather data for available coastal synoptic stations across Kenya, Tanzania and Comoros were acquired from the British Atmospheric Data Centre Global Weather Observation data sets over the period 1984–2014. The sampling time-step for the stations was normally every 3 h but quality and completeness of the data sets varied considerably among sites, with marked improvements from the late 1990s onward.

Hydrogeology and groundwater resources

Hydrogeological investigations were carried out to assess the current hydrogeological conditions and the aquifer vulnerability to seawater intrusion across each area.

Comprehensive baseline surveys were undertaken to establish the existing water well infrastructure. Existing datasets were used to map the location of wells and compile their technical characteristics, their use, pumping rate and records of changing salinity. A strong collaboration with local water services was initiated to ensure a continuous exchange and updating of borehole databases.

Baseline geophysical investigations, using 2D Electrical Resistivity Tomography (ERT), were carried out at each site, to characterise the aquifer structure and the current extent of saltwater intrusion. One to two profiles were undertaken at each site, perpendicular to the shore, using boreholes or wells with existing hydrogeological information to inform the interpretation.

Additional well monitoring was carried out to provide an indication of the groundwater variability in both space and time. Spatially, water levels, Electrical Conductivity (EC) and temperature profiles of wells were measured across a sub-set (dependent on density and accessibility) at each site, for a number of campaigns (2 per year corresponding to the dry and wet seasons). Temporally, in 1–3 strategic boreholes/wells at each site, a borehole datalogger was installed to acquire high-resolution temporal data on temperature, EC and water level, to investigate recharge processes and aquifer diffusivity.

RESULTS

Societal—community and stakeholder perspectives

Results from analysis of census data

Analysis of the available census data, from at least 2 decadal censuses, for each site provides an overview of demographic change (Fig. 3) and associated pressures on groundwater infrastructure (Table 1).

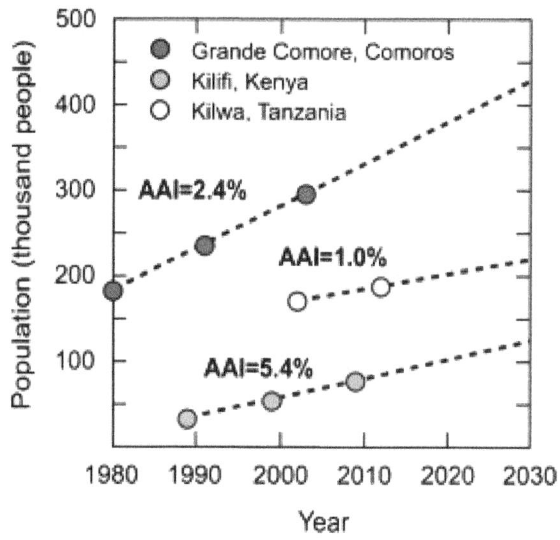

Figure 3: Population changes in the three study areas over the last three decades (from the most recent available census data). AAI = average annual increase; dashed lines are linear projections by 2030.

Table 1: Main source of water used in the three regions as compared to the whole country and change over time.

% (Rounded to upper%)	Remotely piped water	Protected well/spring	Unprotected well/spring	River/pond/dam	Vendors (e.g. trucks)	Rainfall tank/Jabias	Other
Tanzania (excl. Zanzibar) 2012	34	20	31	14	2		
Tanzania (excl. Zanzibar) 2002	37	18	29	11	5		
Lindi Region 2012 (Tanzanian coast)	19	14	53	10	4		
Lindi Region 2002 (Tanzanian coast)	13	17	55	15	1		
Kenya 2009	30	36		27	8		1
Kenya 1999	30	33		33	2		3
Kilifi County 2009	57	24		19	1		1
Kilifi County 1999	50	25		24	1		2
Grande Comore 2014	0.0	20a/2b	0.0	2	5	59	18c
Grande Comore 2000	0.0	15a/2b	0.0	3	3	70	10c
Comoros Union 2014	0.0	33a/3b	0.0	3	No data	30	29c
Comoros Union 2000	0.0	20a/2b	0.0	4	No data	40	24c

Kenya values from KNBS.

A UNDP deep wells.

B Private wells.

C Public fountains supplied by groundwater (protected wells).

Demographic changes

In Kilifi, Kenya, the analysis focused on the township to the north of the estuary and found that population and household densities have grown rapidly. The population in this area more than doubled over the 20 year period, increasing from 255 persons km^{-2} in 1989–535 persons km^{-2} in 2009, and a household density of 54 households km^{-2} in 1989–111 households km^{-2} in 2009.

Population growth in Kilwa district, Tanzania was weaker by contrast; increasing by 10% between 2002 and 2012. The area is still predominantly rural with a population density of 13 people km^{-2}. The urban population increased from 8% to 10% of the total over the decade. In the urban centre of Kilwa Masoko, which has seen some growth in tourism and the development of the port for servicing the gas pipelines to the North.

In Grande Comore (Ngazidja) the population increased by 62% over the 23 years between 1980 and 2003 (last census). The population density was about 250 people km^{-2} in 2003. In 2005 72% of the population was rural, against 76% in 1991. The capital Moroni's population (13.5% of the island's population in 2003) more than doubled between 1980 (17,267 people) and 2003 (40,050 people) and is currently estimated at over 50,000 people. Tourism is as yet embryonic in the Comoros but has potential and is expected to develop in the next decades (Shaaban et al., 2013) due to the many natural and cultural assets.

Water use and evolution of infrastructures

Data on water use (Table 1) were obtained from census records in each country (Kenya, Comoros Union and Tanzania) at the highest resolution available for each study area (respectively, Kilifi County, Grande Comore island and Lindi Province in which Kilwa district is located).

In Kenya, reliance on piped water supply is high in Kilifi with 56.8% of households surveyed in 2009 using piped water over other supplies. There was an increase of 7.1% in piped supplies from 1999 and a corresponding decrease in the use of well or spring supplies. The Kilifi region has no permanent rivers and the piped supplies are abstracted from River Tana and Athi-Sabaki rivers originating from the Kenyan highlands that flow into the Indian Ocean. Currently the supply from inland is inadequate to meet demand and is supplemented by groundwater abstraction, supplying a combined total of 130,000 m^3 daily towards an estimated demand of 200,000 m^3. The decrease in groundwater as a supply (24.5–23.2% between 1999 and 2009) may be linked to the high salinity of coastal aquifers in the area combined with water pollution, environmental degradation and recurrent droughts of wells (Marete, 2006 and Onyancha et al., 2010); which make piped supplies preferable where they can be afforded. Similar patterns were observed in the county during the surveys and FGDs undertaken as part of this study. The use of water vendors and supply trucks has increased by 6.5% in the county over the decade (up to 2009) while the use of water from rivers, ponds and dams directly by households declined by 6.5%.

Piped water is less commonly used in Lindi Region, of which Kilwa is the easternmost district, accounting for 19.1% of use in 2012 (Table 1). This is still low in comparison to the percentage of households with access to piped water at national level (Tanzanian mainland), which was 33.5% in 2002 and reflects the rural character of the area. The increase in piped water use in Lindi from 2002 is attributed to some improvement in infrastructure and a larger urban population in proximity to existing piped networks. Potentially linked to this is the decline in households obtaining water from boreholes, wells and springs from a combined total (unprotected and protected) of 72.1 to 66.9%. Some 79% of all wells/springs in Lindi were considered unprotected in 2012 compared to 76% in 2002. Growth in use of alternative water supplies such as rainwater and supply by water vendors from 0.4% in 2002 to 7.0% in 2012 may reflect the growing urbanisation of the population and increasing demand driving alternatives to traditional well sources as well as a growing perception of water vending as a rewarding business opportunity.

Water supply in Grande Comore Island relies exclusively on local aquifers and on rainfall harvesting. Supply from boreholes and wells more than doubled between 1985 and 2010 as a result of the commissioning of 54 new wells drilled in the early 1980s (UNDP, 1987). Groundwater is currently estimated as supplying just over a third of the island's population with safe water. Alternative water supplies such as rainfall harvesting have decreased as a consequence, from 70% to 59% between 2000 and 2014. Water supply by tanker trucks from operating wells has strongly increased (from 3% in 2000 to 5% in 2014) particularly in the last decade and has become a highly lucrative business (Mohamed, 2012).

Results from household surveys, water stakeholder and community meetings

Further to the available census data, local observations at each of the study areas were undertaken through a combination of household questionnaires, stakeholder and community meetings. For brevity a summary of the key findings is presented here; the detail of the socio-economic work programme will be covered elsewhere.

Water accessibility

In terms of water access, results from the household survey (Fig. 4) established that less than half of households in Kilifi, Kilwa and Comoros used between 20 and 100 L day^{-1}(mean household size = 6 persons). This falls short of the basic water requirements of 20 L person^{-1} day^{-1} recommended by the World Health Organisation (WHO, 2003). Furthermore, less than a third of households used between 120 and 200 L day^{-1} which is classed as 'reasonable access' but still falls below the 50 L person^{-1} day^{-1} which is considered adequate for domestic use (Gleick, 1996). The primary factors contributing to poor water access across all three areas based on the household surveys and community discussions were: lack of finances to purchase greater quantities or purchase storage/harvesting equipment, insufficient storage capacity, water budgeting/rationing in certain areas or during the dry seasons and a failing or inadequate infrastructure.

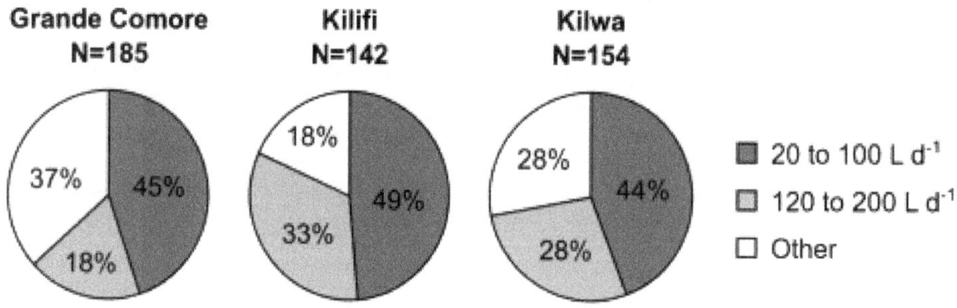

Figure 4: Daily volume of water used per household (mean size of 6 persons) in the three regions. *N* = number of households surveyed.

Across all three sites the common issues raised in community and stakeholder meetings and discussions were inadequate and intermittent supply, water scarcity, increased salinity, poor quality and high cost. The causes of these issues were attributed to drought (lack of rainfall/changing weather patterns), poor water sector management and governance, sea water intrusion and failing infrastructure. Issues with water quality result primarily from contamination of unprotected wells and poor source protection, with unregulated dumping highlighted as an issue in all areas. In Comoros there is a problem with unregulated use of agricultural fertilisers and sediment erosion due to deforestation.

Participation in water resources management

Across the three study sites the stakeholder and community discussions raised issues with governance and disproportionate resource allocation, highlighting failures at district and national levels. Effective water governance requires active participation of all the stakeholders in management of water resources. This can be determined by the proportion of the local community which actively contributes to water resource decision making through membership of social groups and being involved in leadership roles contributing to the decision making process. Findings from the survey indicate that less than half of the households surveyed were members of social groups relating to water resources (Fig. 5a) or involved in meetings for decision making in relation to water resources (Fig. 5b). Furthermore, less than 25% in Kilifi and Grande

Comore and less than 5% in Kilwa had received any training relating to groundwater or surface water management, water quality or water use (Fig. 5c). Additionally, 10–20% of households surveyed identified poor water sector management as a major water problem (Fig. 5d), which needs to be addressed.

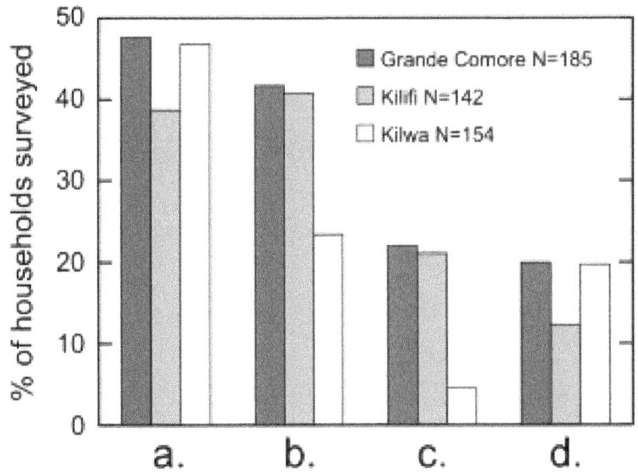

Figure 5: Participation of end-user households in water resources management. (a) Membership of a social group relating to water access; (b) participation in community meetings for decision-making relating to water; (c) water related training received; (d) identification of poor water sector management as a major water problem.

Strategies for management

In summary it was noted that community members were aware that groundwater resources are being depleted and that salinity is a key water quality issue. The community members are willing and ready to engage in water resources management and to pay for water of good quality.

Several strategies were proposed for sustainable management of existing groundwater resources during the meetings including the implementation of effective water resources management to provide water of good quality that is properly and the provision of appropriate infrastructure for water storage, especially for rainwater harvesting. Furthermore, there should be proper siting of boreholes and wells, with regular maintenance programmes, that should involve women as major stakeholders at micro level and encourage alternative sources of water supply. Traditional knowledge systems should be integrated with modern technologies. The need for awareness and training on water issues is still critical (see Fig. 5c). Micro-finance and social networks are also vital as avenues for capacity building of community for effective management of water resources. From the visioning conducted with the community members the desired state for selected aspects of water resources management was captured in Table 2.

Table 2: Summary of community aspirations based on focussed group discussions on water resource management across the 3 study areas.

Current state	Desired state
Not all community members receive high quality water	All community members receive high quality drinking water
Not all community members have adequate supply of water particularly shortages in dry season	All community members have adequate supply of water in relation to the demand during all seasons
Not all community members can afford to pay for water—prohibitive cost	Affordable water pricing
Waiting too long to fetch water	Sufficient water points per unit of population
Long distances to water source	Water source near dwelling or piped water—enhance connection of water
Community not aware of the water issues, particularly the poor	Participation of community in water governance
Poor water resource management	Effective water resources management

Environmental change

Regional climatic data

Seasonal rainfall varies across the region. The rainy seasons occur during boreal spring and autumn; the long rains occurring in spring and short rains in autumn. The strength of the rainy seasons varies with latitude and changes in circulation patterns over the Pacific and Indian Oceans. Orographic influences are particularly strong in Comoros where volcanic massifs create a topographic high that results in high variability across the island.

From the regional synoptic stations (Fig. 6), average annual rainfall in Mombasa, 20 km south of Kilifi, is 1800 mm (average over 2007–2013) with the majority falling in May and June, and generally some rainfall in all months, though little in January and February (<25 mm in 2013). Seasonal rainfall decreases to the north with an average annual rainfall in Lamu, the most northerly district on the Kenyan coast, of 1700 mm over the 2007–2013 period, for which the record is almost complete. Since the 2005–2008 period, during which Kenya experienced extremes of drought and rainfall (Hastenrath et al., 2010) with 3160 mm recorded in 2008 at Mombasa, annual rainfall has not exceeded 2000 mm, with 2011 particularly dry at 1140 mm. The low rainfall pattern has been linked to variability in both the Indian and Pacific Oceans and to the El Niño-Southern Oscillation (Ummenhofer et al., 2009 and Clark et al., 2003). The effect of prolonged droughts places increases pressure on groundwater resources in the region as surface water dries up.

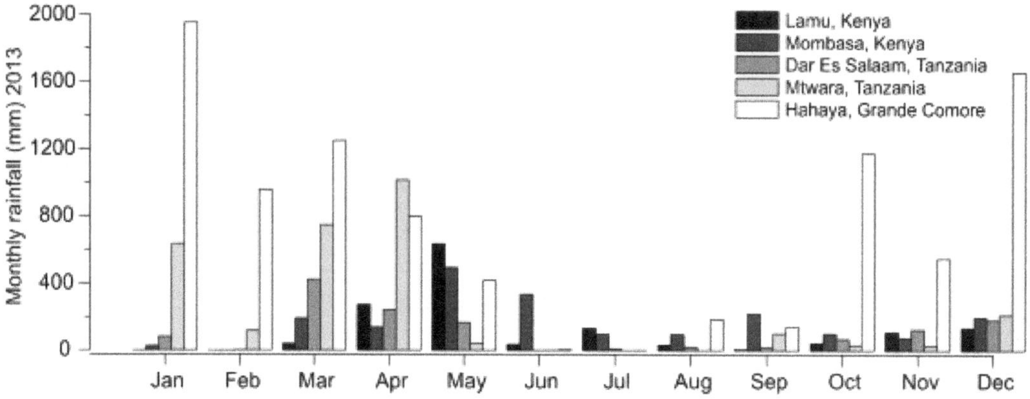

Figure 6: Comparative monthly rainfall totals for 2013 indicating the annual variation in rainfall across seasons from Lamu in northern Kenya to Mtwara in southern Tanzania and Hahaya on Grande Comore island.

To the south at Dar es Salaam in Tanzania, which lies 100 km north of Kilwa, seasonality is more pronounced with clear rainy seasons and almost no rainfall in June, July, August and September and February (43 mm total over these months in 2013 in Dar es Salaam). Average rainfall recorded in Dar es Salaam over the 2007–2013 period was 2570 mm. At Mtwara 170 km south of Kilwa, Tanzania the average annual rainfall was 2430 mm over the 2007–2013 period. In Grande Comore, an average of 4395 mm rainfall was recorded over the same period.

Observations at the weather stations installed at each of the study sites agree well with the patterns at the national synoptic stations overall (Fig. 7). Over the period 25 September 2013–5 July 2014, for which a complete record exists for all three sites, 800 mm of rainfall was recorded in Kilifi (maximum rainfall = 40.8 mm h^{-1}, median rainfall 0.8 mm h^{-1} with rain recorded 5% of time), 1056 mm in Kilwa (maximum rainfall = 46.2 mm h^{-1}, median rainfall 1 mm h^{-1}, with rainfall recorded 4.5% of the time) and 3688 mm in Vouvouni in Grande Comore (maximum rainfall 97.6 mm h^{-1}, median rainfall 1.4 mm h^{-1} and rain recorded 11% of time).

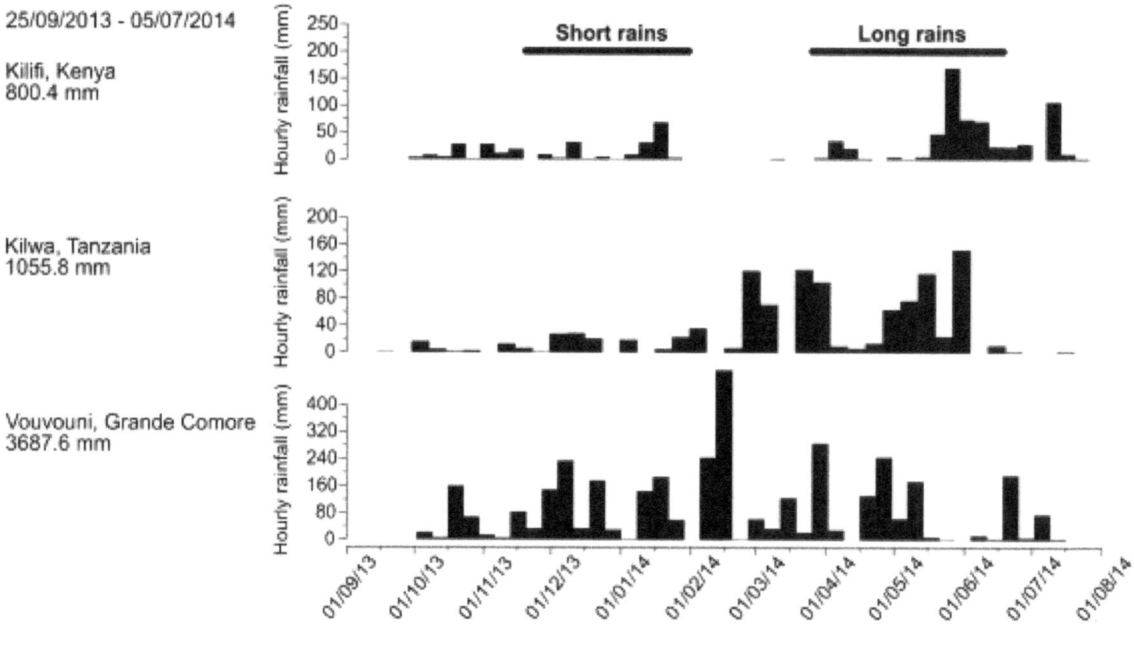

Figure 7: Rainfall over the period 25/09/2013–05/07/2014 at the weather stations at Kilifi, Kenya; Kilwa, Tanzania; and Vouvouni, Grande Comore.

In Comoros, and particularly in Tanzania, which in late April 2014 experienced heavy rainfall and extensive flooding across the north east of the country, the number and magnitude of large rainfall events in recent years was a frequent comment in the social surveys undertaken in the study areas. An analysis of rainfall intensities based on the 3 hourly synoptic station data supports this (Fig. 8) with increasing magnitude events recorded in the last decade at sites (Hahaya, Dar es Salaam and Mombasa) for which the data record is almost complete and where potentially erroneous measurements were filtered based on the Quality Check scores attached to the record. Due to large data gaps and omissions the period prior to 1995 was not included in the analysis. An increase in the extremes of rainfall is apparent from 2005 onward for Hahaya in Comoros and Dar es Salaam, Tanzania. There is less indication of change for Mombasa, which experiences lower annual rainfall.

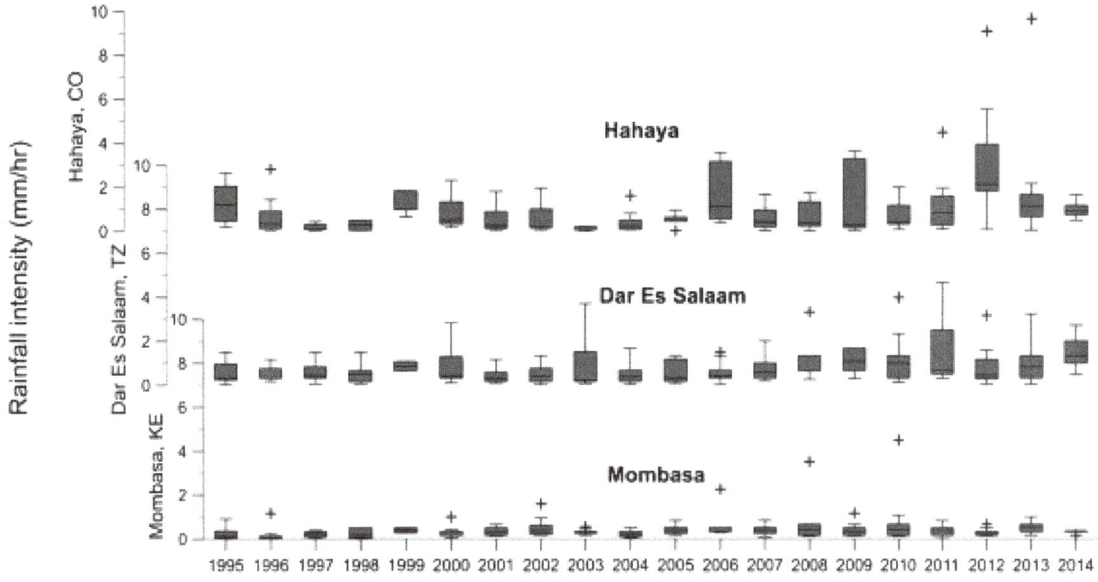

Figure 8: Box whisker plots of monthly rainfall intensities calculated for the closest synoptic stations in each country.

Many projections exist for the impact of climate change on Global Mean Sea-level (GMSL). The rate of change has been widely debated, with recent work (Watson et al., 2015) which corrected for errors in satellite altimetry data over recent decades, indicating that the actual rate of sea-level rise has accelerated. At a regional level, tidal data (Fig. 9) were obtained from the University of Hawaii Sea Level Center (http://uhslc.soest.hawaii.edu/data/download/rq) for the Western Indian Ocean region (gauges at Mombasa, Kenya; Lamu, Kenya; Zanzibar, Tanzania and Pointe des Galets, Réunion Island). Data gaps are present in almost all data sets but from 1986 to 2012 a comparable section from each time series was extracted and a linear fit applied to data during this period for each station. Not accounting for issues with gauge maintenance and instrument drift a clear positive trend is common to all with estimated annual increases for the gauges ranging from 3 to 9 mm. The potential implications for the coastal communities in this area are clear: accelerated coastal erosion, inundations of low-lying land and resulting inland extension of saltwater intrusion which will impact on the availability of freshwater resources.

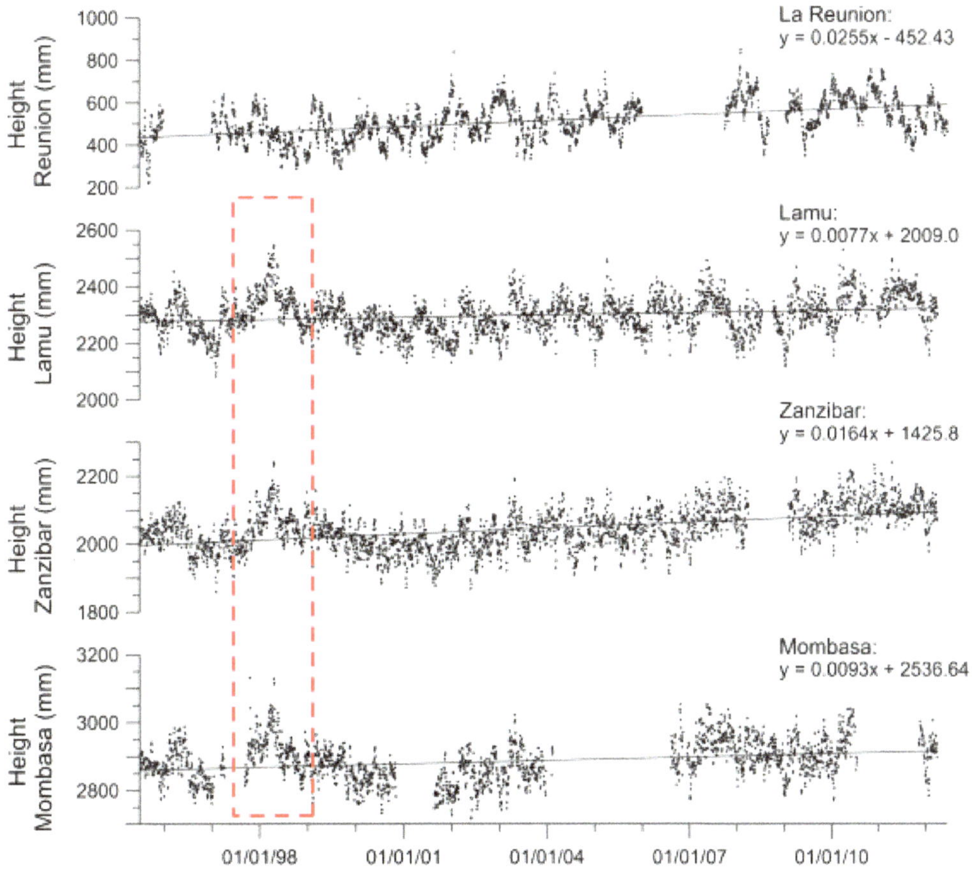

Figure 9: Tidal gauge data for the Western Indian Ocean region proximal to the study area. An upward trend is noted for all sites excepting a peak (red dash box) recorded at Lamu, Zanzibar and Mombasa circa 1998.

Land use change

Change in land use across each study area provided an indication of the effects of demographic change and the potential impact on recharge and groundwater quality. Clearance of vegetation for agriculture affects evapotranspiration and recharge and where bare earth remains there is increased potential for soil erosion during intense rainfall events. Increased sediment entering surface waters through erosion affects their use as water sources and with intensification of agriculture comes a risk of contamination from pesticides and herbicides, animal faecal material and chemical fertilisers. Pressure for space in urban areas leads to dispersion and enlargement of the urban periphery. New houses require water and where piped water supplies are not available new wells are necessary or the amount of time spent per day on collecting water increases.

Mapping was undertaken within defined areas at each study area. The focus was on mapping ground cover (extent of vegetation) and urban areas. Buildings were digitised from aerial imagery in order to provide an indication of changing housing and population patterns in the study areas. All buildings within selected areas, including houses, business establishments and public buildings were mapped. Land use and housing density change is presented for Kilwa, Tanzania over the period 2006–2013 (Fig. 10 and Fig. 11). The comparatively rural situation in this study area makes alterations in the landscape easier to map; the quality of imagery available was high.

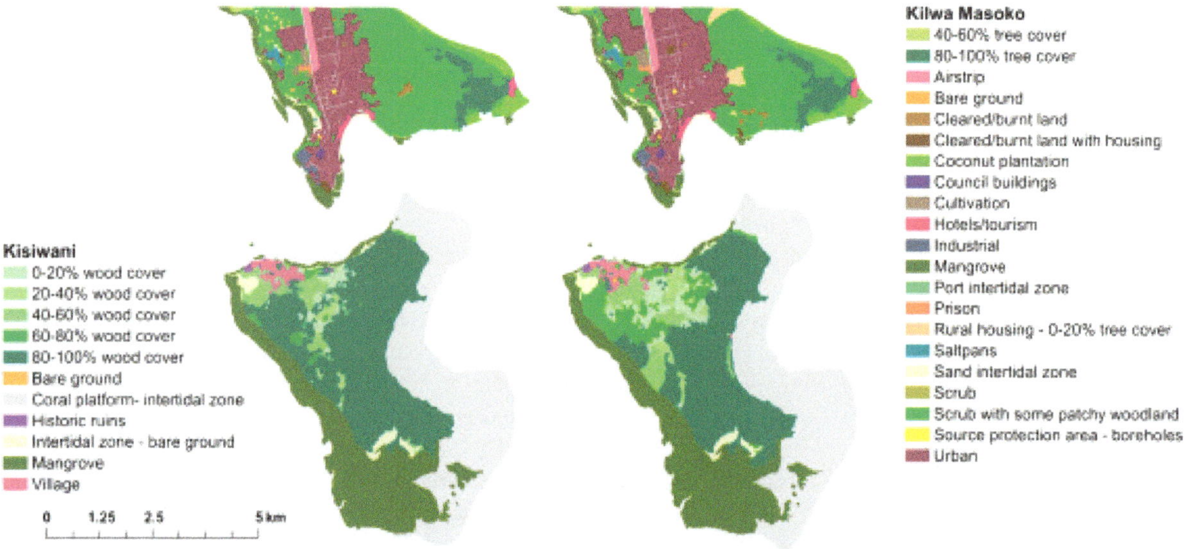

Figure 10: Land use change in Kilwa Masoko (mainland) and Kisiwani (island), Tanzania between 2006 (left) and 2013 (right).

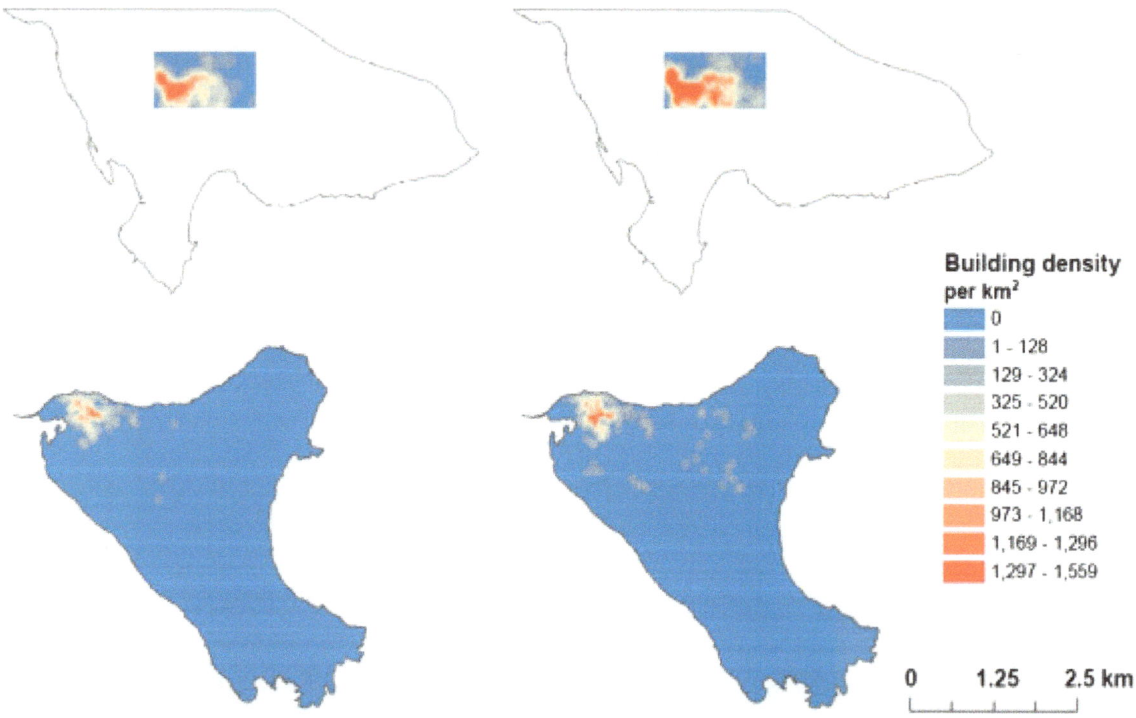

Figure 11: Building density in Kilwa Kisiwani (island) and for a sub-area on the urban periphery of Kilwa Masoko, Tanzania in 2006 (left) and 2013 (right). Density is expressed as number per km^2.

Within dense urban areas, such as Kilifi, the quality of imagery available limited the extent to which change could be mapped. However across the area an increase in land enclosure was noted.

In Kilwa Kisiwani clearance of land for cultivation and to provide fuel has reduced tree cover on the island over the 2006–2013 period (Fig. 10). Areas of formerly dense woodland to the south and east of the village have been thinned and cut for firewood and to provide land for cultivation, with new houses established. Areas with wood cover in the range 80–100%, which made up 35% of the island area in 2006 had reduced to 27% in 2013. The density of housing in the village increased over the period (Fig. 11) and traditional huts were replaced with block built houses. On the mainland, in Kilwa Masoko, the built area increased by 8% over the period, with large areas to the north of the town on both sides of the road cleared for construction. Housing densities increased during the period in formerly low density areas. New housing is largely block-built.

Groundwater resources

Geological structure and spatial patterns of groundwater salinity

The hydrogeological investigations provided baseline information on aquifer characteristics and on groundwater quality across the three study areas (Fig. 2 and Fig. 12).

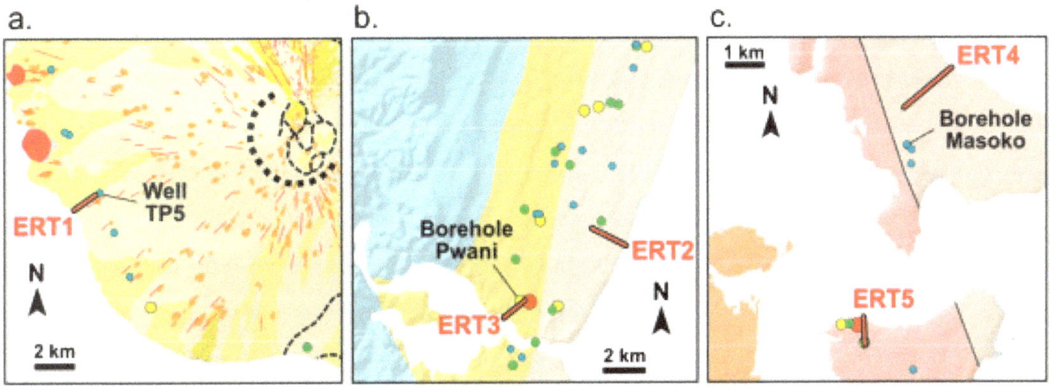

Figure 12: Areas of more detailed hydrogeological investigations, including geophysical profiling (ERT) and high frequency temporal groundwater monitoring in (a) well TP5, Vouvouni area, Grande Comore island; (b) Pwani borehole, Kilifi area and (c) Masoko borehole, Kilwa Masoko. See Fig. 2 for geological legend and borehole colour legend. ERT profiles 1–5 are presented on Fig. 13.

In Grande Comore island (Fig. 2a), aquifers are composed of young volcanic rocks organised in three volcanic massifs of different ages (Bachelery and Coudray, 1993). The Mbadjini massif is the oldest of these (Miocene) and outcrops at limited locations in the south-east of the island. It is characterised by deep weathering, which makes it difficult to distinguish the individual lava flows. The La Grille massif is of intermediate age (Mid-Pleistocene) and outcrops in the north of the island. The degree of weathering varies from moderate to low depending on the chronology of flow emplacement. The Karthala massif is the youngest (Quaternary), and active, volcano on the island characterised by a low degree or absence of weathering of the lava flows.

Water wells on the island are mostly deep (50–100 m), hand-dug, large-diameter wells excavated in the 1980s. Very large spatial variations in groundwater salinities are observed among the wells, which can be related to some extent to geological heterogeneity. In Grande Comore, all wells are located within a few kilometres of the coast due to the much higher topographic gradients compared to hydraulic gradients. This necessitates a large increase in well depths with distance from the coast to reach water, and much higher associated costs. In the Mbadjini massif, groundwater salinity is generally low (<2 g L^{-1}) due to the intensity of weathering resulting in reduced aquifer permeability and, therefore, reduced seawater intrusion due to higher hydraulic gradients. In addition, the weathering of the Mbadjini massif promotes the development of perched aquifers within more recent lava flows that overlie and are disconnected from the basal aquifer which has been subjected to seawater intrusion (Bourhane et al., 2015). In the La Grille and Karthala massifs, water wells have higher salinities (generally above 2 g L^{-1}) except on the western flank of the Karthala massif which experiences higher rainfall. Seawater intrusion in those two massifs is promoted by the high permeabilities of the lava flows and limited weathering; it is reduced by the rainfall intensity notably higher on the western flank of the Karthala volcano due to its exposition to dominant south westerly winds and its high elevation. Important spatial variations of salinities can be noted and attributed to the complexity of imbrication of lava flows of different ages resulting in different degrees of weathering and associated permeabilities.

The Kilifi coast in eastern Kenya (Fig. 2b), is characterised by a 5–15 km wide band of generally highly productive Tertiary and Quaternary sediments overlying less productive Jurassic or older sediments. The Tertiary (Pliocene) and lowest units of Quaternary (Pleistocene) sediments are composed of relatively homogeneous fine-grained marine or lagoon sands. They are overlain by, or for the uppermost units laterally pass into, Pleistocene coral sandstones and limestones displaying variable degrees of karstification. All those formations are covered by Holocene sand dunes along a narrow fringe of about 1 km width along the coastline. Most of the water wells in the coastal zone are shallow (typically <20m) large-diameter wells intersecting late Tertiary and Quaternary aquifers. Groundwater is mainly used for irrigation and as drinking water. Measured salinities display large spatial variations, but remain generally lower than those measured in Grande Comore (2/3 of the surveyed wells in Kilifi region have salinities lower than 2 g L^{-1} and 1/3 lower than the recommended WHO limit of 1 g/L; Fig. 2b). A gradient of increasing salinity also occurs from south to north reflecting the decrease in rainfall northwards in coastal Kenya.

Kilwa district, Tanzania, is composed of Tertiary to possibly Quaternary sediments comprising shale and silt (Nicholas et al., 2006) that are of little hydrogeological interest. However, calcarenite and coral limestone intercalations offer useful yields and are common in the Masoko formation (mid Eocene) that constitutes the western side of Kilwa Masoko peninsula and most of Kilwa Kisiwani island (Fig. 2c). A fault separates the Masoko formation on the west and the more recent (Miocene and younger) sediments of the Pande formation on the east. The moderately productive public boreholes providing drinking water for Kilwa Masoko are drilled in the Pande formation close to this fault. Measured salinities in Kilwa region are generally low, with 70% of wells providing water with less than 1 g/L of salt. This is the case for the majority of boreholes intersecting poorly productive shale and silt aquifers in the Kivinje and Pande formation. Wells drilled in the Masoko formation all occur on Kisiwani Island. They are mostly shallow large diameter hand-dug wells drilled in proximity to habitations on the northwest coast. They have much higher salinities (1–9 g L^{-1}) except for one borehole in the centre of the island.

Fig. 12 shows more detailed, enlarged views of the areas investigated in the three countries. More detailed investigations included geophysical surveys and high frequency temporal monitoring of groundwater in selected wells.

Geophysical investigations

Electrical Resistivity Tomography (ERT) was applied to all three study areas using similar acquisition methodologies. Several 2D resistivity cross-sections were obtained from selected transects run perpendicular to the coast; results are plotted on Fig. 13. All cross-sections show both the spatial organisation of different types of geological units and the spatial patterns of saltwater distribution in the aquifers. All results highlight the relatively high degree of aquifer salinization associated with the interface between freshwater and saltwater occurring at relatively shallow depths with a low angle of dip away from the coast. Such a low angle is consistent with a low hydraulic gradient resulting from the relatively large permeabilities of the coastal aquifers. Significant differences in mean resistivities are also observed between sites. Both the volcanic aquifers of Grande Comore (ERT1) and coral limestones/sandstones of Kilifi (ERT2) have much higher overall resistivities than in the marine/lagoon sands of Kilifi (ERT3) and clastic sediments of Kilwa (ERT4 and 5), which have lower resistivities overall due to a higher aquifer clay content. Because of these lower mean resistivities, the contrast between freshwater and saltwater is not as clear for those latter sites (ERT3, 4 and 5)

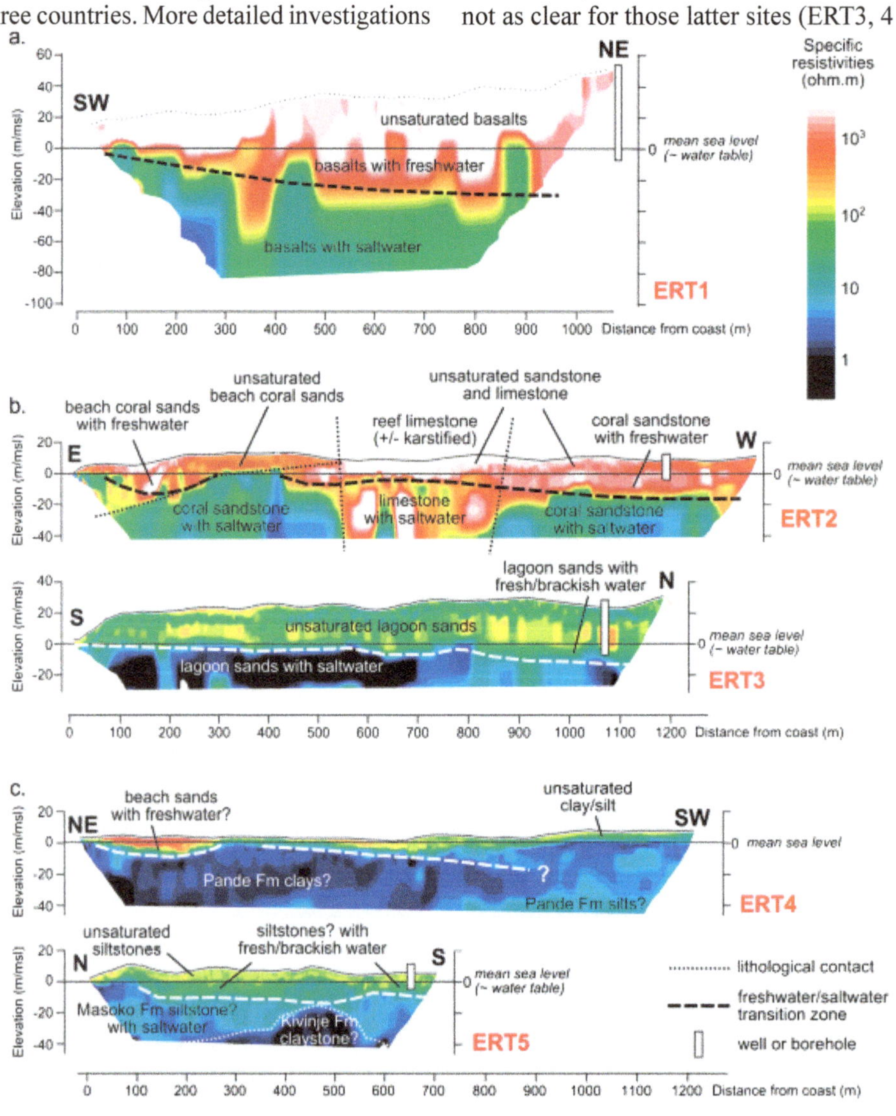

Figure 13: Results of ERT investigations in (a) Grande Comore; (b) Kilifi area and (c) Kilwa Masoko (top) and Kilwa Kisiwani (bottom).

In the Vouvouni area, south west of Grande Comore Island (Fig. 13a), the volcanic aquifer is characterised by a thick (from 10 m at the sea cliffs up to 50 m at 1 km from the shoreline) and highly resistive (>1000 Ω m) unsaturated zone. The transition zone between freshwater-saturated and seawater-saturated basalts (50–1000 Ω m) dips towards the centre of the island at a low angle. Freshwater thickness reaches a maximum of 30 m at 1 km from the coastline. This is consistent with the average low salinity measured in well TP5 (~0.2 g L^{-1}) located at a few hundred meters upslope of the north east end of the section and drilled within about 5–10 m below the water table. The Vouvouni area has the highest rainfall on Grande Comore (more than three times the average rainfall); therefore the freshwater thickness is less elsewhere on the areas of the island that are also underlain by the same young Karthala basalt flows.

Around Kilifi (Fig. 13b and c) resistivity signatures are different depending on the geological nature of the aquifer. Mean resistivities are generally higher in the Quaternary coastal units of the Pleistocene coral limestones and the Holocene sand dunes (Fig. 13b top) compared to the inland Tertiary marine/lagoon sands (Fig. 13b bottom). Coastal quaternary formations display a relatively high degree of heterogeneity. The Holocene coral sands (resistivities of ~20–1000 Ω m depending on the salinity of pore water and degree of saturation) overlie the Pleistocene basement with thickness increasing toward the sea (up to 30–40 m thick at the studied location) from about 500 m from the coastline.

The Pleistocene basement is composed of variability consolidated coral sandstones and breccias (2–1000 Ω m depending on salinity and saturation) with local occurrences of highly resistive reef limestone (10–10000 Ω m). A reef formation occurs between about 500 and 900 m from the coastline in the cross section of Fig. 13b (top). Strong internal variations in resistivities suggest karstification. Superimposed on the geological structure are different degrees of seawater intrusion.

Within the coastal Holocene sands, a freshwater lens (resistivity of 200–2000 Ω m) occurs reaching up to 20 m thickness at 200 m from the coastline. The maximum width of the lens is about 200–300 m. Seawater intrusion is reaching a greater distance inland in the underlying coral sandstones/breccias due to their higher permeability. A freshwater wedge (100–1000 Ω m) starts to develop at about 400 m from the coastline with thickness slowly increasing with distance, reaching about 20 m at 1400 m at the end of the ERT profile. Within this formation (500–900 m), the reef limestone displays extremely heterogeneous resistivities suggesting seawater intrusion is taking place within the karstified structures (10–100 Ω m) while the compact limestone remains relatively unaffected due to its low storativity (100–1000 Ω m). Generally, freshwater thickness appears lower within the karstified limestone than within the coral sandstone/breccias. The hydrogeological structure observed in this profile extends laterally, with similar structural patterns within Holocene/Pleistocene formations observed along the coastline north and south of Kilifi town.

The profile ERT3 (Fig. 13b bottom) trends north east from Kilifi creek within the Pleistocene/Holocene marine and lagoon sand units. Aquifer resistivities are much lower, relatively, than those observed in the Quaternary coral units. This is due to a finer granulometry and higher clay content. The unsaturated zone displays resistivities ranging between 10 and 200 Ω m because of horizontal lithological stratification within the sands. Below sea level, the aquifer displays very low resistivities: seawater intruded sands have values generally lower than 10 Ω m and lower than 1 Ω m in the most southern part of the profile close to the creek. There is a thin freshwater/brackish water lens of resistivity values ranging 10–100 Ω m reaching a thickness of only about 15 m at 1100 m from the creek. Presence of fresh/brackish water at this distance is confirmed by a well where groundwater electrical conductivity has been measured at 2500 microS/cm (corresponding to a salinity of about 1.4 g L^{-1}).

In Kilwa, south east Tanzania, the silt-clay dominant lithology of Tertiary sediments is responsible for low resistivities, typically less than 100 Ω m (Fig. 13c). Only coastal sands occurring on a narrow band of about 200 m along the western coastline have higher resistivities of up to 1000 Ω m (ERT4, Fig. 13c top). No clear resistivity contrasts are observable in the Pande Formation (ERT4, Fig. 13c top) except for the transition between the unsaturated (10–100 Ω m) and the saturated zone (<10 Ω m). Because of such low resistivities in the aquifer›s saturated zone, it is difficult to distinguish between freshwater and seawater. However, a general decrease in resistivity towards the coast suggests diffuse saline intrusion. Kilwa Masoko public water boreholes drilled in this formation a few hundred metres south of the south west end of this profile provide freshwater with salinity of about 0.4 g L^{-1}; this suggests that low resistivities are primarily attributed to the clay/silt lithology rather than the saltwater content. This profile also suggests the presence of a thin freshwater lens of 5–10 m thick within or below the coastal sands.

Within the Masoko formation, profile ERT5 (Fig. 10c bottom) is characterised by slightly higher resistivities due probably to a more silty/carbonate lithology. The unsaturated zone clearly shows resistivities ranging 50–200 Ω. In the saturated zone, a fresh/brackish water lens of about 15 m thickness is suggested by slightly lower resistivities of 20–100 Ω m. The underlying seawater-saturated aquifer has resistivities lower than 10 Ω m. At the base of the profile, the even lower resistivities (<1 Ω m) could be attributed to the underlying Kivinje claystone formation, a clay facies within the Masoko formation, both saturated with saltwater. The existence of a well and a borehole at distance of about 600 m, where salinities were measured at 1.2 and 4.3 g L^{-1} is consistent with a brackish nature of the lens rather than freshwater.

Controls of geology, distance to sea and rainfall on salinity distribution in groundwater

Resistivity profiles clearly reveal a geological control on spatial patterns of seawater distribution in the various coastal aquifers. Other factors such as the distance to the coast and the average rainfall of the area also exert a control on distribution of saltwater within a specific geological unit. Both an increase in distance to the coast and increased rainfall should correspond to a decrease in groundwater salinity. Fig. 14 illustrates such correlations between the combination of distance to coast and amount of rainfall and the salinity observed in wells and boreholes. When all geological types are considered together across the three countries, a poor correlation is observed (Fig. 14a). However, in Grande Comore island (Fig. 14b), strong correlations are observed within the two dominant volcanic units of Karthala and La Grille, respectively with an exponential relationship with $R^2 = 0.68$ and a power law relationship of $R^2 = 0.85$. This suggests relative aquifer homogeneity within individual geological units with the magnitude of seawater intrusion predominantly controlled by the aquifer recharge from rainfall and distance to sea.

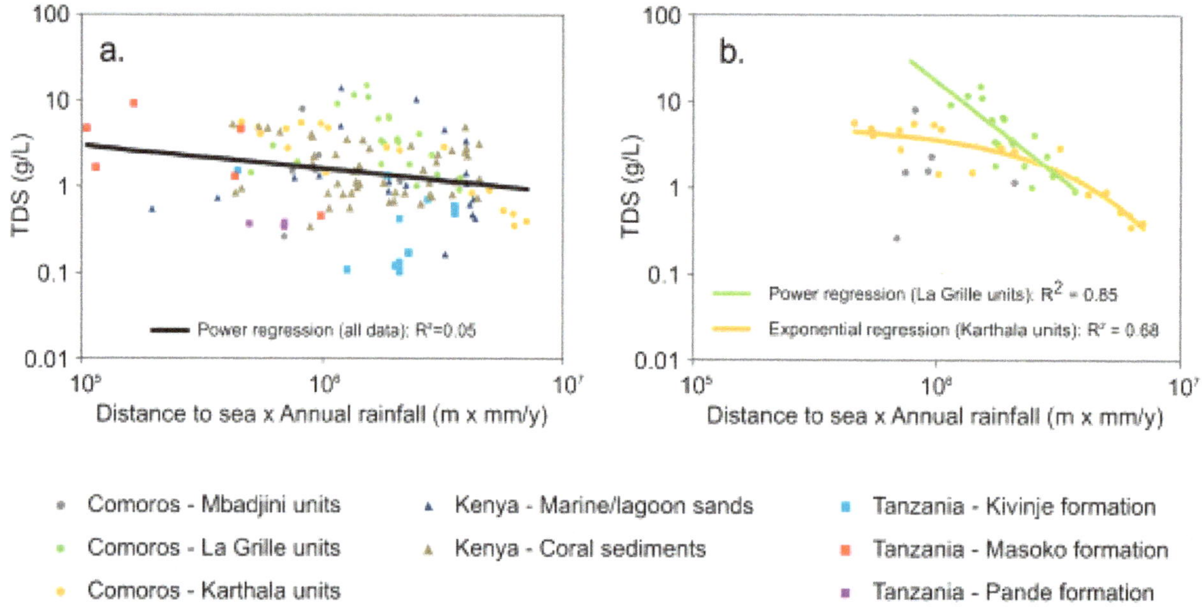

Figure 14: Correlations between borehole/well salinity, geology, distance to coast and rainfall; (a) all data across the three countries; (b) Grande Comore.

In contrast to Grande Comore, the absence of clear correlation for the aquifers of Kilifi and Kilwa can be explained by three main factors: (1) a significant intra-formation aquifer heterogeneity; (2) the large variations of well/borehole depths resulting in large variations of salinity primarily due to the strong vertical salinity gradients in the aquifers; and (3) a large variability in abstraction rates resulting in different degrees of seawater upconing. However, for Kilwa and Kilifi, an analysis of the influence of well depth and pumping rate on salinity is not possible due to lack of available data.

Groundwater behaviour

Locations of temporally monitored boreholes in Comoros, Kenya and Tanzania are shown on Fig. 12. High frequency groundwater time-series recorded in those boreholes reveal large differences in temporal groundwater behaviour in the three studies sites (Fig. 15). In TP5 well in Vouvouni (Fig. 15a), both groundwater heads and electrical conductivity fluctuations are primarily controlled by semi-diurnal tidal fluctuations. The high water Table during the high tide is systematically accompanied by high electrical conductivity levels. This strong tidal control is promoted by high permeabilities and diffusivities in the volcanic aquifers (Bourhane et al., 2015). The short-term influence of rainfall is not visible, reflecting the large thickness of the unsaturated zone that results in diffuse recharge over time. Seasonal variations of heads and electrical conductivity can be noted with slightly higher water tables and slightly lower electrical conductivity during the wet season as compared to the dry season. TP5 is one of the most intensively pumped wells in Grande Comore. Pumping schedules are designed to minimise drawdowns (about 10 cm, due to high permeability and large diameter of the wells). More interestingly, pumping failures result in a sharp increase in salinity. This is a similar effect to the tidal fluctuationsm—a general rise in the water column results in a rise of more mineralized water from the base of the well. Improving access to groundwater of acceptable quality might require considering different abstraction and management approaches (e.g. widespread low-rate shallow wells or horizontal tunnels). Currently, two programmes funded by the African Development Bank [ADB] and the French Development Agency [AFD], are prospecting for new groundwater resources in Grande Comore. The latter is considering drilling large diameter wells to supply rural communities in both coastal and mountain areas (Join et al., 2013).

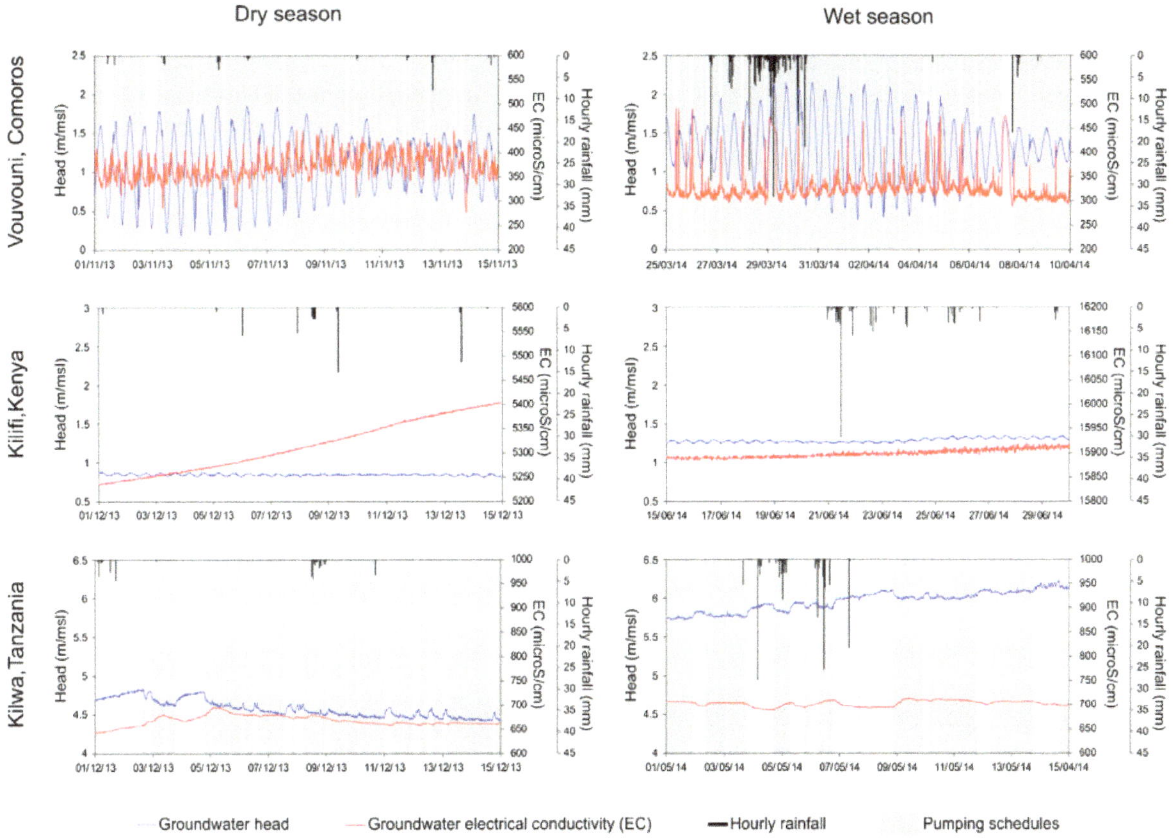

Figure 15: time series of groundwater heads and electrical conductivities measured in monitored wells (Fig. 12) along with hourly rainfall and pumping schedules for dry and wet seasons.

Pwani University boreholes in Kilifi (Fig. 15b) also show the influence of semi-diurnal tides on groundwater heads and salinity fluctuations, but to a much lower extent than TP5 in Grande Comore. However, hourly rainfall does not show a visible effect on groundwater heads and salinity. As for Grande Comore, the thickness of the unsaturated zone (20–30 m) is likely responsible for diffuse recharge over time during the rainy season, i.e. a more seasonal impact of rainfall than in Grande Comore. This seasonality is seen during the dry season showing a progressive increase in electrical conductivity. Prolonged rainfall is also responsible for a slight rise in the water Table (a few cm) during the wet season. The datalogger had to be moved from one borehole to another nearby between the two dry/wet periods because a pump was installed in the original one. The second borehole was previously pumped causing seawater intrusion but pumping was stopped before the datalogger was installed. This explains the much higher EC recorded there during the wet season.

Groundwater records on the Kilwa Masoko boreholes (Fig. 15 bottom) do not show any clear influence from tides. Instead, the influence of pumping is clearly visible on both heads and electrical conductivity: pumping periods are characterised by a drop in head, of about 20 cm, and a rise in electrical conductivity of up to 50 microS/cm. This simultaneous response of both heads and salinity can be attributed to seawater upconing below the boreholes; the drawdown of the water Table is accompanied by a rise in the freshwater/seawater transition zone. Seasonality of rainfall also impacts both heads and salinity, with a progressive decrease in head/increase in salinity during the dry season and increase in head/decrease in salinity during the wet season. Short-term rainfall also shows a slight impact on heads, with highest hourly rains causing a water Table rise of a few cm (e.g. on 4 May 2014).

In summary, the groundwater (heads and salinity) temporal behaviour in the volcanic aquifers of Grande Comore appears to be primarily controlled by the tidal fluctuations and secondary by the pumping regime. In the Kilifi marine/lagoon sands, the same parameters appear to be primarily controlled by seasonality and secondly by tides (monitored boreholes were not pumped at the time of monitoring). In the clayey/calcarenite sediments of Kilwa Masoko, they are primarily controlled by the pumping regime and secondly by both short-term rainfall and seasonality.

OVERALL SYNTHESIS, DISCUSSION AND STATE OF UNDERSTANDING

The results of the various investigations carried out in the three countries provide an overview of the status of the groundwater resources and demand for water as well as the challenges faced with regard to expected environmental and societal changes. They also highlight strong similarities as well as notable differences across the three areas. The status of the groundwater resources and water supply is summarised as follows:

- Groundwater resources are important, and often safer sources of domestic and agricultural water across the whole region; used by over 50% of the population in Tanzania and Comoros (including direct supply from groundwater or indirect supply through water truck delivery) and over 20% in Kilifi region, Kenya (in Kilifi about half of the water supply is imported from other catchments and dams on remote major rivers);

- The observed increase in population of 1.0–5.4% in average annually regards predominantly urban centres and is causing an increase in demand; which is not met by increases in water availability and supply infrastructure;
- There is widespread salinization of wells and boreholes in all areas studied across the region. This is the main limiting factor on the quantity of groundwater availability and was highlighted as a key water quality issue by the community members and stakeholders;
- Fresh groundwater resources in coastal areas systematically occur as thin lenses or wedges; with the freshwater/salt water transition zone is inclined at a low angle so that salt water occurs at relatively shallow depths for a considerable distance inland. This limits the depths to which wells can be bored and the abstraction rates that are possible if fresh water supplies are to be maintained;
- The spatio-temporal variability of groundwater quality is well-correlated with the spatio-temporal distribution of rainfall and geology; in particular, variations in geology are reflected in the temporal response of wells to drivers such as tidal fluctuations (volcanic aquifers of the Comoros), pumping schedules (lower productive sedimentary aquifers of Tanzania), or seasonality (coral aquifers of Kenya);
- Long-term monitoring data for groundwater are lacking across the region and in cases where data are available there are often inconsistencies due to the methodologies used at different times so that comparisons are difficult and trends unclear;
- Well construction strongly influences water quality and long-term sustainability. Traditionally constructed wells are wider and shallower, have better water quality (lower salinity) and are more sustainable (less collapse/failure and more easily refurbished); however unprotected open wells are more likely to have faecal contamination due to dirty collection buckets/ropes (Tanzania and Kenya) and ingress of debris such as animal waste;
- The time-constrained nature of funding and of groundwater development projects are issues, often causing a disconnect where deliverables are required within the short (2–3 years) duration of a project. This results in the prioritisation of delivering a set number of mechanically-drilled production boreholes and rapid speed of installation regardless of the known benefits of hand dug, large diameter wells;
- The importance of local researchers with expertise/experience in coastal hydrogeology and having established links with local water stakeholders is recognised by researchers, stakeholders and communities in order to facilitate literature and data compilation necessary to conduct regional resource assessments.

The environmental drivers (climate and land use) that have potential to affect groundwater resources and their accessibility are:

- An overall change in climate across the region, that is indicated in long-term hydrometric observations and reported by local people who are noticing changes in weather patterns, seasons, dynamics of plants and animal behaviour;
- A clear increase in the extremes of rainfall in the last decade is recorded in Comoros and Tanzania, but is not as clear in Kenya which is drier and less seasonal;
- A clear increasing trend of sea level was recorded over the last two decades ranging from 3 to 9 mm/year across the region and comparable with global estimates for the Indian Ocean region;
- Increasing urbanisation is leading to land clearance for housing and increased demand for wood as a fuel. In areas subject to high intensity rainfall events (Tanzania) this leads to increased potential for runoff and soil erosion and can reduce infiltration to groundwater;
- Tourism developments and modern plumbed housing place strains on existing water infrastructure through the volume used compared to local housing
- Demand for land in coastal areas is leading to increased enclosure of land (particularly in Kenya) with loss of grazing ground for pastoralists and increased stocking densities on land which is accessible.

The socio-economic drivers and governance affecting groundwater resources and access are:

- Low levels of education, income and limited livelihood characterises the region; all indicate a need for strategies to address poverty in the region;
- Communities feel strongly that it is the responsibility of government to solve water issues and identify shortcomings in governance;
- There are low levels of participation to decision making community meetings and training/education related to water resources management, although community members are keen to engage and voice their opinions; they want to be actively involved in water resource management, identify it as a critical issue and a direct link with daily life;
- Community members are critical of researchers who commonly do not return/share project findings;
- Gender issues and equality are getting increased recognition: women are the main stakeholders at community-level, but rarely involved in discussion, decision and future activities;
- Most users are willing to pay for water, however some resist because water quality is poor (i.e. salty);
- Researchers, managers and communities recognise that technical decisions should be bottom up by local communities and incorporate local knowledge and skills; international/national funding agencies are culpable for perpetuating unadapted and unsustainable groundwater infrastructures;

- Student training is viewed as key for water management as they will be the future decision makers in many areas;
- Particularly in Kenya and Tanzania, which are much larger than the Comoros Islands, the lack of centralisation of basic information on wells and boreholes is pointed out, which causes difficulties in accessing crucial hydrogeological literature and reports.

Overall, stakeholders have recognised the need to understand how the available groundwater resource works so that sustainable management and development plans can be adopted. Also, the investigations as well as the dissemination events strongly pointed to the important role of engagement both with the stakeholder and community and the importance of education as to how coastal groundwater and aquifers work and why they need stakeholder safeguarding.

Recent and specific figures for integrated management of groundwater and coastal aquifers within IWRM frameworks in Sub-Saharan Africa are not yet readily available, though projects relevant to East Africa are in progress (e.g. UNEP, 2012 and Mehta and Movik, 2014). Governance aspects of groundwater management in East Africa lack the mechanism for coordination and for fostering cross-sector linkages. For example, deficiencies have been identified in key areas under existing policy, legal and institutional frameworks in Kenya (Mumma et al., 2011). Given the lack of consideration of groundwater, as pointed out by previous studies (e.g. Braune and Xu, 2010; Knüppe, 2011), one can expect progress to be made on managing coastal groundwater resources in East Africa.

The results of this study provide a detailed regional perspective that complements the previous continent-scale review of coastal aquifers by Steyl and Dennis (2010). It confirms the fundamental importance of coastal groundwater resources for the development of the region and the urgent need for matching the rate of water development with the rate of demographic change, including urban expansion and economic growth. Despite the geological diversity of the region, the sustainable provision of safe groundwater is limited in all areas by both seawater intrusion (rather than aquifer productivity) and the lack of adequate local to national management plans and a policy framework adapted to the specificities of these coastal situations. The positive development of national and inter-African consciousness and leadership with regard to water development mentioned by Steyl and Dennis (2010) must however be accompanied by a realisation of the fundamental importance of implementing coastal groundwater development strategies that are different to those of other physical settings where borehole productivity is the target.

International development projects still fail to provide sustainable groundwater infrastructure and fresh (i.e. non salty) groundwater in coastal regions. Drilling campaigns are often deemed successful once productive boreholes with good quality water are obtained without consideration to future salinisation due to over-pumping. Those strategies, which systematically target a limited number of deep and productive boreholes rather than promoting the development of multiple shallow, large diameter wells, have proven unsuccessful in the long-term. Sustaining the use of large numbers of shallow wells through, for example, modernisation, rehabilitation or deployment of traditional wells however poses significant management challenges requiring a strong collaboration of stakeholders at different levels of society, from communities to providers, consultants, NGOs and regulators. Those collaborative aspects must be embedded or strongly enhanced in the design and implementation of future development projects. In particular, the developing, and lucrative, water trucking business in Kenya and Comoros proves that communities are willing to pay for quality water, which suggests that community/associative management frameworks based on paying for quality water from local wells have potential as a local alternative for water management.

CONCLUSION

The coastal aquifers of East Africa are major water resources for domestic supply, industry and agriculture. They currently require development programmes to support the high rate of population increase in the region, including strong urban development. The region is also experiencing clear trends in long-term environmental change, common to all three countries and probably also shared by many other African coastal countries. Those changes include an increase in rainfall intensity without a clear increase in average rainfall together with significant land clearance and urbanisation. This is expected to have a negative long-term impact on groundwater recharge. Sea level is also showing a consistent rise during the last two decades that may increase coastal erosion and lead to seasonal flooding of low-lying areas, which ultimately will decrease fresh groundwater availability.

Notwithstanding, geological controls across the three areas, groundwater quality from wells and even more from narrow boreholes is broadly poor due to saltwater intrusion, and freshwater availability is low despite highly productive aquifers. Fresh groundwater is often limited to thin freshwater lenses/layers requiring a large number of shallow wells with low abstraction rates. The experience shows that traditional (or modernised) shallow large diameter wells are less saline and are more sustainable than deep boreholes. Observed trends of changes in demography, climate, sea level and land use are not expected to improve the availability of fresh groundwater. Recent and current development projects do not account for hydrogeological requirements for drilling more and shallower wells rather than deep boreholes. Communities as wells as stakeholders are often left with wells having inadequate water quality that are abandoned, opening up business for costly water trucking. Communities are willing to pay for water and to engage in local groundwater management framework provided that groundwater is of appropriate quality.

The recent progress in social consciousness and education on groundwater vulnerability in African coastal regions must be strengthened and rapidly accompanied by the establishment/improvement of national strategies for both groundwater development and management. Those strategies will need to take better account for the specificities of coastal groundwater resources as well as their anticipated changes and will also require enhancing the involvement of communities in the management processes. They also have to be supported by improved knowledge of coastal aquifers and long-term monitoring data, which are lacking in the region.

ACKNOWLEDGMENTS

This research was funded by the UK Natural Environment Research Council NERC, the Department for International Development DFID-UKAID, the Economic and Social Research Council ESRC; UPGRO Grant No NE/L001888/1: towards groundwater security in coastal East Africa. Geophysical data acquisition in the Comoros was funded by the French Development Agency AFD: project GECEAU CKM1068. We acknowledge stakeholders and community members from Grande Comore, Kilwa district and Kilifi Region for their active and fruitful participation in the focussed group discussions and questionnaire surveys. We are grateful to Bonventure Obeka and Stanley Wanjala (MA Sociology students at Pwani University) for assistance throughout the project, to postgraduate and undergraduate students from the Universities of the Comoros, Dar Es Salaam, Pwani and Kenyatta and interns from the Comoros Water Directorate DGEME for assistance in field data acquisition. We thank Isaac Marobhe (University of Dar Es Salaam) for assistance and the loan of field equipment, as well as Peter Karanja Gathu (Kilifi District Water Agency), Magai Kakuru (Kilwa District Water Department), Mahabadi Boinali (Comoros Water Directorate DGEME), Youssouf Ben Abdallah (Comoros Community Representative CIGEO) and Simon Onywere (Kenyatta University) for contributions to the final workshop discussions. We also acknowledge the assistance of the British Institute in Eastern Africa (BIEA) and particularly the late Joseph Mutua for his technical and logistical expertise and untiring assistance through fieldwork in Tanzania and Kenya. We are finally grateful to the Reviewers and Editorial Board of the journal for their valuable comments and suggestions that helped improving the quality of the manuscript.

REFERENCES

1. Adelana, S.M.A., 2009. Monitoring Groundwater Resources in Sub-saharan Africa: Issues and Challenges. IAHS Red Book Publication, pp. 103–113.
2. Adelana, S.M.A., Abiye, T.A., Nkhuwa, D.C.W., Tindimugaya, C., Oga, M.S., 2008. Urban groundwater management and protection in Sub-Saharan Africa. In:
3. Adelana, S.M.A., MacDonald, A.M. (Eds.), Applied Groundwater Studies in Africa, IAH Selected Papers on Hydrogeology 13. CRC Press/Balkema, Leiden, The Netherlands, pp. 1–7.
4. Ashton, P.J., Turton, A.R., 2009. Water and security in Sub-Saharan Africa: emerging concepts and their implications for effective water resource management in the Southern African Region. In: Brauch, H.G., Spring, U.O., Grin, J., Mesjasz, C., Kameri-Mbote, P., Behera, N.C., Chourou, B., Krummenacher, H. (Eds.), Facing Global Environmental Change. Springer-Verlag, Berlin, pp. 661–674, http://dx.doi.org/10.1007/978-3-540-68488-6 50.
5. Bachelery, P., Coudray, J., 1993. Carte Géologique Des Comores: Carte Volcanologique De La Grande Comore Au 1/50 000ème Et Carte Volcano-tectonique, Avec Notice Explicative. Ministère Franc¸ ais de la Coopération, 39 p.
6. BGR, Cap-Net, Waternet, WA-Net, 2007. Capacity Building for Groundwater Management in West and Southern Africa. BGR, Hanover, Germany, pp. 65 p.
7. Bourhane, A., Comte, J.-C., Join, J.-L., Ibrahim, K., 2015. Groundwater prospection in Grande Comore Island: joint contribution of geophysical methods, hydrogeological time-series analysis and groundwater modelling. In: Bachelery, P., Lénat, J.-F., Di Muro, A., Michon, L. (Eds.), Active Volcanoes of the
8. Southwest Indian Ocean: Piton de la Fournaise and Karthala. Springer-Verlag, Berlin, pp. 385–401, http://dx.doi.org/10.1007/978-3-642-31395-0 24.
9. Braune, E., Xu, Y., 2008. Groundwater management issues in Southern Africa: an IWRM perspective. Water SA 34 (6), 699–706.
10. Braune, E., Xu, Y., 2010. The role of ground water in sub-Saharan Africa. Groundwater 48, 229–238, http://dx.doi.org/10.1111/j.1745-6584.2009.00557.x.
11. Cai, W., Santoso, A., Wang, G., Weller, E., Wu, L., Ashok, K., Masumoto, Y., Yamagata, T., 2014. Increased frequency of extreme Indian Ocean Dipole events due to greenhouse warming. Nature 510, 254–258, http://dx.doi.org/10.1038/nature13327.
12. Clark, C.O., Webster, P.J., Cole, J.E., 2003. Interdecadal variability of the relationship between the indian ocean zonal mode and east african coastal rainfall anomalies. J. Clim. 16, 548–554, http://dx.doi.org/10.1175/1520-0442(2003)016.
13. Comte, J.-C., Join, J.-L., Banton, O., Nicolini, E., 2014. Modelling the response of fresh groundwater to climate and vegetation changes in coral islands. Hydrogeol. J. 22 (8), 1905–1920, http://dx.doi.org/10.1007/s10040-014-1160-y.
14. Döll, P., 2009. Vulnerability to the impact of climate change on renewable groundwater resources: a global-scale assessment. Environm. Res. Lett. 4 (3), 035006, http://dx.doi.org/10.1088/1748-9326/4/3/035006.
15. Edmunds, M., 2012. Limits to the availability of groundwater in Africa. Environ. Res. Lett. 7, 021003, http://dx.doi.org/10.1088/1748-9326/7/2/021003.
16. Foster, S., Tuinhof, A., Garduno, ˜ H., 2008. Groundwater in sub-Saharan Africa: a strategic overview of developmental issues. In: Adelana, S.M.A., MacDonald, A.M. (Eds.), Applied Groundwater Studies in Africa, IAH Selected Papers on Hydrogeology 13. CRC Press/Balkema, Leiden, The Netherlands, pp. 9–21.
17. Foster, S., Tuinhof, A., van Steenbergen, F., 2012. Managed groundwater development for water-supply security in sub-Saharan Africa: investment priorities. Water SA 38 (3), http://dx.doi.org/10.4314/wsa.v38i3.1.
18. Gleick, P.H., 1996. Basic water requirements for human activities: meeting basic needs. Water Int. 21 (2), 83–92, http://dx.doi.org/10.1080/02508069608686494.
19. Global Water Partnership, 2000. Integrated water resources management, Technical Advisory Committee (TAC) Background Papers, No. 4, 71 p. Available at: http://www.gwp.org/Global/ToolBox/Publications/Background%20papers/04%20Integrated%20Water%20Resources%20Management%20(2000)%20

English.pdf.

20. Global Water Partnership, 2002. The Policy Guidance and Operational Tools on IWRM. Hastenrath, S., Polzin, D., Mutai, C., 2010. Diagnosing the droughts and floods in equatorial east Africa during boreal autumn 2005–08. J. Clim. 23 (3), 813–817, http://dx.doi.org/10.1175/2009JCLI3094.1.
21. Hinkel, J., Brown, S., Lars Exner, Nicholls, R.J., Vafeidis, A.T., Kebede, A.S., 2012. Sea-level rise impacts on Africa and the effects of mitigation and adaptation: an application of DIVA. Reg. Environ. Change 12 (1), 207–224, http://dx.doi.org/10.1007/s10113-011-0249-2.
22. IPCC, 2014a. Climate Change 2014a: Impacts, Adaptation, and Vulnerability. Part B: Regional Aspects. Contribution of Working Group II to the Fifth Assessment Report of the Intergovernmental Panel on Climate Change. In: Barros, V.R., Field, C.B., Dokken, D.J., Mastrandrea, M.D., Mach, K.J., Bilir, T.E., Chatterjee, M., Ebi, K.L., Estrada, Y.O., Genova, R.C., Girma, B., Kissel, E.S., Levy, A.N., MacCracken, S., Mastrandrea, P.R., White, L.L. (Eds.). Cambridge University Press Cambridge, United Kingdom and New York, NY, USA, 688 p.
23. IPCC, 2014b. Climate Change 2014b: Impacts, Adaptation, and Vulnerability. Part A: Global and Sectoral Aspects. Contribution of Working Group II to the Fifth Assessment Report of the Intergovernmental Panel on Climate Change. In: Field, C.B., Barros, V.R., Dokken, D.J., Mach, K.J., Mastrandrea, M.D., Bilir, T.E., Chatterjee, M., Ebi, K.L., Estrada, Y.O., Genova, R.C., Girma, B., Kissel, E.S., Levy, A.N., MacCracken, S., Mastrandrea, P.R., White, L.L. (Eds.). Cambridge University Press Cambridge, United Kingdom and New York, NY, USA, p. 1132 p.
24. Join, J.-L., Comte, J.-C., Bourhane, A., Soule, H., 2013. Tests méthodologiques en géophysique en vue de l'implantation de forages d'eau sur l'île de La Grande Comore.P rojet Pilote de gestion du service public de l'eau en milieu rural sur l'île de La Grande Comore. Unpubl. Geophys. Report, Laboratory Geosciences, University of Reunion Island, 64 p.
25. Knüppe, K., 2011. The challenges facing sustainable and adaptive groundwater management in South Africa. Water SA 37 (1), 67–79.
26. MacDonald, A.M., Bonsor, H.C., Dochartaigh, B.É.Ó, Taylor, R.G., 2012. Quantitative maps of groundwater resources in Africa. Environ. Res. Lett. 7 (2), 024009, http://dx.doi.org/10.1088/1748-9326/7/2/024009.
27. Mapfumo, P., Adjei-Nsiah, S., Mtambanengwe, F., Chikowo, R., Giller, K.E., 2013. Participatory action research (PAR) as an entry point for supporting climate change adaptation by smallholder farmers in Africa. Environ. Dev. 5, 6–22.
28. Marete, J.M., 2006. An assessment of water resources and integrated water resources management (IWRM) strategy in Bahari sub-basin, Kilifi District. In: Unpubl. MSc Thesis. University of Nairobi.
29. Marobhe, I., Songo M., 2006. Report on the study of groundwater resources at Kilwa Kisiwani, Kilwa district, and Lindi region, Unpubl. Geol. Report, Dept. of Geology, University of Dar Es Salaam.
30. Mehta, L., Movik, S., 2014. Flows and practices: integrated Water Resource Management (IWRM) in African contexts, IDS Working Paper, 2014, No.438, 34 p.
31. Mohamed, I., 2012. L'eau en Grande Comore : étude de cas d'un petit espace insulaire abondamment arrosé mais en situation de pénurie d'eau. In: PhD Thesis. University of Reunion Island, 353 p.
32. Mohamed, S.H., Othman, S.A., 2006. Etude causale sur la disponibilité de la ressource en eau et la sécurité de l'approvisionnement en eau potable aux Comores, Unpubl. Report, UNDP, Moroni, 42 p.
33. Mukherjee, A., Saha, D., Harvey, C.F., Taylor, R.G., Ahmed, K.M., Bhanja, S.N., 2015. Groundwater systems of the Indian sub-continent. J. Hydrol.: Reg. Stud. 4 (A), 1–14, http://dx.doi.org/10.1016/j.ejrh.2015.03.005.
34. Mumma, A., Lane, M., Kairu, E., Tuinhof, A., Hirji, R., June 2011. Kenya groundwater governance case study. In: Water Papers. Water and Sanitation Program (WSP), The World Bank, Washington, D.C, pp. 2011, available online at URL: http://www.worldbank.org/water.
35. Munga, D., Mwangi, S., Ong'anda, H., Kitheka, J.U., Mwaguni, S.M., Mdoe, F., Barongo, J., Massa, H.S., Opello, G., 2006. Vulnerability and pollution of groundwater in Kisauni, Mombasa, Kenya. In: Xu, Y., Usher, B. (Eds.), Groundwater pollution in Africa. Taylor and Francis (Balkema), The Netherlands, pp. 213–228.
36. Musingi, J.K., Kithiia, S.M., Wambua, B.N., 1999. The urban growth of Mombasa coastal town and its implication for surface and groundwater resources. In: Ellis, J.B. (Ed.), Impacts of Urban Growth on Surface Water and Groundwater Quality. Proceedings of an International Symposium Held During IUGG
37. 99, the XXII General Assembly of the International Union of Geodesy and Geophysics. at Birmingham, UK 18–30 July 1999. IAHS Press (No. 259).
38. NEMA, 2010. Integrated Coastal Zone Management Action Plan for Kenya, 2011–2015. National Environment Management Authority, Nairobi, Kenya, pp. 76 p.
39. Nicholas, C.J., Pearson, P.N., Bown, P.R., Jones, T.D., Huber, B.T., Karega, A., Lees, J.A., McMillan, I.K., O'Halloran, A., Singano, J.M., Wade, B.S., 2006.
40. Stratigraphy and sedimentology of the Upper Cretaceous to Paleogene Kilwa Group, southern coastal Tanzania. J. Afr. Earth Sci. 45 (4), 431–466, http://dx.doi.org/10.1016/j.jafrearsci.2006.04.003.
41. Onyancha, C., Khaemba, A., Sabuni, B., 2010. Aquifer storage and recovery and surface basins for a greener Kilifi District. Nile Water Sci. Eng. J. 3 (3), 26–33.
42. Pretty, J.N., Vodouhê, S.D., 1997. Using rapid or participatory rural appraisal. In: Swanson, B. (Ed.), Improving Agricultural Extension. FAO, Rome, pp. 47–55.
43. Robins, N.S., Davies, J., Farr, J.L., Calow, R.C., 2006. The changing role of hydrogeology in semi-arid southern

and eastern Africa. Hydrogeol. J. 14 (8), 1483–1492, http://dx.doi.org/10.1007/s10040-006-0056-x.

44. Robins, N.S., 2013. A review of small island hydrogeology: progress (and setbacks) during the recent past. Q. J. Eng. Geol. Hydrogeol. 46 (2), 157–165, http://dx.doi.org/10.1144/qjegh2012-063.

45. Shaaban, I.A., Ramzy, Y.H., Sharabassy, A.A., 2013. Tourism as a tool for economic development in poor countries: the case of Comoro Islands. Afr. J. Bus. Econ. Res. 8 (1), 127–145.

46. Steyl, G., Dennis, I., 2010. Review of coastal-area aquifers in Africa. Hydrogeol. J. 18 (1), 217–225, http://dx.doi.org/10.1007/s10040-009-0545-9.

47. Ummenhofer, C.C., Sen Gupta, A., England, M.H., Reason, C.J.C., 2009. Contributions of Indian Ocean sea surface temperatures to enhanced East African rainfall. J. Clim. 22, 993–1013, http://dx.doi.org/10.1175/2008JCLI2493.1.

48. UNEP, 2012. The UN-Water Status Report on the Application of Integrated Approaches to Water Resources Management. UNEP Division of Communications and Public Information, Nairobi, Kenya, 119 p., ISBN: 978-92-807-3264-1.

49. United Nations, 2011. World Population Prospects: The 2010 Revision. UN Department of Economic and Social Affairs, Population Division, New York.

50. United Nations Development Plan UNDP, 1987. Recherche et mise en valeur des eaux, Comores: perspectives de mises en valeur des eaux souterraines pour l'alimentation en eau des agglomerations de l'ile de Ngazidja, Unpubl. Techn. Report No. DP/UN/COI-79-005, DP/UN/COI-86-001, New York, 60 p.

51. Vafeidis, A., Neumann, B., Zimmerman, J., Nicholls, R.J., 2011. MR9: analysis of land area and population in the low-elevation coastal zone (LECZ). GB, Foresight, Government Office for Science, London.

52. Van Niekerk, L., Van Niekerk, D., 2009. Participatory action research: addressing social vulnerability of rural women through income-generating activities. J. Disaster Risk Stud. 2 (2), 127–144.

53. Vincent, L.A., Aguilar, E., Saindou, M., Hassane, A.F., Jumaux, G., Roy, D., Booneeady, P., Virasami, R., Randriamarolaza, L.Y.A., Faniriantsoa, F.R., Amelie, V., Seeward, H., Montfraix, B., 2011. Observed trends in indices of daily and extreme temperature and precipitation for the countries of the western Indian Ocean. J. Geophys. Res.: Atmos. 116 (D10), 1961–2008, http://dx.doi.org/10.1029/2010JD015303 (1984–2012).

54. Walraevens, K., Mjemah, I.C., Mtoni, Y., Van Camp, M., 2015. Sources of salinity and urban pollution in the Quaternary sand aquifers of Dar es Salaam, Tanzania. J. Afr. Earth Sci. 102, 149–165, http://dx.doi.org/10.1016/j.jafrearsci.2014.11.003.

55. Watson, C.S., White, N.J., Church, J.A., King, M.A., Burgette, R.J., Legresy, B., 2015. Unabated global mean sea-level rise over the satellite altimeter era. Nat. Clim. Change 5, 565–568, http://dx.doi.org/10.1038/nclimate2635.

56. Watto, M.A., Mugera, A.W., 2015. Econometric estimation of groundwater irrigation efficiency of cotton cultivation farms in Pakistan. J. Hydrol.: Reg. Stud. 4 (A), 193–211, http://dx.doi.org/10.1016/j.ejrh.2014.11.001.

57. Werner, A.D., Bakker, M., Post, V.E.A., Vandenbohede, A., Lu, C., Ataie-Ashtiani, B., Simmons, C.T., Barry, D.A., 2013. Seawater intrusion processes, investigation and management: recent advances and future challenges. Adv. Water Res. 51, 3–26, http://dx.doi.org/10.1016/j.advwatres.2012.03.004.

58. World Bank, 2014. Climate change knowledge portal, http://sdwebx.worldbank.org/climateportal/index.cfm, consulted 10/06/2014.

59. World Bank, 2014. Low income economies classification, http://data.worldbank.org/about/country-classifications, consulted 10/06/2014.

60. World Health Organisation WHO, 2003. Total dissolved solids in drinking-water. Background document for preparation of WHO Guidelines for drinking-water quality No. WHO/SDE/WSH/03.04/16, WHO, Geneva, 8 p.

61. World Water Council, 2006. Water resources development in Africa, Africa Regional Document, 4th World Water Forum, Mexico 2006.

CHAPTER 11

TOWARD IMPROVED SIMULATION OF RIVER OPERATIONS THROUGH INTEGRATION WITH A HYDROLOGIC MODEL

Eric D. Morway [a], Richard G. Niswonger [b], Enrique Triana [c]

[a] Nevada Water Science Center, U.S. Geological Survey, Carson City, NV 89701, USA

[b] National Research Program, U.S. Geological Survey, Menlo Park, CA 94025, USA

[c] MWH Global, Inc., 3665 JFK Parkway Building 1, Suite 206, Fort Collins, CO 80525, USA

ABSTRACT

Advanced modeling tools are needed for informed water resources planning and management. Two classes of modeling tools are often used to this end–(1) distributed-parameter hydrologic models for quantifying supply and (2) river-operation models for sorting out demands under rule-based systems such as the prior-appropriation doctrine. Within each of these two broad classes of models, there are many software tools that excel at simulating the processes specific to each discipline, but have historically over-simplified, or at worse completely neglected, aspects of the other. As a result, water managers reliant on river-operation models for administering water resources need improved tools for representing spatially and temporally varying groundwater resources in conjunctive-use systems. A new tool is described that improves the representation of groundwater/surface-water (GW-SW) interaction within a river-operations modeling context and, in so doing, advances evaluation of system-wide hydrologic consequences of new or altered management regimes.

INTRODUCTION

Efforts to address the growing divide between water supply and water demands are complicated by uncertain supplies resulting from climate variability, growing demands, shifting societal values, over-allocation of water resources, trans-boundary compacts, institutionalized inefficiencies, and complex legal entanglements rooted in conjunctive use, among others. The predictive capabilities of simulation models may be the only viable means by which to forecast the potential impacts of alternative management decisions (Singh, 2010). Thus, models are frequently relied upon for identifying potential trade-offs among alternative management strategies. Yet the potential complications resulting from an alternative management strategy can scarcely be appraised with a suitable level of accuracy if the hydrologic consequences caused by those decisions are over-simplified. An important example is the application of operations models that do not properly represent finite groundwater supplies and temporally varying groundwater and surface-water (GW-SW) exchanges. Due to the time delays of groundwater responses to surface-water and groundwater management actions, the effects of a new management decision may range from days to decades as it propagates through the connected river-aquifer system. Oversimplified assessment of the impacts of water-use decisions can interfere with meaningful reform of current water-use practices.

Two pertinent examples of the importance of groundwater return flow in the execution of river operations are found in Colorado, though many other examples are available

Citation: Eric D. Morway, Richard G. Niswonger, Enrique Triana, Toward improved simulation of river operations through integration with a hydrologic model, Environmental Modelling & Software, Volume 82, August 2016, Pages 255-274, ISSN 1364-8152, http://dx.doi.org/10.1016/j.envsoft.2016.04.018.

(e.g.,Pulido-Velázquez et al., 2006, Rassam, 2011, Rassam et al., 2013, Shannon et al., 2000 and Valerio et al., 2010). In *Kansas v. Colorado* [1] (Abrams, 2004), the simplifying assumptions regarding GW-SW interaction were at the focal point of deliberations. The U.S. Supreme Court ruled that irrigation wells installed in Colorado (CO), located upstream of Kansas (KS), after the execution of the CO-KS interstate compact in the late 1940s did in fact "materially deplete" Arkansas River flows in direct violation of the governing compact (Colorado Revised Statutes, 1949), thereby injuring Kansas surface-water users (Hobbs, 2009 and Willis et al., 2008). The Special Master's[2] findings, informed by the Hydrologic-Institutional ("H-I") model simulations, were affirmed by the U.S. Supreme Court (Luecke, 2010) and scores of irrigators had their pumping curtailed (Littleworth, 2003, pgs. 8–9).

The second example is from the Platte River basin, which is the next major river basin to the north. A 'substitute water supply plan' (SWSP) approved by the Colorado State Legislature and implemented by the Colorado State Engineer's Office allowed junior-righted irrigation wells to operate provided they augment surface-water flows in sufficient amount to prevent injury to senior surface-water right holders, a calculation that acknowledges the influence of GW-SW connections and thereby requires a detailed hydrologic model (Howe, 2008). The SWSP had been operating successfully for nearly 30 years before the Colorado Supreme Court ordered the curtailment of pumping in nearly 400 irrigation wells starting in 2006, a move that recognized the impact irrigation wells were having on the already low surface-water flows, and, by extension, on the senior surface-water right holders (Ross, 2014). However, after the court order, water tables rebounded more than expected, negatively impacting home-owners by flooding basements and raising questions regarding curtailment of pumping in the irrigation wells (Waskom, 2014). In these and other examples, the influence of GW-SW exchange on in-stream flows, and therefore on the effective execution of river operations, cannot be overstated.

Rassam (2011) points out that failure to account for GW-SW interaction during periods of low flow limits the forecasting ability of an operations model. Heavy overdraft of the aquifer during average to dry conditions not only impacts GW-SW interaction in the near term, but may sap river water for years, especially during extended droughts. In most hydrogeologic systems, pumping will intercept regional groundwater flow that historically returned to the river, thereby resulting in reduced base flow and consequently reduced river flows that can persist for decades after pumping has ceased (Barlow and Leake, 2012). Conversely, conjunctive use of groundwater and surface-water resources during average to wet years may reduce senior surface-water users' reliance on surface water. As patterns in surface-water usage emerge in this context, the onset of dry conditions may give rise to conflict under the "use it, or lose it" principal that accompanies the prior appropriation doctrine (Wilds, 2010). Finally, though not the focus of this paper, a significant consideration for future integrated hydrologic and operations-modeling studies is the role GW-SW interaction plays on water quality (Sophocleous, 2002 and Winter et al., 1998), a facet of river-operations modeling that will likely receive increased scrutiny in the future. Thus, accounting for the impact of pumping on river flows is critically important for river-operations modeling in the 21st century (Brooks et al., 2012 and Simonovic, 2000).

In addition to the benefits that hydrologic models provide to operations models, operations models can greatly benefit common applications of hydrologic models, such as for water availability studies (Gleick, 1987). For example, where measurements of historical reservoir releases and river diversions are limited, operations models can be used to estimate anthropogenic components of water distribution. The ability to simulate river operations within basin-scale hydrologic models provides an accurate portrayal of the state of the basin in simulations of future conditions. That is, an integrated operations-hydrologic model requires no a priori specification of actual diversion amounts; the operational framework adhered to by the operation-side of the integrated code will dynamically allocate the yet unknown resource at model runtime, subject to all of the priorities and constraints specified by the user. Operations models provide a method to simulate the demand for water in conjunction with the hydrologic model that limits the availability of water supply within the hydrologic environment. If water users have supplemental groundwater rights, reasonable estimates of pumping can be calculated as the difference between the estimated water demand provided by the operations model and the surface-water shortfall provided by the hydrologic model (Schmid and Hanson, 2009). Historical conjunctive use of surface water and groundwater is typically unknown for many basins, and, as a result, simulation can improve the understanding of historical and projected supply versus demand within these systems.

A review of the literature indicates a lack of truly integrated operations-hydrologic codes. Currently, many simplifications are used within river operations models to approximate the underlying groundwater resource. Approaches for representing groundwater in operations models include response coefficients calculated using a calibrated groundwater model (Fredericks et al., 1998) or by simplified one-dimensional equations such as the Glover solution (Glover, 1977); a single feed-forward iteration with output calculated by a distributed-parameter model such as MODFLOW (Valerio et al., 2010), PARFLOW (Condon and Maxwell, 2013), or WEAP21 (Sieber and Purkey, 2007 and Yates et al., 2005); analytical solutions that assume linearity to approximate the cumulative effect of individual stresses (Rassam, 2011 and Welsh et al., 2013); and the use of surrogate models such as artificial neural networks (ANNs) trained by models such as MODFLOW for emulating GW-SW interactions under variable management (Triana, 2008). In the present work, we analyze the importance of integrating operations and hydrologic models, such that the models iterate multiple times within a time step to determine a balanced solution between the available water supply simulated by the hydrologic model and the subsequent distribution of that supply among the demands specified within the operations model. We demonstrate that without convergence between the operations and hydrologic models, erroneous diversion amounts are simulated and significant mass-balance errors are accumulated owing to inaccurate representation of GW-SW interaction.

An integrated operations-hydrologic model that iterates between the respective codes within each time step until specified convergences are attained provides a modeling platform that more accurately accounts for nonlinear feedbacks between water operations and water supply. Because groundwater is an integral component of most alluvial aquifer systems, it will strongly influence the predicted response to potential alternative management scenarios in many river systems, including operational changes (e.g., land fallowing), demand management (e.g., crop type), infrastructure changes (e.g., converting flood to sprinkler irrigation), and conjunctive-use strategies. Due to the profound impact that water operations have on water use and distribution, and, likewise, the impact that water supply has on water operations, integration of these aspects of water resources clearly points toward integrations of software that achieve these capabilities. Thus, an integrated approach to river operations and hydrologic modeling is provided here for exploring sustainable management of groundwater and surface-water supplies.

OBJECTIVES

A 'what-if' scenario in the context of water-resources simulation is an approach that tries to evaluate both obvious and unforeseeable outcomes associated with changes in water use or water supply. The value of 'what-if' scenarios common to most river operations and planning simulations is diminished when these simulations do not represent the finite storage of aquifers, head dependent GW-SW exchanges, and capture of natural groundwater discharge, as these processes control flow in rivers (Winter et al., 1998). Although it is well recognized that groundwater plays a profound role on surface-water supplies, representation of groundwater in operations models has failed to keep up with technology developed in the GW-SW literature (Fairbanks et al., 2001, LaBolle et al., 2003, Markstrom et al., 2008 and Panday and Huyakorn, 2004). Recent drought in the western US has exposed the limitations of water supplies, especially with regard to groundwater supplies (Castle et al., 2014). As drought continues to apply stress on finite groundwater supplies, it is clear that realistic representation of groundwater resources is necessary for managing water resources in the 21st century.

Therefore, the primary objective of this paper is to present a new approach wherein an operations/planning model is integrated with a hydrologic model capable of simulating GW-SW exchange at the time-step iteration level. An important distinction between the loosely-coupled (sometimes referred to as "sequentially-coupled") models already available (Rassam, 2011, Sieber and Purkey, 2007, Valerio et al., 2010 and Welsh et al., 2013) and the iteratively coupled models described here, is that in the iteratively coupled approach, neither model advances in time until differences in simulated values over an iteration fall below convergence criteria specified for each model individually and for quantities common to both models simultaneously. Using a hypothetical river basin modeled after a real river basin characteristic of semi-arid conditions in the western United States, scenarios are presented that highlight the value of integrating a river-planning/operations model (hereafter referred to as an operations model) with a GW-SW model as a tool for water-resources management. The integrated MODSIM-MODFLOW model is presented as an advancement over previously available codes because of its ability to simulate a much wider class of water-resources problems. We highlight the importance of iterative coupling over sequential coupling to achieve an accurate and mass-conservative solution that is robust over a wide range of management, climatic, and hydrologic conditions.

In this work, the Newton formulation (MODFLOW-NWT; Niswonger et al., 2011) of the groundwater-flow model MODFLOW (Harbaugh, 2005) is coupled to the operations/planning model MODSIM (Labadie, 2010a). Although there are a wide-array of hydrologic and river-operations models available in the literature, MODFLOW and MODSIM were chosen for the work presented herein because they are popular and freely available, are numerically efficient and stable, and have a software architecture that promotes enhancement, such as coupling to the watershed runoff model PRMS (GSFLOW; Markstrom et al., 2008). Other hydrologic models that might have been used include MODHMS (Panday and Huyakorn, 2004) and HydroGeoSphere (Therrien et al., 2006), whereas WaterWare (Jamieson and Fedra, 1996), RiverWare (Zagona et al., 1998) (reference), WEAP (Yates et al., 2005), or WRAP (Wurbs, 2013), might have been used to represent water-resource operations. MODSIM customization capabilities allow the user to write code that is compiled with the MODSIM solver, making possible low-level integration with other models. Moreover, widely-varying hydrogeologic settings coupled with disparate sets of legal doctrines necessitate the use of generalized tools that are both customizable and defensible. As to the former, both MODFLOW and MODSIM are offered as generalized software that can be customized to specific river systems. Because of this, both have found widespread acceptance and application to many of the world's most important river basins.

Management capabilities currently exist in MODFLOW; two examples are MODFLOW-GWM (Ahlfeld et al., 2005) and MODFLOW-FMP (Schmid and Hanson, 2009), with other versions of MODFLOW, such as MODFLOW-CDSS (Banta, 2011), compiled for use by river-operations models like StateMod (Alvarado and Bennett, 2005 and Bennett, 2012). However, there are important distinctions between the MODSIM-MODFLOW model developed herein and these other MODFLOW based management codes. MODFLOW-GWM is a management tool that optimizes water use to achieve balances between meeting water demands and satisfying constraints related to water supply and other environmental factors. It is important to note that GWM does not include capabilities to represent rule and priority based water-allocation systems or reservoir operations. Conversely, MODSIM-MODFLOW is focused on representing laws that govern water use, which is distinct from the simulation-optimization capabilities of MODFLOW-GWM. This distinction also applies to the difference between MODSIM-MODFLOW and MODFLOW-FMP, which was recently re-released with added functionality as MODFLOW-OWHM (Hanson et al., 2014 and Schmid et al., 2014). MODFLOW-FMP does not provide capabilities for representing rule and priority-based water allocation. Rather, water allocation is demand based. It should be noted that MODFLOW-GWM and MODFLOW-FMP offer capabilities that are not available in MODSIM-MODFLOW, and, as such, may be the most appropriate code for a particular water-management problem. MODSIM-MODFLOW is intended as a simulation tool for analyzing water-resources problems in basins that are governed by rule and priority controls on water allocation, as well as reservoir operations.

METHODS

Integrated river operations and hydrologic modeling framework

MODSIM is a generic river-basin management decision-support system capable of simulating complex, large-scale surface-water networks, and excels in the area of administering water in systems governed by water rights, administrative constraints, and agreements. However, MODSIM standard functionality approximates stream capture by wells using simplified one-dimensional equations or estimated stream-depletion factors. Stream seepage rates can be specified prior to model execution, but must be determined outside of the simulation process and do not account for the transient behavior of GW-SW interactions or the effects of climate variability. A more robust approach for simulating changes in groundwater storage and GW-SW interaction affected by river-system management changes is to use a three-dimensional groundwater-flow model such as MODFLOW. Accordingly, weaknesses in MODSIM are bolstered by the strengths of MODFLOW, and vice versa. For example, MODSIM does not simulate spatially and temporally varying GW-SW interaction due to changes in pumping, climate, or other groundwater-recharge or groundwater-discharge processes. MODFLOW can simulate these processes through a wide variety of boundary-condition packages that account for interactions with surface-water features (Hughes et al., 2012, Merritt and Konikow, 2000 and Niswonger and Prudic, 2005) and the unsaturated zone (Niswonger et al., 2006) and the effects of groundwater pumping (Harbaugh, 2005 and Konikow et al., 2009) However, options currently available for simulating streamflow diversions with the SFR2 streamflow routing package (Niswonger and Prudic, 2005) and (or) releases from reservoirs with the LAK3 package (Merritt and Konikow, 2000) are not capable of accounting for prior-appropriation water rights and reservoir operations, and lack the functionality to divert water based on priorities (or rules) that are not in downstream order. Moreover, it is not possible to use MODFLOW alone to explore "what-if" water-management scenarios such as permanent or temporary water-right transfers from agriculture to municipal uses or among alternative agricultural uses, or the creation of reservoir storage accounts that affect river flows.

MODSIM is designed to allocate river flows and stored water based on physical and legal availability, while also capable of simulating complex operations such as river exchanges (i.e., water-market type trades), minimum instream-flow requirements, and multi-reservoir systems. Thus, MODSIM and MODFLOW are integrated such that they reach a converged solution for every time step by satisfying the water-right rules that are constrained by water availability simulated by MODFLOW. The newly developed code will provide water planners and managers in appropriated systems with a more robust decision-making support tool than either model could achieve when applied independently.

Modsim

MODSIM is short for MODified SIMyld and is freely available for download at http://modsim.engr.colostate.edu/index.shtml. Its predecessor, SIMYLD, was an open-source network-flow model developed in the early 1970s by the Texas Water Development Board (Shafer and Labadie, 1978). MODSIM uses a minimum-cost network-flow optimization algorithm to allocate run-of-the-river flows and stored volumes among a specified set of demands according to the institutional framework governing the distribution of water. Moreover, the solution technique to the network-flow problem solves the continuity equation and ensures conservation of water. Unregulated inflows to the model domain, consumptive and instream-flow demands, reservoir and channel evaporation, precipitation, exchanges with the groundwater system, reservoir storage rights and exchanges, and reservoir operating targets are each simulated as network elements within MODSIM (Shourian et al., 2008).

MODSIM takes advantage of the linear-network structure of the modeled system and minimizes the following objective function for each time period t = 1, …, T (Labadie, 2010a):

$$Minimize \sum_{l \in A} c_l q_l \quad (1)$$

Subject to:

$$\sum_{j \in O_i} q_j - \sum_{k \in I_i} q_k = b_{it}(\vec{q}); \text{ for all } i \in N \quad (2)$$

$$l_{lt}(\vec{q}) \le q_l \le u_{lt}(\vec{q}); \text{ for all } l \in A \quad (3)$$

where c_l are the costs, weighting factors, or water-right priorities per unit of flow rate within link l (unitless); q_l is the integer-valued flow rate in link l (L^3/T); A is the set of all links in the network; O_i is the set of all links originating at node i (i.e., outflow links); I_i is the set of all links terminating at node i (i.e., inflow links); b_{it} is gain (positive) or loss (negative) at node i in time step t; N is the set of all nodes; l_{lt} and u_{lt} are the lower and upper bounds on the flow in link l (L^3/T) at time step t, respectively; and the vector $\vec{q}$ is the unknown set of network flow rates (L^3/T) determined at each model time step. Allowing the link constraints l_{lt} and u_{lt} and node supplies b_{it} to vary as functions of $\vec{q}$ enables non-network constraints to affect the solution, including nonlinearities related to the interdependence of reservoir surface area on evaporation, precipitation, and reservoir exchange with groundwater. Equations (2) and (3) represent constraints on mass balance (continuity) and flow-dependent diversions, such as minimum in-stream flows or maximum conveyances. Although the magnitude of the costs, c_l, are arbitrarily set by the user, their relative values must be ordered such that the most senior priority has the most negative cost. In this way, junior surface-water demands are the first to be curtailed during water limited periods.

Equations (1), (2) and (3) are solved by the RELAX-IV Lagrangian relaxation algorithm (Bertsekas and Tseng, 1994), which provides a solution that is up to two orders of magnitude faster than the revised simplex method of linear programming (Labadie, 2010a). The set of flows $\vec{q}$ found by the RELAX-IV algorithm are subsequently used to update the estimates of parameters b_{it}, l_{lt}, and u_{lt}, and the network-flow solution continues iterating until convergence is achieved. Additional explanation of advanced features as well as the solution algorithm can be found in the MODSIM manual, appendices, and tutorials (Labadie, 2010b), available from the MODSIM homepage. Detailed inter-model comparisons between MODSIM and other river operations models can be found in Winchester, 2008, Wurbs, 2012 and Sulis and Sechi, 2013, and Johnson (2015). In each of the four studies, MODSIM was well ranked among the other models considered in the study. Table 1 provides a list of selected MODSIM applications from around the world.

Table 1: MODSIM applications and their treatment of GW-SW interaction.

River basin or irrigation district	Country	Citation	Considers groundwater?
Geum	South Korea	Labadie et al. (2007)	Imported from external model, not synchronized
Han	South Korea	Shim et al. (2002)	No, though conjunctive use not the focus of study
Sirvan	Iran	Shourian et al. (2008)	No
Karkheh	Iran	Jamali et al., 2013 and Vaghefi et al., 2013	No, hydropower generation focus of the study
Awash	Ethiopia	Berhe et al. (2013)	No
Upper Pampanga	Philippines	Faux et al. (1986)	No
Nile	Egypt	El-Beshri and Labadie (1994)	Uses aquifer response functions
Piracicaba	Brazil	de Azevedo et al. (2000)	No
Paraguaçu	Brazil	Srdjevic et al. (2004)	No
Nizao	Dominican Republic	Labadie (1993)	Unknown
Rio Grande	USA	Graham et al. (1986)	No
Upper Colorado	USA	Law and Brown (1989)	Unknown
Upper Snake	USA	Briand et al., 2008, Frevert et al., 1994,Larson and Spinazola, 2000 and Miller et al., 2003	Uses aquifer response functions
Clearwater	USA	Lanini et al. (2013)	No, hydropower generation focus of the study
Platte	USA	Fredericks et al., 1998, Houk et al., 2007 and Shafer et al., 1981	Uses aquifer response functions
Arkansas	USA	Triana (2008)	Imported from external model, not synchronized
Klamath	USA	Campbell et al. (2001)	No
Deschutes	USA	LaMarche, 2001 and Larson et al., 2014	Imported from external model, not synchronized
Imperial Irrigation District	USA	Triana and Labadie (2012)	No
San Joaquin	USA	Marques et al. (2006)	Uses loosely-coupled approach

Modflow

Developed and maintained by the U.S. Geological Survey and made freely available to the public, MODFLOW is a three-dimensional finite-difference groundwater-flow model. Through a robust modular structure that facilitates plug-in of new or enhanced functionality, additional simulation capabilities from a wide array of contributors has continually expanded upon its original scope.

MODFLOW solves a cell-centered finite-difference approximation of the groundwater-flow equation (Harbaugh, 2005):

$$\frac{\partial}{\partial x}\left(K_{xx}\frac{\partial h}{\partial x}\right)+\frac{\partial}{\partial y}\left(K_{yy}\frac{\partial h}{\partial y}\right)+\frac{\partial}{\partial z}\left(K_{zz}\frac{\partial h}{\partial z}\right)+W=S_s\frac{\partial h}{\partial t} \quad (4)$$

where K_{xx}, K_{yy}, and K_{zz} are values of hydraulic conductivity along the *x*, *y*, and *z*coordinate axes, (L/T); *h* is the groundwater head (L); *W* is a volumetric flux per unit volume representing sources and/or sinks of water, (T^{-1}); S_s is the specific storage of the porous material (L^{-1}); and *t* is time (T).

Streams and lakes/reservoirs having arbitrary geometry and are hydraulically connected with the underlying groundwater system are represented in MODFLOW using the SFR2 streamflow routing package (Niswonger and Prudic, 2004) and a modified version of the Lake (LAK3) Package (Markstrom et al., 2008, Merritt and Konikow, 2000 and Prudic et al., 2004), respectively. Seepage from channels or lakes is calculated according to a Darcy-type formulation:

$$Q_{gw}=\frac{KwL}{m}(h_s-h_{aq}) \quad (5)$$

where Q_{gw} is the seepage rate through a streambed or lake cell and may be positive (aquifer discharge to the stream or lake) or negative (stream or seepage loss to the aquifer) depending on the head differential between the groundwater and channel or lake; K is the hydraulic conductivity of the streambed or lakebed/aquifer sediments; $w$is a representative width of the stream or lake cell; L is the length of stream or lake cell corresponding to a volume of aquifer (sometimes taken as the length of meandering streams within a finite-difference cell); m is the thickness of the streambed or lake deposits; hs is the head in the stream or the sum of the stream stage and the specified elevation of the streambed or lakebed; and haq is the calculated groundwater head in the finite-difference cell in which the stream reach or lake cell is simulated. Prudic et al. (2004) describe channel-geometry options that are available–for example, an 8-point cross-section or a user-supplied depth-discharge or width-discharge relationship. In addition, SFR2 accommodates any user-specified channel length within a grid cell, and can simulate steady or transient surface-flow conditions. Where the stream network flows in and out of lakes (or reservoirs), the LAK3 package may be used to exchange surface-water flows with the SFR2 package.

Integration

As discussed by Barlow and Leake (2012), the geology, basin and stream geometry, distribution of natural groundwater discharge, distribution of groundwater heads, and the location of wells all impact streamflow. Thus, the three-dimensional characterization of the hydrology and hydrogeology is important for realistically simulating the effects of GW-SW interactions on streamflow. In this work, the stream network is represented by the SFR2 package of MODFLOW and is made identical to its counterpart stream network represented in MODSIM to assure parallel representation of river-tributary-ditch connectivity in both MODFLOW and MODSIM. For the integration of MODSIM-MODFLOW, channel losses and gains passed to MODSIM are calculated by MODFLOW on the basis of surface-water stages relative to the calculated groundwater heads. Within the newly integrated MODSIM-MODFLOW code, diversions and releases (determined by MODSIM and sent to MODFLOW) and, conversely, GW-SW exchanges (calculated by MODFLOW and sent to MODSIM) are passed via computer memory, thereby allowing for a Picard nonlinear iterative-solution scheme to determine flows and potentiometric heads throughout the model domain and during each time increment. In summary, groundwater gains and losses affected by spatially and temporally varying pumping, riparian ET, and bank storage—that is, processes that influence surface flows—are fully accounted for when MODSIM solves for daily diversion amounts. Table 2 summarizes the role of each model regarding surface-water features that are common to both codes.

Table 2: Description of surface-water features represented by both MODSIM and MODFLOW.

Feature	MODSIM	MODFLOW
Reservoir	Simulates storage operations (accrual of direct flows into 'owned' storage accounts and the subsequent release of flow from storage accounts).	Simulates GW-SW interaction and surface-area dependent gains and losses due to precipitation and evaporation through the use of the LAK3 lake package.
River and tributaries	Routes surface flows and storage releases through the MODSIM stream network after being informed of spatially- and temporally-varying GW-SW exchange rates calculated by MODFLOW.	Equipped with diversions and releases calculated by MODSIM, simulates river (and tributary) stage and the GW-SW exchanges resulting from those stages using the SFR2 streamflow-routing package.
Diversions	Calculated based on water rights and water availability.	Uses MODSIM diversion amounts at the points of diversion within the SFR2 network for routing of flows via delivery ditches to points of demand within the MODFLOW model.

Integration at the time-step iteration level is easily accommodated using the MODSIM customization capabilities. That is, before MODSIM-MODFLOW advances to the next simulation time step, simulated flows by MODFLOW must be consistent with those simulated by MODSIM. Because MODSIM does not calculate potentiometric heads, calculated diversions (and lake or managed reservoir releases, when pertinent) are the sole quantities passed from MODSIM to MODFLOW. MODFLOW then re-solves the groundwater-flow equation and the GW-SW exchanges corresponding to the newest values of operationally compliant diversions and reservoir releases passed from MODSIM. The solution process for the integration of MODSIM-MODFLOW for a single time step is enumerated in the following steps:

1. Upon model initiation, MODSIM is run to calculate initial diversions and reservoir releases based on conditions specified in the MODSIM input file assuming no GW-SW exchanges among all elements in the link-node network
2. MODFLOW executes using diversions and reservoir releases calculated in step 1
3. MODSIM is rerun using updated GW-SW exchanges from the MODFLOW solution in the last iteration
4. MODFLOW is rerun using updated reservoir releases and diversions calculated in step 3
5. Steps 3 and 4 are repeated until convergence is reached
6. The simulation moves to the next time step, whereby the iterative process begins again at step 3, in which MODSIM uses the GW-SW exchanges calculated in the final MODSIM-MODFLOW iteration from the previous time step.

Prior to initiating the integrated code, it is recommended that the user develop a steady-state MODFLOW simulation using long-term average boundary conditions (Reilly and Harbaugh, 2004). Once satisfied that the steady-state solution roughly approximates aquifer conditions at the start of the transient solution, the steady-state groundwater head solution, soil-moisture distribution in the unsaturated zone, and lake stages should be inserted into the corresponding input files for use by the first transient time-step in the integrated code.

Model convergence is reached when changes in all diversion and reservoir-release rates (and/or flow in any additional stream segment, depending on user preferences) are below a specified tolerance between two successive iterations. The L^2Norm (Sun, 2013) is used for the convergence criteria:

$$L^2\,Norm = \sqrt{\sum_i \left(Q_{i,k} - Q_{i,k-1}\right)^2} < \delta \qquad (6)$$

where $Q_{i,k}$ is the MODSIM calculated flow rate for each managed diversion or release i and iteration k ; $Q_{i,k-1}$ is the MODSIM calculated flow rate for each diversion or release i and iteration $k - 1$; and δ is the L^2 *Norm* convergence criteria (Sun, 2013).

In this work, MODSIM and MODFLOW are called by the .NET integrator (Fig. 1). To facilitate the integration, MODFLOW is compiled as a dynamic-link library (DLL) and called multiple times within a given time step by the .NET integrator that implements the built-in custom-code interface (Fig. 2). Within Fig. 2, names appearing in the blue boxes on the left (e.g., "IterationTop" or "IterationConverge", etc.) correspond to the 'handlers' within the MODSIM custom code module. Code written into the handlers is compiled with the MODSIM executable and subsequently executed as part of the MODSIM source code. It is through the custom-code interface that MODSIM users are able to access nearly all of the internal variables used by MODSIM to initialize, retrieve, and modify parameter values. Because of this, it is not necessary to access the full MODSIM source code for integration with MODFLOW.

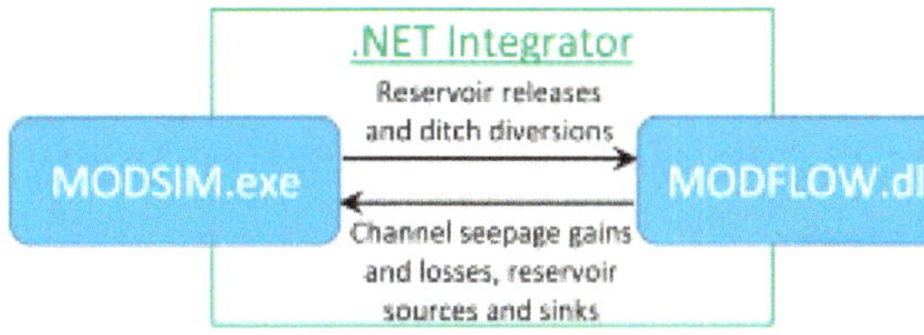

Figure 1: Diagram showing the information that is passed between the integrated codes.

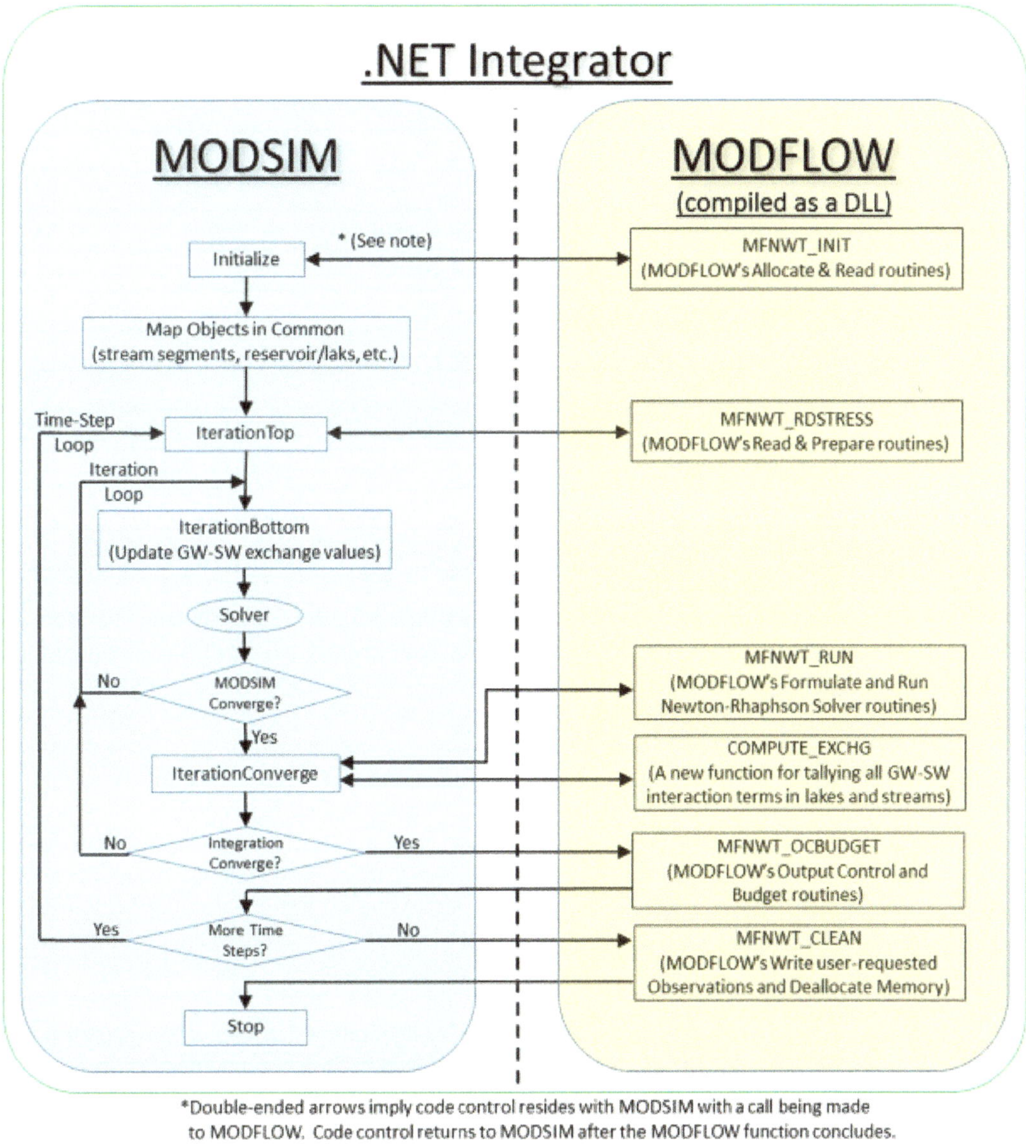

Figure 2: Integration of MODSIM and MODFLOW relied on reorganization of MODFLOW such that similar functions were grouped into a single function callable by the .NET integrator. Thus, the generalized functions to the right of the dashed line are the only visible functions in the MODFLOW DLL. The OnMessage and OnError handlers are used to display internal progress and error messages from the MODSIM model.

There are seven handler locations within MODSIM: (1) OnInitialize, (2) OnMessage (3) OnError, (4) IterationTop, (5) IterationBottom, (6) IterationConverge, and (7) OnFinished. By their names, users can get a sense of when their custom code will be executed during the model run. However, for more information regarding the specific details of the custom-code functionality, readers are referred to the MODSIM manual and Tutorial C available with the MODSIM download (Labadie, 2010b).

Compiling MODFLOW as a DLL required that 'themed' sections of the MODFLOW code be wrapped (or bundled) in a unique function name, for example, 'MFNWT_RUN,' which contains all of the MODFLOW Formulate routines [that is, functions with "FM" appearing in their header within the MODFLOW source code (Harbaugh, 2005)] responsible for calculating the terms sent into the linear matrix solver. Once the code lying within MFNWT_RUN is completed, program control is returned to the. NET integrator, where subsequent calls to MODFLOW are made and checks for convergence are carried out. The code continues to iterate within a given time step as shown by Fig. 2, which shows the complete MODSIM-MODFLOW code flow.

HYPOTHETICAL TEST MODEL OF AN AGRICULTURAL RIVER BASIN

The integrated MODSIM-MODFLOW code is applied to a hypothetical test model of an agricultural river basin located in a semi-arid environment (approximately 23 cm [9 in] of annual precipitation), where most of the water is supplied by upstream snowmelt. The test problems conceptually represent a shallow (37 m thick) alluvial aquifer underlain by an impermeable bedrock. Aquifer properties are uniform throughout the model domain and are summarized in Table 3.

Table 3: Specified aquifer properties and boundary conditions in the MODFLOW model.

Model parameter	Value
Saturated-zone model parameters	
Horizontal hydraulic conductivity	10 m d^{-1}
Ratio of vertical to horizontal hydraulic conductivity	0.1
Specific yield	0.28
Unsaturated-zone model parameters	
Saturated hydraulic conductivity of unsaturated zone	0.18 m d^{-1}
Brooks-Corey epsilon	7.10
Saturated water content	0.42
Extinction water content	0.08
Boundary conditions	
Precipitation rate	0.0–0.07 m d^{-1} (0–2.75 in d^{-1})
Irrigation rate (i.e., application rate)	0.01–0.84 m d^{-1} (0.4–33.1 in d^{-1})
River flow entering model	23,200–16,147,200 m^3 d^{-1} (9.5–6600 ft^3 s^{-1})
Tributary flow entering model	1030–484,400 m^3 d^{-1} (0.4–198 ft^3 s^{-1})

In the two scenarios presented below, the resulting impacts to both the hydrologic solution and operational decisions are assessed to highlight the advantages of an integrated operations-hydrologic simulation at the time-step level. For example, comparison of the changes in groundwater levels during the simulation period illustrates how river and reservoir operations affect groundwater sustainability in agricultural regions that conjunctively use groundwater and surface-water supplies.

Although the size and complexity of this example problem is simple relative to many real-world agricultural systems, it illustrates the complexities associated with conjunctive-use agricultural systems. Nonlinear processes that occur in this hypothetical test problem are consistent with those in larger real-world systems, and, as such, the problem provides an exploration of the feedbacks between water supply and water demand in an adjudicated river basin.

Although the MODSIM-MODFLOW integration offers many important advantages for a robust exploration of water resource allocation, there are some noteworthy challenges. Experienced river-operation modelers may be unfamiliar with MODFLOW and therefore find it difficult to navigate the myriad input files required by a MODFLOW simulation. Conversely, researchers and practitioners experienced with distributed-parameter hydrologic modeling may find representation of legal frameworks governing river administration in a modeling environment unfamiliar. That is, imparting real-world operating rules in a river-operations model may not be intuitive to newcomers of this field. Thus, it is important to recognize that water-resource investigations are increasingly multi-disciplinary, requiring a diverse set of expertise to design and implement complex solution strategies – the aforementioned 'what-if' scenarios (Brekke et al., 2009 and Yates et al., 2005) – and should therefore be approached as a collaboration between river-operations specialists and experienced hydrologic modelers, as was done for this present work.

MODSIM model construction

MODSIM attributes common to both investigated scenarios include four agricultural diversions from the main-stem river, sometimes referred to as "ditch-diversions." Priorities are assigned to each of the four ditch-diversion locations; farm-level priorities are not represented, although the generalized nature of MODSIM and MODFLOW would allow representation of farm-level priorities. Each of the four diversions are hereafter referred to as 'Command Area 1' (CA1) through 'Command Area 4' (CA4), with CA1 being the most upstream, CA2 the next downstream diversion, and CA3 and CA4 the next two downstream diversions, respectively (Fig. 3). Each ditch was assigned two water rights of staggered priority. The tiered priorities are provided in Table 4. The more negative a cost in MODSIM is, the more senior the priority (variable c_1 in Equation (1)); thus, the minimum instream-flow requirement mandated for the main-stem river just downstream of the confluence with the large tributary (Fig. 3) is the most senior right on the river, with a cost of −1100.

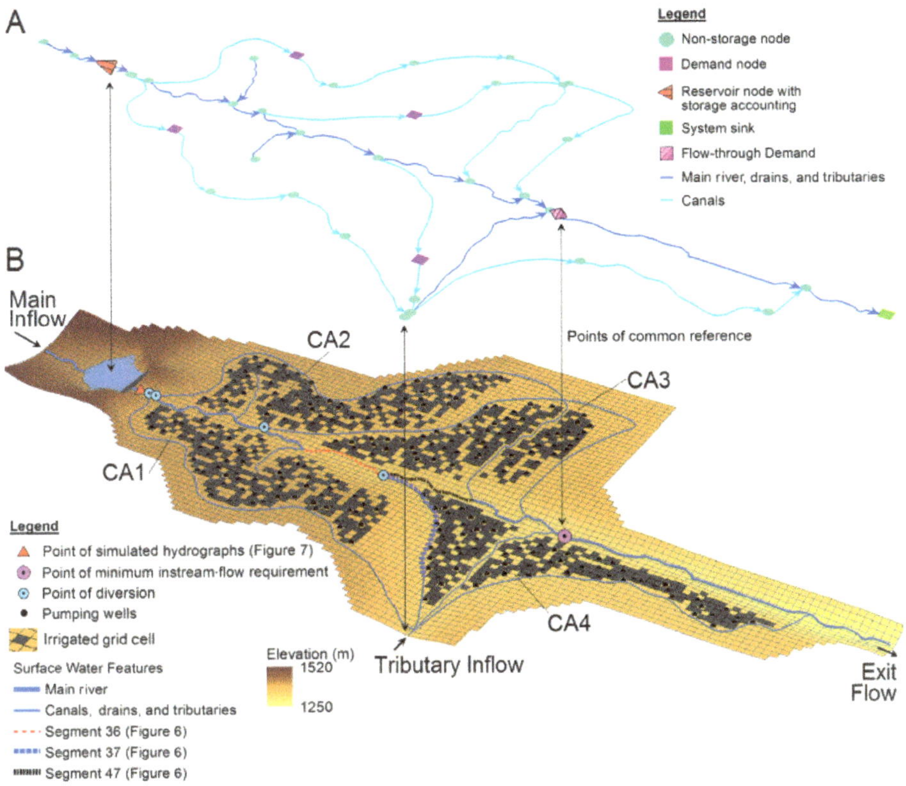

Figure 3: Hypothetical test model layout. Direction of the river flow is left to right. (A) The schematic network of MODSIM links and nodes and (B) the MODFLOW grid calculating groundwater flow terms in response to river operations. Command areas (CA1, etc.) are delineated by the ditches delivering their water and the main river.

Table 4: Hierarchy of priorities and maximum flow amounts enforced by MODSIM. The more negative costs reflect increasing seniority among the demands.

Water right global priority[a]	Water right type	MODSIM cost	Flow allotment m^3 d^{-1} (ft^3 s^{-1})	Storage volume m^3(ac ft)	User/owner
1	Direct flow	−1100	24,466 (10.0)		Minimum instream flow requirement[b]
2	Direct flow	−1000	80,000 (32.7)		CA1
3	Direct flow	−900	80,000 (32.7)		CA2
4	Direct flow	−800	80,000 (32.7)		CA3
5	Direct flow	−700	80,000 (32.7)		CA4
6	Direct flow	−500	25,000 (10.2)		CA1
7	Direct flow	−400	25,000 (10.2)		CA2
8	Direct flow	−300	25,000 (10.2)		CA3
9	Direct flow	−200	25,000 (10.2)		CA4
10	Storage	−190		12,000,000 (9730) 36,000,000 (29,190)	CA1 & CA2
11	Storage	−180		26,008,500 (21,090)	CA3

a More negative numbers indicate higher priority corresponding to higher MODSIM costs.

b See Fig. 3 for location of minimum instream flow requirement.

Minimum instream-flow requirement

Due to its location, tributary flows at the confluence with the main river will, at times, satisfy the minimum instream-flow requirement. Under this circumstance, MODSIM will attempt to satisfy the maximum allocation allotment for each ditch, diverting all of the available river flow starting with the most senior priority. When tributary inflow fails to satisfy the minimum instream-flow requirement, ditch diversions will be curtailed, beginning with the lowest-priority ditch first. Depending upon water availability, MODSIM will turn off all diversions during the irrigation season in an attempt to satisfy the higher-priority minimum instream-flow requirement of 24,466 m^3 d^{-1} (10 ft^3 s^{-1}).

Reservoir storage accounts

One of the simulated scenarios, described in Section 4.3.2, takes advantage of reservoir storage capabilities provided by MODSIM. The LAK3 package simulates lake-aquifer interaction, but offers no functionality for simulating regulated operations (i.e., releases) from storage accounts within the reservoir other than a priori specification of releases. Thus, as streamflow, precipitation, overland runoff, and groundwater enters the reservoir, it is accrued in reservoir storage accounts specified by the user, provided senior water-right priorities elsewhere in the simulation are satisfied. Conversely, as water is removed from the reservoir (i.e., reservoir releases and/or direct withdrawals) and delivered to the owner through the surface-water network, it is deducted from the appropriate account(s). Other reservoir losses, including evaporation and seepage, are distributed equally among all of the storage accounts. That is, based on the feedback from MODFLOW regarding the interaction of the reservoir with the groundwater system, storage accounts are re-balanced in such a way that the losses are partitioned equally among all of the storage accounts, although user-specified preferences within MODSIM could direct the code to allocate these losses to a particular combination of accounts. Additional detail regarding the setup of storage accounts in the hypothetical test model is provided in Section 4.3.

MODFLOW model construction

The MODFLOW component of the test model spans the same period of time as the MODSIM model, from water year (WY) 1990 through WY 2006 (10/01/1989–09/30/2006). The MODFLOW-side of the simulation also uses the same daily stress periods as MODSIM, where a stress period in MODFLOW represents the period of time over which boundary conditions remain constant. Longer time steps could be chosen, so long as they are of equal length in both models. The total number of daily stress periods in the model is 6,210, and each stress period consists of a single daily time step. Spatially, the MODFLOW simulation is discretized into 64 rows, 133 columns, and 4 layers, representing the 37 m thick alluvial aquifer. The surface-water network is located within the upper model layer, which is a uniform 7 m across the model domain. Layers 2 through 4 are each 10 m thick. Horizontal discretization is 400 m square, resulting in 16 ha (~40 acre) grid cells that correspond one-to-one with irrigated fields, as shown by the discretely shaded grid cells in Fig. 3. There are no specified-head boundaries within the model domain; lateral boundaries in all four layers of the model are specified as no-flow. A uniform, constant recharge of 9.6 m^3 d^{-1} is applied to layer 4 grid cells that touch the upper (northern) and lower (southern) perimeter of the active domain and represents natural recharge from surrounding (non-modeled) lands. Total recharge, excluding recharge from agricultural inefficiencies and reservoirs and stream seepage losses, was set to 2870 m^3 d^{-1} (1.2 ft^3 s^{-1}).

Surface-water network

Streamflow entering the model domain occurs in the main channel (left-edge of the model domain) as well as in a tributary crossing midway along the lower-edge of the model domain (Fig. 3). At these two locations, the same inflow must be specified within both of the appropriate model-input files. Daily inflow hydrographs for the river and tributary are derived from continuously gaged records collected on the West Fork of the Carson River, located near Woodfords, CA, and Clear Creek, a tributary to the Carson River located near the south side of Carson City, respectively. Fig. 4 shows the annual total flow entering each of these locations. The simulation period roughly corresponds to a dry period (WY1990 – WY1994, with WY1993 the only exception), followed by five years of above average conditions (WY1995 – WY1999), followed by a five-year below average period (WY2000 – WY2004), and ends with a two year period of normal to above-average conditions (WY2005 – WY2006). All surface-water flow, including connections with the reservoir and alluvial aquifer, are simulated by use of the streamflow routing (SFR2) package (Niswonger and Prudic, 2005). More complex problems, including problems with overbank (2-dimensional) flow could be solved with the Surface-water Routing (SWR) package (Hughes et al., 2015) written for MODFLOW; however, the first release of MODSIM-MODFLOW does not facilitate use of the SWR package. Streambed hydraulic conductivities are 0.25 m d^{-1} in the river and its tributaries and 0.04 m d^{-1} in the delivery ditches. Hydraulic conductivity values for delivery ditches represented in the model are constant in time and the simulated seepage from them ranges between 10%–20% of the diverted amount, which is consistent with seepage values published for the region (Allander et al., 2014).

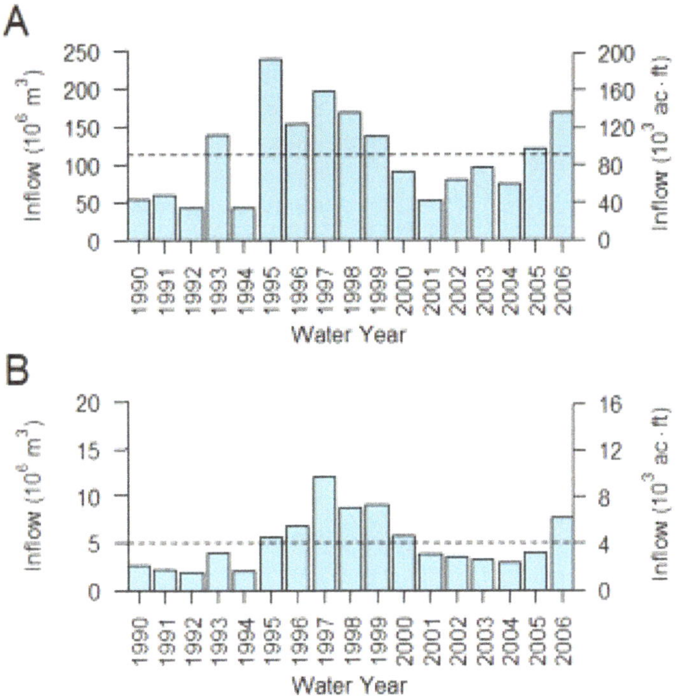

Figure 4: Annual inflow entering the model domain through (A) the main river channel and (B) tributary. Dashed lines show the average annual inflow during the modeled period, approximately 108 million m^3 (88,000 ac ft) and 5 million m^3 (4000 ac ft) in the river and tributary, respectively.

Once diverted, the SFR2 package calculates seepage (or groundwater return flow) using Equation (5) for each stream reach. Diverted water plus or minus ditch seepage is distributed among 'cropped' fields (i.e., MODFLOW grid cells) as irrigation water applied to land surface via the Unsaturated-Zone Flow (UZF1) package (Niswonger et al., 2006). Irrigation events are described in Section 4.3.3.

Evapotranspiration

UZF1 partitions the applied irrigation water, whether derived from surface-water diversions or pumped groundwater, into ET, recharge, and unsaturated-zone storage changes. To accomplish this, the user specifies potential ET (pET), which varies by stress period (i.e., daily) within the UZF1 input file. pET arrays are based on values (i.e., solar radiation, dew point, wind speed, etc.) measured at a weather station situated in Carson Valley, Nevada, nearby where the streamflow hydrographs were recorded. Using guidelines described in ASCE-EWRI (2005) and Huntington and Allen (2010), the ASCE standardized grass reference ET (ET_o) was calculated for each day (i.e., stress period) of the simulation. *ETo* is resolved into actual ET (ET_a) for five crop types typical of Carson Valley, NV, including (1) alfalfa, (2) corn, (3) grass hay, (4) highly managed pasture, and (5) lightly managed pasture, which are randomly assigned to the cropped cells within each command area (Fig. 3). To convert ET_o to ET_a, a dual crop coefficient approach is used:

$$ET_a = K_c ET_o \quad (7)$$

$$K_c = K_{cb} + K_e \quad (8)$$

where K_c is the crop-specific coefficient for converting ET_o to ET_a, K_{cb} is the basal crop coefficient and K_e is the soil evaporation coefficient. K_{cb} is the ratio of ET_a toET_o when the soil surface is dry and transpiration is derived solely from root zone soil moisture under non-stressed conditions (Huntington and Allen, 2010). Fig. 5A highlights the variation of K_c in response to precipitation and irrigation, whereasFig. 5B depicts the impact the dual crop coefficient approach has on ET_o, resulting inET_a.

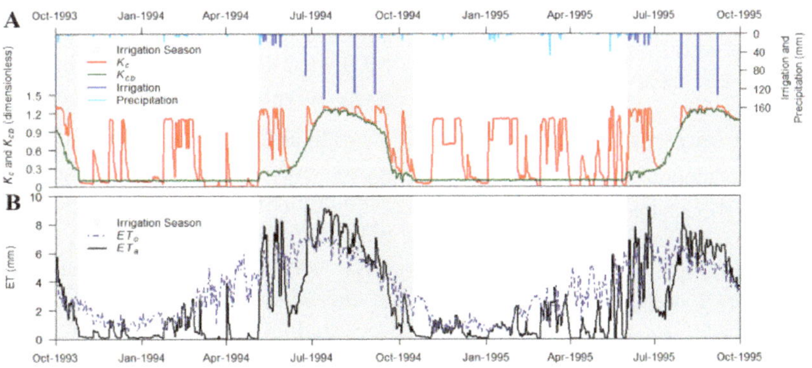

Figure 5: An example time series of (A) irrigation and precipitation rates and crop coefficients, K_c and K_{cb}, and (B)ET_o and ET_a for corn during WY1994 (dry year) and WY1995 (wet year). Time series of ET_a for the other 4 simulated crop types used in the hypothetical model are not shown, but use a similar approach.

Precipitation, irrigation applications, and pumping

Precipitation (Fig. 5) is based on a rain-gage record collected in Carson Valley, Nevada, and varies with each stress period (i.e., daily). Because crop-water needs far exceed precipitation inputs, irrigation frequency for each cropped field is such that well-water conditions are maintained (i.e., crop-water stress is avoided). Crop water requirements are first satisfied by surface-water supplies delivered via the SFR2 package. When surface-water supplies fall short of maintaining well-watered conditions, pumping wells simulated using the well (WEL) package (Harbaugh, 2005) are activated by the integrated code to offset surface-water delivery shortfalls. Additionally, groundwater pumping is reduced or turned off as the groundwater head drops below the well screen (Niswonger et al., 2011). Approximately 62 pumping wells are located within each command area (Fig. 3); thus, roughly four fields (i.e., cells) are served by a single well. Within the MODFLOW framework, pumped volumes are amended to the applied water array in the UZF1 package (the parameter 'FINF'), where it then flows back to the stream network as tail-water runoff, is lost as ET, contributes to deep percolation leading to recharge, or is stored in the unsaturated zone.

Reservoir configuration

A broad and shallow reservoir with a capacity of 74 million m^3 (60,000 ac·ft) is simulated using the lake (LAK3) package (Merritt and Konikow, 2000). Prior to this integration, the ability to simulate distinct storage accounts within a MODFLOW lake (or reservoir) did not exist.

The reservoir is incised into layer 1 and, as such, is in hydraulic connection with both layer 1 (around the perimeter of the lake) when it is near full-stage and layer 2 (below the lakebed) during any stress period in which it is not empty. Lakebed conductance is set to a constant value of 0.01 m d^{-1} that is typical of fine sediment that settles at the bottom of reservoirs. In circumstances when the reservoir is at full stage and inflow exceeds the MODSIM determined release amount, reservoir water spills through an SFR2 segment with a bed elevation equal to the crest stage of the reservoir. A 151-point stage-capacity-surface area lookup table is provided in an ancillary support file used by MODFLOW to determine the amount of surface-area dependent inflow and outflow from precipitation and evaporation, respectively.

Investigation of alternative management scenarios

No reservoir scenario

River operations simulated in the first scenario allocate run-of-the-river water into the ditches serving CA1-CA4. As the scenario name indicates, MODSIM does not simulate reservoir operations in this simulation, and the LAK package is not active in MODFLOW. Relative priorities among the ditches, as well as the minimum instream-flow requirement, are the only management features bearing upon the diversion amounts. As flow entering the model ebbs (i.e., spring runoff transitioning into lower summer flows), MODSIM curtails junior water-right diversions to deliver the available water supply to the senior water rights. In so doing, MODSIM simulates "calls on the river." This scenario is hereafter referred to as NR (Table 5).

Table 5: Description of simulated scenarios selected to illustrate the importance of tightly-coupling river-operation models to groundwater-simulation models.

Management-scenario abbreviated name	Management-scenario description
NR	"No reservoir." Scenario allocates run-of-the-river water among 5 priorities, including CA1 – CA4 and the minimum instream-flow requirement.
WR	"With reservoir." Same as NR, but includes an on-stream reservoir located upstream of the first diversion (Fig. 3). All other model settings, including relative ditch priority, minimum instream-flow requirement, and aquifer parameters remain the same. Junior storage rights are added to the system to store any unused direct flow.

With reservoir scenario

The second scenario investigates the impact of a reservoir on the total ditch deliveries and investigates hydrologic-operations feedback via the groundwater system. In this scenario, hereafter referred to as "WR" (Table 5), CA1 – CA3 are provided reservoir-storage accounts with junior storage rights, as summarized in Table 4. Owing to the junior status of the storage rights in this scenario, water rights senior to the storage rights (water rights 10 and 11 in Table 4) should not be injured by diversions to surface-water storage accounts. Thus, even though CA4 does not have reservoir storage, it should not be negatively impacted by reservoir-storage operations due to its senior status. It is possible that under this scenario, CA4 may benefit from additional groundwater returns caused by increased irrigation associated with storage releases. The WR scenario, and, specifically, the differences in CA4's diverted amount, serves as an important check that the MODSIM-MODFLOW coupling is working properly, as CA4 should not be negatively impacted by reservoir capture.

Model-input modifications for simulating the addition of a reservoir were minimal. Because the MODSIM GUI readily accommodates the insertion of new nodes anywhere in the network, a reservoir node was added to the stream-tributary-ditch network with no further modification anywhere else in the MODSIM model. Likewise, the LAK package was activated in MODFLOW for representing reservoir-aquifer interaction. Simulated gains and losses in the LAK3 package, including stream inflow, surface-area dependent values such as precipitation and evaporation, and GW-SW exchanges, are passed to MODSIM. MODSIM determines the release amount that is then passed back to MODFLOW and the two codes continue to iterate until convergence according to Equation (6).

In the prior-appropriation doctrine, storage rights facilitate accrual of run-of-the-river water to storage accounts according to their global priority. Storage rights are defined by annual limits and, in some cases, are accompanied by a flow-rate limit. Once the storage right has stored the annual limit, it is out of priority until the next water year. Water in storage accounts is available to its owners as a supplemental water source at any time based on the owner's needs. In this example, a dual-storage ownership is used to simulate shared storage ownership for the Global Priority 10 (Table 4) between CA1 and CA2. In this modeling construct, each owner is entitled to a portion of total storage. In this case, one-quarter of the storage is assigned to CA1 and three-quarters of the storage is assigned to CA2. MODSIM uses the percent of storage ownership to split the storage accrual among the storage accounts. Furthermore, MODSIM simulates storage releases as a supplemental supply source based on demand shortages that remain after allocation of the direct flows. Within this storage-ownership modeling mode, each storage account is operated independently, based on the supply needs for each owner. MODSIM operates releases from storage accounts and guarantees deliveries to the corresponding storage owners to meet supply needs.

RESULTS AND DISCUSSION

Because there is a computational cost due to iteration between MODSIM and MODFLOW, it is important to evaluate the degree of nonlinearity among diversions and GW-SW exchanges that are reflected by differences in the solution between the first and the last iteration. It is to be expected that the degree of nonlinearity, and therefore the sensitivity of the solution to iterations, would vary from one problem to the next. For example, Niswonger et al. (2014) found nonlinear feedbacks between reservoir releases, diversions, and water flowing to a terminal lake to be significant for the Walker Basin in Nevada. Nonetheless, as the example problem tested herein is typical of semi-arid alluvial basins, the degree of nonlinearity found in this example likely is representative of many agricultural settings around the world. The degree of nonlinearity expressed in this example also provides an evaluation of previous studies that have relied on explicit formulations (e.g., Valerio et al., 2010). To explore this question, iterative results (Fig. 6) are examined for three stream segments located near the center of the model (Fig. 3). We note here that neglecting nonlinearities results in incorrect simulated deliveries to farms, and is therefore indicative of overall mass-balance errors that arise in explicit, versus implicit, formulations.

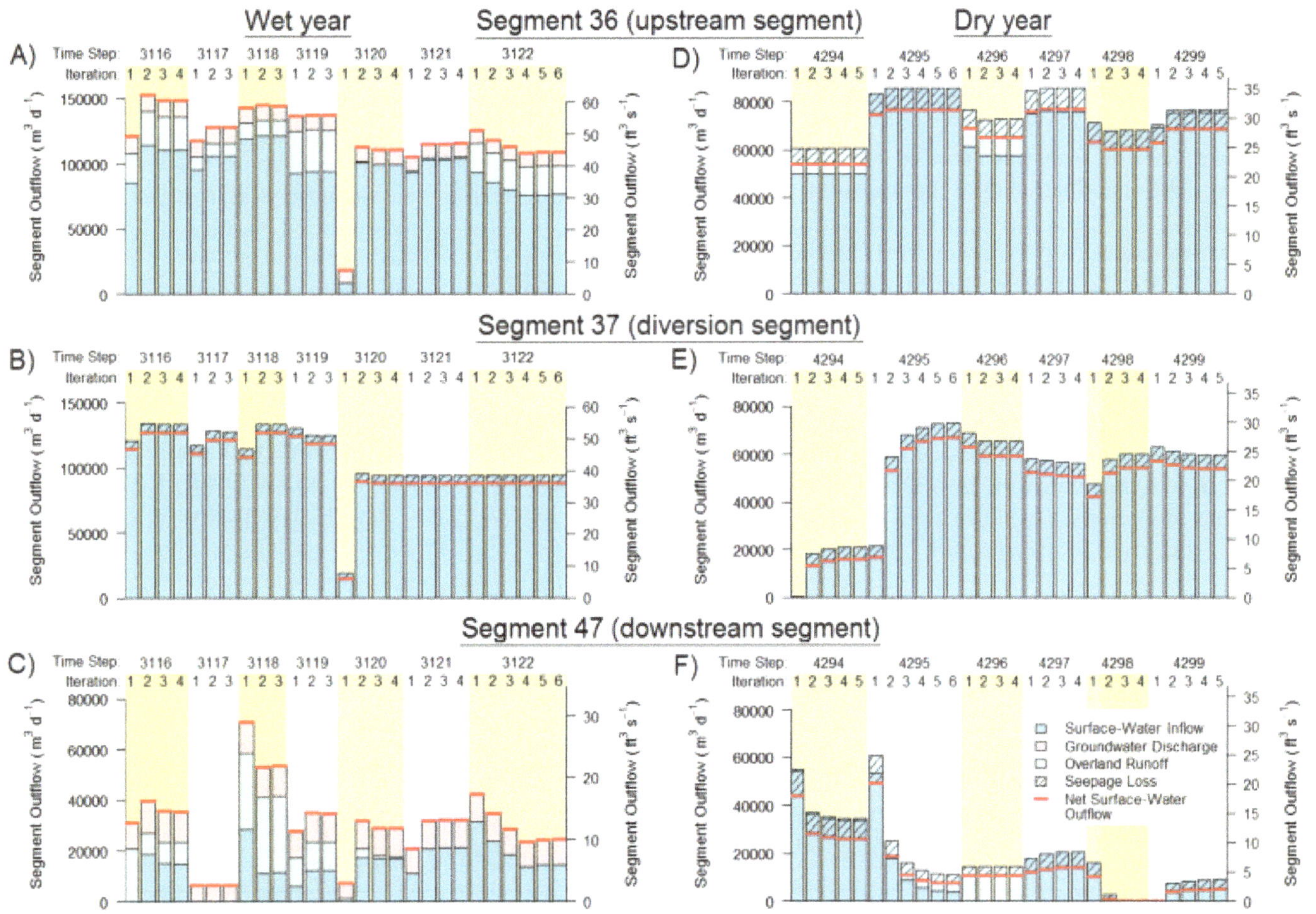

Figure 6: Changes in model-calculated inflows and outflows to segments 36, 37, and 47 (highlighted in Fig. 3) with each MODSIM-MODFLOW iteration. The stacked bars comprise the total inflow to the segment from all sources (surface-water inflow, groundwater discharge, and overland runoff), whereas the red lines represent the net surface-water outflow from the segment, which in many cases are less than the height of the stacked bars because of seepage losses. In sub-figures B and E, the height of the blue bar represents the diverted amount as calculated by MODSIM. (For interpretation of the references to color in this figure legend, the reader is referred to the web version of this article.)

Investigating operations responses to inclusion of hydrologic model feedbacks

Fig. 6 highlights changes in the diverted amounts as iterations progress. Surface flows are shown for the segment immediately upstream of the CA4 diversion (Fig. 6A, D), the flow diverted into CA4 (Fig. 6B, E), and the segment immediately downstream (Fig. 6C, F) of the CA4 diversion. Locations of the segments are highlighted in Fig. 3. Results are presented for both a wet year (Fig. 6A–C) and a dry year (Fig. 6D–F) for the NR scenario. An important difference between these two years is the nature of the GW-SW interaction in segments 36 (Fig. 6A, D) and 47 (Fig. 6C, F), upstream and downstream of the diversion, respectively. During the selected wet year (1998, 3rd wettest year of the simulation; Fig. 4), groundwater discharges to the stream, whereas during the selected dry year (2001, 3rd driest year of the simulation; Fig. 4), a net loss of surface water to groundwater in segments 36 and 47 occurs (note the cross-hatching in Fig. 6D, F). In addition to the GW-SW interaction, other sources of flow in the segments include total surface flow entering that segment from an upstream segment and runoff derived from irrigation events or excess precipitation. The relative contribution of each source of water is highlighted by the stacked bars. Horizontal red lines in Fig. 6 reflect the net surface-water outflow from that segment for every iteration. Thus, Fig. 6 qualitatively depicts MODSIM-MODFLOW model convergence when the value of the red line stabilizes for two consecutive iterations. Note that Fig. 6 shows flow conditions above, within, and below only one of the four diversion points within the NR scenario; the L^2 *Norm* convergence criteria also was satisfied for the other three diversions. Also note that inFig. 6 the displayed values may appear to have converged between two iterations that are not the last two iterations (e.g., time-step 3121, iterations 2 and 3 in Fig. 6A–C), yet the model continued to iterate. This is due to the influence of the other diversions not plotted in Fig. 6 that have not converged.

Over the entire simulated period, the integrated code required an average of approximately 3.2 MODSIM-MODFLOW iterations to converge each time step. Considering only the growing season (April 1st – October 31st), the only dates when diversions are allowed in either scenario, the integrated code used an average of 4.7 iterations to converge each time step. To contrast the integrated approach with a single-iteration feed-forward approach (Sieber and Purkey, 2007, Valerio et al., 2010 and Yates et al., 2005), Table 6 provides statistics related to the percent change in the simulated diversions (CA1–CA4) between the first and last iterations for which diversions occur.Results from time steps such as 3120, 3122 (Fig. 6A–C), and 4295 (Fig. 6D–F) highlight the potential pitfall of using a single feed-forward approach. That is, had the lagged MODFLOW results been used by the river-operations model without further iteration, flow downstream (net surface-water outflow) of segment 47 would have been under predicted by 74% in time step 3120 (change between iterations 1 and 4) and over predicted by 74% in time step 3122 (change between iterations 1 and 6) and 560% in time step 4295 (change between iterations 1 and 6). Alternatively, had one iteration between the operations and hydrologic model been completed within each of these three time steps, outflow from segment 47 would have been over predicted by a more modest, but still substantial, 10%, 41%, and 149% in time steps 3120, 3122, and 4295, respectively. The need to iterate between an operations and hydrologic model is clearly demonstrated by the magnitude of the results reported in Table 6. In CA3, 5% of the time steps with non-zero diversions decreased (less than the 5th percentile) by as much as −35.7%, equating to reductions in flow equal to or below −0.21 $m^3 s^{-1}$ (−7.6 $ft^3 s^{-1}$). Conversely, 5% of CA3's time steps with non-zero diversions increased (greater than the 95th percentile) by 39.2%, a flow amount equal to or above 0.21 $m^3 s^{-1}$ (7.6 $ft^3 s^{-1}$). For CA4, the 5th and 95th percentile change in flows were −27.0% and 48.8%, respectively, meaning that 10% of the time steps (time steps below the 5th percentile and above the 95th percentile) with non-zero diversion amounts were reduced by at least −0.18 $m^3 s^{-1}$(6.4 $ft^3 s^{-1}$) or increased by at least 0.26 $m^3 s^{-1}$ (9.2 $ft^3 s^{-1}$) between the first and last iterations. Were the differences between the first and last iterations accumulated as volumes, these differences would be equivalent to 5.0 million m^3 (4060 ac·ft), 4.7 million m^3 (3810 ac·ft), 17.3 million m^3 (14,060 ac·ft), and 26.1 million m^3 (21,120 ac·ft), in CA1–CA4, respectively. These volumes are equivalent to depths spread over the entire 4000-ha (9900-ac) irrigated area (agricultural fields) of each Command Area, CA1–CA4, of 0.13 m (0.4 ft), 0.12 m (0.4 ft), 0.43 m (1.4 ft), and 0.65 m (2.1 ft), respectively.

Table 6: Statistics of the percent change in total inflow for the identified diversions between the indicated iterations for the NR scenario. (TS, time step).

		First-to-last iteration changes, entire simulation				
Ditch[a]	No. of TS with non-zero diversions[b]	Minimum	5th %ile	Mean	95th %ile	Maximum
CA1	1428	−99.9%	−20.2%	3.7%	17.1%	1020%
CA2	1645	−94.6%	−11.4%	5.5%	15.9%	3170%
CA3	2463ss	−97.9%	−35.7%	6.6%	39.2%	6500%
CA4	3519	−98.1%	−27.0%	9.7%	48.8%	3020%
		First-to-last iteration changes in Fig. 6A–C				
Segment	No. of TS considered	Minimum		Mean		Maximum
36	7	−13.5% (TS 3122)		74.3%		490% (TS 3120)

37	7	−4.1% (TS 3119)		78.8%		519% (TS 3120)
47	7	−42.4% (TS 3122)		44.5%		288% (TS3120)
		First-to-last iteration changes in Fig. 6D–F				
Segment	No. of TS considered	Minimum		Mean		Maximum
36	6	−5.2% (TS 4296)		0.5%		8.9% (TS 4299)
37	5[c]	−5.7% (TS 4299)		46.7%		217% (TS 4295)
47	5[d]	−100% (TS 4298)		−41.6%		18.1% (TS 4297)

a Values for each ditch are calculated using the first segment of the ditch, which are approximately one quarter of the total length of each ditch.

b A non-zero diversion is defined as having a non-zero diversion amount for all iterations within a time step. That is, time steps that started with 0 diversion in the first iteration and increased to a non-zero value, or, alternatively, started with a non-zero value in the first iteration, but then went to zero–an example of which can be seen in time step 4298 (Fig. 6F)–are not factored in to the reported statistics due to division by zero errors.

c Time step 4294 not included in the analysis owing to initial diversion amount of zero.

d Time step 4299 not included in the analysis owing to initial diversion amount of zero.

Results shown in Fig. 6B and E highlight the smoothly varying changes in the calculated diversion amounts between each iteration within a time step. While it cannot be mathematically proven that the final solution within each time step is the global optimum given the nonlinear feedbacks between the iterating models, the observed behavior (generally monotonic convergence) provides confidence that the solution space is dominated by a global solution.

Investigating hydrologic responses to inclusion of river operations

The test model is representative of a typical basin in the semi-arid western United States. As these basins are typically supply limited due to the persistence of below-average years of precipitation separated by infrequent above-average precipitation, groundwater overdraft is becoming a growing problem as populations increase in the west (Dettinger et al., 2011). Developing additional water supplies will be an important component (although challenging) for increasing society's resilience to future drought. As demonstrated here, the new MODSIM-MODFLOW model provides a powerful tool for evaluating options for increasing water supply, changes in irrigation efficiency, and various administrative decisions including evaluation of the potential benefits of design options as well as possible negative impacts to subsets of the stakeholder community (e.g., third-party impacts). An important concept of the water-supply problem is the assessment of sustainability in conjunctive-use systems. For example, in regions where groundwater extraction is unsustainable, as illustrated by steady long-term downward trends in groundwater levels exhibited by many basins in the western United States (Faunt et al., 2009), it is important to evaluate how enhanced water supply increases sustainability of groundwater supply. However, in order to address this problem, aquifer storage must be represented in the modeling framework. Without explicit consideration of groundwater through a model such as MODFLOW, groundwater sustainability cannot be assessed in the context of operations modeling. Thus, a new and powerful capability is provided by the integrated MODSIM-MODFLOW model that allows practitioners to connect reservoir and river operations to groundwater sustainability. In the following examples, we present simulations that compare water resources within the test model for cases with no reservoir and with a reservoir, designated as NR and WR, respectively.

Design of the NR and WR scenarios (Table 5) facilitates a more robust exploration of the hydrologic impacts resulting from development of new water storage and change in irrigation practices. The input files for the investigated scenarios are very similar, with the notable exception that the reservoir capacity was set equal to zero in the MODSIM model in the NR scenario. Thus, despite being present in the NR scenario, the zero capacity assigned to the reservoir forced the network-flow optimization solution in MODSIM to 'release' all of the water flowing into the reservoir.

Among the most notable hydrologic impacts resulting from inclusion of the reservoir are the capture of excess streamflow during the irrigation off-season months of November through March (Fig. 7A), the build-up of storage within the reservoir that generally occurs from November through March (Fig. 7B), and subsequent release of stored water during the spring recession from April through June (Fig. 7A). In Fig. 7A, the red hydrograph generated by the NR scenario is representative of natural flow at the location identified by the red dot in Fig. 3, whereas the blue hydrograph reflects the capture (zero flow) and release (blue cross-hatching) of stored water. Owing to the model set up, 100% of the natural streamflow is captured during winter, hence the zero flow associated with the blue hydrograph in winter. Despite the unrealistic nature of the simulated conditions, a minimum instream-flow requirement downstream of the dam was not enforced to provide a convenient check of the integrated code (that is, zero wintertime releases). During these periods, storage in the two reservoir accounts is accrued (Fig. 7B) based on the storage priority shown in Table 4. Because no storage targets were used, flows will bypass the reservoir (spill) when the storage accounts are full (Fig. 7B). Although reporting specific flows like these has little meaning in this hypothetical context, it points to the importance of runtime decision-making by a river-operations model integrated with a hydrologic model. That is, comparison of these two scenarios would be nearly impossible in a hydrologic-model only (i.e., MODFLOW-only) simulation due to the fact that storage releases and diversions require a priori specification when total storage volumes are not yet known.

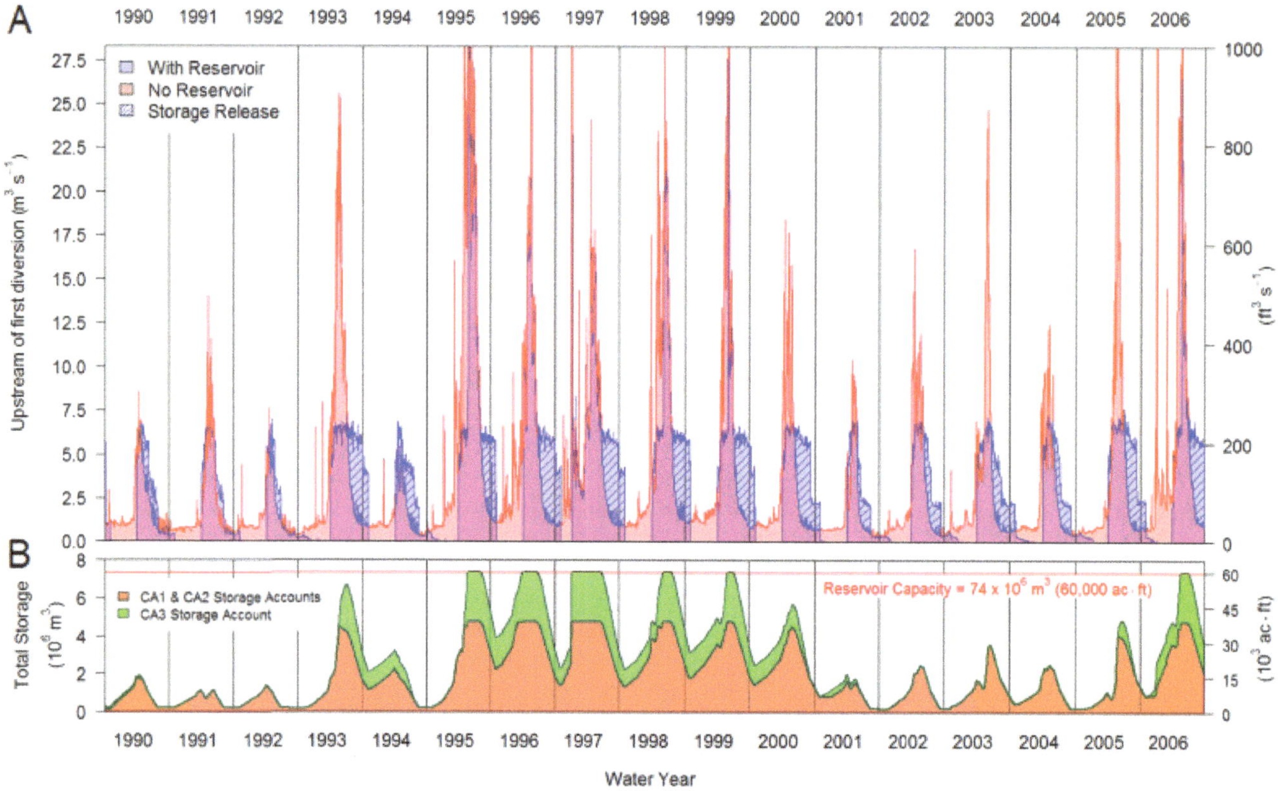

Figure 7: (A) Simulated hydrograph in the stream segment immediately downstream of the dam and (B) the volumes stored in reservoir accounts resulting from simulated operations of the dam by MODSIM. Vertical lines delineate water years, which extend between October 1st and September 30th each year.

The WR scenario highlights an important aspect of river operations that is not typical of hydrologic models, with important implications for large real-world problems: Even though a single flow rate is calculated for each stream reach, two classifications of water are routed downstream by the integrated model. The first, previously alluded to as run-of-the river water, is perhaps more commonly referred to as "natural flow." The second class of water, sometimes referred to as "storage water," is water that is stored in an account, owned by a water user, and delivered based on that owner's need. As such, this stored water is divorced from the prior-appropriation rules governing allocation of natural flows. Thus, in the WR scenario, MODSIM directs MODFLOW to release stored water to be routed along with natural flows, and guarantees that storage water is delivered to its owner through manipulation of flows at diversion points, regardless of its location in the basin.

Many additional hydrologic effects resulting from river operations are experienced downstream. For example, simulating the dynamically-operated reservoir within the hydrologic model to capture, store, and release peak flows during hot summer months when natural flows typically wane led to the increased diversions shown in Fig. 8. Boxplots to the right of each time series summarize the annual residuals between the two scenarios. In general, appreciable increases, averaging roughly 35%, 52%, and 26% in CA1, CA2, and CA3, respectively, result from the inclusion of the reservoir. This, in turn, leads to additional seepage losses along the delivery ditches, more recharge from surface water, and less supplementary pumping. The larger increases in CA1 and CA2 result from their senior storage water rights relative to the account assigned to CA3. The lack of increase in the CA4 diversion amounts is due to the assumption that CA4 does not take stake in the reservoir storage, which is something to be expected from a senior water-right holder. Interestingly, the addition of a reservoir resulted in an average reduction of 2% by the CA4 diversion, with the largest delivery decreases occurring in the first five years of the simulation, a relatively dry period (Fig. 4). An un-anticipated result like this, commonly referred to as a 'third-party effect' (Gould, 1988 and Pease, 2012), may be enough to stop a new reservoir project from launching owing to the injuries caused to senior rights not involved in the project. A closer examination of the flows upstream of the CA4 diversion showed the integrated code attempting to deliver just enough water to meet the full right of CA4 while also filling reservoir accounts during the spring runoff, thereby preventing peak flows from moving through the system. Under this type of management, the integrated hydrologic model resulted in slightly smaller river flows reaching CA4 from the beginning of July through the end of the irrigation season, possibly due to altered GW-SW interaction upstream of CA4 following consistently lower flows associated with reservoir capture (i.e., altered bank storage and release associated with diminished peak flows).

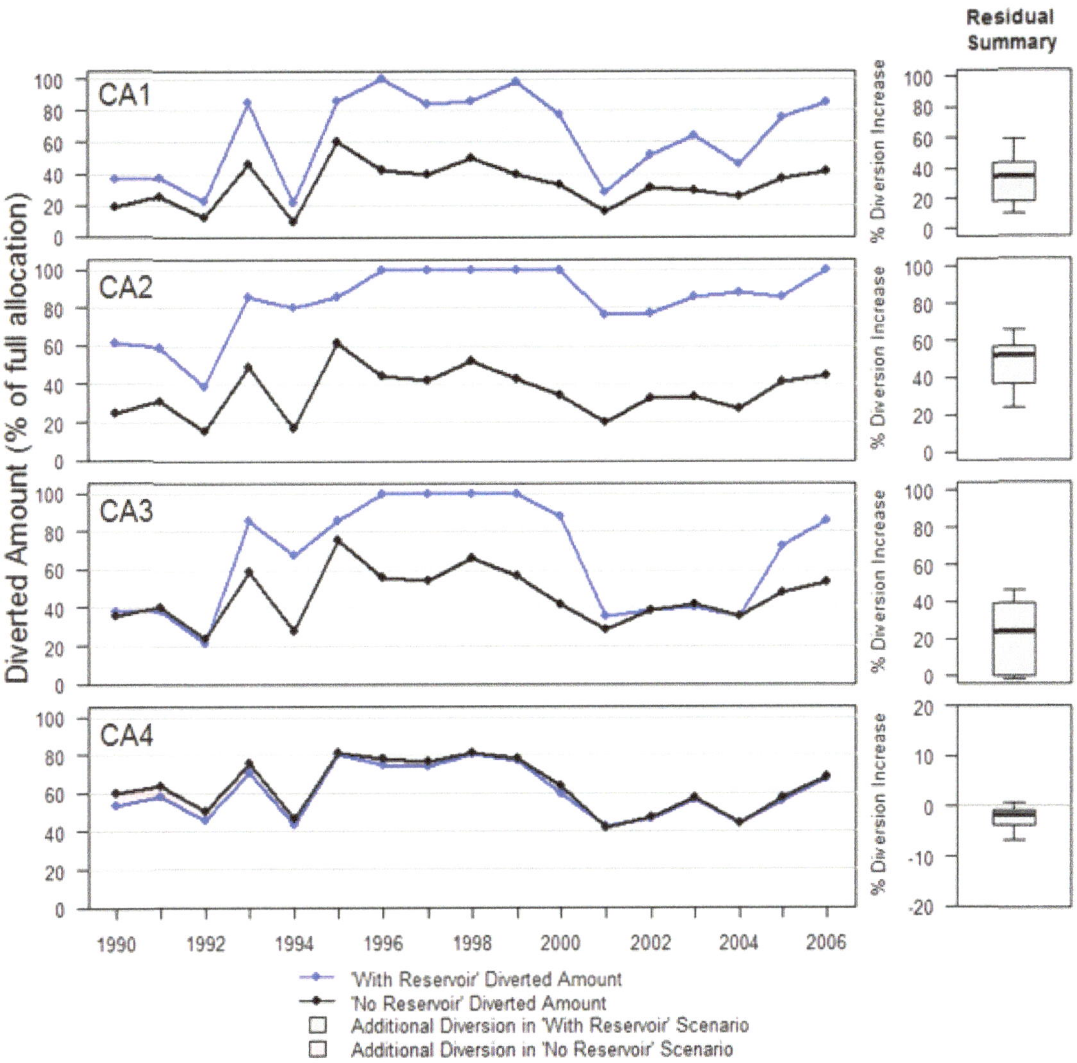

Figure 8: Total diversion amounts for the four command areas in both the NR and WR scenarios expressed as a percentage of full allocation, where full allocation would be the amount of water delivered when the ditch is running at capacity for the entire irrigation season (April 1st–October 31st). Boxplots to the right of the graphs show statistical summaries of the difference between the 'No Reservoir' and 'With Reservoir' simulations.

To reinforce the importance of accurately accounting for GW-SW exchange within river-operations modeling, Fig. 9 shows the GW-SW interaction on the same arbitrary day for two very different years in the NR scenario. Upward pointing blue bars highlight cells with groundwater discharge; downward pointing red bars show cells with seepage that recharges the alluvial aquifer. Bar height depicts the relative magnitude of the GW-SW exchange. Among the most noticeable differences between the two years is along the tributary stream segment, where significantly more groundwater discharged during the wet year (Fig. 9A). As a result, greater groundwater discharge in the tributary bolsters streamflow at the confluence of the main stem and tributary, the site of the minimum instream-flow requirement. With the additional streamflow from the tributary satisfying the minimum instream-flow requirement, flows in the main stem previously designated for satisfying the minimum instream-flow requirement can instead be diverted to support irrigation demands. Thus, the impacts of climate on GW-SW interactions that are accounted for in MODFLOW provide a more realistic representation of climate impacts on water operations. The increased seepage and recharge associated with the enhanced diversions will in turn bolster future groundwater discharge to streams, and in this way, the 'memory' of the groundwater system is more accurately and appropriately accounted for in the simulation of river operations.

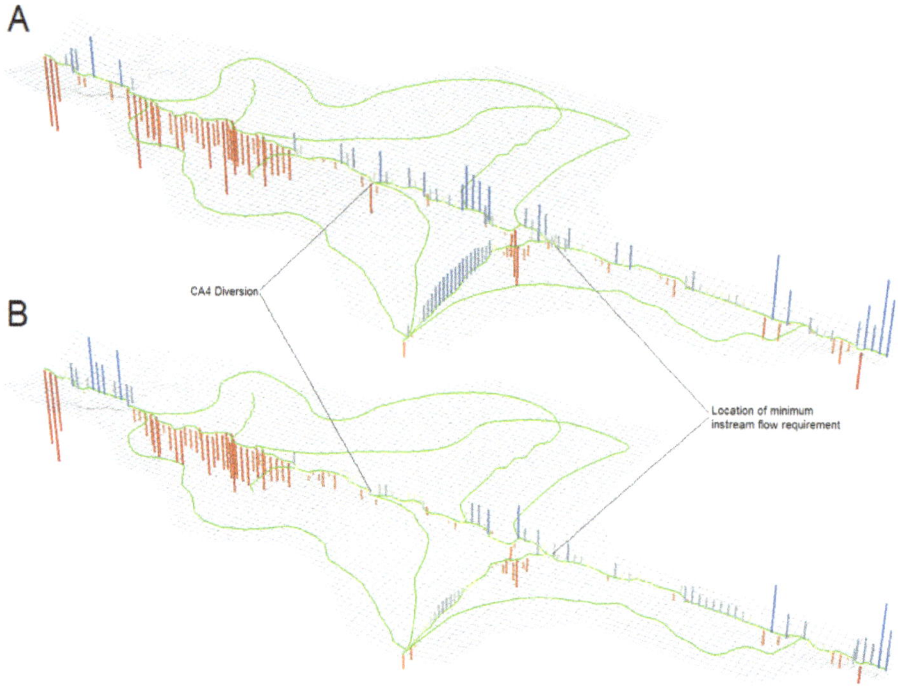

Figure 9: Perspective plots showing the spatially distributed (cell-by-cell) GW-SW interaction on an arbitrary day (April 21st) during a (A) wet year (1998) and (B) dry year (2001). GW-SW interaction is only shown for the main stem river and tributary and are from the NR scenario. Blue bars indicate groundwater discharge to the river or tributary; red bars indicate streamflow seepage to the aquifer. On the day for which results are shown, the net GW-SW exchange during the wet year resulted in a loss of 16,900 m^3 (13.7 ac ft) along the main stem, an amount equivalent to approximately 4% of the surface flow exiting the model. For the same day but during the dry year, the main stem lost a net of 9300 m^3 (7.5 ac ft), equivalent to about 11% of the surface flow exiting the model on that day. In the tributary, the net groundwater discharge for the day was 8500 m^3 (6.9 ac ft) and 1100 m^3 (0.9 ac ft) during the wet and dry years, respectively, amounts that are equivalent to 9.5% and 5.4% of the flow in the tributary at the confluence with the main stem. (For interpretation of the references to color in this figure legend, the reader is referred to the web version of this article.)

Increased groundwater discharge in the upstream area of the model during the dry year (Fig. 9B) is likely a result of lower river stage during the dry year, thereby allowing more groundwater to discharge back to the river. Very similar seepage losses between the wet and dry years are seen just downstream of the reservoir. Because these results are from the NR scenario, the reservoir accounts for boosting CA1 and CA2 diversions are not available. Limitations in surface-water supplies lead to increased supplementary pumping in the upper reaches of the model, which in turn leads to the relatively large surface-flow losses seen in both Fig. 9A and B. Streamflow capture by wells is significant even during wet years when supplementary pumping is low owing to the long-term effects of pumping on streamflow capture. Alluvial basins, as represented in the test model, often exhibit multi-decadal cumulative response to groundwater pumping, and drawdown recovery is similarly a very slow process (Barlow and Leake, 2012 and Konikow and Kendy, 2005). Accordingly, it is important to represent these aquifer systems using the three-dimensional groundwater flow equation (e.g., MODFLOW).

Fig. 10 highlights the spatio-temporal interdependence between the hydrologic system and the water-management operations. Whereas both of the simulated scenarios start out with the same average depth to the water table (*Dwt*) under each of the four command areas, the effects of increased supplementary pumping in the NR scenario versus additional recharge with surface-storage supplies in the WR scenario on the average *Dwt*are immediately recognized in CA1–CA3. While these results illustrate the benefits of reservoir storage, they also highlight the ability of the integrated code to simulate both surface and groundwater supplies. As the preponderance of below-average years of precipitation stack up during the latter part of the simulation period, downward trends in groundwater storage develop, and suggest that this system, similar to many basins in the west, cannot sustain historical water use without a significant increase in precipitation above what has occurred during the period 2000–2006. Under the NR scenario, for example, the average pumping increase for CA1–CA4 with respect to the WR scenario was 27%, 51%, 29%, and −2%, respectively. As a result, groundwater levels under CA3 clearly switched from an apparently sustainable 'dynamic equilibrium' to an unsustainable downward trend in groundwater levels. Similarly, groundwater-level declines in CA1 and CA2 are accelerated during the second half of the simulation period due to increased pumping in the NR scenario. Under CA4, the *Dwt* remained relatively the same in both scenarios. The slight increase in pumping under the WR scenario is related to the slight decreases in delivered surface water as shown in Fig. 8. An interesting aspect of Fig. 10 is that the pumping rates, as well as the *Dwt* levels, are the synchronized response based on the simulated river operations. We note here that the results shown inFig. 10 could not be attained without the integrated code, and that these results illustrate, in a way that previous codes could not, the benefits to the sustainability of groundwater of developing a surface reservoir in this conjunctive-use basin. Thus, the integrated MODSIM-MODFLOW code facilitates a fuller assessment of the hydrologic response to any arbitrary management decisions made with surface-water or groundwater supplies, and associated feedbacks.

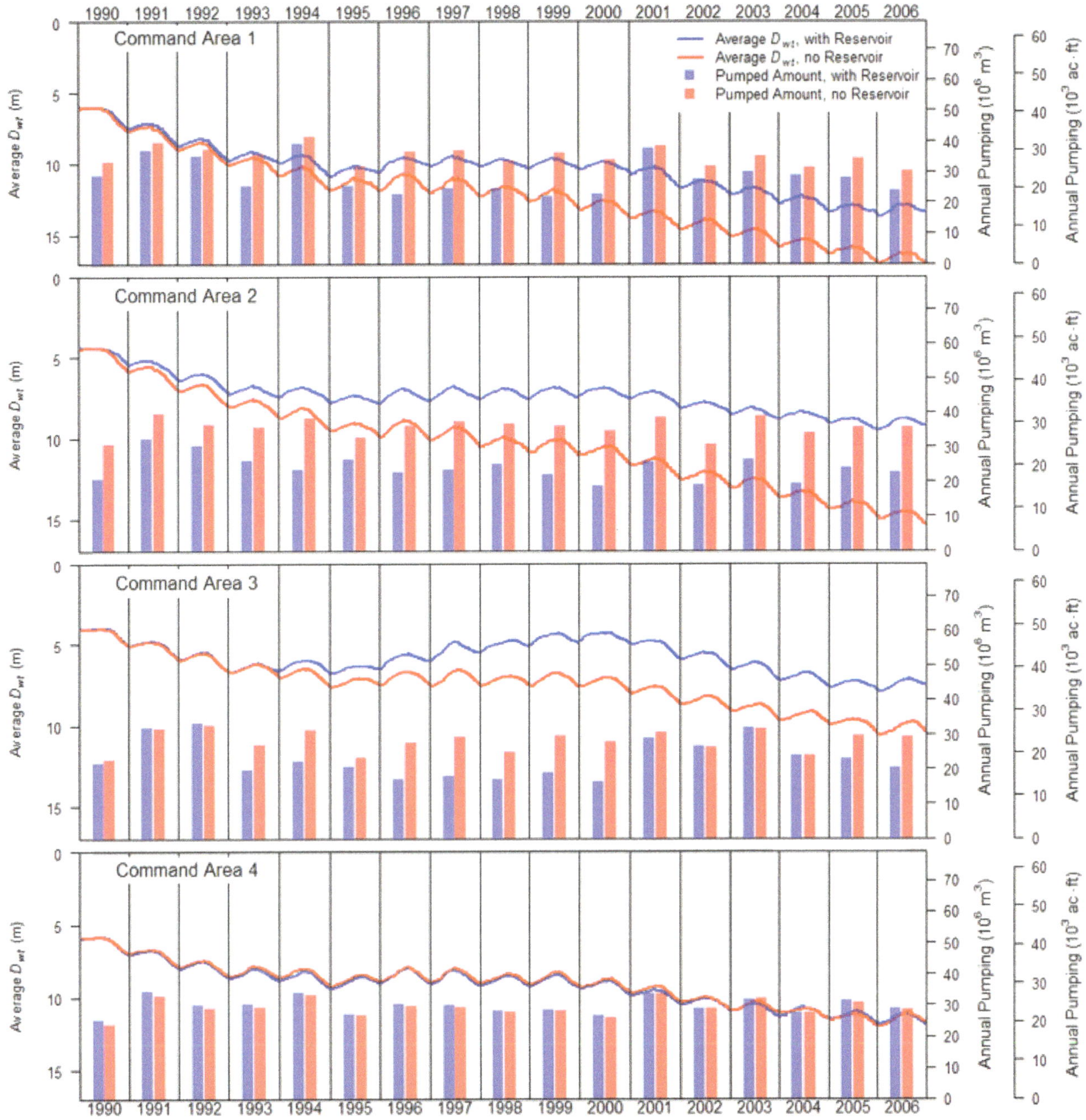

Figure 10: The impact of altered operations on prevailing hydrologic conditions as predicted by the MODSIM-MODFLOW integration. Lines show the average water-table depth under each of the four command areas (depth scales shown on left-hand y axes); bars show the total annual pumping under each of the command areas for the indicated scenario (pumping scales shown on right-hand y axes).

Another aspect of this simulation that is important is the assumption that groundwater pumping is reduced as groundwater levels fall below the well screens, which are assumed to extend to the bottom of the layer from which each well pumps. Thus, within these simulations, groundwater pumping is curtailed as wells dry, and the simulations do not consider drilling of new deeper wells. However, in some regions, such as the Central Valley of California, water users have drilled deeper wells to continue pumping after their original wells have become dry. Thus, groundwater depletion by wells in these simulations is conservative relative to regions where wells are drilled to accommodate deeper groundwater levels. Fig. 11 shows the cumulative pumping curtailment for both scenarios and for each CA. In the NR scenario (Fig. 11A), pumping curtailment in CA1 was approximately 23 million m^3 [~18,000 ac ft, approximately enough water to cover CA1 in 0.6 m (2 ft) of water] by the end of the simulation. With the benefits of reservoir storage, however, pumping curtailment within CA1 was lowered by roughly 87% to 3 million m^3 (~3000 ac ft) in the WR scenario (Fig. 11B). In addition to reducing supplemental pumping requirements, recharge is increased by virtue of the increased surface-water supply. In both scenarios, pumping in CA4 was curtailed by roughly 18 million m^3 (15,000 ac ft) owing to CA4's lack of access to surface water storage. Although this is only a hypothetical example, results such as this further demonstrate the benefit of integrating river-operations and hydrologic models at the time-step level. That is, assessments regarding the state of the hydrologic system, particularly the groundwater system, reflect the management of the river that was heretofore strained by explicit-type coupling between operations and hydrologic models or simpler representations of systems where surface water and groundwater is used conjunctively.

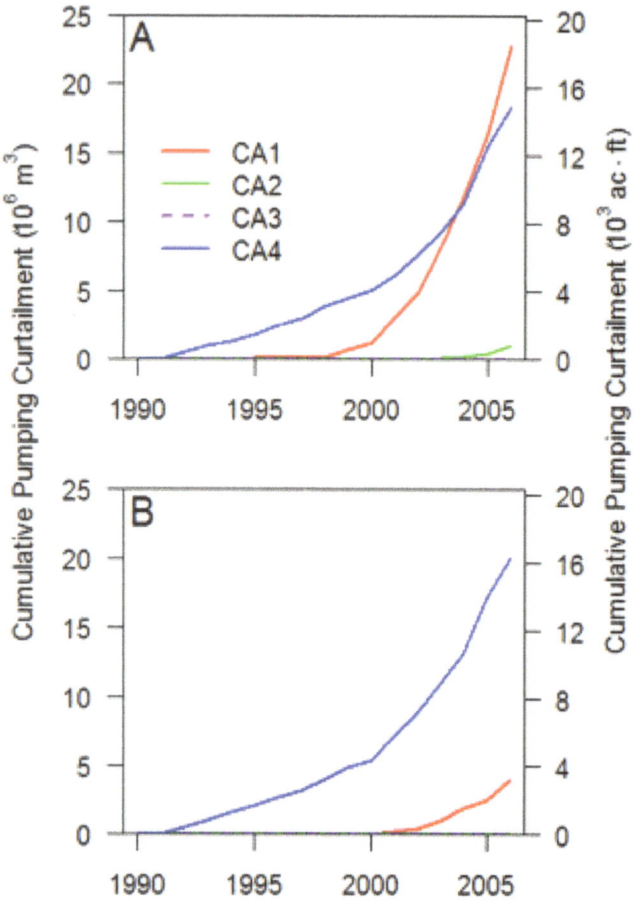

Figure 11: Cumulative pumping curtailment under each Command Area for the (A) NR and (B) WR scenarios.

CONCLUSIONS

While this analysis emerges from a hypothetical test model, it demonstrates the impact that physically-based distributed-parameter modeling has on river-operations modeling, and vice versa. The spatio-temporal representation of GW-SW exchanges as simulated by a hydrologic model, in this case MODFLOW, has a clear impact on simulated river operations. Previously, less accurate approximations of GW-SW exchanges, including loosely-coupled models (e.g., "feed-forward" approaches), analytical solutions with their various assumptions, complete omission of groundwater, or simply lumping GW-SW exchanges in with all other unmeasured gains and losses and assuming the same GW-SW exchanges for different management conditions, were commonly applied. However, through simulation of GW-SW interaction with a physically-based distributed-parameter hydrologic model, a river-operations model is equipped to more accurately account for finite water resources, and in particular, finite groundwater resources in conjunctive-use systems. Other approaches, such as those named above, are in many situations unsuitable for examining water resources in water-scarce, over-appropriated, conjunctive-use basins that water-resource engineers must now manage.

A distinct advantage of the MODSIM-MODFLOW integration, beyond the fact that it takes advantage of two long established models within their respective genres, is that the movement of information between the codes is via computer memory in compiled codes and can occur several times within each simulated time step, making the tool presented computationally efficient. Thus, simulated operational decisions are based on surface flows that are synchronized with GW-SW interactions resulting from the current, as well as all previous, management (i.e., diversions and releases) decisions. Through the synchronized solutions, 'what-if' scenarios and/or future forecasts run with an integrated operations-hydrologic model like MODSIM-MODFLOW more accurately account for the non-linear feedbacks and the cumulative effects of those feedbacks on yet unrealized operational decisions. Measured or inferred aquifer property heterogeneities, spatially and temporally variable stresses such as pumping, and, finally, their collective impact on river operations are vetted without relying on analytical approximations or loosely-coupled (i.e., out-of-synchronization) approaches.

In addition, the particular code pairing adopted herein provides the MODFLOW community with access to new and novel functionality, allowing existing MODFLOW applications to be readily expanded to include the prior-appropriation framework (or a similar rule-based system). Conversely, MODSIM users can readily expand applications with the world's most widely used groundwater-modeling software for more accurate representation of GW-SW exchange based on simulated conjunctive use practices.

While this integration effectively integrates hydrologic and operational considerations with detailed accounting of GW-SW interactions by the operations model, this effort does not address integration with economic, social, ecological, or other dimension of water-resource planning and management. However, given the integrated approach described herein, it is possible that one of the aforementioned modeling platforms (e.g., economic, ecological, etc.) could be integrated with MODSIM-MODFLOW such that the within time-step iteration includes a third type of model (e.g., an economic or ecologic model).

ACKNOWLEDGEMENTS

We wish to thank Toby Welborn and Sue Buto with U.S. Geological Survey at the Nevada Water Science Center for their assistance helping to set up the hypothetical test model. We also wish to thank Wil de Jong and Stuart Mentzer for their help sorting out Fortran/C# communications. We would like to thank Paul Barlow with the U.S. Geological Survey Office of Groundwater, and three anonymous reviewers, for their timely and insightful colleague review. We gratefully acknowledge the assistance of Justin Huntington with the Desert Research Institute for providing sensor-based estimates of ET for use by the hypothetical test model and Josh Lee of the Nevada Water Science Center for assisting with the graphics. Research supported by grant from the Water Sustainability & Climate Program jointly funded by the National Science Foundation (1360506) and U.S. Department of Agriculture/National Institute of Food & Agriculture(1360507). Support was also provided by the U.S. Geological Survey's Groundwater Resources Program.

REFERENCES

1. Abrams, R.H., 2004. Arkansas river: one hundred years later, Kansas and Colorado still are at odds. In: Preview U.S. Supreme Court Cases, pp. 29e32.

2. Ahlfeld, D.P., Barlow, P.M., Mulligan, A.E., 2005. GWM-A Ground-water Management Process for the U.S. Geological Survey Modular Ground-water Model (MODFLOW-2000). U.S. Geological Survey, Open-File Report: 2005-1072.

3. Allander, K.K., Niswonger, R.G., Jeton, A.E., 2014. Simulation of the Lower Walker River Basin Hydrologic System, West-central Nevada, Using PRMS and MODFLOW Models, p. 93. U.S. Geological Survey Scientific Investigations Report 2014-5190.

4. Alvarado, R., Bennett, R.R., 2005. State of Colorado's stream simulation model (StateMod). In: Singh, V.P., Frevert, D.K. (Eds.), Watershed Models. CRC Press.

5. ASCE-EWRI, 2005. The ASCE Standardized Reference Evapotranspiration Equation, ASCE-EWRI Standardization of Reference Evapotranspiration Task Committee Report. Available at: http://www.kimberly.uidaho.edu/water/asceewri/.

6. Banta, E.R., 2011. MODFLOW-CDSS, a Version of MODFLOW-2005 with Modifications for Colorado Decision Support Systems, p. 19. U.S. Geological Survey OpenFile Report: 2011-1213.

7. Barlow, P.M., Leake, S.A., 2012. Streamflow Depletion by Wells: Understanding and Managing the Effects of Groundwater Pumping on Streamflow, p. 84. U.S. Geological Survey, Circular 1376.

8. Bennett, R.R., 2012. State of Colorado's Water Supply Model (StateMod), Ver. 13.00.00. Colorado's Decision Support Systems. Colorado Water Conservation Board. Accessed Sept. 1, 2015, Available at: http://cwcbweblink.state.co.us/ WebLink/ElectronicFile.aspx?docid¼158584&searchid¼eb29ce30-08c0-4578- ab80-bf44869d73f0&dbid¼0.

9. Berhe, F., Melesse, A., Hailu, D., Sileshi, Y., 2013. MODSIM-based water allocation modeling of Awash River Basin, Ethiopia. Catena 109, 118e128.

10. Bertsekas, D.P., Tseng, P., 1994. RELAX-IV: a Faster Version of the RELAX Code for Solving Minimum Cost Flow Problems. Completion report for NSF Grant CCR-9103804. Dept. of Electrical Engineering and Computer Science, Massachusetts Institute of Technology, Cambridge.

11. Brekke, L., Kiang, J., Olsen, J., Pulwarty, R., Raff, D., Turnipseed, D., Webb, R., White, K., 2009. U.S. Geological Survey Circular. Climate Change and Water Resources ManagementdA Federal Perspective, vol. 1331, p. 65.

12. Briand, G., Schuck, E.C., Holland, D.W., 2008. Effects of flow augmentations in the snake river basin on farms profitability1. JAWRA J. Am. Water Resour. Assoc. 44 (2), 360e366.

13. Brooks, K.N., Ffolliott, P.F., Magner, J.A., 2012. Hydrology and the Management of Watersheds. John Wiley & Sons.

14. Campbell, S.G., Hanna, R.B., Flug, M., Scott, J.F., 2001. Modeling Klamath River system operations for quantity and quality. J. Water Resour. Plan. Manag. 127 (5), 284e294.

15. Castle, S.L., Thomas, B.F., Reager, J.T., Rodell, M., Swenson, S.C., Famiglietti, J.S., 2014. Groundwater depletion during drought threatens future water security of the Colorado River Basin. Geophys. Res. Lett. 41 (16), 5904e5911.

16. Colorado Revised Statutes, 1949. Arkansas River Compact, Sections 37-69-101-37- 69-106. Colorado General Assembly.

17. Condon, L.E., Maxwell, R.M., 2013. Implementation of a linear optimization water allocation algorithm into a fully integrated physical hydrology model. Adv. Water Resour. 60, 135e147.

18. de Azevedo, L.G.T., Gates, T.K., Fontane, D.G., Labadie, J.W., Porto, R.L., 2000. Integration of water quantity and quality in strategic river basin planning. J. Water Resour. Plan. Manag. 126 (2), 85e97.

19. Dettinger, M.D., Ralph, F.M., Das, T., Neiman, P.J., Cayan, D.R., 2011. Atmospheric rivers, floods and the water resources of California. Water 3 (2), 445e478.

20. El-Beshri, M., Labadie, J., 1994. Optimal conjunctive use of surface and groundwater resources in Egypt. In: Proc. of the 8th IWRA World Congr. on Water Resour: Ministry of Public Works and Water Resources, Cairo, Egypt, Nov. 21e25.

21. Fairbanks, J., Panday, S., Huyakorn, P., 2001. Sept. 16-19, 2001, Vol. 1, pp. 356e361. In: Poeter, E., Hill, M., Doherty, J., Zheng, C. (Eds.), Comparisons of Linked and Fully Coupled Approaches to Simulating Conjunctive Surface/subsurface Flow and Their Interactions. Int. Ground Water Modeling Center, Colorado School of Mines, Golden CO., pp. 356e361

22. Faunt, C.C., Hanson, R.T., Belitz, K., Rogers, L., 2009. California Central Valley Groundwater Study, a Powerful New Tool to Assess Water Resources in California's Central Valley, p. 4. U.S. Geological Survey Fact Sheet, FS2009e3057.

23. Faux, J.C., Labadie, J.W., Lazaro, R.C., 1986. Improving performance of irrigation/ hydro projects. J. Water Resour. Plan. Manag. 112 (2), 205e224.

24. Fredericks, J.W., Labadie, J.W., Altenhofen, J.M., 1998. Decision support system for conjunctive stream-aquifer management. J. Water Resour. Plan. Manag. 124 (2), 69e78.

25. Frevert, D.K., Labadie, J.W., Larson, R.K., Parker, N.L., 1994. Integration of water rights and network flow modeling in the Upper Snake River Basin. In: Fontane, D.G., Tuvel, H.N. (Eds.), 21st Annual Conf., Water Resour. Plng. and Mgmt Div. ASCE. ASCE, Denver, CO.

26. Gleick, P.H., 1987. The development and testing of a water balance model for climate impact assessment: modeling the Sacramento basin. Water Resour. Res. 23 (6), 1049e1061.

27. Glover, R., 1977. Transient Groundwater Hydraulics. Water Resources Publications, LLC., Highland Ranch, CO.

28. Gould, G.A., 1988. Water rights transfers and third-party effects. Land Water Law Rev. 23 (1).

29. Graham, L., Labadie, J., Hutchison, I., Ferguson, K., 1986. Allocation of augmented water supply under a priority water rights system. Water Resour. Res. 22 (7), 1083e1094.

30. Hanson, R.T., Boyce, S.E., Schmid, W., Hughes, J.D., Mehl, S.M., Leake, S.A., Maddock, T., Niswonger, R.G., 2014. One-water Hydrologic Flow Model (MODFLOW-OWHM). U.S. Geological Survey, Techniques and Methods 6eA51, p. 120. Doi: 110.3133/tm3136A3151, Available at:

31. Harbaugh, A.W., 2005. MODFLOW-2005: the U.S. Geological Survey Modular Ground-water Modelethe Ground-water Flow Process, p. 253. U.S. Geological Survey Techniques and Methods 6-A16.

32. Hobbs, G., 2009. Citizen's Guide to Colorado Water Law. Colorado Foundation for Water Education.

33. Houk, E.E., Frasier, M., Taylor, R., 2007. Evaluating water transfers from agriculture for reducing critical habitat water shortages in the Platte Basin. J. Water Resour. Plan. Manag. 133 (4), 320e328.

34. Howe, C.W., 2008. Water law and economics: an assessment of river calls and the South Platte well shut-down. Univ. Denv. Water Law Rev. 12, 181.

35. Hughes, J.D., Langevin, C.D., Chartier, K.L., White, J.T., 2012. Documentation of the Surface-water Routing (SWR1) Process for Modeling Surface-water Flow with the U.S. Geological Survey Modular Ground-water Model (MODFLOW-2005), p. 113. U.S. Geological Survey Techniques and Methods book 6, chap. 40 (version 1.0).

36. Hughes, J.D., Langevin, C.D., White, J.T., 2015. MODFLOW-based coupled surface water routing and groundwater-flow simulation. Ground Water 53 (3), 452e463.

37. Huntington, J., Allen, R., 2010. Evapotranspiration and Net Irrigation Water Requirements for Nevada. Nevada State Engineer's Office Publication, Nevada Division of Water Resources, Carson City, Nevada, p. 288. Available at: http:// water.nv.gov/mapping/et/Docs/ Evapotranspiration_and_Net_Irrigation_ Requirements_ for_Nevada_Compiled.pdf.

38. Jamali, S., Abrishamchi, A., Madani, K., 2013. Climate change and hydropower planning in the Middle East: implications for Iran's Karkheh hydropower systems. J. Energy Eng. 139 (3), 153e160.

39. Jamieson, D., Fedra, K., 1996. The 'WaterWare' decision-support system for riverbasin planning. 1. Conceptual design. J. Hydrol. 177 (3), 163e175.

40. Johnson, J., 2015. MODSIM versus RiverWare: a Comparative Analysis of Two River Reservoir Modeling Tools. U.S. Bureau of Reclamation, Pacific Northwest Region. Final Report 2014.3669. Boise, Idaho Accessed on 16.06.15. at. http://www.usbr. gov/research/projects/ detail.cfm?id¼3669.

41. Konikow, L.F., Hornberger, G.Z., Halford, K.J., Hanson, R.T., 2009. Revised Multi-node Well (MNW2) Package for MODFLOW Ground-water Flow Model, p. 67. U.S. Geological Survey Techniques and Methods 6eA30.

42. Konikow, L.F., Kendy, E., 2005. Groundwater depletion: a global problem. Hydrogeol. J. 13 (1), 317e320.

43. Labadie, J., 1993. Combining Simulation and Optimization in River Basin Management, Stochastic Hydrology and its Use in Water Resources Systems Simulation and Optimization. Springer, pp. 345e371.

44. Labadie, J.W., 2010a. MODSIM 8.1: River Basin Management Decision Support System. User manaul and documentation.

45. Labadie, J.W., 2010b. MODSIM 8.1: River Basin Management Decision Support System: Tutorials and Example Networks. Dept. of Civil Eng., Colo. State Univ., Ft. Collins, CO.

46. Labadie, J.W., Fontane, D.G., Lee, J.H., Ko, I.H., 2007. Decision support system for adaptive river basin management: application to the Geum River basin, Korea. Water Int. 32 (3), 397e415.

47. LaBolle, E.M., Ahmed, A.A., Fogg, G.E., 2003. Review of the integrated groundwater and surface-water model (IGSM). Ground Water 41 (2), 238e246.

48. LaMarche, J., 2001. Upper and Middle Deschutes Basin Surface Water Distribution Model. Surface Water Open File Report #SW02-001. Oregon Water Resources Department. Available at: http://www.oregon.gov/owrd/ pubs/docs/reports/ sw02-001.pdf.

49. Lanini, J.S., Dozier, A.Q., Furey, P.R., Kampf, S.K., 2013. Stochastic method for examining vulnerability of hydropower generation and reservoir operations to climate change: case study of the Dworshak reservoir in Idaho. J. Water Resour. Plan. Manag. 140 (9).

50. Larson, K.B., Ham, K.D., Niehus, S.E., Geerlofs, S.H., Tagestad, J.D., Richmond, M.C., 2014. The Integrated Basin-Scale Opportunity Assessment Initiative: Pilot Assessment for the Deschutes River Basin. Pacific Northwest National Laboratory, Richland, Washington. Available at: http://basin.pnnl.gov/Content/reports/ Deschutes/PNNL-23197_Deschutes_final_report.pdf.

51. Larson, R., Spinazola, J., 2000. Conjunctive management analyses for endangered species flow augmentation alternatives in the Snake River. In: Proceedings of Watershed Management and Operations Management.

52. Law, J.E., Brown, M.L., 1989. Development of a Large Network Model to Evaluate the Yield of a Proposed Reservoir. In: Computerized Decision Support Systems for Water Managers. ASCE, pp. 621e631.

53. Littleworth, A.L., 2003. State of Kansas v. State of Colorado. Fourth Report. Supreme Court of the United States.

54. Luecke, D.F., 2010. Hydrologic models in the Courtroom, 47. Idaho L. Rev, p. 113. Markstrom, S.L., Niswonger, R.G., Regan, R.S., Prudic, D.E., Barlow, P.M., 2008. GSFLOW, Coupled Ground-water and Surface-water Flow Model Based on the Integration of the Precipitation-runoff Modeling System (PRMS) and the Modular Ground-water Flow Model (MODFLOW-2005), p. 240. U.S. Geological Survey Techniques and Methods 6-D1.

55. Marques, G.F., Lund, J.R., Leu, M.R., Jenkins, M., Howitt, R., Harter, T., Hatchett, S., Ruud, N., Burke, S.,

2006. Economically driven simulation of regional water systems: Friant-Kern, California. J. Water Resour. Plan. Manag. 132 (6), 468e479.

56. Merritt, M.L., Konikow, L.F., 2000. Documentation of a Computer Program to Simulate Lake-aquifer Interaction Using the MODFLOW Ground-water Flow Model and the MOC3D Solute-transport Model, p. 146. U.S. Geological Survey Water-Resources Investigations Report 00e4167.

57. Miller, S.A., Johnson, G.S., Cosgrove, D.M., Larson, R., 2003. Regional scale modeling of surface and ground water interaction in the snake river basin. J. Am. Water Resour. Assoc. 39, 517e528.

58. Niswonger, R., Panday, S., Ibaraki, M., 2011. MODFLOW-NWT, a Newton Formulation for MODFLOW-2005, p. 44. U.S. Geological Survey Techniques and Methods 6eA37.

59. Niswonger, R.G., Allander, K.K., Jeton, A.E., 2014. Collaborative modelling and integrated decision support system analysis of a developed terminal lake basin. J. Hydrol. 517, 521e537.

60. Niswonger, R.G., Prudic, D.E., 2004. Modeling variably saturated flow using kinematic waves in MODFLOW. In: Hogan, J.F., Phillips, F.M., Scanlon, B.R. (Eds.), Groundwater Recharge in a Desert Environment. The Southwestern United States. Am. Geophysical Union, Washington, D.C, pp. 101e112.

61. Niswonger, R.G., Prudic, D.E., 2005. Documentation of the Streamflow-routing (SFR2) Package to Include Unsaturated Flow beneath StreamseA Modification to SFR1, p. 48. U.S. Geological Survey Techniques and Methods 6eA13.

62. Niswonger, R.G., Prudic, D.E., Regan, R.S., 2006. Documentation of the Unsaturatedzone Flow (UZF1) Package for Modeling Unsaturated Flow between the Land Surface and the Water Table with MODFLOW-2005, p. 62. U. S. Geological Survey Techniques and Methods 6eA19.

63. Panday, S., Huyakorn, P.S., 2004. A fully coupled physically-based spatially-distributed model for evaluating surface/subsurface flow. Adv. Water Resour. 27 (4), 361e382.

64. Pease, M., 2012. Water transfer laws and policies: tough questions and institutional reform for the western United States. J. Nat. Resour. Policy Res. 4 (2), 103e119.

65. Prudic, D.E., Konikow, L.F., Banta, E.R., 2004. A New Streamflow-routing (SFR1) Package to Simulate Stream-aquifer Interaction with MODFLOW-2000, p. 96. U.S. Geological Survey Open-File Report 2004e1042.

66. Pulido-Velazquez, M., Andreu, J., Sahuquillo, A., 2006. Economic optimization of conjunctive use of surface water and groundwater at the basin scale. J. Water Resour. Plan. Manag. 132 (6), 454e467.

67. Rassam, D.W., 2011. A conceptual framework for incorporating surfaceegroundwater interactions into a river operationeplanning model. Environ. Model. Softw. 26 (12), 1554e1567.

68. Rassam, D.W., Peeters, L., Pickett, T., Jolly, I., Holz, L., 2013. Accounting for surfaceegroundwater interactions and their uncertainty in river and groundwater models: a case study in the Namoi River, Australia. Environ. Model. Softw. 50, 108e119.

69. Reilly, T.E., Harbaugh, A.W., 2004. Guidelines for Evaluating Ground-water Flow Models, p. 30. U.S. Geological Survey Scientific Investigaions Report 2004-5038.

70. Ross, A., 2014. Banking for the future: prospects for integrated cyclical water management. J. Hydrol. 519, 2493e2500.

71. Schmid, W., Hanson, R., Leake, S., Hughes, J.D., Niswonger, R.G., 2014. Feedback of land subsidence on the movement and conjunctive use of water resources. Environ. Model. Softw. 62, 253e270.

72. Schmid, W., Hanson, R.T., 2009. The Farm Process Version 2 (FMP2) for MODFLOW- 2005: Modifications and Upgrades to FMP1. U.S. Geological Survey, Techniques and Methods: 6eA32.

73. Shafer, J.M., Labadie, J.W., 1978. Synthesis and Calibration of a River Basin Water Management Model. Colorado Water Resources Research Institute. Completion Report No. 89. Available at: http://cwi.colostate.edu/publications/cr/89.pdf.

74. Shafer, J.M., Labadie, J.W., Jones, E.B., 1981. Analysis of firm water supply under complex institutional constraints. J. Am. Water Resour. Assoc. 17 (3), 373e379.

75. Shannon, T., Labadie, J., Baldo, M., Larson, R., 2000. Integration of GIS and river basin network flow modelling. In: Proceedings of 20th Annual ESRI International User Conference, San Diego. In: http://gis.esri.com/library/userconf/proc00/ professional/papers/PAP628/p628.htm. Citeseer.

76. Shim, K.-C., Fontane, D.G., Labadie, J.W., 2002. Spatial decision support system for integrated river basin flood control. J. Water Resour. Plan. Manag. 128 (3), 190e201.

77. Shourian, M., Mousavi, S., Tahershamsi, A., 2008. Basin-wide water resources planning by integrating PSO algorithm and MODSIM. Water Resour. Manag. 22 (10), 1347e1366.

78. Sieber, J., Purkey, D., 2007. Water Evaluation and Planning System User Guide for WEAP21. Stockholm Environment Institute, U.S. Center.

79. Simonovic, S.P., 2000. One view of the future. Water Int. 25 (1), 76e88.

80. Singh, A., 2010. Decision support for on-farm water management and long-term agricultural sustainability in a semi-arid region of India. J. Hydrol. 391 (1), 63e76.

81. Sophocleous, M., 2002. Interactions between groundwater and surface water: the state of the science. Hydrogeol. J. 10 (1), 52e67.

82. Srdjevic, B., Medeiros, Y., Faria, A., 2004. An objective multi-criteria evaluation of water management scenarios. Water Resour. Manag. 18 (1), 35e54.

83. Sulis, A., Sechi, G.M., 2013. Comparison of generic simulation models for water resource systems. Environ. Model. Softw. 40, 214e225.

84. Sun, N.Z., 2013. Inverse Problems in Groundwater Modeling, vol. 6. Springer Science & Business Media.

85. Therrien, R., McLaren, R., Sudicky, E., Panday, S., 2006. HydroGeoSphere: a Threedimensional Numerical Model Describing Fully-integrated Subsurface and Surface Flow and Solute Transport. Groundwater Simul. Group, Waterloo, Ont., Canada.

86. Triana, E., 2008. A Spatial Decision Support System for Basin Scale Assessment of Improved Management of Water Quantity and Quality in Stream-aquifer Systems (PhD thesis). Colorado State University, Fort Collins, CO.

87. Triana, E., Labadie, J.W., 2012. GIS-based decision support system for improved operations and efficiency conservation in large-scale irrigation systems. J. Irrig. Drain. Eng. 138 (10), 857e867.

88. Vaghefi, S.A., Mousavi, S.J., Abbaspour, K.C., Srinivasan, R., Arnold, J.R., 2013. Integration of hydrologic and water allocation models in basin-scale water resources management considering crop pattern and climate change: Karkheh River Basin in Iran. Reg. Environ. Change 1e10.

89. Valerio, A., Rajaram, H., Zagona, E., 2010. Incorporating groundwater-surface water interaction into river management models. Ground Water 48 (5), 661e673.

90. Waskom, R., 2014. Improving Groundwater Management in the South Platte Alluvial Aquifer, 2014 NGWA Groundwater Summit. National Groundwater Association. Available at: https://ngwa.confex.com/ngwa/2014gws/webprogram/ Paper9659.html.

91. Welsh, W.D., Vaze, J., Dutta, D., Rassam, D., Rahman, J.M., Jolly, I.D., Wallbrink, P., Podger, G.M., Bethune, M., Hardy, M.J., 2013. An integrated modelling framework for regulated river systems. Environ. Model. Softw. 39, 81e102.

92. Wilds, L.J., 2010. Water Politics in Northern Nevada: a Century of Struggle. University of Nevada Press, Reno.

93. Willis, D.B., Hamilton, J.R., Robison, M.H., Whittlesey, N.K., 2008. Secondary damages in interstate water compact litigation. Nat. Resour. J. 48, 679.

94. Winchester, J.N., 2008. Predicting when the well will go dry: a review of commonlyused models for water supply planning and management. Am. Water Res. Assoc. Hydrol. Watershed Manag. Tech. Comm. 6 (3), 11. Accessed on 15.08.13., available at: http://www.awra.org/committees/techcom/watershed/pdfs/AWRA %20 H&WM_2008_3_watersupplymodel.pdf.

95. Winter, T.C., Harvey, J.W., Franke, O.L., Alley, W.M., 1998. Ground Water and Surface Water e a Single Resource. U.S. Geological Survey, Circular 1139.

96. Wurbs, R., 2013. Water Rights Analysis Package (WRAP) Modeling System Reference Manual: Technical Report No. 255, tenth ed. Texas Water Resources Institute, The Texas A&M University System, College Station, Texas. Available at: https:// ceprofs.tamu.edu/rwurbs/wrap.htm.

97. Wurbs, R.A., 2012. Reservoir/river system development and management. Tex.

98. Water J. 3 (1), 26e41.

99. Yates, D., Sieber, J., Purkey, D., Huber-Lee, A., 2005. WEAP21dA demand-, priority-, and preference-driven water planning model: part 1: model characteristics. Water Int. 30 (4), 487e500.

100. Zagona, E., Fulp, T., Goranflo, H., Shane, R., 1998. RiverWare: a general river and reservoir modeling environment. In: Proceedings of the First Federal Interagency Hydrologic Modeling Conference, pp. 19e23.

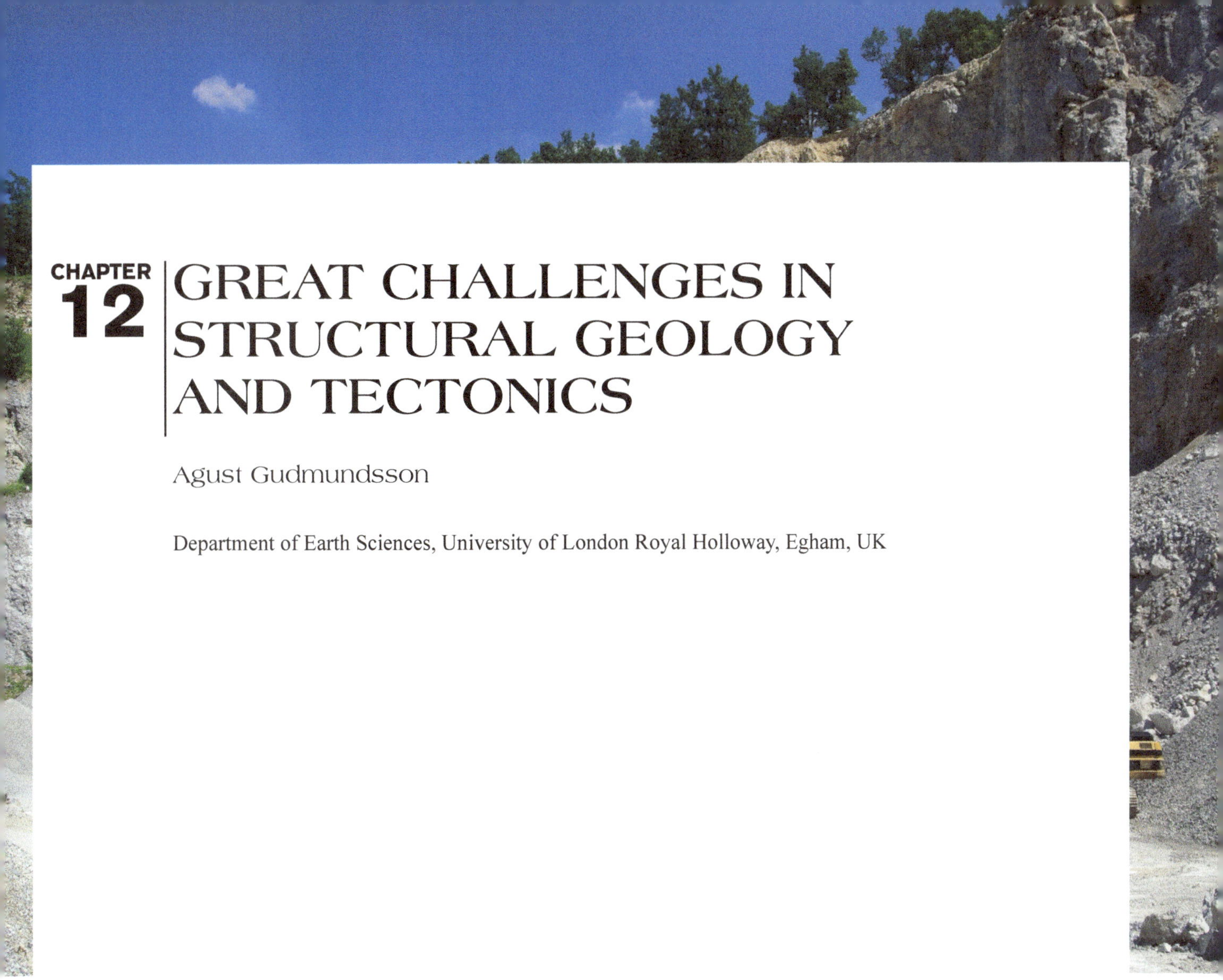

CHAPTER 12 GREAT CHALLENGES IN STRUCTURAL GEOLOGY AND TECTONICS

Agust Gudmundsson

Department of Earth Sciences, University of London Royal Holloway, Egham, UK

Successful scientific theories provide theoretical understanding for reliable deterministic or probabilistic forecasting of events. In the past decades the volume of data obtained within the fields of structural geology and tectonics has increased enormously, but the theoretical understanding of the processes that generate these data has not advanced at the same rate. For example, after decades of research we still do not have successful theories explaining the size distribution of the Earth's tectonic plates or the processes that form, drive, and destroy them. Subduction zones certainly contribute to these processes; yet we do not know how these zones initiate. While the largest earthquakes occur in subduction zones, much of the fault slip is aseismic ("creep"). We do not know the conditions that determine whether slip on a fault will be seismic or aseismic, or small or large. Similarly, with dense geodetic and seismic networks on many active volcanoes, we should be able to make a reliable forecast of the propagation path of an injected dyke—but normally we cannot. Dyke propagation and fault propagation depend on the energy available to drive these fractures. Our present knowledge of the energy budget of volcanic and seismic zones, however, is so limited that we cannot assess the maximum energies available for driving dyke-fed eruptions and earthquake faults. The topics listed here are among the great challenges in structural geology and tectonics. They all relate to the physics behind the observational data. Answering satisfactorily any of these challenges would constitute a major breakthrough in structural geology and tectonics.

INTRODUCTION

Structural geology and tectonics are closely related topics. Structural geology focuses primarily on the geometric description and the analysis of the spatial arrangement of structural features. It is also concerned with the forces or stresses or pressures that generate these structures—the result of brittle and/or ductile deformation—but to a lesser degree than tectonics. Tectonics focuses more on structures and processes on a larger scale, such as plate movements and mountain formation, as well as active processes and their physical causes. Its subfields include neotectonics, volcanotectonics, and seismotectonics.

There has been an enormous progress in obtaining high-quality tectonic/structural data in the past decade. This is largely due to greatly improved geodetic technique, in particular GPS and InSAR (Burgmann et al., 2000, 2006; Hooper et al., 2004; Wright et al., 2004; Fukushima et al., 2005;Sturkell et al., 2006; Delouis et al., 2010). The technique, described in detail by Dzurisin (2007) and Segall (2010), allows us to monitor active deformation within plate boundaries, including active fault zones and volcanoes worldwide.

Despite the new technique and wealth of data, we still cannot answer many of the basic questions related to tectonic processes, such as (Figure 1):

Citation: Agust Gudmundsson, Great challenges in structural geology and tectonics, http://dx.doi.org/10.3389/feart.2013.00002

- What is the size distribution of Earth's tectonic plates, and how many are the plates?
- What are the main forces/processes that form, drive, and destroy the plates?
- How do subduction zones initiate?
- What physical conditions favor aseismic slip or creep ("slow earthquakes") on active faults?
- What conditions favor large seismic slips?
- Why is the fault-length/slip ratio so much larger than the fault-length/displacement ratio on the same faults?
- What is the likely path of an injected dyke during an unrest period in a volcano?
- Which energy sources are available to drive seismic and aseismic fault slip?
- Which energy sources are available to drive volcanic eruptions?

Figure 1: Plate-tectonic driving forces. We can estimate the stresses needed to form a fault at a divergent plate boundary, such as here, but we do not know how the forces that cause those stresses and move the plates. Aerial view southwest, part of a large mixed-mode (opening and vertical displacement) normal fault in a Holocene pahoehoe lava flow of the rift zone in Southwest Iceland. The maximum opening of the fault is about 60 m. Buses and houses provide a scale.

Other topics of great importance in structural geology and tectonics include (1) the conditions for the formation of transform faults vs. overlapping spreading centers; (2) how oblique-spreading ridge segments are initiated and maintained; (3) whether the mechanism of plate movements is primarily through fracturing or through flow; (4) the exact conditions for ductile vs. brittle deformation during crustal deformation; (5) general mountain-chain development; and (6) how fracture-related permeability in rocks is generated and maintained.

Here, however, the focus is on the bullet points listed above. These are all major challenges in structural geology and tectonics. Satisfactory answers to these challenges imply theories that are advanced enough to have testable implications and to make reliable forecasts as to likely scenarios for given boundary conditions and physical properties. The purpose of the paper is to outline and discuss these challenges.

PLATES AND THEIR DRIVING FORCES

The theory of plate tectonics was formulated in the 1960s, that is, close to half a century ago. It was formulated as a kinematic theory—focusing on plate movements without being too much concerned with the causes of the movements—and it still is. The kinematic details of the theory are now much better understood through the reconstruction of previous plate movements (e.g., Hall, 2002; Torsvik et al., 2008). Current plate motions are monitored by satellites and well established (DeMets et al., 1994, 2010). The forces or stresses that supposedly drive the plates, however, are still poorly understood, and so are the geometric characteristics of the plates themselves.

Consider first the plates. Originally, the plates were regarded as "rigid" parts of the Earth's lithosphere within which essentially no deformation took place; nearly all the deformation was supposed to take place at the plate boundaries. The boundaries were classified as divergent (primarily ocean ridges), convergent (primarily trenches and mountain chains), and transform (primarily transform faults—and their fracture-zone extensions). Initially, the main plates were defined as 7, with about 12 additional smaller plates.

But solid fragments commonly follow power-law size distributions, and there is little reason why that sort of distribution should not hold for the fragmented lithosphere as well. In fact, recent studies show that the plates that have so far been identified follow roughly a power-law size (area) distribution. This was suggested by Bird (2003) and analyzed further bySornette and Pisarenko (2003) and Morra et al. (2013), the latter showing that the abrupt change or "kink" in the scaling exponent on the log-log frequency-size plot of plates has been a permanent feature for the past 200 Ma. Similar changes in scaling exponents—resulting in different scaling regimes—are common; for example for the lengths and apertures of rock fractures (Gudmundsson and Mohajeri, 2013). The kinks relate to mechanical constraints and different fracture types. Short fractures (with small apertures) are mostly tension fractures and confined to single mechanical layers, whereas longer fractures are mostly normal faults and penetrate many mechanical layers. Similar constraints may occur as regards the plates—the small ones being more constrained laterally (Bird, 2003; Morra et al., 2013) and/or vertically. But this needs to be explored.

How many are the plates? In the original plate-tectonic theory there were, as said, 7 main plates and about 12 smaller plates. In the analysis of Sornette and Pisarenko (2003) there are 42 plates, but the revised number used by Bird (2003) is 52 plates. In what is presumably the most up-to-date counting (List of tectonic plates, Wikipedia, 2013) there are 7 primary plates, 8 secondary plates, and 57–58 tertiary plates, bringing the total to about 73 plates. If plate-size distribution indeed follows a power law, there is no particular reason why the minimum plate size should be limited to tertiary plates. The exact number would depend on the definition of a tectonic plate which, in turn, depends on our theoretical understanding of the formation, destruction, and driving forces of plates.

While the power-law distribution of plate sizes is not surprising, its demonstration by actual data is a step forward in understanding the mechanism of plate formation and evolution. Power laws are very common in natural and social sciences and constitute a class of heavy-tailed distributions (Turcotte, 1997; Barabasi and Albert, 1999; Sornette, 2006; Clauset et al., 2009; Pisarenko and Rodkin, 2010). Generally, it is agreed that power-law size distributions occur when the sizes of events/objects are dependent whereas, by contrast, normal distributions occur when the sizes of events/object sizes are independent. Several mechanisms have been proposed for power laws, including entropy maximizing, self-organized criticality, and preferential attachment. However, none of these mechanisms really explains why certain processes and objects and events follow power-law size distributions, whereas others do not.

While the plate areas thus seem belong to the class of objects whose size-distributions follow power laws, the forces/stresses moving the plates are still largely unknown (Figure 1). Many possible forces have been suggested, including those related to mantle convection and the Earth's rotation, but the main forces are still considered to be "ridge push" and "slab pull." Ridge push has normally been primarily attributed to elevation difference between ocean ridges and the surrounding ocean floor. Gravity sliding of a plate away from an elevated ridge generates a push, a horizontal compressive force or stress, in the surrounding part of the lithosphere (Frisch et al., 2011). However, an additional stress, usually ignored in calculations of ridge push, is generated by the magmatic overpressure of the dykes injected within the active zone of rifting at mid-ocean ridges (the neovolcanic zone). The overpressure can reach as much as tens of mega-pascals (e.g., Becerril et al., 2013). The contribution of overpressure associated with dyke injection to ridge push is well established through horizontal compressive stresses (generated by the overpressured dykes) and associated excess spreading rates in recent rifting episodes in Iceland and the East African Rift (Wright et al., 2012). Slab pull, however, is commonly regarded as the main mechanism for driving the plates. The pull is supposed to be generated by the down-going (sinking) slabs in the subduction zones. A slab sinks because of its comparatively high density and it is thought to drag the associated plate with it, that is, to generate a tensile force or stress, hence the name plate pull. But so long as we do not know how subduction zones initiate, it is difficult to assess this force and its likely changes in the near future.

SUBDUCTION-ZONE INITIATION

Subduction zones are a key element in the theory of plate tectonics (Figure 2). It has been known for a long time that the plates velocities tend to be much higher for plate whose margins are partly subduction zones (Forsyth and Uyeda, 1975; Schellart et al., 2007). Consequently, it is thought that subduction zones contribute to the plate motion through the drag or pull effect of the sinking slab. Individual subduction zones range in length or strike dimension from 250 to 7400 km (Schellart et al., 2007). By implication, the associated thrust faults have similar strike dimensions and reach dip dimensions (widths in seismology) of at least hundreds of kilometres. The cumulative length of the subduction zones is about 55,000 km (Stern, 2002;Frisch et al., 2011), whereas that of the mid-ocean ridges is variously estimated at 60–80,000 km (e.g., Macdonald, 1982; Stern, 2002; Mid-ocean ridge, Wikipedia, 2013). The active spreading portion of the mid-ocean ridge system is regarded as somewhat shorter—perhaps 55–60,000 km. The cumulative length of a fracture system, however, depends on its irregularity and the scale and method used for the length measurement; the "exact" cumulative lengths of these structures thus cannot be determined.

Figure 2: Subduction zones. Subduction zones generally migrate (e.g., through retreat of the trench and rollback of the slab) and change their geometries through time. The South-American subduction zone (trench), indicated here schematically (by a yellow line), is one of the most stable of all the subduction zones (Schellart et al., 2007). While subduction-zone geometries and evolution are well known, we still do not know how subduction zones initiate (Image from Google Earth).

Given that plate pull is regarded as the main driving force of plate movements, understanding the initiation and evolution of subduction zones is of great importance. A new subduction zone normally changes the plate configuration, the plate movements, and the associated forces. Thus, we can understand the long-term future plate movements, as well as past movements, only if we have a theoretical framework that makes it possible to understand and forecast subduction zone initiation.

There has been considerable work on subduction-zone initiation in the past decade (e.g., Stern, 2002, 2004; Hall et al., 2003; Niu et al., 2003; Gurnis et al., 2004; Gerya et al., 2008; Duarte et al., 2013). The initial idea (when plate-tectonics was established) was as follows: As the oceanic lithosphere moves from the ocean ridges and becomes older, its thickness (partly through cooling of the upper asthenosphere) and density increase until its density exceeds that of the asthenosphere and the plate starts to sink. The sinking was supposed to generate the subduction zone. Mere aging of the lithosphere, however, does not seem to be a plausible explanation for subduction-zone initiation; other factors are needed (Cloetingh et al., 1989). Adding sediment load and water-weakening of the lithosphere may possibly help to initiate subduction at passive margins (Regenauer-Lieb et al., 2001), but this idea has still not been formulated as a compelling theory.

Another widely discussed idea is that subduction zones initiate on existing weaknesses, discontinuities of some sort, in the lithosphere. Discontinuities identified as potential or actual sites for subduction-zone initiation include fracture zones, back arcs, spreading centers, and former (extinct) subduction zones (Gurnis et al., 2004; Duarte et al., 2013). The most widely considered discontinuities for subduction-zone initiation are transform faults/fracture zones. The mechanism of changing a fracture into a subduction zone has been analyzed numerically (Hall et al., 2003; Gurnis et al., 2004). The results indicate that:

1. At least 100 km of forced convergence is needed before the stage of self-sustaining subduction is reached.
2. The fracture zone must be subject to considerable horizontal compressive force or stress in order to initiate subduction. This stress was initially estimated at 80 MPa by McKenzie (1977), but may be somewhat lower depending on the boundary conditions.

The two main models of subduction initiation, namely (a) increase in lithosphere density and sedimentary load with age, resulting in instability and subsidence, and (b) forced convergence, particularly at fracture zones, are both possible and not necessarily exclusive. Both models, however, face certain theoretical and observational difficulties that need to be overcome before reliable forecasts based on these models could be made.

For model a, the subduction zones could be expected to be at a certain, roughly constant, distance from the nearby ocean ridges, but these distances vary widely. One explanation for the variation in distance is that both ridges (Gudmundsson, 1995) and subduction zones (Schellart et al., 2007, 2011) migrate with time. But migration of ridges and subduction zones does not explain the great variation in distances between these two. If the sedimentary load is added, then a certain critical load, for a given lithosphere strength, might be expected to initiate subduction, but there is little evidence for this actually happening—witness, for example, the lack of subduction zones at the continental margins of the Atlantic Ocean.

As for model b, one difficulty is lack of specification: which of the many discontinuities with a potential of changing into subduction zones actually do so and why? There are many fracture zones that could potentially change into subduction zones once a new ridge segment forms that is parallel with the fracture zones in a given area. Also, there are numerous ridge-parallel discontinuities in any tectonic plate that are subject to ridge push and yet do not, as a rule, change into subduction zones. Ridge-parallel discontinuities include large normal faults, some with displacements of hundreds of meters or several kilometers and cutting through the entire lithosphere. With dips of 50–70°, it should be easier to transform them into subduction zones than vertical strike slip faults in fracture zones (fracture zones also contain many normal faults, however). But there is no clear evidence that any of the large normal faults in the ocean floor or on the continental shelf have changed into subduction zones. Similarly, the contacts between continental crust and the oceanic crust are discontinuities that are subject to horizontal ridge-push stresses; yet few develop into subduction zones. Thus, while these and related models certainly indicate several possibilities as to subduction imitation, it is fair to conclude that we do not yet have a general quantitative theory—one with a clear explanatory power—as to subduction-zone initiation.

SEISMIC AND ASEISMIC SLIP

In the past few decades, the importance of aseismic slip has become recognized as fundamental in active fault zones (Figure 3). In fact, aseismic slip is so common that it is estimated that in subduction zones and transform faults, for example, some 50% of the slip may be aseismic (Stein and Wysession, 2003; Peng and Gomberg, 2010). Slip is referred to as aseismic when it is comparatively slow and not accompanied by an ordinary earthquake. There are many other terms used in seismotectonics for aseismic slip such as "slow earthquake," "creep" in the uppermost part of fault zones, and "slow-slip phenomena" (Sleep and Blanpied, 1992; Galehouse and Lienkaemper, 2003; Jordan, 2003; Rolandone et al., 2008; McFarland et al., 2009; Peng and Gomberg, 2010). Aseimic slip occurs partly through "afterslip" and partly through creep, but in this overview the term used for all these processes is aseismic slip.

Figure 3. Seismic and aseismic slip. Fault zones are composed of a core and a damage zone. View west, the core and the damage zone of a part of the Husavik-Flatey transform fault, which is partly exposed on land in North Iceland (Gudmundsson, 1995, 2000). The core is 10 m thick and the lava flows in the damage zone, originally sub-horizontal, now dip 40°NW. The core is composed of soft breccia and presumably subject to much aseismic slip, whereas the fault as a whole is seismically highly active.

Why is part of the displacement on an active fault generated through aseismic slip and part through seismic slip? More generally: Why do surface (and subsurface) rocks sometimes respond to generally similar boundary conditions through flow and sometimes through fracture? At present, we do not know the detailed answers to these questions. We know that strain rate, mechanical properties (and asperities), and temperature affect the response of rocks to loading. However, even when all the conditions are generally similar, the rock response may sometimes vary from ductile to brittle; for example, in many mountain chains and in individual fault zones (Figure 3).

One possible explanation for the common aseismic slip is that it is a partly the result of a "correction" or adjustment of the slip to that which would have occurred if the dynamic Young's modulus were equal to the static Young's modulus (Gudmundsson et al., 2013). These moduli are very different; the dynamic modulus, which controls co-seismic slip, is commonly 10-times and may be 100-times larger than the static modulus of the core and the innermost part of the damage zone of a fault zone (Figure 3). The static modulus of the core and innermost part of the damage zone then controls the aseismic slip. The rock behavior may be regarded as analogous to an elastic-plastic model in continuum mechanics where to an initial elastic strain or displacement (the seismic slip) a ductile strain (the aseismic slip) is gradually added. This model may also account for the 1–2 order of magnitude difference between the length-slip ratios in individual earthquakes and the length-cumulative displacement in the same fault zones.

Whether or not this model is along fruitful lines, understanding the conditions for seismic and aseismic slip is of fundamental importance in tectonics and structural geology in general and in earthquake mechanics in particular. The magnitude of aseismic slip has come as a surprise to many, and the topic relates to the question of conditions that favor ductile over brittle deformation. Furthermore, this topic relates directly to the conditions for the initiation of fault slip, aseismic and seismic, and therefore to earthquake mechanics.

LARGE FAULT SLIPS AND EARTHQUAKES

Fault zones are three-dimensional structures composed of fault cores and damage zones (Figure 3; Bruhn et al., 1994; Caine et al., 1996; Shipton and Cowie, 2003; Kim et al., 2004; Faulkner et al., 2006). Because the mechanical properties of the damage zone and, in particular, the core, are normally widely different from those of the host rock, the fault zone develops a local stress field that may be very different from the surrounding regional stress field (Gudmundsson et al., 2010). The local stress in a fault zone determines if and when slip occurs and, presumably, whether the slip will be seismic or aseismic. Simple elastic dislocation models miss entirely the internal structures and properties of fault zones, namely the factors that, to large degree, control earthquakes and aseismic slip (and fault-zone permeability). Such models, however, may be useful for matching the general co-seismic surface deformation (Segall, 2010).

Earthquake fractures are dynamic cracks, for which considerable body of knowledge exists (Freund, 1998; Ravi-Chandar, 2004; Shulka, 2006). Much of this literature, however, focuses on the propagation of mode I cracks (extension fractures such as tension fractures and dykes in rift zones) rather than shear fractures (mode II, mode III, or mixed-mode cracks), as earthquake faults are. There are also specific monographs on earthquake source mechanics (Kostrov and Das, 2005).

We know that earthquakes occur on faults, and that the faults develop from existing weaknesses such as joints—which are commonly generated when the rock forms (Figure 4). In the simplest terms, earthquake initiation can be understood through considerations of stresses and material properties. For example, we know that increasing the pore-fluid pressure in the crust, such as during injection of water into reservoirs or the building of dams, triggers earthquakes. These triggering effects are reasonably well understood through elementary stress analysis and presentable by Mohr circles, as is routinely done in textbooks on structural geology and tectonics. The main unsolved problem, however, is to explain why some earthquake slips become large, generating large earthquakes, whereas most remain small (e.g., Main and Naylor, 2012). This brings us again to the power-law size distribution that we discussed in connection with tectonic plate-size distributions.

Figure 4: Fault initiation and earthquakes. Most faults originate from existing weaknesses that, on an outcrop-scale, are commonly joints generated during the formation of the rock. View east, columnar (cooling) joints in a basaltic intrusion in South Iceland, with a common column diameter (width) of 20–40 cm.

The Gutenberg-Richter frequency-magnitude relation for earthquakes is among the best known of all the power laws in physical science. The Omori relation for the rate of aftershock production with time since the main shock, and the Bath's relation, indicating that the difference in magnitude between the main shock and the largest aftershock is nearly a constant, are also well known earthquake relations (Turcotte et al., 2007). Other scaling laws refer to the geometric aspects of the seismogenic faults, perhaps the most remarkable being the great difference between the rupture length-slip ratios and the fault length-displacement ratios. Commonly, the rupture length-slip ratios are of the order of 10^{3-4} (Wells and Coppersmith, 1994; Leonard, 2010) whereas the fault length-displacement ratios are of the order of 10^{1-2} (e.g., Clark and Cox, 1996; Schlische et al., 1996). These ratios thus differ by factors of the order of 10^{1-2}.

None of the relations discussed above explains why earthquake magnitudes follow power law size distributions—why some earthquakes are very large while most are small. It is known from studies of fault zones and fissure swarms, however, that fault-size distributions also follow power laws (e.g.,Gudmundsson and Mohajeri, 2013). Since fault slip is a function of fault size (the controlling dimension, which is either the strike dimension or the dip dimension), and since earthquake moments (energies) and related magnitudes depend on slip and rupture area, it could be argued that when the fault sizes follow power laws, so must the magnitudes. More specifically, if we could explain the power-law size distribution of fault dimensions, the power laws in earthquake energy releases and magnitudes might follow. At the moment, however, no power laws have been satisfactory explained in terms of well-developed physical theory (Turcotte, 1997; Barabasi and Albert, 1999;Sornette, 2006; Clauset et al., 2009; Pisarenko and Rodkin, 2010). The development of such a theory is one of the most important tasks in science in general, and in tectonics and earthquake mechanics in particular.

FEEDER-DYKE PROPAGATION

Volcanic activity is often marginalized, or excluded altogether, in mainstream research in structural geology and tectonics. For example, many textbooks on structural geology hardly mention dykes, sills, and inclined sheets at all. Yet, all these are rock fractures; more specifically, fluid or magma-driven fractures (Figure 5) and thus part of general brittle deformation. Most importantly, hardly any eruptions occur unless a magma-driven fracture is able to form and propagate to the Earth's surface. Dyke propagation is essentially that of a mode I crack, that is, an extension fracture; the same crack model is also used for all tension fractures in rift zones and at divergent plate boundaries as well as for most joints (Gudmundsson, 2011). Dyke (and sheet) propagation is and should be regarded as an important part of structural geology and tectonics—in fact, a major part of the fields of neotectonics and volcanotectonics.

Figure 5: Dyke propagation. Dyke paths are complex and cannot, at this stage of theoretical understanding, be reliably forecasted during unrest periods in volcanoes. Many dykes become arrested, particularly at contacts between mechanically dissimilar rocks, like this basaltic dyke in Tenerife (Canary Islands). The soft rock hosting the dyke is altered breccia, whereas the stiff rock arresting the dyke is an older, inclined sheet.

Dyke propagation is generally simpler than seismogenic fault propagation. This follows partly because dyke propagation is essentially quasi-static mode I crack propagation. By contrast, earthquake fracture is a dynamic mode II or mode III, and commonly mixed mode, shear-fracture propagation. Also, many dyke fractures are geometrically simple (Figure 5). Each dyke segment is normally composed of essentially one fracture, and its geometry can be analyzed in detail because it is filled with solidified magma. Dyke propagation does normally not generate much of a damage zone or core, partly because most dykes are not subject to reactivation (they are single injections, Figure5). By contrast, an earthquake fault is commonly reactivated many times and generates a fault core and damage zone (Figure 3). Some damage zones are composed of many layers or subzones (Gudmundsson et al., 2010) and contain a very large number of extension fractures and small faults in addition to the main-fault segments.

Despite dyke-fractures being comparatively simple, their propagation paths are commonly anything but. This is demonstrated by direct field observations (Geshi et al., 2010; Gudmundsson, 2011, 2012) which show many dykes being deflected into sills or inclined sheets along parts of their paths. In addition, in any particular dyke area, most of the dykes become arrested and thus never reach the surface to supply magma to eruptions (Figure 5). The complex paths of magma-driven fractures are also indicated by detailed geodetic and seismic studies of volcano unrest periods. In many very well-monitored recent volcano unrest periods, with wealth of seismic and geodetic data, it has proven impossible to provide reliable forecasts as to either the path of the dyke or the likelihood of its erupting at the surface. Monitoring indicates that many dykes become arrested, others change into sills, and then back to dykes, in agreement with field observations. The likelihood as to the dyke making it to the surface is generally difficult to assess. When a dyke eventually reaches the surface, there are still no reliable methods for forecasting the likely duration and size of the resulting eruption (Gudmundsson, 2012), both of which depend on the energy available to drive the eruption.

ENERGY SOURCES DRIVING EARTHQUAKES AND ERUPTIONS

The formation and propagation of every rock fracture, including dykes (Figure5) and slipping faults (Figures 1, 3, 6), requires energy. More specifically, the propagation of a fracture requires surface energy, that is, energy to rupture the rock and form the two fracture surfaces (Anderson, 2005; Gudmundsson, 2011). Thus, for a magma-driven fracture and an earthquake fracture to propagate, energy inputs into the volcano and fault zone are needed.

Figure 6: Energy sources driving earthquakes. Major fault zones receive much energy (primarily elastic energy) through plate movements. View east, part of the Husavik-Flatey transform fault in North Iceland (Figure 3, Gudmundsson, 1995,2000) showing a sag pond (Lake Botnsvatn) formed along normal faults. We still do not know of much of the energy stored in fault zones is available to drive large-scale seismogenic faulting.

There is work done on, and therefore increase in internal energy of, fault zones and volcanic zones by plate-tectonic forces (Figures 1, 6); in addition, there is heat input through the general heat flow as well as magma intrusion. The heat received by volcanic zones is primarily from magma chambers and reservoirs, as well as from dyke and sill and sheet injections. Some fault zones, such as transform faults and fissure swarms at divergent plate boundaries, also receive heat through magma injection, primarily in the form of dykes (e.g., Gudmundsson, 1995, 2000).

The elastic energy released or transformed during individual earthquakes is well known. Using elastic dislocation theory, this energy is routinely calculated from the seismic moment (Aki, 1966; Aki and Richards, 2009) and the related moment-magnitude scale (Hanks and Kanamori, 1979). A small fraction of the released energy is radiated as seismic waves, but the greatest part is transformed into heat (partly because of friction) and surface energy needed for the seismic rupture (the earthquake fault) to propagate. The exact proportion of the energy released into seismic waves is not really known (e.g.,Stacey and Davis, 2008), but is thought to be somewhere around 10% of the total. Similarly, the surface energy needed to propagate a dyke to the surface and supply magma to an eruption can be estimated (Gudmundsson, 2012). Because of the logarithmic scales of the earthquake magnitudes, as well as of the explosive index for volcanic eruptions, the largest earthquakes/eruptions entirely dominate the energy release or transformation in a given zone.

How much of the energy received by fault zones is available to drive seismogenic faulting is not well known (Figure 6). For example, we do not know how much of the energy received by a major fault zone is transformed into aseismic slip and general ductile deformation, small earthquakes, extension fracturing (hydrofracture or mineral-vein formation), and dissipated as heat. Nor do we know how much of the energy—primarily elastic energy—in the fault zone just before a potential major earthquake is actually available to drive the fault propagation. These energy considerations, which must be formulated in terms of thermodynamic principles, have not received the attention they deserve. At present, these energy proportions are poorly constrained but need to be well constrained in order to understand the potential of an active fault zone for generating large to great earthquakes.

Stress drops, the driving shear stresses for seismogenic faulting, are mostly small, commonly 1–10 MPa (Aki and Richards, 2009), and constitute only a fraction of the total stress in the fault zone. Since slip on a fault relates to the driving stress and partly determines the energy (moment) of the earthquake, the driving stress is not an indication of the total stress or the total energy in a fault zone. Also, the boundary conditions of the fault zone partly determine the energy available during the earthquake-fault propagation. If the boundaries are rigid (so-called fixed grip) then they do not contribute to the available energy. By contrast, if the boundaries are flexible, then their movement (that is, work) during the fault propagation may provide significant energy input and increase the chances of the generation of large to great earthquakes (Gudmundsson, 2012). In general, the energy budget and available energy in seismogenic zones remain poorly known.

Similarly, for volcanic zones and individual volcanoes, the energy available to propagate a feeder dyke to the surface and then squeeze out magma from the chamber remains poorly constrained (Gudmundsson, 2012). Heat is continuously transported into volcanoes, particularly through their magma chambers, but its transformation into mechanical energy and work is inefficient and difficult to estimate. Elastic energy is stored in the volcano during magma-chamber inflation. In addition, the flanks of the volcano, or the boundaries of the associated volcanic zone, may be flexible, in which case energy can be brought into the system (through work) during feeder-dyke propagation. In order to assess volcanic risks when a volcano is subject to unrest and inflation, particularly as regards the likely size and duration of the eventual eruption, we must know the energy available to propagate the feeder dyke and sustain the eruption. At the moment, we have no detailed theoretical framework to make accurate estimates of this energy. The development of such a framework as well as a framework for estimating the available energy in seismogenic zones are among the main challenges in structural geology and tectonics.

ACKNOWLEDGMENTS

I thank the Frontiers reviewer and the associate editor for helpful comments.]

REFERENCES

1. Aki, K. (1966). Generation and propagation of G waves from the Niigata Earthquake of June 16, 1964. Part 2. Estimation of earthquake moment, released energy, and stress-strain drop from the G wave spectrum. *Bull. Earthq. Res. Inst.* 44, 73–88.
2. Aki, K., and Richards, P. G. (2009). *Quantitative Seismology*. 2nd Edn. Herndon, VA: University Science Books.
3. Anderson, T. L. (2005). *Fracture Mechanics*. 3rd Edn. New York, NY: CRC Press.
4. Barabasi, A. L., and Albert, R. (1999). Emergence of scaling in random networks. *Science* 286, 509–512. doi: 10.1126/science.286.5439.509
5. Becerril, L., Galindo, I., and Gudmundsson, A. (2013). Depth of origin of magma in eruptions. *Sci. Rep.* 3, 2762. doi: 10.1038/srep02762
6. Bird, P. (2003). An updated digital model of plate tectonics. *Geochem. Geophys. Geosyst.* 4. doi: 10.1029/2001GC000252
7. Bruhn, R. L., Parry, W. T., Yonkee, W. A., and Thompson, T. (1994). Fracturing and hydrothermal alteration in normal fault zones. *Pure Appl. Geophys.* 142, 609–644. doi: 10.1007/BF00876057
8. Burgmann, R., Hilley, G., Ferretti, A., and Novali, F. (2006). Resolving vertical tectonics in the San Francisco Bay area from permanent scatterer InSAR and GPS analysis. *Geology* 34, 221–224. doi: 10.1130/G22064.1
9. Burgmann, R., Rosen, P. A., and Fielding, E. J. (2000). Synthetic aperture radar interferometry to measure Earth's surface topography and its deformation. *Annu. Rev. Earth Planet Sci.* 28, 169–209. doi: 10.1146/annurev.earth.28.1.169
10. Caine, J., Evans, J. P., and Forster, C. B. (1996). Fault zone architecture and permeability structure.*Geology* 24, 1025–1028. doi: 10.1130/0091-7613(1996)024<1025:FZAAPS>2.3.CO;2
11. Clark, R. M., and Cox, S. J. D. (1996). A modern regression approach to determining fault displacement-scaling relationships. *J. Struct. Geol.* 18, 147–152, 1996. doi: 10.1016/S0191-8141(96)80040-X
12. Clauset, A., Chalizi, R. C., and Newman, M. E. J. (2009). Power-law distributions in empirical data.*Soc. Ind. Appl. Math.* 51, 661–703. doi: 10.1137/070710111
13. Cloetingh, S., Wortel, R., and Vlaar, N. J. (1989). On the initiation of subduction zones. *Pure Appl. Geophys.* 129, 7–25. doi: 10.1007/BF00874622
14. Delouis, B., Nocquet, J. M., and Vallee, M. (2010). Slip distribution of the February 27, 2010 Mw = 8.8 Maule Earthquake, central Chile, from static and high-rate GPS, InSAR, and broadband teleseismic data. *Geophys. Res. Lett.* 37. doi: 10.1029/2010GL043899
15. DeMets, C., Gordon, R. G., and Argus, D. F. (2010). Geologically current plate motions. *Geophys. J. Int.* 181, 1–80. doi: 10.1111/j.1365-246X.2009.04491.x
16. DeMets, C., Gordon, R. R., Argus, D. F., and Stein, S. (1994). Effects of recent revisions to the geomagnetic reversal time scale on estimates of current plate motions. *Geophys. Res. Lett.* 21, 2191–2194. doi: 10.1029/94GL02118
17. Duarte, J. C., Rosas, F. M., Terrinha, P., Schellart, W. P., Boutelier, D., Gutscher, M. A., et al. (2013). Are subduction zones invading the Atlantic. Evidence form the southwest Iberia margin. *Geology*41, 839–842. doi: 10.1130/G34100.1
18. Dzurisin, D. (2007). *Volcano Deformation: New Geodetic Monitoring Technique*. Berlin: Springer Verlag.
19. Faulkner, D. R., Mitchell, T. M., Healy, D., and Heap, M. J. (2006). Slip on weak fault by the rotation of regional stress in the fracture damage zone. *Nature* 444, 922–925. doi: 10.1038/nature05353
20. Forsyth, D., and Uyeda, S. (1975). On the relative importance of the driving forces of plate motion.*Geophys. J. R. Astr. Soc.* 43, 163–200. doi: 10.1111/j.1365-246X.1975.tb00631.x
21. Freund, L. B. (1998). *Dynamic Fracture*. Cambridge: Cambridge University Press.
22. Frisch, W., Meschede, M., and Blakey, R. (2011). *Plate Tectonics: Continental Drift and Mountain Building*. Berlin: Springer Verlag.
23. Fukushima, Y., Cayol, V., and Durand, P. (2005). Finding realistic dike models from interferometric synthetic aperture radar data: the February 2000 eruption at Piton de la Fournaise. *J. Geophys. Res.* 110. doi: 10.1029/2004JB003268
24. Galehouse, J. S., and Lienkaemper, J. J. (2003). Inferences drawn from two decades of alinement array measurements of creep on faults in the San Francisco Bay region. *Bull. Seismol. Soc. Am.* 93, 2415–2433. doi: 10.1785/0120020226
25. Gerya, T. V., Connolly, J. A. D., and Yuen, D. A. (2008). Why is terrestrial subduction one-sided.*Geology* 36, 43–46. doi: 10.1130/G24060A.1
26. Geshi, N., Kusomoto, S., and Gudmundsson, A. (2010). Geometric difference between non-feeder and feeder dikes. *Geology* 38, 195–198. doi: 10.1130/G30350.1
27. Gudmundsson, A. (1995). Stress fields associated with oceanic transform faults. *Earth Planet. Sci. Lett.* 136, 603–614. doi: 10.1016/0012-821X(95)00164-8
28. Gudmundsson, A. (2000). Dynamics of volcanic systems in Iceland: example of tectonism and volcanism at juxtaposed hot spot and mid-ocean ridge systems. *Annu.*

Rev. Earth. Planet. Sci. 28, 107–140. doi: 10.1146/annurev.earth.28.1.107

29. Gudmundsson, A. (2011). *Rock Fractures in Geological Processes*. Cambridge: Cambridge University Press. doi: 10.1017/CBO9780511975684
30. Gudmundsson, A. (2012). Strengths and strain energies of volcanic edifices: implications for eruptions, collapse calderas, and landslides. *Nat. Hazards Earth Syst. Sci.* 12, 2241–2258. doi: 10.5194/nhess-12-2241-2012
31. Gudmundsson, A., De Guidi, G., and Scudero, S. (2013). Length-displacement scaling and fault growth. *Tectonophysics*. doi: 10.1016/j.tecto.2013.06.012
32. Gudmundsson, A., and Mohajeri, N. (2013). Relations between the scaling exponents, entropies, and energies of fracture networks. *Bull. Geol. Soc. Fr.* 184, 377–387. doi: 10.2113/gssgfbull.184.4-5.373
33. Gudmundsson, A., Simmenes, T. H., Larsen, B., and Philipp, S. L. (2010). Effects of internal structure and local stresses on fracture propagation, deflection, and arrest in fault zones. *J. Struct. Geol.* 32, 1643–1655. doi: 10.1016/j.jsg.2009.08.013
34. Gurnis, M., Hall, C., and Lavier, L. (2004). Evolving force balance during incipient subduction.*Geochem. Geophys. Geosyst.* 5. doi: 10.1029/2003GC000681
35. Hall, C. E., Gurnis, M., Sdrolias, M., Lavier, L. L., and Dietmar Muller, R. (2003). Catastrophic initiation of subduction following forced convergence across fracture zones. *Earth Planet. Sci. Lett.* 212, 15–30. doi: 10.1016/S0012-821X(03)00242-5
36. Hall, R. (2002). Cenozoic geological and plate tectonic evolution of SE Asia and the SW Pacific: computer-based reconstructions, model and animations. *J. Asian Earth Sci.* 20, 353–431. doi: 10.1016/S1367-9120(01)00069-4
37. Hanks, T. C., and Kanamori, H. (1979). A moment magnitude scale. *J. Geophys. Res.* 84, 2348–2350. doi: 10.1029/JB084iB05p02348
38. Hooper, A., Zebker, H., Segall, P., and Kampes, B. (2004). A new method for measuring deformation on volcanoes and other natural terrains using InSAR persistent scatterers. *Geophys. Res. Lett.* 31. doi: 10.1029/2004GL021737
39. Jordan, T. H. (2003). *Living on an Active Earth: Perspectives on Earthquake Science*. National Research Council of the National Academies. Washington, DC: The National Academic Press.
40. Kim, Y. S., Peacock, D. C. P., and Sanderson, D. J. (2004). Fault damage zones. *J. Struct. Geol.* 26, 503–517. doi: 10.1016/j.jsg.2003.08.002
41. Kostrov, B. V., and Das, S. (2005). *Principles of Earthquake Source Mechanics*. Cambridge: Cambridge University Press.
42. Leonard, M. (2010). Earthquake fault scaling: self-consistent relating of rupture length, width, average displacement, and moment release. *Bull. Seismol. Soc. Am.* 100, 1971–1988. doi: 10.1785/0120090189
43. Macdonald, K. C. (1982). Mid-ocean ridges: fine scale tectonic, volcanic and hydrothermal processes within the plate boundary zone. *Annu. Rev. Earth Planet. Sci.* 10, 155–190. doi: 10.1146/annurev.ea.10.050182.001103
44. Main, I., and Naylor, M. (2012). Extreme events and predictability of catastrophic failure in composite materials and in the Earth. *Eur. Phys. J. Special Topics* 205, 183–197. doi: 10.1140/epjst/e2012-01570-x
45. McFarland, F. S., Lienkaemper, J. J., and Caskey, S. J. (2009). "Data from theodolite measurements of creep rates on San Francisco Bay region faults, California, 1979-2009," in *USGS Open-File Report 2009-1119*. Available online at: http://pubs.usgs.gov/of/2009/1119/
46. McKenzie, D. P. (1977). "The initation of trenches: a finite amplitude instability," in *Island Arcs, Deep Sea Trenches and Back-Arc Basins. Maurice Ewing Ser, Vol. 1*, eds M. Talwani and W. C. Pitman (Washington, DC: American Geophysical Union), 57–61. doi: 10.1029/ME001p0057
47. Morra, G., Seton, M., Quevedo, L., and Dietmar Muller, R. (2013). Organization of the tectonic plates in the last 200 Myr. *Earth Planet Sci. Lett.* 373, 93–101. doi: 10.1016/j.epsl.2013.04.020
48. Niu, Y., O'Hara, M. J., and Pearce, J. A. (2003). Initiation of subduction zones as a consequence of lateral compositional buoyancy contrast within the lithosphere: a petrological perspective. *J. Petrol.* 44, 851–866. doi: 10.1093/petrology/44.5.851
49. Peng, Z., and Gomberg, J. (2010). An integrated perspective of the continuum between earthquakes and slow-slip phenomena. *Nat. Geosci.* 3, 599–607. doi: 10.1038/ngeo940
50. Pisarenko, V., and Rodkin, M. (2010). *Heavy-Tailed Distributions in Disaster Analysis*. Berlin: Springer Verlag. doi: 10.1007/978-90-481-9171-0
51. Ravi-Chandar, K. (2004). *Dynamic Fracture*. Amsterdam: Elsevier.
52. Regenauer-Lieb, K., Yuen, D. A., and Branlund, J. (2001). The initation of subduction: criticality by addition of water. *Science* 294, 578–580. doi: 10.1126/science.1063891
53. Rolandone, F., Burgmann, R., Agnew, D. C., Johanson, I. A., Templeton, D. C., d'Alessio, M. A., et al. (2008). Aseismic slip and fault-normal strain along the central creeping section of the san andreas fault. *Geophys. Res. Lett.* 35, L14305. doi: 10.1029/2008GL034437
54. Schellart, W. P., Freeman, J., Stegmann, D. R., and May, D. (2007). Evolution and diversity of subduction zones controlled by slab width. *Nature* 446, 308–311. doi: 10.1038/nature05615
55. Schellart, W. P., Stegman, D. R., Farrington, R. J., and Moresi, L. (2011). Influence of lateral slab edge distance on plate velocity, trench velocity, and subduction partitioning. *J. Geophys. Res.* 116. doi: 10.1029/2011JB008535
56. Schlische, R. W., Young, S. S., Ackermann, R. V., and Gupta, A. (1996). Geometry and scaling relations of a population of very small rift-related normal faults. *Geology* 24, 683–686. doi: 10.1130/0091-7613(1996)024<0683:GASROA>2.3.CO;2

57. Segall, P. (2010). *Earthquake and Volcano Deformation*. Princeton, NJ: Princeton University Press.

58. Shipton, Z. K., and Cowie, P. A. (2003). A conceptual model for the origin of fault damage zone structures in high-porosity sandstone. *J. Struct. Geol.* 25, 333–344. doi: 10.1016/S0191-8141(02)00037-8

59. Shulka, A. (2006). *Dynamic Fracture Mechanics*. Singapore: World Scientific Publishing.

60. Sleep, N. H., and Blanpied, M. (1992). Creep, compaction and weak rheology of major faults. *Nature*359, 687–692. doi: 10.1038/359687a0

61. Stacey, F. D., and Davis, P. M. (2008). *Physics of the Earth*. 4th Edn. Cambridge: Cambridge University Press. doi: 10.1017/CBO9780511812910

62. Stern, R. J. (2002). Subduction zones. *Rev. Geophys.* 40. doi: 10.1029/2001RG000108

63. Stern, R. J. (2004). Subduction initiation: spontaneous and induced. *Earth Planet. Sci. Lett.* 226, 275–292. doi: 10.1016/j.epsl.2004.08.007

64. Sornette, D. (2006). *Critical Phenomena in Natural Sciences: Chaos, Fractals, Selforganization and Disorder: Concepts and Tools*. Berlin: Springer Verlag.

65. Sornette, D., and Pisarenko, V. (2003). Fractal plate tectonics. *Geophys. Res. Lett.* 30. doi: 10.1029/2002GL015043

66. Stein, S., and Wysession, M. (2003). *An Introduction to Seismology, Earthquakes, and the Earth Structure*. Oxford: Blackwell.

67. Sturkell, E., Einarsson, P., Sigmundsson, F., Geirsson, H., Olafsson, H., Pedersen, R., et al. (2006). Volcano geodesy and magma dynamics in Iceland. *J. Volcanol. Geotherm. Res.* 150, 14–34. doi: 10.1016/j.jvolgeores.2005.07.010

68. Torsvik, T. H., Muller, R. D., Van der Voo, R., Steinberger, B., and Gaina, C. (2008). Global plate motion frames: toward a unified model. *Rev. Geophys.* 46. doi: 10.1029/2007RG000227

69. Turcotte, D. L. (1997). *Fractals and Chaos in Geology and Geophysics*. Cambridge: Cambridge University Press. doi: 10.1017/CBO9781139174695

70. Turcotte, D. L., Shcherbakov, R., and Rundle, J. B. (2007). "Complexity and earthquakes," in*Treatise on Geophysics*, Vol. 4, Chapter 23, ed H. Kanamori (Amsterdam: Elsevier), 675–700. doi: 10.1016/B978-044452748-6.00085-7

71. Wells, D., and Coppersmith, K. (1994). New empirical relationships among magnitude, rupture length, rupture width, rupture area, and surface displacement. *Bull. Seismol. Soc. Am.* 84, 974–1002.

72. Wright, T. J., Parsons, B. E., and Lu, Z. (2004). Toward mapping surface deformation in three dimensions using InSAR. *Geophys. Res. Lett.* 31. doi: 10.1029/2003GL018827

73. Wright, T. J., Sigmundsson, F., Pagli, C., Belachew, M., Hamling, I. J., Brandsdottir, B., et al. (2012). Geophysical constraints on the dynamics of spreading centres from rifting episodes on land.*Nat. Geosci.* 5, 242–250. doi: 10.1038/ngeo1428

CHAPTER 13

COSTS AND BENEFITS OF SATELLITE-BASED TOOLS FOR IRRIGATION MANAGEMENT

Francesco Vuolo, Laura Essl and Clement Atzberger

Institute of Surveying, Remote Sensing and Land Information, University of Natural Resources and Life Sciences, Vienna, Austria

This paper presents the results of a collaborative work with farmers and a cost-benefit analysis of geospatial technologies applied to irrigation water management in a semi-arid agricultural area in Lower Austria. We use Earth observation (EO) data to estimate crop evapotranspiration (ET) and webGIS technologies to deliver maps and irrigation advice to farmers. The study reports the technical and qualitative evaluation performed during a demonstration phase in 2013 and provides an outlook to future developments. The calculation of the benefits is based on a comparison of the irrigation volumes estimated from satellite vs. the irrigation supplied by the farmers. In most cases, the amount of water supplied was equal to the maximum amount of water required by crops. At the same time high variability was observed for the different irrigation units and crop types. Our data clearly indicates that economic benefits could be achieved by reducing irrigation volumes, especially for water-intensive crops. Regarding the qualitative evaluation, most of the farmers expressed a very positive interest in the provided information. In particular, information related to crop ET was appreciated as this helps to make better informed decisions on irrigation. The majority of farmers (54%) also expressed a general willingness to pay, either directly or via cost sharing, for such a service. Based on different cost scenarios, we calculated the cost of the service. Considering 20,000 ha regularly irrigated land, the advisory service would cost between 2.5 and 4.3 €/ha per year depending on the type of satellite data used. For comparison, irrigation costs range between 400 and 1000 €/ha per year for a typical irrigation volume of 2000 cubic meters per ha. With a correct irrigation application, more than 10% of the water and energy could be saved in water-intensive crops, which is equivalent to an economic benefit of 40–100 €/ha per year.

INTRODUCTION

Irrigated agriculture is the main user of freshwater resources (WWAP, United Nations World Water Assessment Programme, 2015). Recent studies show that many large aquifer located in agricultural regions are exposed to stress conditions due to intense exploitation combined to a slower rate of recharge (Richey et al., 2015b); our knowledge about the total amount of water stored and the resilience of these systems is often limited (Richey et al., 2015a). Hence, efficient irrigation management is vitally important, especially within the growing context of challenges such as increased use of groundwater and water quality degradation (OECD, 2012). While agricultural decisions are often made solely on the basis of a farmer's experience and at individual level, today the farming sector is expected to deliver food security, environmental sustainability, and create new economic opportunities (WEF, 2013). This requires not only improved efficiency at farm scale but also tools that allow water management and monitoring on a larger scale. With a 70 per cent share of worldwide water withdrawals, improved

Citation: Francesco Vuolo, Laura Essl and Clement Atzberger, Costs and benefits of satellite-based tools for irrigation management, http://dx.doi.org/10.3389/fenvs.2015.00052

efficiency of water application and management is expected to result in a major step toward more sustainable water usage (WEF, 2013).

The potential of Earth observation (EO) techniques to support irrigation water management is now widely recognized: high spatial resolution satellite images are well-suited to monitor the spatial and temporal distribution of crop development and to derive crop evapotranspiration (ET) at field scale with a regional coverage (Choudhury et al., 1994; Bausch, 1995; Courault et al., 2005; Verstraeten et al., 2008; McMahon et al., 2013). However, the adoption of these technologies and the integration into the day-to-day routine operations of farmers is a complex process and favorable conditions depend on several technical, social, and economic factors (Baptista, 2006).

A commonly used approach to produce maps of ET in operational frameworks relies on the methodology proposed by FAO in 1998 (Allen et al., 1998; Task Committee on Standardization of Reference Evapotranspiration, 2005). The approach requires local agro-meteorological data for the direct application of the Penman-Monteith (P-M) equation, with EO-based estimates of albedo (r) and Leaf Area Index (LAI) (D'Urso and Menenti, 1995; D'Urso, 2001; D'Urso et al., 2010; Vuolo et al., 2015a). The minimum satellite data requirements to apply this approach are multispectral observations in the visible (VIS) and near-infrared (NIR) spectrum. Such data are available with the appropriate frequency and spatial resolutions. Suitable satellite platforms are for instance Landsat-8 (U.S. Geological Survey—USGS—and NASA) which provides free data within less than 24-h of acquisition. In an operative scenario, Landsat-8 is often combined with commercial satellite data to achieve a temporal resolution of about 7–10 days between acquisitions, which is adequate to monitor the various phases of the crop development throughout the growing season. The availability of free and open access data for commercial use is expected to improve with the Sentinel-2 mission (Drusch et al., 2012), developed by European Space Agency (ESA) within the Copernicus initiative.

Several research projects have facilitated the advancement of readiness levels pertaining to the application of space technologies to water resource management. The first concept of satellite-based irrigation advisory services was designed in the context of the DEMETER EU-project in 2005 (D'Urso and Belmonte, 2006; Osann Jochum, 2006). It was successively improved and automatized to meet the requirements of individual farmers with personalized weekly irrigation advices by using Short Message Services on mobile phones[1] (Vuolo et al., 2006; De Michele et al., 2009). The development of advisory services based on webGIS platforms has been initially developed within the PLEIADeS[2] EU-project and further tested in SIRIUS[3]. The methodologies have been validated during several experimental field campaigns (Rubio, 2006; D'Urso et al., 2010; Vuolo et al., 2013) showing a strong potential and good transferability to different contexts.

In Lower Austria, the region of Marchfeld is one of the major crop production areas. Water resource management is a real and tangible issue in this region. Due to the shortage of precipitation and the semi-arid regional climate, after the Second World War farmers began to intensify crop production by irrigation with groundwater. Today most of the cultivated land in Marchfeld is equipped with irrigation infrastructures. Groundwater is shared with urban and industrial sectors, leading to high pressure on the quantity and quality of resources. The increased demand for high quality food, sustainable local production, and market development strategies is stimulating the application of tools for more efficient and effective irrigation water use (Cepuder, 2004). Tests have been carried out to regulate irrigation based on soil water content measurements performed at plant or field scale (Cepuder and Nolz, 2007). However, the widespread application of soil water content sensors that can support the decision making process is hampered by investment and maintenance costs. Generally, the information available to help in the decision process is based only on agro-meteorological data that does not account for the large spatial variability of crop development, soil conditions, and agronomic practices.

This work presents the results of a demonstration campaign in the Marchfeld region (conducted in May–September, 2013) and evaluation with users. First, we compared the irrigation volumes estimated from satellite and the irrigation supplied by the farmers to calculate the efficiency of water use. Second, during and after the irrigation season, we assessed the service with the users based on questionnaires and meetings. The results of this qualitative evaluation provided an indication of the usefulness of the information in relation to the user needs. Finally, we developed a cost and benefit analysis and evaluated the willingness to pay for the information generated by the project. This study reports preliminary results of the evaluation. It also introduces the information technologies and webGIS solutions used to provide the service and presents future steps for implementation.

MATERIAL AND METHODS

Pilot Area

Implementation and evaluation activities were carried out in the years 2012–2013 in the framework of a case study in the Marchfeld region (Lat. 48.20°N, Long. 16.72°E), Lower Austria (Figure 1). In this region of about 60,000 hectares, irrigation water is sourced mainly from private well via electric or diesel pumps. Irrigation water accounts for up to 60 per cent of the total freshwater use[4]. Irrigation is regularly performed (using sprinkler systems) during the cropping period for intensive crops (i.e., sugarbeet, potatoes, and various vegetables), and during the dry periods for semi-intensive crops (such as cereals, oilseed, and maize). The climate in Marchfeld is semi-arid with annual precipitations of less than 550 mm (about 250–300 mm from May to September). The soil conditions are characterized by a high spatial variability, including soils with low to moderate water-storage capacity (Rischbeck, 2007;Thaler and Eitzinger, 2008; Eitzinger et al., 2009).

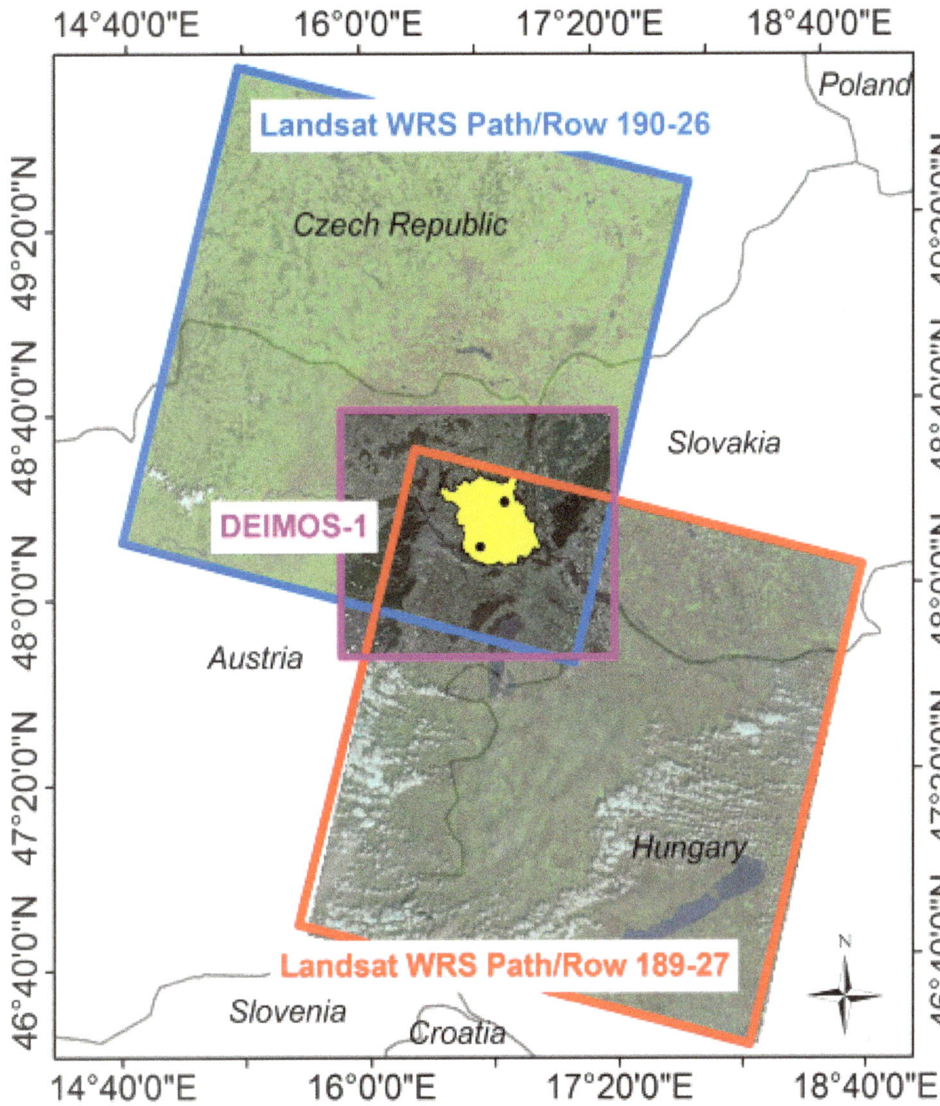

Figure 1: Configuration of the Landsat-8 and DEIMOS-1 acquisitions over the test site of Marchfeld (yellow polygon). The black dots indicate the location of the two weather stations.

Methodological Background for the Estimation of Crop Water Requirements

The approach for the calculation of crop ET using satellite observations was developed by (D'Urso and Menenti, 1995), based on Penman–Monteith (P–M) equation as implemented in the FAO-56 procedure (Allen et al., 1998). A detailed description of the methodologies for the generation of maps of crop ET and crop water requirements (CWR) was recently reported in a comparative study in different pilot areas (Vuolo et al., 2015a). The model considers the canopy as a "big leaf," with a surface area expressed by the LAI, a crop height hc and a hemispherical spectrally integrated albedo r. LAI and albedo (r) maps can be obtained using atmospherically corrected multi-spectral images. To calculate crop ET, the model also required daily values of main agro-meteorological variables (i.e., temperature, solar radiation, wind speed, and air humidity) obtained from local weather stations. The ratio of crop ET to the reference ET (ET of the hypothetical grass reference crop i.e., hc = 0.12 m; r = 0.23, and LAI = 2.88) leads to the crop coefficient (k_c), which is often used in irrigation management. The CWR is calculated by subtracting the rainfall from crop ET. Depending on user requirements, the calculated CWR is further aggregated over different temporal (e.g., weekly) and spatial (e.g., individual parcels) scales. The information is then transferred to farmers for irrigation management using webGIS and mobile phone technologies.

The most critical point in the data processing chain is the successful acquisition of cloud-free satellite images. A first estimation of the availability of adequate data over our pilot area was carried out using the revisit time map created for the Sentinel-2 mission planning (Martimort et al., 2007). According to this map, the constellation of the two Sentinel-2 satellites (the first satellite was successfully launched in June 2015) would be able to acquire cloud-free images in summer every 5 days in areas with clear-sky conditions (as in Southern Europe) and every 15–30 days in cloudy areas. For our geographic location, cloud-free images would be available approximately every 10–15 days (with two satellites in operation). Sentinel-2 data will be available with global coverage as well as free and open access for the European and international users, therefore there will be a high potential for application in Europe and in other regions. Moreover, the combined use of Sentinel-2 and

Landsat-8 will further increase the opportunities to acquire cloud-free scenes with adequate temporal resolution.

Waiting for a fully operational Sentinel-2 mission, our development and initial operations activities have been based on existing free and commercial satellite sensor data. In particular, the Marchfeld region is located in the overlapping area between two Landsat orbits (Figure 1); within this area, Landsat achieves a revisit time of 8 days, which is usually not sufficient to produce cloud-free scenes with the required temporal interval. Therefore, our approach was to use the commercial satellite DEIMOS-1 (featuring an image swath of 630 km and a ground sampling distance of 22 m)[5] in a virtual constellation of satellites with Landsat-8. DEIMOS-1 acquisitions were tasked during the campaign to complement Landsat-8 in case of cloud cover. Thanks to its large image swath, DEIMOS-1 can theoretically achieve a revisit time of 1–3 days. Based on this strategy, we obtained cloud-free acquisitions every 15 days (average value during the irrigation season) in 2013. An overview of the images acquired in the year 2013 is provided in Figure 2. By using Landsat-8 images in addition to DEIMOS-1 data, substantial cost reductions were achieved.

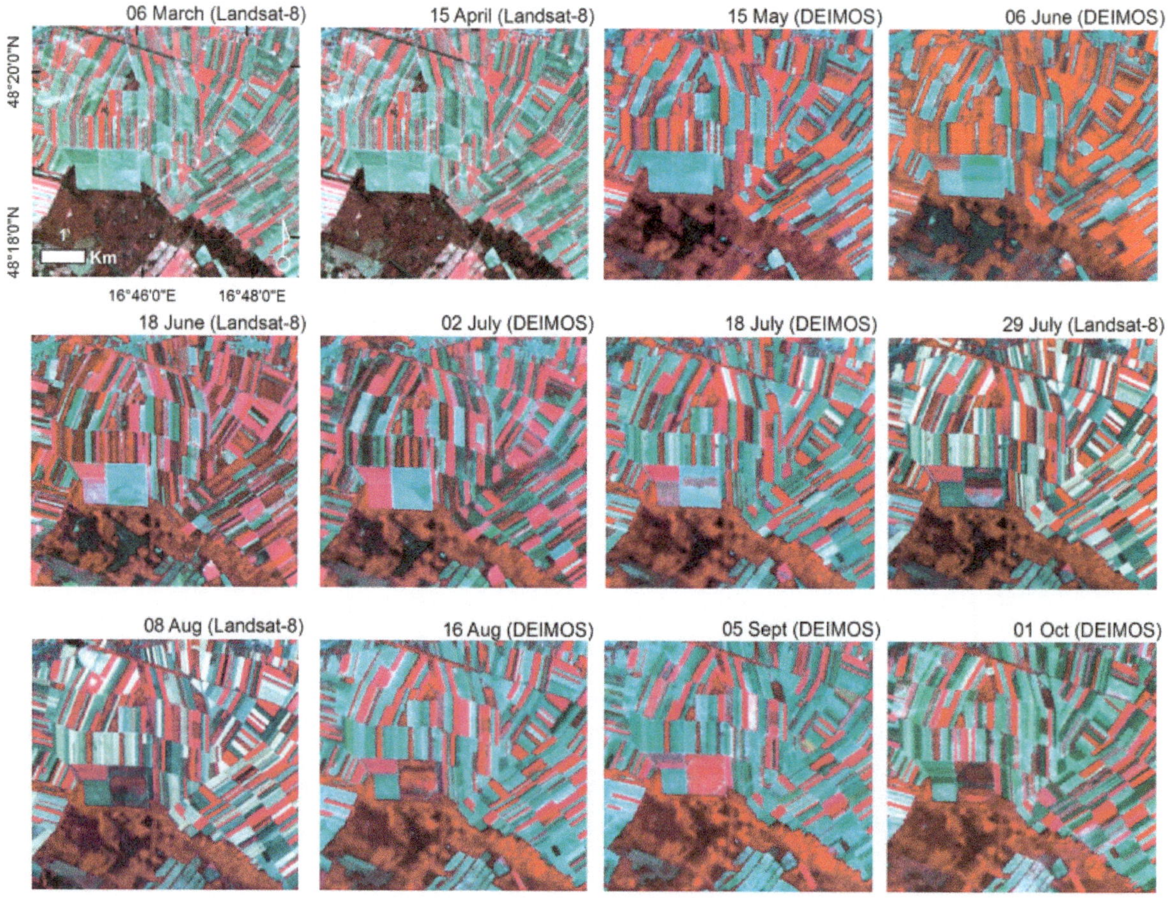

Figure 2: Near-infrared color composite of a time series of DEIMOS-1 and Landsat-8 images over Marchfeld in the year 2013. Four more cloud-free Landsat-8 scenes were available on May 1st, September 6th, October 1st and 8th.

Agro-meteorological variables were measured using wireless weather stations at two different sites located in the southern and northern part of the pilot area (black dots in Figure 1). Reference ET was calculated according to the FAO-56 standards and made available twice a day: we provided a first forecast at midday and updated it with the actual value at midnight. In this way, farmers could calculate the water requirement already for the current day. All meteorological data were recorded and stored on a 15-min basis; reference ET was calculated on an hourly basis and aggregated to daily values.

Delivery of the Information

Experience gained in past projects revealed that decision support tools have to be easy-to-use, reliable and correctly embedded into irrigation routine operations of farmers. Otherwise, it is not possible to achieve successful and effective use of maps and other provided data. Several information technology solutions are possible, based on spatial online data analysis and visualization systems with dynamically interactive graphics display for decision-support or personalized information sheet. We used an existing and consolidated webGIS solution implemented in past projects in Italy in the context of an operative irrigation advisory service (Irrisat) and made available for the purpose of the project. The webGIS system is the main repository for the maps and irrigation advices, which are published on a timely basis, with access restricted to individual farmers for their parcels. Figure 3 shows the mapping panel of the webGIS that provides tools to browse and query maps and to display the boundaries of the irrigation units. A data frame (right panel, Figure 3), provides the plot details (crop type, irrigation method) and the temporal series of crop ET, effective rainfall, and irrigation requirements. The dictionary, units and style of maps of the existing webGIS system were adapted to the local conditions and deployed for demonstration in Marchfeld.

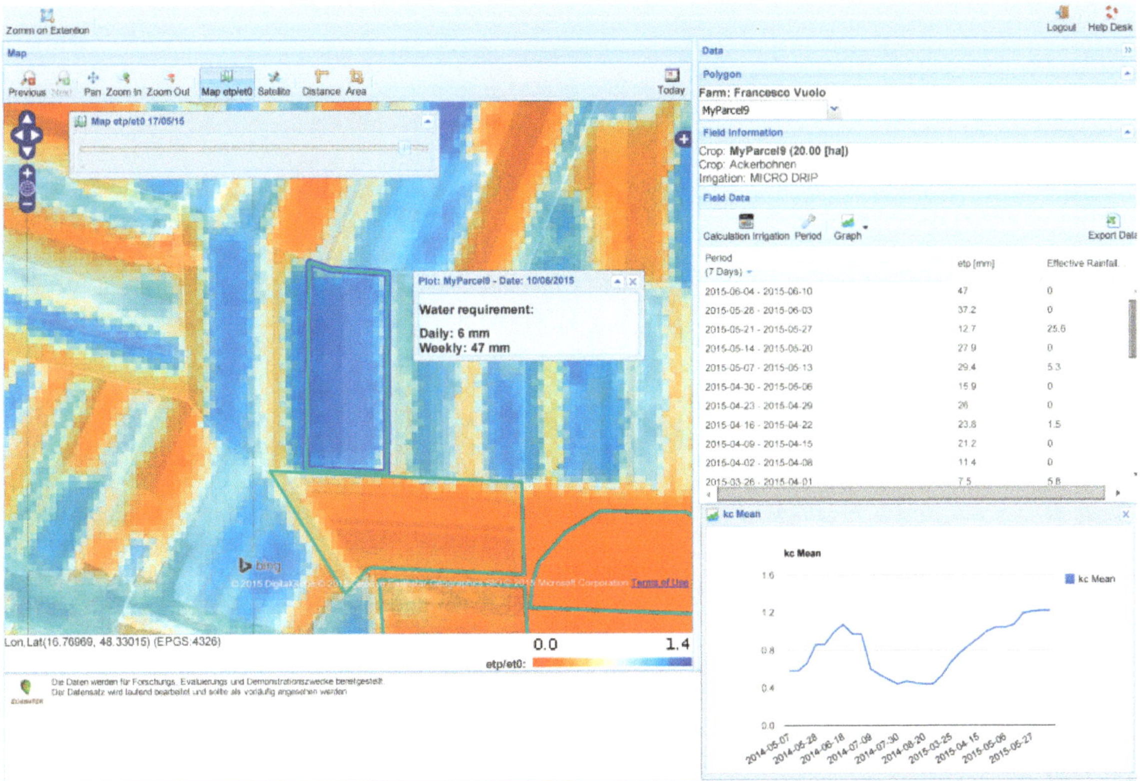

Figure 3: Example of webGIS interface used to deliver the irrigation advice. The map displays the ratio of crop ET to reference ET (i.e., crop coefficient image, k_c, where red tones indicate low k_c-values); the data and chart panel on the right side display the weekly crop ET, the effective rainfall values, and the k_c chart of the selected parcel.

Evaluation

The demonstration campaign conducted in the year 2013 provided the framework for evaluation that was achieved at different levels.

First, we compared the satellite-derived irrigation volumes with the irrigation supplied by the farmers. The volumes of supplied water for each irrigation unit were directly reported by the farmers based on their measurements. In total, we received 46 irrigation reports. These volumes were aggregated to an adequate temporal scale (based on the time of irrigation) and compared to the CWR estimated from satellite. For this comparison, we used the time interval between the first and last irrigation during the period covered by satellite images. These data were further summarized to calculate the Total Crop Water Use Indicator (IP) that expresses the adequacy of water supplied compared to the water requirements (water supply/water demand). A value greater than 1 indicates that the amount of water supplied is higher than the potential water needed.

Second, during and after the irrigation season, a qualitative evaluation with the users was achieved. The main aim was to get insight into the user's perspective and gain knowledge about the expectations, needs and problems of the farmers relating to the use of the information and tools provided. Our approach was to turn existing issues into opportunities to further improve the functionality of the system. Therefore, we worked closely with a core group of farmers who tested the system during a demonstration campaign (May–September, 2013). Personal contacts were established before the campaign by visiting farmers at their place or by keeping regular contact by telephone. During the irrigation season, farmers had different options to provide their feedback and to get help in case of problems: they could either call the project team or participate in one of the monthly group meetings. If the problem was not solved by phone, personal visits were conducted. At the end of the irrigation season (October 2013), all farmers were visited again personally to collect their final feedback and their perception of problems and benefits. For this purpose, a semi-structured questionnaire was developed and compiled during face-to-face interviews with each of the farmers. It aimed at capturing information connected to the use of the system (frequency of use, benefit, problems), but was kept open to leave space for personal comments.

Finally, we developed a cost and benefit analysis for different data cost scenarios and evaluated the willingness to pay for the information generated by the project. The benefits were classified in two main categories according to the approach proposed by (Wichelns and Oster, 2006), which differentiates between farm-level benefits and the social cost-benefit of irrigation management. In this paper, we focus only on farm-level benefits. A qualitative description of the benefit was based on the analysis to the semi-structured questionnaire and measured by the willingness of the farmers to pay for the service. Table 1 summarizes the benefits and possible valuation methods.

Table 1: Benefits and possible valuation methods.

Benefit	Impact	Valuation method	Explanation
Save water and money	Optimization of irrigation management	Market price method	Calculation of cost of saved water due to improved irrigation management
Save time	Increased knowledge and quality of work	Contingent valuation	Willingness to pay for increased knowledge
Improve decision	Information on regional water use	No valuation	Further collaboration with irrigation associations in the future
Improve water use transparency	Allocation of resources and water rights	No valuation	Further collaboration with the local water authorities in the future

RESULTS

A total number of 358 irrigation units were registered into the webGIS system and 30 different farmers contributed in various ways to the demonstration and evaluation of the project activities. The participating irrigation units covered an area of approximately 2000 ha (~10% of total irrigated area in Marchfeld) and were distributed over 40 different administrative districts. The average field size was 5 hectares (ha) and the crop inventory comprised 33 different crop species. Twenty-two farmers were engaged in regular face-to-face meetings and 17 farmers provided irrigation volumes for a total number of 46 fields or irrigation units (comprising different crop types, sowing dates, soil characteristics, and irrigation techniques).

Evaluation of Crop Water Requirements vs. Water Supply

The comparison was performed for all the irrigation volumes for the 46 fields collected during the campaign in the year 2013 and included various crops such as Sugarbeet, Maize, Potatoes, Soya, Onion, Carrots, Spinach, and Peas. Figure 4 shows (A) an example of satellite-based crop coefficient curves for different crops, (B) the temporal evolution of the CWR estimated from satellite vs. the irrigation supplied and (C) the total water supplied (irrigation + effective rainfall) vs. the water demand. Figure 5 provides a synthesis of the results. Figure 6 shows the Total Crop Water Use Indicator (IP). We can notice that the average value of the IP is close to 1, indicating that the amount of water supplied is equal to the maximum amount of water required by crops in the reference time frame. However, a high variability can be observed for the different irrigation units and crop types. For example, results indicate that sugarbeet (number of irrigation units, $n = 16$) received less water than the potential requirements (from 0.83 to 0.91 times the maximum requirements), while Onion ($n = 3$), Maize ($n = 9$), and other vegetables ($n = 5$) (Carrots, Spinach, and Peas) received more water than required. For the mentioned vegetables, IP ranged between 1.47–1.58, 1.15–1.2, and 1.38–1.64. Soya ($n = 5$) and Potatoes ($n = 8$) presented an IP range of 0.98–1.12, indicating that the maximum water requirements were fully satisfied. The data indicate that some economic benefit could be directly achieved by reducing the irrigation volumes. However, these results are limited to one irrigation season only, which was particularly dry from middle June to the end of August. The total reference ET (from May 15th to September 30th 2013) ranged between 480 and 530 mm, with a total rainfall of 300 mm. A more detailed analysis will require considering the irrigation and the water requirements over multiple seasons before drawing conclusions on irrigation efficiency in the study area. Regarding the validation of the crop ET maps, we evaluated the performance of the atmospheric correction over multiple seasons and years (Vuolo et al., 2015b) and the accuracy in the retrieval of LAI over a wide range of crops (Vuolo et al., 2013, 2015a). In both cases we achieved good results. Additionally, the crop ET was directly validated using actual ET measurements based on eddy covariance data for a maize field during the irrigation season 2014; results (not shown) demonstrated a very good agreement between measured and satellite-based crop ET.

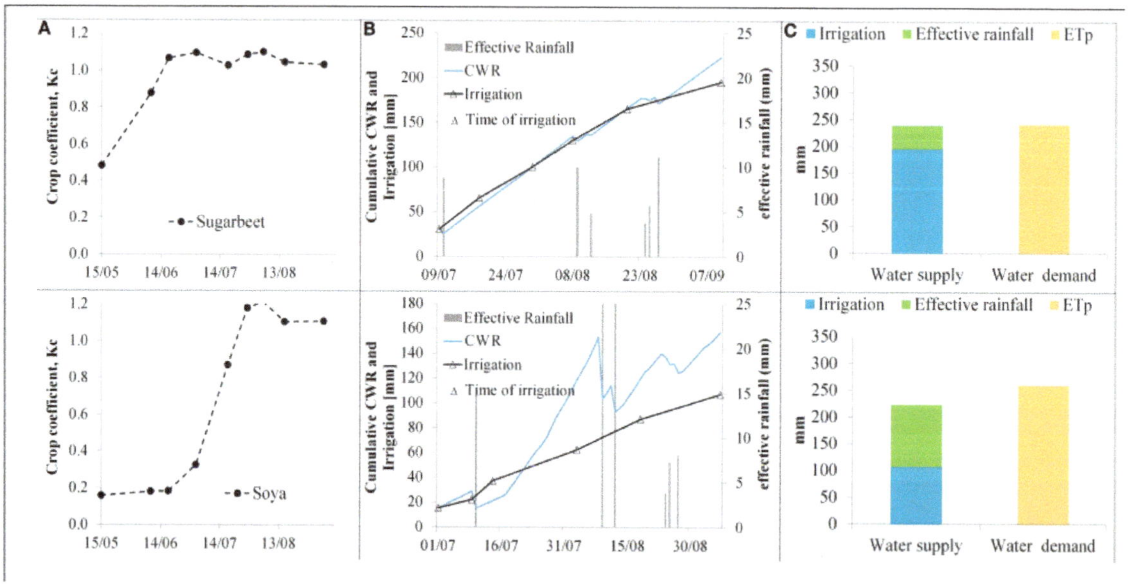

Figure 4: Examples of (A) satellite-based crop coefficient, (B) Crop Water Requirements (CWR) estimated from satellite vs. supplied irrigation volumes provided by the farmers, and (C) Total water supplied (Irrigation + effective rainfall) vs. water demand (crop ET).

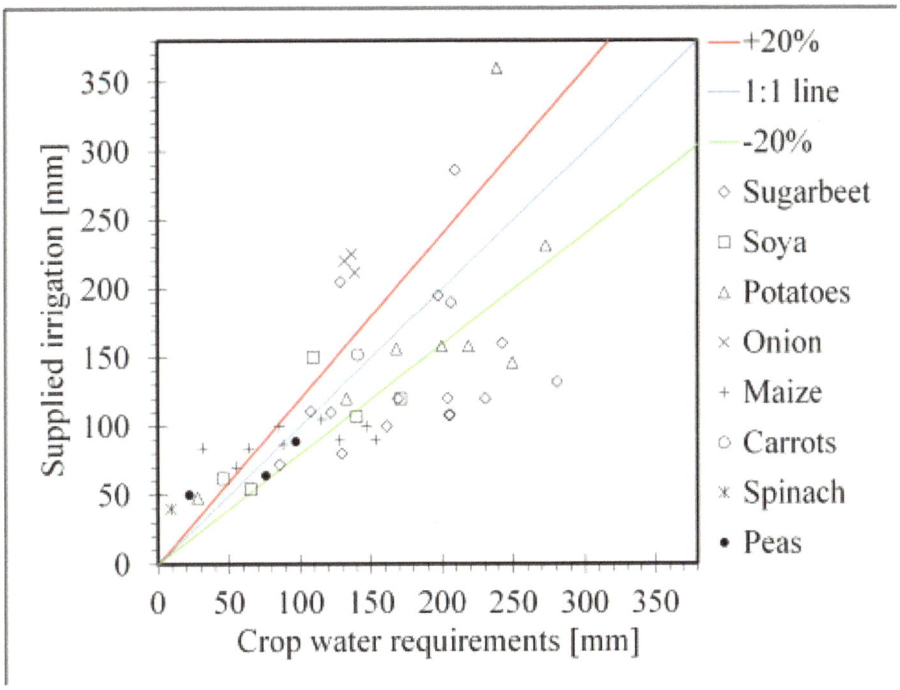

Figure 5: Crop water requirements (CWR) vs. supplied irrigation volumes for Sugarbeet (number of irrigation units, $n = 16$), Soya ($n = 5$), Potatoes ($n = 8$), Onion ($n = 3$), Maize ($n = 9$), and other crops (Carrots, Spinach and Peas) ($n = 5$).

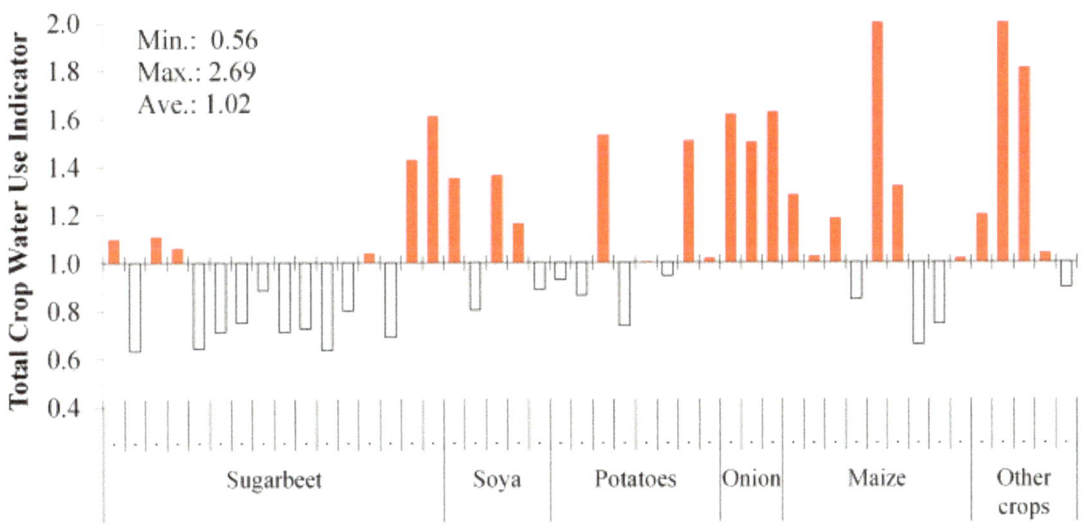

Figure 6: Total Crop Water Use Indicator (Water supply/Potential Water Demand) for Sugarbeet ($n = 16$), Soya ($n = 5$), Potatoes ($n = 8$), Onion ($n = 3$), Maize ($n = 9$), and other crops ($n = 5$).

QUALITATIVE EVALUATION

Most farmers expressed a positive interest in the information (mainly crop evapotranspiration that helps them to make decision on irrigation) and the majority of them (54%) also expressed a general willingness to pay, either directly or via cost sharing with complementary services. Farmers were asked about the expected benefits and the value-added information and multiple answers were possible. The most often mentioned reason for taking part in the demonstration and evaluation activities was (i) the determination of the best time for irrigation, (ii) the interest in new technologies, and (iii) the information about crop evapotranspiration. Another incentive was the possibility to observe—by using crop vigor maps—differences in crop development within a parcel or between parcels planted with the same crop and under different management practices.

Based on the visualization and interpretation of the crop vigor maps, farmers expected to prove their long professional experience and to save money on energy costs for irrigation, on travels and time necessary to survey all their parcels. Expected benefits that were mentioned less often included the possibility: (i) to derive information on soil characteristics, (ii) to improve the sustainable use of water within the farm, and (iii) to get local weather information.

The most frequently mentioned issues were that the units and terminology were difficult to understand. Some farmers noted that the system is not compatible with their management structure as they need a period of a few days to irrigate one field and therefore cannot react immediately on changing water requirements of the crops. Other farmers had (minor) technical problems related to lost passwords and the type of internet browser they used. Distribution of rainfall in Marchfeld strongly varies and the two weather stations were considered insufficient to cover the whole area. Sometimes, satellite images with a higher spatial resolution were expected.

Concerning the decision making process for irrigation, results indicate that farmers base their decisions on various criteria: the most frequently listed factors were the actual weather conditions and the weather forecast. Farmers use various online platforms to get information about precipitation and wind direction. In addition, many farmers have their own precipitation measurement stations within the farm or in field. Another environmental criterion is the actual condition of the plants, which is checked personally on site.

Economic criteria also play an important role in the decision to irrigate crops. The price of the seeds determines the initial investment and crops with expensive seeds (e.g., onions) are more frequently irrigated (resulting often in over-irrigation). The expected market price of the product is another factor to decide which crops are irrigated. Irrigation itself is expensive with prices between 0.2–0.5 € per cubic meter depending on the type of irrigation system. Fuel prices are subject to high variations and are therefore also a relevant factor in the decision-making process. To give a practical example, we can consider that 1 hectare (ha) of onion receives about 200 mm of water, which is equivalent to 2000 cubic meters per ha. This leads to a total cost of 400–1000 € per ha only for irrigation. In addition, farmers have obligations to fulfill production contracts, which is only possible with irrigation. For certain crops, they have to assure that irrigation on the fields is possible and therefore these crops will be irrigated first. Another relevant aspect is the personal experience. Farmers recognized that their habits are difficult to change. They receive suggestions from friends concerning irrigation and decisions of their respective neighbors influence their own irrigation management practices. The capacity of infrastructure is another constraint. Farmers usually do not have the capacity to irrigate all their fields, so they have to decide at the beginning of the season where to place the irrigation infrastructure.

Some of the expectations were completely fulfilled: (i) farmers confirmed their observations about differences between different fields with the spatial variability depicted in the satellite-based maps; (ii) the information on plant development confirmed or even improved their experience. Some expected benefits were not achieved within the timeframe of this analysis: the determination of the best time to irrigate was not permitted by the system as the farmers had to concentrate on other limiting factors (mainly technical) due to the dry summer months experienced in the year 2013.

ECONOMIC ANALYSIS

Benefits

Farmers were asked whether further optimization of their irrigation management practices was possible. Only 23% stated that there were no more margins for optimization whereas 50% said that further improvements could be achieved by optimizing the total amount of water requirements and the temporal frequency and distribution of individual irrigation events. No general tendency could be found as some farmers claimed to irrigate too much while others complained about water deficits of their crops. Carrots and soya were mentioned to be especially sensitive to water scarcity and therefore require special care in the future.

To support the decision whether to irrigate or not, farmers listed various factors: (i) they would like to know the actual status of the crops on the field without visiting every single plot, (ii) be able to compare their own experiences with farmers who are irrigating less, (iii) to know the water need of the plant in different phases of growing season, or (iv) to have access to more exact weather forecasts.

Improved irrigation management also has substantial economic impacts. The running costs for irrigation vary between around 0.1–0.2 € per cubic meter for irrigation systems powered by electricity to around 0.3–0.5 € per cubic meter for diesel systems. The amount of water irrigated therefore has a direct economic impact on savings. If water is reallocated between the plots, indirect benefits through higher yields can also be expected.

The willingness to pay for the provided service was used as a measure of the value of the system. The majority (54%) of the farmers indicated a general willingness to pay, either directly or via cost sharing funding options; 32% did not answer and 14% expressed no willingness to pay. One quarter of the farmers indicated a willingness to pay 3 € per ha. A small percentage would be willing to pay more (5 €/ha) for higher spatial resolution images and others would be willing to pay only if costs are shared with complementary services (e.g., fertilizer and pesticide consultancy services).

Costs

The costs of the service include the initial set-up of the webGIS and farmer's database and the cost for satellite data acquisition, processing and delivery of final maps. These costs can be grouped in two categories: variable costs (that vary with the number of hectares and users served) include the initial farmer's database implementation and a dedicated customer service. Fixed costs, for a given service coverage/area, are the main component of the operative costs and include the satellite data acquisition and processing.

To calculate operative cost items we consider three categories: (1) personnel, (2) satellite data, and (3) other costs. Personnel-effort account for the entire satellite processing chain that includes: EO data procurement, atmospheric correction and generation of products, quality control and calibration/validation activities, agro-meteorological data elaboration, and maintenance of geo-infrastructures. The personnel-effort is a fixed cost given a certain region of interest, duration of the operation and number of images to process. Other costs include field travel, software licenses, and satellite data.

Satellite data cost depends on the type of data, minimum order size and required spatial resolution. Depending on field size different spatial resolutions are required (i.e., for smaller field size higher spatial resolution is needed). We use four satellite data cost scenarios. Table 2 provides an overview of current commercial satellites (and data costs) that would be able to provide data for the service in our region of interest. Considering the minimum order size and length of the irrigation season (10 images), the cost of the satellite data per irrigation/growing season would range between 15,000 € (DEIMOS-1) and 35,000 € with RapidEye. With the availability of free satellite data from current Landsat-8 (30 m pixel size) and upcoming Sentinel-2 missions (10–20 m pixel size), we foresee minimum need for commercial data, therefore the cost will be notably reduced.

Table 2: Summary of relevant satellite data provides and cost of data.

Data provider	Pixel size (m)	Price per sq. km	Revisit time	Notes	Cost of EO data for growing season (10 images)
DEIMOS-1	22	~0.15 €	1–3 days	Minimum order size 10000 sq. km	15,000 €
SPOT image	10	~2.0 €	2–3 days	Coverage: 900 sq. km - 1/4 scene	18,000 €
RapidEye	5	~1.0 €	1 day	Minimum order size: 3500 sq. km; multiples days might be needed to achieve full coverage	35,000 €

Using commercial satellite data at 10–30 m pixels size, the service cost ranges between 65,000 and 68,000 € per growing season. The cost of the satellite data would comprise about 25% of the total cost of the service. If very high spatial resolution data (e.g., 5 m pixel size) are needed (e.g., to cover higher level of field fragmentation—smaller parcels), the cost of the satellite data would reach around the 40% of the total cost. With the availability of free satellite data from the upcoming Sentinel-2 mission (at 10–30 m spatial resolution), we foresee an average service cost of about 50,000 € per growing season. This cost will be further reduced with the automation of various processing steps as for instance atmospheric correction (Vuolo et al., 2015b).

Using the four data cost scenarios and the extent of the region of interest (Marchfeld), we calculated the total cost of the service for an entire growing season. A summary graph is provided in Figure 7. Taking into account only the 20,000 ha that are regularly irrigated in the region of Marchfeld (Neugebauer and Vuolo, 2014) and depending on the type of satellite data used, the service would have an estimated cost ranging between 2.5 and 4.3 €/ha per year. Instead, considering the total area of the Marchfeld region (40,000 ha) and the use of free data (possible with Landsat-8 and Sentinel-2 in operation), the service would have a cost of approximately 1.3 €/ha per year. In the first case, the cost of the service would be equal or below the willingness to pay expressed by most of the users (3 to 5 €/ha per year). In the second case, the cost of the service would be substantially below the willingness to pay, which allows for revenue generation.

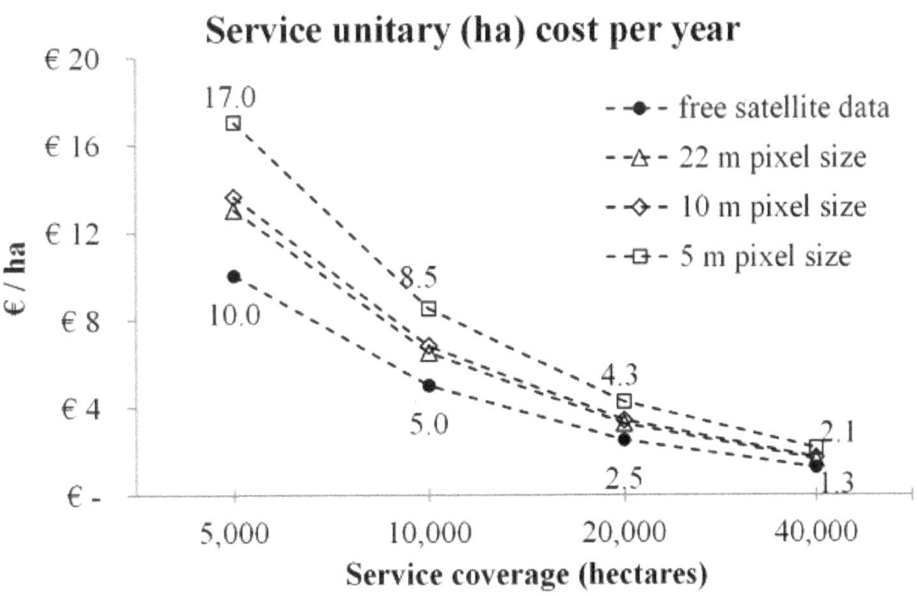

Figure 7: Service cost per hectare (ha) and per growing season.

In comparison: irrigation costs range between 400 and 1000 €/ha per year for a typical irrigation volume of 2000 cubic meters per ha. With a correct irrigation application, more than 10% of the water and energy could be saved in water-intensive crops, which is equivalent to an economic benefit of 40–100 €/ha per year.

It is worth noting that the benefit analysis only includes the farmer's perspective and does not consider the benefit derived from more efficient use of environment and water resources. We did not consider the social and environmental benefits because they can vary widely and their quantification is beyond the scope and resources available within this work. Therefore, we expect even a larger net economic benefit for society than the one estimated here. Social and environmental costs express the value of water allocation to alternative users (e.g., opportunity costs associated with urban and industrial water use) and the impact on water resources, pollution, soil degradation, and landscape modifications due to intense irrigated agriculture, while our financial analysis is based only on the on-farm costs that farmers generally pay for water as well as for any other production costs, such as seed, machinery, labor, and maintenance. The literature indicates that these two categories of costs (social and environmental) are far from being negligible.

CONCLUSIONS

The adoption of better irrigation practices requires information technologies and tools that can be easily embedded into existing systems and that can fully respond to the user needs. Research and demonstration activities (this project and other case studies such as DEMETER, PLEIADeS, and SIRIUS-GMES EU projects), show that satellite-based irrigation advisory services can be successfully implemented in different environments and, once in place, can contribute to assessments of crop water requirements in a spatially, timely, and cost-efficient way. Current information technologies (webGIS and smartphones) support the delivery of irrigation advice to farmers in a user-friendly environment and in a timely manner. The combination of satellite-based mapping tools and webGIS/smartphone devices provides the basis for better and more informed decision over a variety of spatial and temporal scales. The final decision is then taken knowing local system components and management constraints. This implies a strong commitment by users and is contingent on the assumption that irrigation will be managed to achieve a specific target, for example maximum water use efficiency, yield, or profitability. This assumption seems valid in many intensive irrigated areas since farmers are interested in reducing operating constraints and costs, maximize or stabilize yield and, at the same time, maintain sustainability and environmental efficiency.

Our work contributed to evaluate the implementation and functioning process of a satellite-based irrigation advisory service in a very different environment compared to precursors systems running already in Mediterranean countries (e.g., Irrimet[6]), in Australia (e.g., Irrieye[7]) or in California (TOPS[8]). We calibrated and validated the models to derive crop biophysical variables using ground measurements (Vuolo et al., 2013), performed the demonstration and evaluation with a group of farmers and completed the adaptation of mapping and data dissemination technologies; finally, we assessed the benefits and costs of the service. An estimation of the net economic benefit was also obtained and presented in this study. Results confirm that the adoption of new space-based solutions for crop and irrigation water management is a complex process and favorable conditions depend on several aspects; creating the connection with the users and embedding new solutions into standard practices remains the most critical issue. Therefore, our approach has been based on establishing a local network and on intensive work with the users since the very beginning. This way of implementing has contributed to create distinctive results and has contributed to stimulate and build local capacities and, potentially, to create job opportunities in the future.

Feedback from final users is essential to improve the effectiveness of the service in terms of water use optimization, especially during the early stage of service implementation. In our opinion, a continuous effort will still be necessary to transfer, tune, and demonstrate these technologies in different environments and to adapt them to new challenges with the participation of the final users. Only with this approach we expect satellite technologies to become an integral part of existing management systems and bring real benefits.

OUTLOOK

Different irrigation management strategies are possible. For example, one may use satellite-based technologies for management of water-intensive crops. In that case, irrigation is optimized so that the crops produce at their maximum potential. A different irrigation strategy is to seek a compromise between cost of irrigation and loss of production. In this case, farmers irrigate their crops at the limit of the water stress conditions, to reduce costs and limit yield lost (Fereres and Soriano, 2007). This irrigation technique is adopted to optimize economic and technical resources, which are often limited.

In the Marchfeld, the latter irrigation approach is more prevalent. For instance, winter cereals or maize are irrigated only in dry years to avoid excessive yield losses. In these situations, crops grow under limited water availability, and irrigation water is applied only to supplement scarce rainfall. The decision whether to irrigate or not, when and how much, is a critical issue for farmers and it is often based only on personal experience. Monitoring the daily soil water balance is the first steps toward the implementation of these irrigation strategies. The approach necessitates the development of operative frameworks to run a soil water balance model at parcel level and to collect in near-real-time information on irrigation volumes applied by farmers. These technical requirements reflect and complement the requests of the users to adapt the existing webGIS system to (i) manually enter and store in the database the amounts of rainfall and irrigation volumes and use the system as a logbook for their management activities; (ii) access the information via smartphones to save time and have full mobility and (iii) receive alerts via SMS if particular stress conditions are met. To respond to this request, we developed an application for smartphones based on *Sencha Touch*[9], a software development framework for mobile applications that can be deployed on iOS, Android, BlackBerry, Windows Phone systems. The application shares the same mapping and database infrastructure of the desktop version of the existing webGIS and therefore maintenance and operational efforts are minimized. It consists of (i) a main panel for map and data visualization (Figure 8) with the information (crop ET, rainfall and irrigation) aggregated to daily, 7-day and from the start of irrigation; ii) a data input panel (Figure 9), where users can log parcel-specific management activities such as irrigation volumes and crop type; and (iii) 7-day weather forecast, currently based on OpenWeatherMap[10] services. The new system also includes a calculation module to assimilate ET raster maps in a simple soil water balance model (FAO-56), which activates if the farmer inserts the minimum set of inputs (irrigations and crop type) for the parcel under monitoring. The application will be deployed during the growing season 2015 for a first evaluation with users.

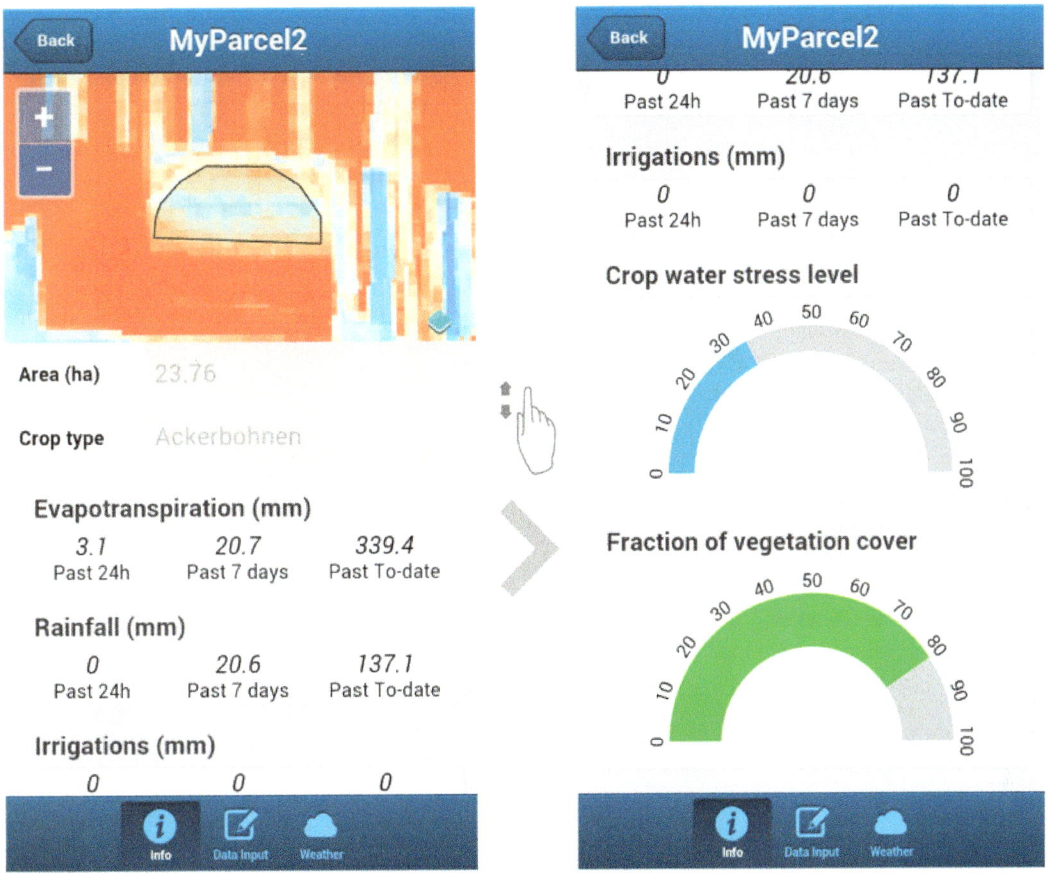

Figure 8. Mobile app: map and data visualization panel. The farmer can connect using user name and password to the list of subscribed parcels and display the latest crop coefficient map (red tones: low values), along with irrigation management parameters (evapotranspiration, rainfall, provided irrigations). The graphs show the crop water stress level based on the FAO-56 soil water balance model (which is activated only if irrigation inputs are provided) and the fraction of vegetation cover for the parcel.

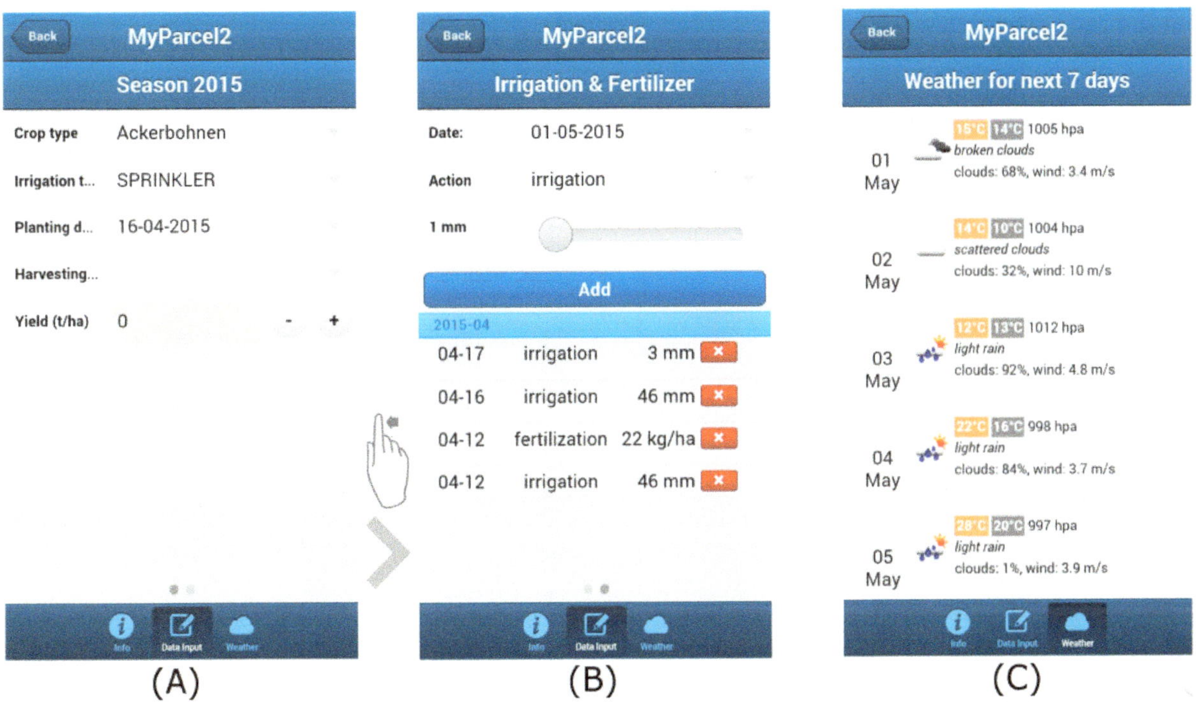

Figure 9. Data input panels (A,B) and weather data (C). Using panel **(A)** farmers can change the crop and irrigation type, dates of planting and harvest, crop yield. In **(B)** farmers can log their irrigation (used for the soil water balance) and fertilizer applications; this also helps to keep the history of the parcel and management. In **(C)** 7-day weather forecast are displayed; we use data obtained from Open Weather Map community but any type of local weather data can be used.

CONFLICT OF INTEREST STATEMENT

The authors declare that the research was conducted in the absence of any commercial or financial relationships that could be construed as a potential conflict of interest.

ACKNOWLEDGMENTS

This research was carried our within the "Earth Observation for water management project" (EO4Water) funded under *Die Österreichische Forschungsförderungsgesellschaft* (FFG) and the Austrian Space Applications Programme (ASAP). We thank Nikolaus Neugebauer from BOKU for help in processing the Earth observation and agrometeorological data. We thank Prof. Guido D›Urso from the University of Naples Federico II for fruitful discussion and feedback on the application of the remote sensing methodologies. We also acknowledge feedback from colleagues involved in the operational development of satellite-based services: C. De Michele, L. Marotta, S. Falanga Bolognesi and M. Colandrea from Ariespace. The smartphone application was developed by Mateusz Żółtak. The framework for the cost analysis was initially developed in the context of the SIRIUS-GMES project and reviewed by Anna Osann from University Castilla La Mancha, Andrew Groom from Geo-Intelligence Airbus Defence and Space and C. De Michele from Ariespace.

FOOTNOTES

1 ^www.irrisat.it

2 ^www.pleiades.es

3 ^www.sirius-gmes.es

4 ^www.marchfeldkanal.at/home.htm

5 ^www.deimos-imaging.com/

6 ^www.irrimet.eu/

7 ^www.irrieye.com/

8 ^ecocast.arc.nasa.gov/dgw/sims/

9 ^www.sencha.com/products/touch/

10 ^openweathermap.org/

REFERENCES

1. Allen, R. G., Pereira, L., Raes, D., and Smith, M. (1998). *Crop Evapotranspiration-Guidelines for Computing Crop Water Requirements-FAO Irrigation and Drainage Paper 56*. FAO, Rome. Available online at:http://www.engr.scu.edu/~emaurer/classes/ceng140_watres/handouts/FAO_56_Evapotranspiration.pdf (Accessed October 4, 2013).
2. Baptista, A. (2006). "Transferability Of DEMETER. A case study of The Irrigation Scheme of Veiga De Chaves," in *AIP Conference Proceedings* (Naples), 84–92. doi: 10.1063/1.2349331
3. Bausch, W. C. (1995). Remote sensing of crop coefficients for improving the irrigation scheduling of corn. *Agric. Water Manag.* 27, 55–68. doi: 10.1016/0378-3774(95)01125-3
4. Cepuder, P. (2004). "Soil water content monitoring for irrigation," in *Forstliche Schriftenreihe, Universität für Bodenkultur Wien, Vol. 18.*, eds R. Jandl, H. Hager, W. Loiskandl, M. Gerzabek, H. Mauser, O. Moog, and H. Richter (Wien: Österr Gesellschaft für Waldökosystemforschung und experimentelle Baumforschung), 46–52.
5. Cepuder, P., and Nolz, R. (2007). Irrigation management by means of soil moisture sensor technologies. *J. Water L. Dev.* 11, 79–90. doi: 10.2478/v10025-000-0007-0
6. Choudhury, B. J., Ahmed, N. U., Idso, S. B., Reginato, R. J., and Daughtry, C. S. (1994). Relations between evaporation coefficients and vegetation indices studied by model simulations. *Remote Sens. Environ.* 50, 1–17. doi: 10.1016/0034-4257(94)90090-6
7. Courault, D., Seguin, B., and Olioso, A. (2005). Review on estimation of evapotranspiration from remote sensing data: from empirical to numerical modeling approaches. *Irrigation Drain. Syst.* 19, 223–249. doi: 10.1007/s10795-005-5186-0
8. Drusch, M., Del Bello, U., Carlier, S., Colin, O., Fernandez, V., Gascon, F., et al. (2012). Sentinel-2: ESA›s optical high-resolution mission for GMES operational services. *Remote Sens. Environ.* 120, 25–36. doi: 10.1016/j.rse.2011.11.026
9. D›Urso, G. (2001). *Simulation and Management of On-demand Irrigation Systems: A Combined Agrohydrological and Remote Sensing Approach.* Wageningen: Wageningen University Press.
10. D'Urso, G., and Belmonte, A. C. (2006). "Operative approaches to determine crop water requirements from earth observation data: methodologies and applications," in *AIP Conference Proceedings* (Naples), 14–25.
11. D›Urso, G., and Menenti, M. (1995). "Mapping crop coefficients in irrigated areas from Landsat TM images," in *Proc. SPIE 2585, Remote Sensing for Agriculture, Forestry, and Natural Resources*, eds E. T. Engman, G. Guyot, and C. M. Marino (Paris: International Society for Optics and Photonics), 41–47.
12. D'Urso, G., Richter, K., Calera, A., Osann, M. A., Escadafal, R., Garatuza-Pajan, J., et al. (2010). Earth Observation products for operational irrigation management in the context of the PLEIADeS project. *Agric. Water Manag.* 98, 271–282. doi: 10.1016/j.agwat.2010.08.020
13. Eitzinger, J., Kersebaum, K. C., and Formayer, H. (2009). *Landwirtschaft im Klimawandel: Auswirkungen und Anpassungsstrategien für die Land- und Forstwirtschaftlichen Betriebe in Mitteleuropa.* (Agrimedia).
14. Fereres, E., and Soriano, M. A. (2007). Deficit irrigation for reducing agricultural water use. *J. Exp. Bot.* 58, 147–159. doi: 10.1093/jxb/erl165
15. Martimort, P., Berger, M., Carnicero, B., Bello, U. E., Arino, O., Biasutti, R., et al. (2007). Sentinel-2, the optical high-resolution mission for GMES operational services. *Esa Bull.* 131, 19–23. doi: 10.1109/igarss.2007.4423394
16. McMahon, T. A., Peel, M. C., Lowe, L., Srikanthan, R., and McVicar, T. R. (2013). Estimating actual, potential, reference crop and pan evaporation using standard meteorological data: a pragmatic synthesis. *Hydrol.*

Earth Syst. Sci. 17, 1331–1363. doi: 10.5194/hess-17-1331-2013

17. De Michele, C., Vuolo, F., D›Urso, G., Marotta, L., and Richter, K. (2009). "The irrigation advisory program of campania region: from research to operational support for the water directive in Agriculture," in *33rd International Symposium on Remote Sensing of Environment (ISRSE) Proceedings, May 4-8, 2009* (Stresa).
18. Neugebauer, N., and Vuolo, F. (2014). Crop water requirements on regional level using remote sensing data – a case study in the marchfeld region berechnung des pflanzenwasserbedarfs für sommerfeldfrüchte mittels fernerkundungsdaten. *Eine Fallstud. Marchfeld Region Photogramm. FernerkundungGeoinf.* 2014, 369–381. doi: 10.1127/1432-8364/2014/0230
19. OECD (2012). *OECD Environmental Outlook to 2050: The Consequences of Inaction*. Paris: OECD Publishing. doi: 10.1787/9789264122246-en
20. Osann Jochum, M. A. (2006). "Operational space-assisted irrigation advisory services: overview of and lessons learned from the project DEMETER," in *AIP Conference Proceedings* (Naples), 3–13. doi: 10.1063/1.2349322
21. Richey, A. S., Thomas, B. F., Lo, M.-H., Famiglietti, J. S., Swenson, S., and Rodell, M. (2015a). Uncertainty in global groundwater storage estimates in a total groundwater stress framework. *Water Resour. Res.* doi: 10.1002/2015WR017351. Available online at:http://onlinelibrary.wiley.com/doi/10.1002/2015WR017351/pdf
22. Richey, A. S., Thomas, B. F., Lo, M.-H., Reager, J. T., Famiglietti, J. S., Voss, K., et al. (2015b). Quantifying renewable groundwater stress with GRACE. *Water Resour. Res.* doi: 10.1002/2015WR017349. Available online at:http://onlinelibrary.wiley.com/doi/10.1002/2015WR017349/pdf
23. Rischbeck, P. (2007). *Der Einfluss von Klimaänderung, Bodenbearbeitung und Saattermin auf den Wasserhaus halt und das Ertragspotential von Getreide im Marchfeld*. Available online at:https://zidapps.boku.ac.at/abstracts/download.php?dataset_id=6338&property_id=107&role_id=NONE
24. Rubio, E. (2006). "Golden day comparison of methods to retrieve ET (Kc-NDVI, Kc-analytical, MSSEBS, METRIC)," in *AIP Conference Proceedings* (Naples), 193–200. doi: 10.1063/1.2349344
25. Task Committee on Standardization of Reference Evapotranspiration. (2005). "ASCE Standardized Reference Evapotranspiration Equation," in *ASCE-EWRI Task Committee Report*, eds R. G. Allen, I. A. Walter, R. L. Elliott, T. A. Howell, D. Itenfisu, M. E. Jensen, and R. L. Snyder (Kansas city: American Society of Civil Engineers).
26. Thaler, S., and Eitzinger, J. (2008). "Climate change impacts on selected crops in Marchfeld, Eastern Austria," *28th Conference ...*, 3–7. Available online at:https://ams.confex.com/ams/28Hurricanes/techprogram/paper_138941.htm (Accessed October 13, 2013).
27. Verstraeten, W. W. W., Veroustraete, F., and Feyen, J. (2008). Assessment of evapotranspiration and soil moisture content across different scales of observation. *Sensors* 8, 70–117. doi: 10.3390/s8010070
28. Vuolo, F., D›Urso, G., De Michele, C., Bianchi, B., and Cutting, M. (2015a). Satellite-based irrigation advisory services: a common tool for different experiences from Europe to Australia. *Agric. Water Manag.* 147, 82–95. doi: 10.1016/j.agwat.2014.08.004
29. Vuolo, F., Mattiuzzi, M., and Atzberger, C. (2015b). Comparison of the landsat surface reflectance Climate Data Record (CDR) and manually atmospherically corrected data in a semi-arid European study area. *Int. J. Appl. Earth Obs. Geoinf.* 42, 1–10. doi: 10.1016/j.jag.2015.05.003
30. Vuolo, F., De Michele, C., and Lazzaro, U. (2006). "Success stories: analysis and evaluation of the italian case-study of DEMETER in the Destra Sele Plain," in *AIP Conference Proceedings* (Naples), 33–42. doi: 10.1063/1.2349325
31. Vuolo, F., Neugebauer, N., Bolognesi, S., Atzberger, C., and D›Urso, G. (2013). Estimation of leaf area index using DEIMOS-1 data: application and transferability of a semi-empirical relationship between two agricultural areas. *Remote Sens.* 5, 1274–1291. doi: 10.3390/rs5031274
32. WEF (2013). *Putting the New Vision for Agriculture into Action: A Transformation is Happening*. Available online at: http://www.weforum.org/projects/new-vision-agriculture.
33. Wichelns, D., and Oster, J. D. (2006). Sustainable irrigation is necessary and achievable, but direct costs and environmental impacts can be substantial. *Agric. Water Manag.* 86, 114–127. doi: 10.1016/j.agwat.2006.07.014
34. WWAP (United Nations World Water Assessment Programme) (2015). *The United Nations World Water Development Report 2015: Water for a Sustainable World*. Paris: UNESCO. Available online at: http://unesdoc.unesco.org/images/0023/002318/231823E.pdf.

CHAPTER 14

VIEWS ON GRAND RESEARCH CHALLENGES FOR QUATERNARY GEOLOGY, GEOMORPHOLOGY AND ENVIRONMENTS

Steven L. Forman[1] and Gary E. Stinchcomb[2]

[1]Department of Geology, Baylor University, Waco, TX, USA

[2]Department of Geosciences and Watershed Studies Institute, Murray State University, Murray, KY, USA

INTRODUCTION

This missive presents one view of the grand challenges, broadly for disciplines within the geosciences, atmospheric, and ocean sciences, biosciences, chemistry, physics, engineering, and archaeology concerned with Quaternary geology, geomorphology, and associated environmental changes. This area of inquiry encompasses the global earth system on 10^{-3}–10^{6} year timescales addressing pressing questions such as tectonic and climatic dynamics, hominid evolution, and past, current, and future sustainability. Outlining "grand challenges" for this interdisciplinary science is a daunting task. Though often attempted, no one scientist or group of scientists has the depth of knowledge or insights to fully justify research challenges for a scientific discipline at a particular time. This effort, like many preceding, has unintentional biases based on limited scientific knowledge, reach, and vision, particularly in reference to the range of influencing disciplines. Often advances in a discipline do not originate at the center of inquiry but by those at the margins who bring new perspectives, knowledge, and computational and analytical prowess. Many future frontiers may remain unrealized because the spark of creativity has yet to leap a disciplinary boundary.

A revolution has occurred in the geosciences in the past 25 years with mounting evidence that human activity has and is altering the climate system, challenging the sustainability of the planet (cf. IPCC, 2014). The need to better understand current and future climate dynamics has propelled the international scientific community to study past periods of pronounced climate change, like the transition from glacial to interglacial conditions. The Quaternary, the past 2.6 million years, and more resolutely the past 25 ka is an enduring focus of climate change research because of the wealth of proxy climatic data and identifiable boundary conditions for Earth System Models (ESM) (cf. Ruddiman, 2014). Thus, for many in the community the grandest challenge is a better understanding of the sensitivity and the variable response of the past, present, and future state of the climate system to internal (e.g., greenhouse gases) and external (e.g., solar variability) drivers of planetary change; and the proportional response of and complex feed backs from the biosphere, hydrosphere, cryosphere, lithosphere, pedosphere, and anthrosphere.

As a scientific community we pass on a research legacy, an enduring grand challenge, to future generations to better understand anthropogenic climate change to curtail, adapt, and mitigate unintended harm done for continued sustenance of life on this planet. What follows is an outline of six "grand challenges" that are not necessarily definitive directions, but serve as examples of research frontiers that have and will continue to motivate scientists. This journal welcomes science from these, others and future frontiers, particularly those that are unidentified or unintentionally neglected in this contribution.

Citation: Steven L. Forman and Gary E. Stinchcomb, Views on grand research challenges for Quaternary geology, geomorphology and environments, http://dx.doi.org/10.3389/feart.2015.00047

UNDERSTANDING QUATERNARY ENVIRONMENTS FOR THE EVOLUTION AND MIGRATION OF HOMINIDS

Quaternary geology and geomorphology is a central discipline for understanding the progression and mechanisms of climate variability and associated changes in glacial, periglacial, fluvial, lacustrine, marine, and eolian systems in the past 2.6 million years (cf. Wright and Frye, 1965; Flint, 1971; Ehlers, 1996). This research coupled with paleo-genomics (e.g., Jobling et al., 2014; Ermini et al., 2015) and an expanding archaeological record (e.g.,Madsen et al., 2014; Bretzke and Conard, 2015) informs the community on the paleoenvironments associated with and timing of hominid evolution (e.g.,Scholz et al., 2011) and the spread of our genus across the planet. Outstanding questions remain on the paleoenvironmental context for the evolution of Hominids in East African during the past 800 ka, particularly with groups near large lake and river systems. There is a pressing need to define the multiple pathways and timing for dispersal of our ancestors across Africa, into the Middle East, Europe, and Asia (cf. Pearson, 2013).

In the Americas, the evidence for human migration is much later than in Asia or Europe, post the last glaciation. However, there is recent compelling archaeological evidence for a pre-Clovis culture, 14.5–15 ka old (e.g., Gilbert et al., 2009; Waters et al., 2011; Sistiaga et al., 2014) and possibly older (Lahaye et al., 2013). Large unknowns persist on the routes and timing of human migrations southward into the Americas (cf. Pitblado, 2011; Lahaye et al., 2013). Equally compelling questions remain on the associated paleoenvironmental and paleoclimatic conditions conducive for human migration from Subarctic Asia through North and South America.

INTERGLACIAL CLIMATE VARIABILITY AND MECHANISMS OF CLIMATE CHANGE ON A WARMER PLANET

A clear interdisciplinary focus is elucidating the timing and mechanisms for climate variability in the past 130 ka to place in context the ongoing changes in climate dynamics with the rise of greenhouse gases (cf. Bradley, 2015). The calibration and refining of Earth System models (ESM) to boundary conditions of past periods in the Quaternary is an ongoing effort to improve the predictive capabilities for future climate simulations (cf. Schmidt et al., 2014; Ziemen et al., 2014; Moberg et al., 2015). The integration of Quaternary proxy climatic data with ESMs on the global and the regional scale is needed to provide insight into non-linear response of climate dynamics on various spatial scales, for areas affected by monsoonal systems, and varying modes of ocean circulation, like the El Nino Southern Oscillation or the North Atlantic Oscillation. New knowledge is needed globally on the state of glaciers, ice sheets, northern tree line, tropics, and subtropics and arid lands during past interglacial periods such as the Eemian, ca. 130 ka ago (Dahl-Jensen et al., 2013; Nikolova et al., 2013) and during a shorter warm period centered around 80 ka, when sea level appears to be near current levels (Dorale et al., 2010). Additional insights are critical for the non-analogous climate response of Southern Hemisphere glaciers, lakes, rivers, and eolian systems, particularly in areas influenced by monsoonal variability, which was potentially intensified by ice sheet-induced cooling of the Northern Hemisphere oceans (Kanner et al., 2012; Rhodes et al., 2015).

ABRUPT CLIMATE AND ENVIRONMENTAL CHANGE

Studies of ice cores from Greenland that span the last glacial period revealed astonishing, abrupt transitions, spanning a few years, from warm to pronounced cooling periods, like during the Younger Dryas chronozone (Steffensen et al., 2008). Equally astonishing is that these rapid climate changes may be forced by mechanisms operating on comparatively gradual timescales that trigger a rapid change in system state (i.e., threshold response). The inferred driver for these rapid climate changes during the last glacial period is ice sheet modulation of Atlantic meridional overturning circulation and associated effects on CO2 levels (Clement and Peterson, 2008;Landais et al., 2015).

Pressing questions remain if abrupt shifts in the climate system can occur during interglacial conditions. Two time periods in the Holocene at ca. 8.2 ka and 4–5 ka appear to be intervals of rapid climate change, though the 8.2 ka event is associated with the last global effects of the retreating Laurentide ice sheet (Alley et al., 1997; Barber et al., 1999). Less is known about the 4–5 ka period, which may herald dramatic changes in global hydrology with lasting aridfication of East and North Africa (e.g., Forman et al., 2014; Bloszies et al., 2015) and possible desertification in eastern China (e.g., Yang et al., 2015). North American research on fluvial systems document an episode of incision, 3.5–6 ka, a possible hydrologic response to episodes of mid-Holocene drought (Counts et al., 2015) or a more complex hydrologic response involving sediment supply and changes in climate (Springer et al., 2009; Stinchcomb et al., 2012). Refining the timing and spatial extent of this incision could yield new insights into how drainage basins respond to abrupt climate change. Climatic variability in the past 5000 years on decadal to century timescales remains insufficiently understood with many past climate "surprises," such as sustained multi-decadal long mega-droughts during the Medieval Climate Anomaly in the central USA that exceeded historic drying (e.g., Cook et al., 2007; Stahle et al., 2007). Understanding extreme climate variability for our current interglacial, the Holocene, and associated atmosphere-vegetation-landscape feedbacks in semiarid and temperate areas is critical knowledge for assessing the potential threshold response of earth systems as our planet warms into the twenty-first century.

EARTH SURFACE PROCESSES IN THE (PALEO-) CRITICAL ZONE

A clear research focus within the community is within the so called "Critical Zone" (CZ) which is defined as the, "heterogeneous, near-surface environment in which complex interactions involving rock, soil, water, air, and living organisms regulate the natural habitat and determine the availability of life-sustaining resources" (National Research Council, 2001). Quantifying and modeling the modern CZ is a central challenge for achieving a sustainable planet because the CZ responds to climatic, anthropogenic, and tectonic forcings (Brantley et al., 2007). The CZ and, in particular, Earth's surface and soil is a product of multiple environmental factors that have varied over time, creating a "polygenetic paleosol" (Richter and Yaalon, 2012). This CZ history influences the current and future state of Earth's surface and global carbon cycle; thus, reconstructing this history and its effects on earth's surface is necessary to predict future changes. In this regard, advances in Quaternary geology, geomorphology and environmental reconstruction research that emphasize modern and paleo-CZs stand to make major contributions.

Reconstructions of past CZs in the Middle to Late Quaternary have been achieved by use of optical and cosmogenic radionuclide dating providing estimates of soil production, denudation rates, residence times, and dates of burial (cf. Ahr et al., 2013; Reusser et al., 2015). For example, luminescence dating along the Colorado Front Range show periods of CZ stability, burial and pedogenic rejuvenation over the past 18 ka (Leopold et al., 2011). Measurement of *in situ* cosmogenic ^{10}Be of river sediment on the southeastern Piedmont USA provides background (pre-European settlement) erosion rates that are 100-fold lower than post-settlement erosion rates (Reusser et al., 2015). Grand challenges remain to explore the effects of soil production and erosion rates on influx and storage (residence times) of sediment and carbon in depositional environments under a range of geographic and environmental constraints.

There is compelling research in the modeling of the CZ in the past and into the future with parameterization for earth surface processes, land use changes and pedogenesis (e.g., Nordt et al., 2012, 2013; Nordt and Driese, 2013;Pelletier et al., 2015). Challenges remain on quantifying non-linear, threshold-response interactions, particularly with extreme states, amongst such factors as topography, vegetation, land use, and sediment flux. Research on paleo-CZ has focused on development of transfer function for geochemical proxies from surface soils applied to buried soils that has yielded unique information on past climatic components, such as growing season temperature and mean annual precipitation (e.g., Nordt and Driese, 2010;Klopfenstein et al., 2015). Challenges remain to refine, test, and develop geochemical proxies linked to a wide variety of current pedogenic environments, recognize non-analogous conditions in the paleo-record and forward judicious interpretations of past critical zones.

There is a significant stock of buried carbon in Quaternary peat lands and sediments which may stay sequestered or be released as climate warms. Only recently has the carbon sequestered in Quaternary buried soils been systematically study. In the Holocene and Late Pleistocene the burial of soils, particularly by eolian activity have sequestered large carbon pools and with anthropogenic disturbance may become labile (Marin-Spiotta et al., 2014). Much like quantifying erosion rates, there is a pressing need to quantify the landscape scale variations in carbon storage due to soil burial, erosion, and exhumation (Chaopricha and Marin-Spiotta, 2014).

THE HUMAN IMPRINT ON THE EARTH SYSTEM

Evidence is mounting that human activities continue to affect earth surface processes with non-analog responses compared to Holocene landscape dynamics (Steffen et al., 2015). The Quaternary sciences have anchored this ongoing discussion by investigating the meaning, the initiation and the effects of the Anthropocene; a proposed new geologic epoch dominated by human impacts on earth systems (Crutzen, 2002; Zalasiewicz et al., 2008, in press;Lewis and Maslin, 2015). There is a robust dialog on how and to what extent humans have altered earth surface processes from sediment source to sink in the past decades to millennia in Asia, Europe, Australia, Africa, and the Americas and affected the global greenhouse gas inventory, ecosystem services, and modulated climate variability.

Land-use practices, especially agricultural expansion, across continents since the middle Holocene has been implicated in lowering biodiversity and providing new sources for atmospheric CH_4 and possible CO_2 (Ruddiman, 2007). These changes at the catchment scale also significantly altered upland and hillslope soil erosion rates through increased mass wasting (Notebaert et al., 2011; Reusser et al., 2015), flooding and fluvial sediment storage (Notebaert and Verstraeten, 2010; Hoffmann et al., 2013), regional hydrology (e.g., Contreras et al., 2013; Stinchcomb et al., 2013) and the overall sediment budget. Using empirical Holocene records to tune land surface and hydrological models that predict water and sediment flux; and the sensitivity of landscapes to change remains a challenge.

In turn, the growing foot-print of urbanization since the Nineteenth century has degraded biodiversity of riparian and adjacent environments and enhanced hydrological variability (Walsh et al., 2005; Wohl, 2015). Further advances in geomorphic and hydrologic data collection and modeling are needed to better understand processes that influence variability in sediment, water, and nutrient flux for natural and human-altered landscapes. Pressing questions remain on the role of human activities in desertification in past 5 ka in East and North Africa, across northern Asia and with more recent extreme hydrologic variability in the Americas. Studies show that there is a close correspondence between rapid changes in climate, flooding, and civilization (Macklin et al., 2013; Munoz et al., 2015). Continuing to resolve the timing in these changes will yield new insights into climate-human-water interactions. Although strides have been made toward quantifying anthropogenic impact (e.g., Ackermann et al., 2014; Notebaert and Berger, 2014; Stinchcomb et al., 2014; Vanmaercke et al., 2015), challenges persist in separating natural from anthropogenic drivers in landscape change.

QUATERNARY TECTONIC FRAMEWORK AND SEISMIC HAZARDS

Knowledge of Quaternary geology and geomorphology also yield significant insights in tectonic and volcanic processes for many fault zones and serve to quantify seismic hazards to safe guard urban population centers (cf. Yeats et al., 1997; McCalpin, 2009). The advances in paleoseismology in the past two decades has provided a deeper time perspective on earthquake generation, spanning centuries to multiple millennia, to place in context the shorter seismological and differential-GPS and satellite-based measurements of vertical and horizontal crustal movements. Studies have focused on addressing fault segmentation and geometry changes, recurrence intervals, and inferring earthquake magnitude in a variety of tectonic settings from active and passive plate margins to interplate areas. There remains too many fault systems in proximity to growing cities in Asia, the Caribbean, and Central, and South America in which little is known about seismic source zones, segmentation, slip-rates, and earthquake recurrence. Examples of critical areas of research include the North Anatolian Fault in Turkey, Longmenshen Belt in central China, fault systems in Hispaniola and the San Ramon Fault within the Santiago, Chile metropolitan area. Also, recent research has begun to address the paleo-tsunami record around the Pacific Ocean, with more studies needed (cf. Nelson et al., 2015).

CONCLUSION

It is an opportune time to pursue research in the geosciences, particularly focused on the Quaternary, earth surface processes, and environmental change. Scholarship that addresses these grand challenges should yield needed insights on the origins of our genus and specie, the response of earth systems to past, current, and future climate variability and anthropogenic-sourced change, and how expanding urban populations can sustainably interface with tectonic hazards and rapidly changing hydrologic conditions. An enduring need is the synthesis of proxy environmental information across Quaternary archives such as speleothems, glacial, marine, lacustrine, fluvial, and eolian records to better understand the components of climate variability and the complex response of earth systems. We hope this missive feeds new ideas and alternative and contradictory viewpoints that will motivate research beyond this limited view. These research challenges are not static but as John Bardeen said Science is a field which grows continuously with ever expanding frontiers."

CONFLICT OF INTEREST STATEMENT

The authors declare that the research was conducted in the absence of any commercial or financial relationships that could be construed as a potential conflict of interest.

ACKNOWLEDGMENTS

The views expressed in this contribution have developed through collaboration with many scientists and past students, particularly G. H. Miller, O. Ingolfsson, A. Bettis, M. Waters, A. Tripaldi, D. Wright, X. Yang, J. McCalpin, M. Machette, L. Nordt, R. Cox, J. Hoeffecker, and C. Bloszies. Though any deficiencies are ours alone. The comments of S. Driese and the reviewers are much appreciated.

REFERENCES

1. Ackermann, O., Greenbaum, N., Ayalon, A., Bar-Matthews, M., Boaretto, E., Bruins, H. J., et al. (2014). Using palaeo-environmental proxies to reconstruct natural and anthropogenic controls on sedimentation rates, Tell es-Safi/Gath, eastern Mediterranean. *Anthropocene* 8, 70–82. doi: 10.1016/j.ancene.2015.03.004
2. Ahr, S. W., Nordt, L. C., and Forman, S. L. (2013). Soil genesis, optical dating, and geoarchaeological evaluation of two upland alfisol pedons within the tertiary gulf coastal plain.*Geoderma* 192, 211–226. doi: 10.1016/j.geoderma.2012.08.016
3. Alley, R. B., Mayewski, P. A., Sowers, T., Stuiver, M., Taylor, K. C., and Clark, P. U. (1997). Holocene climatic instability: a prominent, widespread event 8200 yr ago. *Geology* 25, 483–486.
4. Barber, D. C., Dyke, A., Hillaire-Marcel, C., Jennings, A. E., Andrews, J. T., Kerwin, M. W., et al. (1999). Forcing of the cold event of 8,200 years ago by catastrophic drainage of Laurentide lakes.*Nature* 400, 344–348. doi: 10.1038/22504
5. Bloszies, C., Forman, S. L., and Wright, D. K. (2015). Water level history for Lake Turkana, Kenya in the past 15,000 years and a variable transition from the African Humid Period to Holocene aridity.*Glob. Planet. Change* 132, 64–76. doi: 10.1016/j.gloplacha.2015.06.006
6. Bradley, R. S. (2015). *Paleoclimatology: Reconstructing Climates of the Quaternary*. New York, NY: AcademicPress.
7. Brantley, S. L., Goldhaber, M. B., and Ragnarsdottir, K. V. (2007). Crossing disciplines and scales to understand the Critical Zone. *Elements* 3, 307–314. doi: 10.2113/gselements.3.5.307
8. Bretzke, K., and Conard, N. J. (2015). The archaeological signatures of Late Pleistocene populations' dynamics of archaic and modern humans in Arabia and Southwestern Asia. *Am. J. Phys. Anthropol.* 156, 92.
9. Chaopricha, N. T., and Marin-Spiotta, E. (2014). Soil burial contributes to deep soil organic carbon storage. *Soil Biol. Biochem.* 69, 251–264. doi: 10.1016/j.soilbio.2013.11.011
10. Clement, A. C., and Peterson, L. C. (2008). Mechanisms of abrupt climate change of the last glacial period. *Rev. Geophys.* 46:RG4002. doi: 10.1029/2006RG000204
11. Contreras, S., Santoni, C. S., and Jobbagy, E. G. (2013). Abrupt watercourse formation in a semiarid sedimentary landscape of central Argentina: the roles of forest clearing, rainfall variability and seismic activity. *Ecohydrology* 6, 794–805. doi: 10.1002/eco.1302
12. Cook, E. R., Seager, R., Cane, M. A., and Stahle, D. W. (2007). North American drought: reconstructions, causes, and consequences. *Earth Sci. Rev.* 81, 93–134. doi: 10.1016/j.earscirev.2006.12.002
13. Counts, R. C., Murari, M. K., Owen, L. A., Mahan, S. A., and Greenan, M. (2015). Late Quaternary chronostratigraphic framework of terraces and alluvium along the lower Ohio River, southwestern Indiana and western Kentucky, USA. *Quat. Sci. Rev.* 110, 72–91. doi: 10.1016/j.quascirev.2014.11.011
14. Crutzen, P. J. (2002). Geology of mankind. *Nature* 415, 23. doi: 10.1038/415023a
15. Dahl-Jensen, D., Albert, M. R., Aldahan, A., Azuma, N., Balslev-Clausen, D., Baumgartner, M., et al. (2013). Eemian interglacial reconstructed from a Greenland folded ice core. *Nature* 7433, 489–494. doi: 10.1038/nature11789
16. Dorale, J. A., Onac, B. P., Fornós, J. J., Ginés, J., Ginés, A., Tuccimei, P., et al. (2010). Sea-level highstand 81,000 years ago in Mallorca. *Science* 327, 860–863. doi: 10.1126/science.1181725
17. Ehlers, J. (1996). *Quaternary and Glacial Geology*. New York, NY: Wiley.
18. Ermini, L., Der Sarkissian, C., Willerslev, E., and Orlando, L. (2015). Major transitions in human evolution

revisited: a tribute to ancient DNA. *J. Hum. Evol.* 79, 4–20. doi: 10.1016/j.jhevol.2014.06.015

19. Flint, R. F. (1971). *Glacial and Quaternary Geology*. New York, NY: Wiley.

20. Forman, S. L., Wright, D. K., and Bloszies, C. (2014). Variations in water level for Lake Turkana in the past 8500 years near Mt. Porr, Kenya and the transition from the African Humid Period to Holocene aridity. *Quat. Sci. Rev.* 97, 84–101. doi: 10.1016/j.quascirev.2014.05.005

21. Gilbert, M. T. P., Poinar, H., Fiedel, S., King, C., Devault, A., Bos, K., et al. (2009). DNA from pre-Clovis human coprolites in Oregon, North America. *Science* 325, 148–148. doi: 10.1126/science.1168457

22. Hoffmann, T., Schlummer, M., Notebaert, B., Verstraeten, G., and Korup, O. (2013). Carbon burial in soil sediments from Holocene agricultural erosion, Central Europe. *Glob. Biogeochem. Cycles* 27, 828–835. doi: 10.1002/gbc.20071

23. IPCC. (2014). “Climate Change 2014: Synthesis Report,” in *Contribution of Working Groups I, II and III to the Fifth Assessment Report of the Intergovernmental Panel on Climate Change*, eds R. K. Pachauri and L. A. Meyer (Geneva: IPCC), 151.

24. Jobling, M., Hollox, E., Hurles, M., Kivisild, T., and Tyler-Smith, C. (2014). *Origins of Modern Humans, Human, in Evolutionary Genetics, 2nd Edn*. New York, NY: Garland Science.

25. Kanner, L. C., Burns, S. J., Cheng, H., and Edwards, R. L. (2012). High-latitude forcing of the South American summer monsoon during the last glacial. *Science* 335, 570–573. doi: 10.1126/science.1213397

26. Klopfenstein, S. T., Hirmas, D. R., and Johnson, W. C. (2015). Relationships between soil organic carbon and precipitation along a climosequence in loess-derived soils of the Central Great Plains, USA. *Catena* 133, 25–34. doi: 10.1016/j.catena.2015.04.015

27. Lahaye, C., Hernandez, M., Boeda, E., Felice, G. D., Guidon, N., Hoeltz, S., et al. (2013). Human occupation in South America by 20,000 BC: the Toca da Tira Peia site, Piaui, Brazil. *J. Archaeol. Sci.* 40, 2840–2847. doi: 10.1016/j.jas.2013.02.019

28. Landais, A., Masson-Delmotte, V., Stenni, B., Selmo, E., Roche, D. M., Jouzel, J., et al. (2015). A review of the bipolar see-saw from synchronized and high resolution ice core water stable isotope records from Greenland and East Antarctica. *Quat. Sci. Rev.* 114, 18–32. doi: 10.1016/j.quascirev.2015.01.031

29. Leopold, M., Voelkel, J., Dethier, D., Huber, J., and Steffens, M. (2011). Characteristics of a paleosol and its implication for the critical zone development, rocky mountain front range of Colorado, USA.*Appl. Geochem.* 26, S72–S75. doi: 10.1016/j.apgeochem.2011.03.034

30. Lewis, S. L., and Maslin, M. A. (2015). Defining the anthropocene. *Nature* 519, 171–180. doi: 10.1038/nature14258

31. Macklin, M. G., Woodward, J. C., Welsby, D. A., Duller, G. A. T., Williams, F. M., and Williams, M. A. J. (2013). Reach-scale river dynamics moderate the impact of rapid Holocene climate change on floodwater farming in the desert Nile. *Geology* 41, 695–698. doi: 10.1130/G34037.1

32. Madsen, D. B., Oviatt, C. G., Zhu, Y., Brantingham, P. J., Elston, R. G., Chen, F., et al. (2014). The early appearance of Shuidonggou core-and-blade technology in north China: implications for the spread of Anatomically Modern Humans in northeast Asia? *Quat. Int.* 347, 21–28. doi: 10.1016/j.quaint.2014.03.051

33. Marin-Spiotta, E., Chaopricha, N. T., Plante, A. F., Diefendorf, A. F., Mueller, C. W., Grandy, A. S., et al. (2014). Long-term stabilization of deep soil carbon by fire and burial during early Holocene climate change. *Nat. Geosci.* 7, 428–432. doi: 10.1038/ngeo2169

34. McCalpin, J. P. (2009). *Paleoseismology*. New York, NY: AcademicPress.

35. Moberg, A., Sundberg, R., Grudd, H., and Hind, A. (2015). Statistical framework for evaluation of climate model simulations by use of climate proxy data from the last millennium—Part 3: practical considerations, relaxed assumptions, and using tree-ring data to address the amplitude of solar forcing. *Clim. Past* 11, 425–448. doi: 10.5194/cp-11-425-2015

36. Munoz, S. E., Gruley, K. E., Massie, A., Fike, D. A., Schroeder, S., and Williams, J. W. (2015). Cahokia’s emergence and decline coincided with shifts of flood frequency on the Mississippi River.*Proc. Natl. Acad. Sci. U.S.A.* 112, 6319–6324. doi: 10.1073/pnas.1501904112

37. National Research Council. (2001). *Basic Research Opportunities in Earth Science.* Washington, DC: National Academy Press.

38. Nelson, A., Briggs, R., Dura, T., Englehart, S., Gelfenbaum, G., Bradley, L.-A., et al. (2015). Tsunami recurrence in the eastern Alaska-Aleutian arc: a stratigraphic record from Chirikof Island, Alaska. *Geosphere* 11, 1172–1203. doi: 10.1130/GES01108.1

39. Nikolova, I., Yin, Q., Berger, A., Singh, U. K., and Karami, M. P. (2013). The last interglacial (Eemian) climate simulated by LOVECLIM and CCSM3. *Clim. Past* 9, 1789–1806. doi: 10.5194/cp-9-1789-2013

40. Nordt, L. C., and Driese, S. G. (2010). A modern soil characterization approach to reconstructing physical and chemical properties of paleo-vertisols. *Am. J. Sci.* 310, 37–64. doi: 10.2475/01.2010.02

41. Nordt, L. C., and Driese, S. G. (2013). Application of the critical zone concept to the deep-time sedimentary record. *Sedimen. Rec.* 11, 4–9. doi: 10.2110/sedred.2013.3.4

42. Nordt, L. C., Hallmark, C. T., Driese, S. G., and Dworkin, S. I. (2013). “Multi-analytical pedosystem approach to characterizing and interpreting the fossil record of soils,” in *New Frontiers in Paleopedology and Terrestrial Paleoclimatology*, Vol. 104, eds S. G. Driese and L. C. Nordt (SEPM Special Publication), 89–107.

43. Nordt, L. C., Hallmark, C. T., Driese, S. G., Dworkin, S. I., and Atchley, S. C. (2012). Biogeochemical characterization of a lithified paleosol: implications for the interpretation of ancient Critical Zones.*Geochim. Cosmochim. Acta* 87, 267–282. doi: 10.1016/j.gca.2012.03.019

44. Notebaert, B., and Berger, J.-F. (2014). Quantifying the anthropogenic forcing on soil erosion during the Iron Age and Roman Period in southeastern France. *Anthropocene* 8, 59–69. doi: 10.1016/j.ancene.2015.05.004

45. Notebaert, B., and Verstraeten, G. (2010). Sensitivity of West and Central European river systems to environmental changes during the Holocene A review. *Earth Sci. Rev.* 103, 163–182. doi: 10.1016/j.earscirev.2010.09.009

46. Notebaert, B., Verstraeten, G., Ward, P., Renssen, H., and van Rompaey, A. (2011). Modeling the sensitivity of sediment and water runoff dynamics to Holocene climate and land use changes at the catchment scale. *Geomorphology* 126, 18–31. doi: 10.1016/j.geomorph.2010.08.016

47. Pearson, O. M. (2013). "Africa: the cradle of modern people," in *The Origins of Modern Humans: Biology Reconsidered*, eds F. H. Smith and J. C. M. Ahern (Hoboken, NJ: John Wiley & Sons, Inc.), 1–43. doi: 10.1002/9781118659991.ch1

48. Pelletier, J. D., Murray, A. B., Pierce, J. L., Bierman, P. R., Breshears, D. D., Crosby, B. T., et al. (2015). Forecasting the response of Earth's surface to future climatic and land-use changes: a review of methods and research needs. *Earth Future* 3, 220–251. doi: 10.1002/2014EF000290

49. Pitblado, B. L. (2011). A tale of two migrations: reconciling recent biological and archaeological evidence for the pleistocene peopling of the americas. *J. Archaeol. Res.* 19, 327–375. doi: 10.1007/s10814-011-9049-y

50. Reusser, L., Bierman, P., and Rood, D. (2015). Quantifying human impacts on rates of erosion and sediment transport at a landscape scale. *Geology* 43, 171–174. doi: 10.1130/G36272.1

51. Rhodes, R. H., Brook, E. J., Chiang, J. C. H., Blunier, T., Maselli, O. J., McConnell, J. R., et al. (2015). Enhanced tropical methane production in response to iceberg discharge in the North Atlantic. *Science* 348, 1016–1019. doi: 10.1126/science.1262005

52. Richter, D. deB., and Yaalon, D. H. (2012). "The changing model of soil" revisited. *Soil Sci. Soc. Am. J.* 76, 766–778. doi: 10.2136/sssaj2011.0407

53. Ruddiman, W. F. (2007). The early anthropogenic hypothesis: challenges and responses. *Rev. Geophys.* 45, RG4001. doi: 10.1029/2006RG000207

54. Ruddiman, W. F. (2014). *Earth's Climate: Past and Future.* New York, NY: W. H. Freeman.

55. Schmidt, G. A., Annan, J. D., Bartlein, P. J., Cook, B. I., Guilyardi, E., Hargreaves, J. C., et al. (2014). Using palaeo-climate comparisons to constrain future projections in CMIP5. *Clim. Past* 10, 221–250. doi: 10.5194/cp-10-221-2014

56. Scholz, C. A., Cohen, A. S., Johnson, T. C., King, J., Talbot, M. R., and Brown, E. T. (2011). Scientific drilling in the Great Rift Valley: the 2005 lake malawi scientific drilling project—an overview of the past 145,000 years of climate variability in Southern Hemisphere East Africa. *Palaeogeogr. Palaeoclimatol. Palaeoecol.* 303, 3–19. doi: 10.1016/j.palaeo.2010.10.030

57. Sistiaga, A., Berna, F., Laursen, R., and Goldberg, P. (2014). Steroidal biomarker analysis of a 14,000 years old putative human coprolite from Paisley Cave, Oregon. *J. Archaeol. Sci.* 41, 813–817. doi: 10.1016/j.jas.2013.10.016

58. Springer, G. S., Rowe, H. D., Hardt, B., Cocina, F. G., Edwards, R. L., and Cheng, H. (2009). Climate driven changes in river channel morphology and base level during the Holocene and late Pleistocene of southeastern West Virginia RID C-1395-2008. *J. Cave Karst Stud.* 71, 121–129.

59. Stahle, D. W., Fye, F. K., Cook, E. R., and Griffin, R. D. (2007). Tree-ring reconstructed megadroughts over North America since AD 1300. *Clim. Change* 83, 133–149. doi: 10.1007/s10584-006-9171-x

60. Steffen, W., Richardson, K., Rockström, J., Cornell, S. E., Fetzer, I., Bennett, E. M., et al. (2015). Planetary boundaries: guiding human development on a changing planet. *Science* 347, 736. doi: 10.1126/science.1259855

61. Steffensen, J. P., Andersen, K. K., Bigler, M., Clausen, H. B., Dahl-Jensen, D., Fischer, H., et al. (2008). High-resolution Greenland Ice Core data show abrupt climate change happens in few years.*Science* 321, 680–684. doi: 10.1126/science.1157707

62. Stinchcomb, G. E., Driese, S. G., Nordt, L. C., and Allen, P. M. (2012). A mid to late Holocene history of floodplain and terrace reworking along the middle Delaware River valley, USA. *Geomorphology*169–170, 123–141. doi: 10.1016/j.geomorph.2012.04.018

63. Stinchcomb, G. E., Messner, T. C., Stewart, R. M., and Driese, S. G. (2014). Estimating fluxes in anthropogenic lead using alluvial soil mass-balance geochemistry, geochronology and archaeology in eastern USA. *Anthropocene* 8, 25–38. doi: 10.1016/j.ancene.2015.03.001

64. Stinchcomb, G. E., Messner, T. C., Williamson, F. C., Driese, S. G., and Nordt, L. C. (2013). Climatic and human controls on Holocene floodplain vegetation changes in eastern Pennsylvania based on the isotopic composition of soil organic matter. *Quat. Res.* 79, 377–390. doi: 10.1016/j.yqres.2013.02.004

65. Vanmaercke, M., Poesen, J., Govers, G., and Verstraeten, G. (2015). Quantifying human impacts on catchment sediment yield: a continental approach. *Glob. Planet. Change* 130, 22–36. doi: 10.1016/j.gloplacha.2015.04.001

66. Walsh, C. J., Roy, A. H., Feminella, J. W., Cottingham, P. D., Groffman, P. M., and Morgan, R. P. (2005). The urban stream syndrome: current knowledge and the search for a cure. *J. North Am. Benthol. Soc.* 24, 706–723. doi: 10.1899/0887-3593(2005)024

67. Waters, M. R., Forman, S. L., Jennings, T. A., Nordt, L. C., Driese, S. G., Feinberg, J. M., et al. (2011). The Buttermilk creek complex and the origins of clovis at the debra l. friedkin site, Texas.*Science* 331, 1599–1603. doi: 10.1126/science.1201855

68. Wohl, E. (2015). Legacy effects on sediments in river corridors. *Earth Sci. Rev.* 147, 30–53. doi: 10.1016/j.earscirev.2015.05.001

69. Wright, H. E., and Frye, D. G. Jr. (1965). *Quaternary of the United States*. New Jersey, NJ: Princeton University Press.

70. Yang, X. P., Scuderi, L. A., Wang, X. L., Scuderi, L. J., Zhang, D. G., Li, H. W., et al. (2015). Groundwater sapping as the cause of irreversible desertification of Hunshandake Sandy Lands, Inner Mongolia, northern China. *Proc. Natl. Acad. Sci. U.S.A.* 112, 702–706. doi: 10.1073/pnas.1418090112

71. Yeats, R. S., Sieh, K., and Allen, C. A. (1997). *The Geology of Earthquakes*. New York, NY: Oxford.

72. Zalasiewicz, J., Smith, A., Barry, T. L., Coe, A. L., Bown, P. R., Brenchley, P., et al. (2008). Are we now living in the Anthropocene? *GSA Today*, 18, 4–8. doi: 10.1130/GSAT01802A.1

73. Zalasiewicz, J., Waters, C. N., Williams, M., Barnosky, A. D., Cearreta, A., Crutzen, P., et al. (in press). When did the Anthropocene begin? A mid-twentieth century boundary level is stratigraphically optimal. *Quat. Int.* doi: 10.1016/j.quaint.2014.11.045.

74. Ziemen, F. A., Rodehacke, C. B., and Mikolajewicz, U. (2014). Coupled ice sheet-climate modeling under glacial and pre-industrial boundary conditions. *Clim. Past* 10, 1817–1836. doi: 10.5194/cp-10-1817-2014

CHAPTER 15 GEOLOGY OF LONDON, UK

Katherine R. Royse [a], Mike de Freitas [b,c], William G. Burgess [d], John Cosgrove [e], Richard C. Ghail [c], Phil Gibbard [f], Chris King [g], Ursula Lawrence [h], Rory N. Mortimore [i], Hugh Owen [j], Jackie Skipper [k]

[a] British Geological Survey, Keyworth, Nottingham NG12 5GG, UK

[b] First Steps Ltd, Unit 17 Hurlingham Studios, London SW6 3PA, UK

[c] Department of Civil and Environmental Engineering, Imperial College London, London SW7 2AZ, UK

[d] Department of Earth Sciences, University College London, WC1E 6BT, UK

[e] Department of Earth Science and Engineering, Imperial College London, London SW7 2AZ, UK

[f] Cambridge Quaternary, Department of Geography, University of Cambridge, CB2 3EN, UK

[g] 16A Park Road, Bridport, Dorset, UK

[h] Crossrail Ltd. 25 Canada Square, Canary Wharf, London E14 5LQ, UK

[i] University of Brighton & ChalkRock Ltd, 32 Prince Edwards Road, Lewes BN7 1BE, UK

[j] Department of Palaeontology, the Natural History Museum, Cromwell Road, London SW7 5BD, UK

[k] Geotechnical Consulting Group (GCG), 52A Cromwell Road, London SW7 5BE, UK

ABSTRACT

The population of London is around 7 million. The infrastructure to support this makes London one of the most intensively investigated areas of upper crust. However construction work in London continues to reveal the presence of unexpected ground conditions. These have been discovered in isolation and often recorded with no further work to explain them. There is a scientific, industrial and commercial need to refine the geological framework for London and its surrounding area. This paper reviews the geological setting of London as it is understood at present, and outlines the issues that current research is attempting to resolve.

INTRODUCTION

This paper reviews the geological framework for London and its surrounding area. It highlights the complex nature of London's geology and the possible implications for current and future development. Within this review article the London Basin's boundaries are defined by the limit of the Chalk outcrop (Sumbler, 1996). The Basin is described bySumbler (1996), as a broad, gentle synclinal fold (Fig. 1); although its seaward extension is not shown, the structure continues out into the North Sea (King, 2006). The term 'London Basin' was first used on the maps of Smith (1815) and Greenough (1820) to describe the sediments that make up the geology of London (Sheppard, 1917). The outcrop limits of the Chalk were fundamental in defining the original limits of this "basin". However, to understand London's geology, it is necessary to go beyond these present-day geographical limits. It is really only in the later stages of geological history that the idea of a London-centred geology can be considered helpful.

Citation: Katherine R. Royse, Mike de Freitas, William G. Burgess, John Cosgrove, Richard C. Ghail, Phil Gibbard, Chris King, Ursula Lawrence, Rory N. Mortimore, Hugh Owen, Jackie Skipper, Geology of London, UK, Proceedings of the Geologists' Association, Volume 123, Issue 1, January 2012, Pages 22-45, ISSN 0016-7878, http://dx.doi.org/10.1016/j.pgeola.2011.07.005

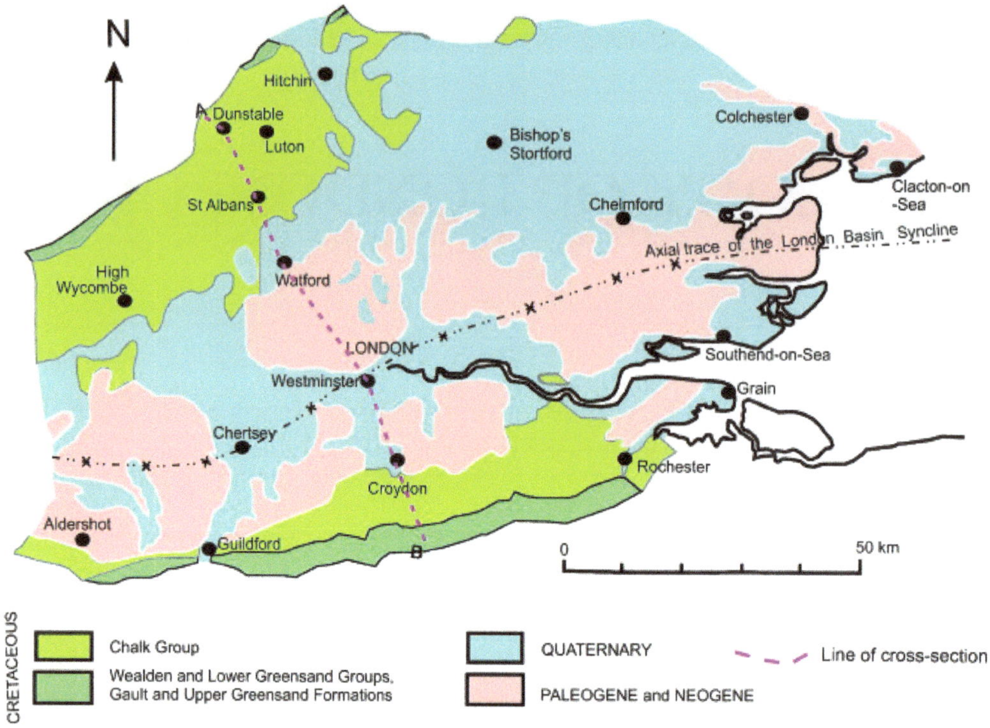

Figure 1: Geological sketch map of the London Basin. Based on Sumbler (1996, Fig. 1).

The British Geological Survey commenced detailed geological mapping of the London Basin in 1861 and soon realised that the correlation of strata across and beneath London was not always as straightforward as it at first appeared (Whitaker, 1872). For example, the substantial and isolated Chalk outcrop at Windsor in the west occurs 5 km south-east of the main Chalk outcrop near Maidenhead. There are smaller but largely unexplained Chalk structures mapped in the east by Wooldridge, 1923 and Wooldridge, 1926 andWooldridge and Linton (1939). However in 1947 the Basin was still presented by the British Geological Survey as a simple unfaulted downwarp on undifferentiated basement (Sherlock, 1947).

Boreholes drilled in the London Basin for oil, coal and gas have proved the extent of a Palaeozoic basement, the London Platform, of folded Silurian and foreland Devonian rocks at depths of ~300 m in central London (Sumbler, 1996). Upon these Palaeozoic rocks are found remnants of strata of Jurassic age, which are themselves covered unconformably by strata of Early Cretaceous age (Table 1). The Gault is the earliest preserved formation to cover the whole area. The London Platform (Fig. 2) is considered to extend to the Worcestershire Basin in the west, to the East Midlands Shelf in the north, to the southern North Sea graben in the east, and to the Weald Basin in the south (Sumbler, 1996). This Platform forms the western part of the London-Brabant massif (Lee et al., 1993). The southern boundary of the Platform and its Western boundary are fault controlled. The faulted southern boundary has been referred to as the "Variscan Front", a tectonic line extending from the Bristol Channel to cross southern England to the Strait of Dover.

Table 1: Summary of the geological strata of the London Basin from Ellison et al., 2004 with Chalk Group thickness updated from Royse et al. (2010).

Period	Group	Formation	Thickness (m)
Paleogene		Bagshot Formation: sand, fine-grained with thin clay beds	10–25
	Thames	London Clay Formation: clay, silty; fine sand clay at base. Claygate member: interbedded sand and clay at top	90–130
		Harwich Formation: sand, clayey fine grained sand and pebble beds	0–10
	Lambeth	Reading, Woolwich And Upnor Formations: clay mottled with fine-grained sand, laminated clay, flint pebble beds and shelly clay	10–20
		Thanet Sand Formation: sand, fine-grained	0–30

Cretaceous	Chalk	Seaford and Newhaven Chalk Formations Undivided: chalk soft, white with flint courses	Up to 70
		Lewes Nodular Chalk Formation: chalk, white with hard, nodular beds	25–46
		New Pit Chalk Formation: chalk white to grey with few flints	30–50
		Holywell Nodular Chalk Formation: chalk white to grey, shelly, hard and nodular	11–18
		Zig Zag Chalk Formation and West Melbury Chalk Formation: chalk, pale grey with thin marls; glaconitic at the base	40–80
		Upper Greensand Formation: sand fine-grained, glauconitic	Up to 17
		Gault Formation: clay, silty	50–70
	Lower	Folkestone Formation: sandstone, fine to medium-grained	60
Concealed strata	Greensand	Sandgate, Hythe and Atherfield Clay Formations: sandstone and mudstone	34
	Wealden	Weald Clay Formation: mudstone	Up to 150
		Hastings Beds: sandstone and mudstone	
Jurassic		Limestone and mudstone	0–c. 750
Silurian and Devonian		Sandstone and siltstone	

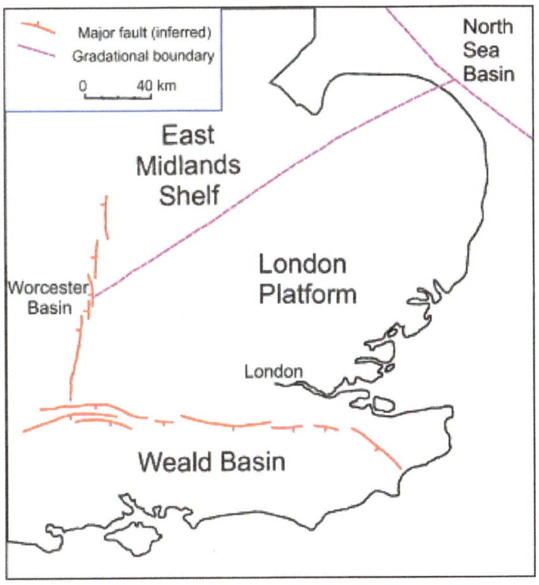

Figure 2: Mesozoic structural setting (Sumbler, 1996, Fig. 8). Sumbler notes that the Southern and western boundaries of the London Platform are defined by basement faults but the northern and eastern boundary with the East midlands Shelf and North Sea Basin are gradational and arbitrary.

The late Variscan pattern of faulting in the Platform had a profound effect on the subsequent location and development of Mesozoic and Cenozoic tectonic structures, because tectonic stresses associated with the rotation of Africa and the opening of the Atlantic reactivated late Variscan block-faulting. At the Jurassic – Cretaceous boundary, these late Variscan faults which had produced a "proto-Wealden Basin" at the end of the Palaeozoic were partly reactivated to form the current Weald Basin. The products of this erosion were redeposited in the Weald and Hampshire Basins during the Early Cretaceous, initially as the freshwater Wealden Group and subsequently as the marine Lower Greensand. The Lower Greensand is present widely below the Gault in borings in the London Platform, but is essentially of Late Aptian and Early Albian age. Later, the Gault Sea covered the Platform, depositing deeper water clays.

The geological structure of the Cretaceous and Paleogene strata which overlie this basement has in the past been considered 'relatively simple' (Ellison et al., 2004); for example, despite the accumulated indirect evidence for brittle structures in the basin, only two faults are shown on the current geological maps for the region: the Wimbledon-Streatham fault and the Greenwich fault (Fig. 3). There is, however, a growing body of direct evidence, particularly from recent deeper engineering projects, such as the Channel Tunnel Rail Link, Thames Water Ring Main (Newman, 2009), Crossrail and the Docklands Light Railway, which demonstrate that there is much more faulting in London and that the structure of London is more complex (Royse, 2010). One aspect of the tectonic history that is of overriding significance to the geological development of the London Basin, and consequently to the application of London's geology to engineering and water supply, is the location of a broad tectonic boundary running approximately east to west beneath the London Basin (Fig. 4). It originates from movements in the Variscan orogeny during and after the Carboniferous and is likely to have been reactivated periodically ever since as a line of crustal weakness.

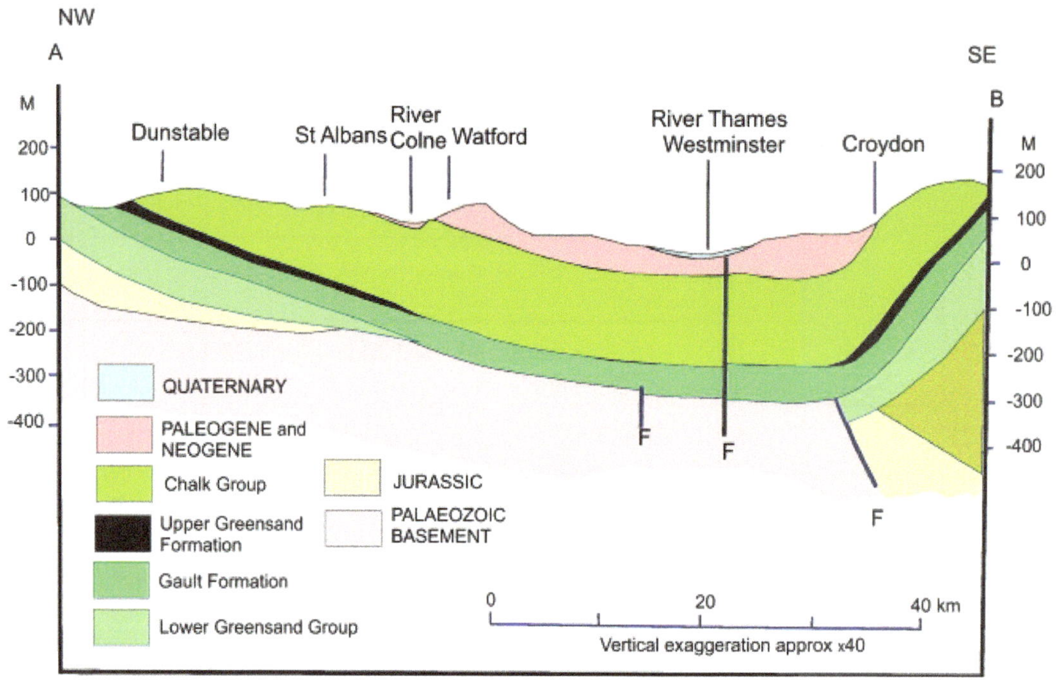

Figure 3: Geological cross-section across the region showing the 'relatively simple' geological structure of region as previously proposed by Sumbler (1996). Section based on Sumbler (1996, Fig. 2).

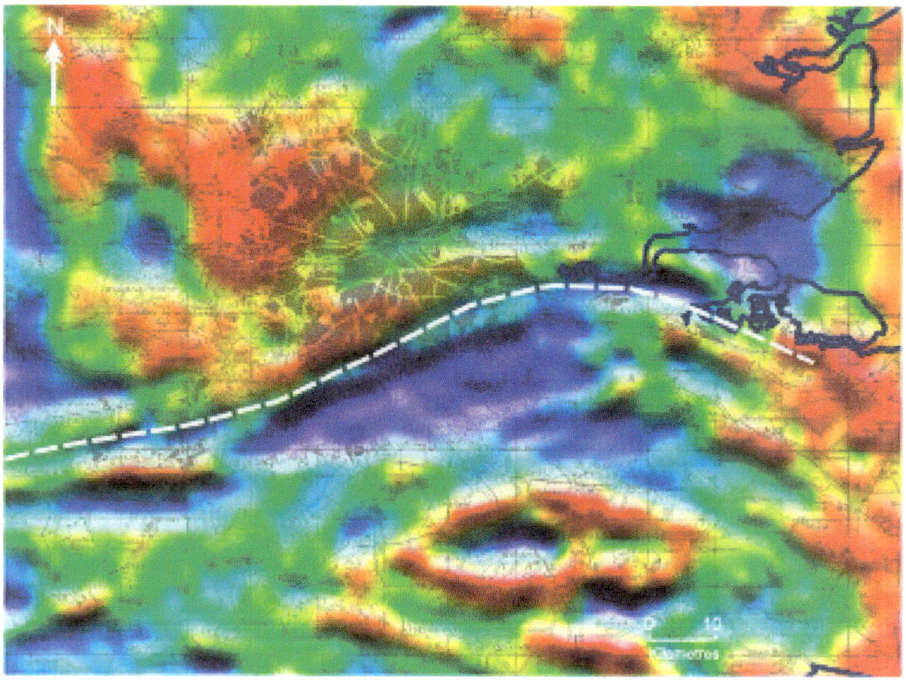

Figure 4: Colour-shaded Bouguer gravity relief map showing location of a broad tectonic boundary running east–west beneath the London Basin (dashed white line) From Ellison et al. (2004). OS data © Crown Copyright. All rights reserved. BGS 100017897/2009.

The Upper Cretaceous Chalk Group is present at subcrop throughout London and comes to the surface along its southern margin (the North Downs) and its northwest margin (the Chiltern Hills); elsewhere it occurs locally at or close to the surface, e.g. along the Greenwich and Purfleet anticlines in East London and also within the Chislehurst 'horst'. The Chalk of the London Basin is typically a fine grained white limestone made largely from organically precipitated aragonite (Hancock, 1975), in contrast to the largely clastic formations below and above it, and represents a period of very high sea-levels. Of the nine Chalk formations defined for mapping in southern England (Bristow et al., 1997), six are present in the London Basin (Fig. 5). These are distinguished by the presence or absence of marl and flint bands, their physical properties including density and porosity and by their colour. However, it is important to appreciate that in the London area the total thickness of the Chalk preserved is only between 170 and 210 m and it generally thins from west to east; the succession within the London Basin is relatively thin compared to that of the more complete Hampshire–Dieppe Basin where younger formations are preserved and the Chalk is over 400 m thick (Ellison et al., 2004). Thus, a substantial history of late and post-Cretaceous uplift and erosion is recorded in the London region.

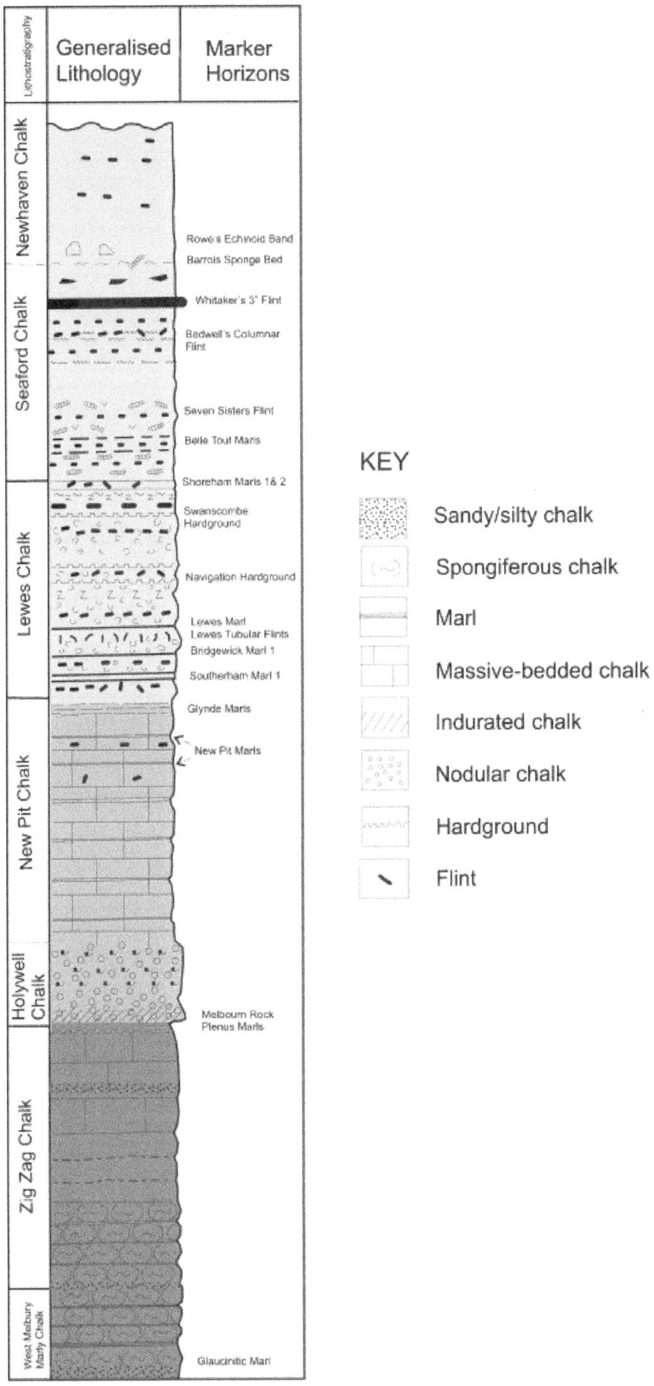

Figure 5: Detailed lithostratigraphy of Chalk in London. From Ellison et al. (2004).

Overlying the Chalk is the oldest Paleogene deposit, the Thanet Sand Formation. This consists of a coarsening upwards succession of fine grained, clean grey sand. The formation reaches a maximum thickness of about 30 m. Above the Thanet Sand Formation lies the Lambeth Group, consisting of the Upnor, the Woolwich and the Reading Formations. The Lambeth Group is between 20 and 30 m thick in London and has a highly variable lithology, containing differing proportions of sands, silts, clays and gravels. Overlying the Lambeth Group are the Eocene sediments of the Thames Group which consist of the Harwich and London Clay Formations. The Harwich (formerly known as the Blackheath or Oldhaven Beds) contain predominantly sand and pebble beds up to 4 m thick. Above this is approximately 90–130 m of the London Clay Formation, a sequence of grey to blue grey, often bioturbated, silty clay. Quaternary deposits are encountered throughout the London Basin and these include evidence of ancient river systems and the development of the present-day River Thames. Deposits include alluvium, peat, brickearth and river terrace deposits, for example the Kempton Park, Taplow and Shepperton Gravels.

One aspect of the post-Cretaceous succession in the London region that is not widely appreciated is the considerable period during which no record of deposition exists. Comparison with the successions of similar age in the Hampshire Basin demonstrates that approximately 50 Ma of deposition is missing from the successions that lie above the Chalk. It is likely that periods of uplift are the cause for these absences and this will be discussed further in the paper. Recent and extensive investigations for engineering excavations in the London region have provided considerable quantities of high quality geological data causing some long established concepts concerning the continuity and extent of sedimentation to be revisited. Currently, work is being undertaken by the London Basin Forum to synthesise, summarise and present an updated version of the geology in a Geological Atlas of the London Basin. The London Basin Forum is a collaborative project, with goescientists from industry, academia and government whose sole aim is to develop an up-dated regional framework within which data from local ground investigations can be interpreted.

To explain the need for such work, selected details from the geological history of London will now be reviewed which illustrate major issues and anomalies within the region. The structure of the London Basin will then be described as understood at present, followed by a review of the implications a better understanding of the geology of the London Basin will have on major engineering projects and long term groundwater management strategies. The paper concludes by describing how technologies such as GIS and 3D modelling are changing the way geological, geotechnical and hydrogeological datasets of the sort available for London are presented and interpreted.

PRE-GAULT AND GAULT SEDIMENTARY FORMATIONS

The mudstones of the Gault Formation are the first sedimentary unit to cover the whole of the London Basin. These mudstones were deposited following a period of intense tectonic activity related to the opening of the northern Atlantic Ocean (Sumbler, 1996). This was followed throughout the Aptian and Albian by periods of erosion, subsidence and sea level rise, which eventually led to the establishment of shallow marine conditions across the whole of the London basin (Table 2).

Table 2: Pre-Cenomanian stratal representation, depositional/erosional phases, principal events and distribution of sediments on the London Platform. Presumed extent of Middle (Callovian) and Late Jurassic sediments based on their development in regions adjacent to the Platform.

Phase	Periods/groups/formations	Principal events/type of sedimentation	Platform
Total	Upper Greensand contiguous with high	Shallowing in the west	South western and southern areas east to west Kent
	Upper Gault		
	Upper Gault	Block faulting in late*auritus* Subzone	Complete cover
Early Late Albian Faulting Along the Thames Axis			
Cover	Lower Gault	Transgressive over Lower Greensand; faulting along Thames Line, removed Lower Gault from Essex and Suffolk by erosion	Probably complete cover originally
Transgressive	Lower Greensand	Transgressive over Wealden and eroded Jurassic and Palaeozoic	Thick succession in western, south-western and southern areas; Woburn Sands and overstepping Carstone in East Anglia
	Wealden	Redeposition of eroded sediment from Jurassic uplands	Essentially in the southern marginal area of the platform in Kent and Surrey
Commencement of the Erosion of the Jurassic Uplands			
Jurassic – Cretaceous Boundary Faulting Along the Southern and North Eastern Margins with Major Uplift and the Inception of the Current Weald Graben			
Shallowing	Purbeckian	Shallowing and oncoming of brackish and freshwater conditions	Probably western area only
	Portlandian	Shallowing marine	Absent from the Platform, but developed to the south of it
Total	Kimmeridge Clay	Relatively deep water mud deposition	Probably covering the whole platform formerly
	Ampthill Clay/Corallian/ West Walton	Shallowing phase	Ampthill Clay and West Walton Beds in west; Corallian in the south
Cover	Oxford Clay	Relatively deep water mud deposition	Present in grabens in east Kent and in north Kent – southern Essex; probably continuous cover
Onlapping	Kellaways	Relatively shallow water sediments	Eroded succession on the W. marginal area and Kent
	Great Oolite	Shallow water carbonates overstepping Bajocian sediments to rest on Palaeozoic	Widespread in the southern area of the Platform and in the western fringing region
	Inferior Oolite	Essentially relatively shallow-water below more extensive Great Oolite	Marginal areas
	Lias	Essentially incomplete mudstone deposition affected by regression In the Toarcian	Marginal below Oolites in S E Kent and in the west
Erosion	Triassic	Erosion of the uplifted platform and redeposition up to the Platform margins	Absent on the Platform and final configuration of the pre-Jurassic surface
	Permian	Major erosion of the Coal Measures, Carboniferous Limestone and earlier Palaeozoic formations	Absent on the Platform

Major Late Variscan Block Faulting with Uplift of the Platform and the Production of the Oxfordshire and Kent Coalfields And Proto-Weald Grabens			
Platform	Coal measures	The succession in the Oxfordshire and Kent Coalfields suggests a former continuous cover.	Kent Coalfield graben
	Carboniferous Limestone	Strong evidence of a continuous sedimentation over southern England	Remnants at Palaeozoic surface in E. Kent, Bedfordshire and beneath Coal Measures. South of the Platform, extensive areas in the proto-Weald graben
Variscan	Devonian	Marine and Old Red Sandstone facies deposition	Extensive areas on the central area of the Platform
End Caledonian Deformation and Subsequent Erosion			
Orogenic	Silurian	Faulted and folded shelf mudstones limestones and sandstones deposition	Extensive areas in East Anglia (Pridoli), in Kent and in the south western region of the Platform
	Ordovician	Deep-water mudstone deposition	Limited to a small area of N. Kent, and the south western region

The mudstones are transgressive across a terrain of eroded Jurassic, Wealden and Lower Greensand sediments and, at the core of the Platform, rest directly on Palaeozoic rocks. In much of the area to the north, in East Anglia, the Gault is underlain immediately by the Carstone Formation, essentially of latest Early Albian and early Middle Albian age (e.g. Gallois and Morter, 1982, Gallois, 1994 and Owen, 1995). To the west, the Carstone Formation is itself underlain in places by the Late Aptian Woburn Sands Formation. In the southern bounding area of the Platform, the Gault is separated from the eroded Jurassic surface by the Lower Greensand Group. Further south, beyond the southern bounding faults of the London Platform, limited areas of Atherfield Clay and Hythe Beds Formations overlie Wealden sediments, which are the product of the early Cretaceous erosion of the Jurassic upland terrain.

Such was the scene at the start of the Gault. A fascinating aspect of the research currently under way for the Geological Atlas of the London Basin is the extent to which a re-interpretation of known deposits against a background of a faulted basement will permit a better understanding of how the basin developed at this time. If that is possible, then some of the anomalous conditions found within the London area may be amenable to rational explanation and indeed prediction. With that in mind, aspects of the overlying succession, i.e. the cover, will be reviewed.

The stratigraphy of the Lower and Upper Gault and contiguous Upper Greensand in south east England has been described in some detail by Owen, 1971a, Owen, 1971b,Owen, 1976 and Owen, 1996 and its subcrop succession in central East Anglia byGallois and Morter (1982), the latter being used extensively in the British Geological Survey Memoirs of the region. The evidence indicates that the Lower Gault originally covered the whole Platform area. However, it is absent over much of northern and western Greater London and Essex, where it was removed as a result of uplift from renewed faulting along older lineations of Variscan and Jurassic–Cretaceous boundary age (Ellison et al., 2004, Owen, 1971a, Owen, 1971b and Sumbler, 1996). This faulting phase and subsequent erosion can be dated as earliest Late Albian (*Dipoloceras cristatum* Zone) age (Owen, 1971a and Owen, 1971b).

The sedimentary character of the Upper Gault is essentially one of the more marly mudstones deposited in deeper water than those of the Lower Gault and replaced progressively from the south west by the arenaceous Upper Greensand facies. The Upper Greensand underlies the Chalk over much of the southern area of the Platform from the Thames axis southward with an eastern termination in west Kent (Owen, 1996). Continued tectonic activity in the mid-Late Albian Stage (*Callihoplites auritus* Subzone) produced minor graben structures found preserved in the Fetcham Mill area of Surrey (Gray, 1965), near Sevenoaks (Owen, 1996) and in the Folkestone area of Kent. Gault Formation sedimentation terminated with the Early Cenomanian transgression, although earliest Cenomanian sediments of Upper Greensand facies are locally present in south Essex and Surrey (Sumbler, 1996).

THE CHALK

For many, the Chalk remains divided at its Formation level as Lower, Middle and upper, but these divisions are of limited value to both pure and applied geology and attention is now given to the relatively recent development of new Formation divisions. This has meant returning to first principles. First, a stratigraphy has been constructed from the many cored boreholes, particularly in east London. From these boreholes, marker beds have been identified that provide reliable horizons for correlation (Mortimore et al., 2011). Most of the marker beds identified through east London are also beds present in the main basinal areas for the Chalk to the south and west. Some of the marker beds, including the main marl seams, have been occluded where major hardgrounds coalesce to form rock bands such as the Chalk Rock and Top Rock of the Chiltern Hills and the Dover Chalk Rock of the North Downs in Kent.

Marker beds are being correlated with borehole geophysical wire-line logs to extend the stratigraphy into areas where no cored boreholes exist. By combining the wire-line logs with cored boreholes and field sections, maps of thickness, lithology and lateral variation are being developed as part of the research that should be published by the London Basin Forum in a Geological Atlas of the London Basin. In addition, the differential uplift and erosion of the Chalk in different parts of the Basin prior to commencement of the Thanet Sand Formation is being mapped from the age of the topmost Chalk.

The main outcomes from these recent and detailed stratigraphical studies of the Chalk in London are an ability to:

1) Locate faults and fault zones previously unidentified.
2) Identify the age of the Chalk immediately beneath the sub-Paleogene surface.
3) Interpret geophysical borehole logs and seismic sections.

These studies are thus providing a new understanding of the relationship between the tectonic structure and chalk sedimentation (Mortimore et al., 2011). The advantage of having a consistent lithostratigraphic interpretation to use when interpreting ground investigations throughout London is an ability to:

1) Identify the presence of major fault complexes displacing the Chalk.
2) Distinguish where lateral variations observed within chalk sediments across the Basin are related to syn-depositional reactivation of folds and faults.
3) Appreciate that styles and frequency of fracturing are stratigraphically significant as stratabound fracture sets (Fig. 6) and that these fracture styles relate to syn-depositional phases occurring within tectonic pulses in the Late Cretaceous.

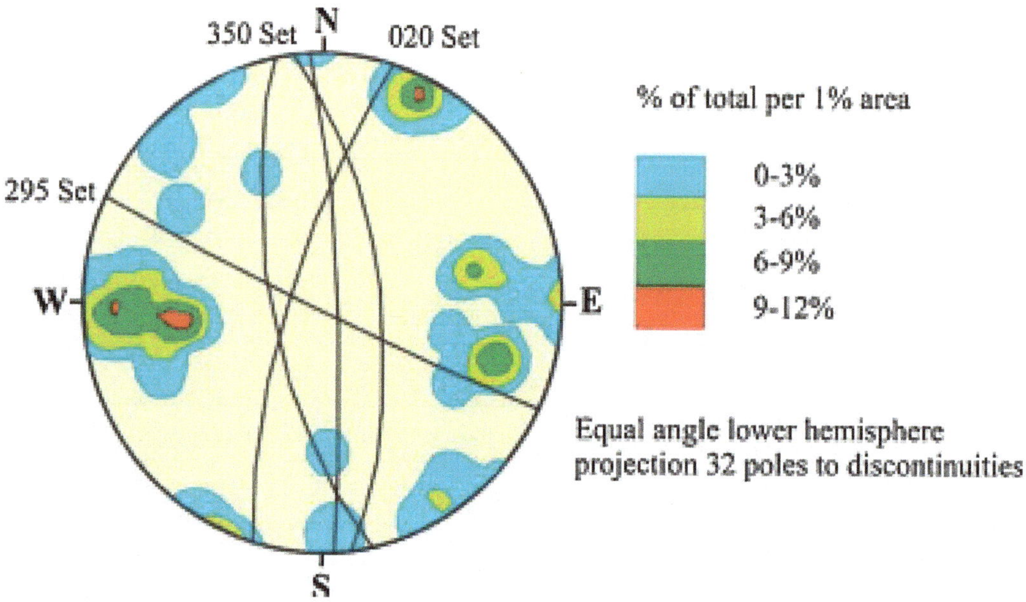

Figure 6: Major persistent primary joints (many sheet-flint filled) and fault orientations in the Seaford Chalk Formation, Northfleet Quarries, north Kent. Adapted from Mortimore et al. (2011).

4) Categorise fractures in the Chalk which are primarily related to tectonic stresses and recognise that these are usually preserved best in Chalk which is buried beneath thick Paleogene deposits and protected from the effects of Quaternary weathering processes.
5) Predict the position of particular and sometimes troublesome layers, such as bands of large flints within the Lewes Nodular Chalk and Seaford Chalk formations.
6) Recognise stratigraphic control of index properties such as intact dry density and porosity, and, to a lesser extent, unconfined compressive strength.
7) Understand that groundwater flow is partly controlled by lithology with flow horizons occurring along sub-horizontal features such as sheet-flints and marl seams.

From these it will be appreciated that four of the seven developments provide insights into the behaviour of the basement during periods of important tectonism within Europe, and indeed the "featureless and uniform" chalk may yield some of the best evidence for reconstructing the history of the London region.

The remaining three developments have practical implications for applied geology. As engineering projects go deeper than ever before beneath London, the more often they are constructed in the Chalk, particularly tunnels (Mortimore et al., 2011). Projects such as the Channel Tunnel Rail Link, Docklands Light Railway, Crossrail, Thames Water Ring Main and the Thames Tideway Tunnel schemes have or will be encountering Chalk. Consequently, there needs to be a good understanding of important elements of the Chalk affecting construction in terms of alignment and ground conditions, including the size and frequency of flints and flint bands and other lithological features such as hard-grounds and marl seams, which might affect ground stiffness, permeability and groundwater flow horizons. Equally important to engineering is the degree to which tectonics and weathering have influenced rock mass fracture characteristics (fracture openness, persistence, frequency, style, fill) and material strength as a function of density (Lord et al., 2002). Features of the material, such as these and of the rock mass, are combined to assess potential engineering behaviour.

This research and its contribution to applied geology through projects of the sort mentioned above illustrate the need for a conceptual ground model in which all boreholes are correlated; in this way, project alignments in both the horizontal and vertical planes can be planned and the impact of subsequent changes assessed in terms of ground conditions and ground response. Such models would incorporate groundwater behaviour; however, the evaluation of groundwater resources in the Chalk aquifer requires its own conceptual model (Buss and Daily, 2009 and Royse et al., 2010; Fig. 7).

Figure 7: 3D model of Chalk Group under London (Royse, 2010). OS data and Crown Copyright. All rights reserved. BGS 100017897/2011.

LOWER PALEOGENE

The tectonic activity recorded in Chalk times was followed by a period of erosion and quiescence that, by the start of the Paleogene, placed London on the edge of a sedimentary basin incorporating a large proportion of the North Sea and probably extending as far east as Poland (Ellison et al., 2004). In this marginal position, Paleogene deposits were laid down during a period of repeated transgressions and regressions driven by global sea-level changes and or by tectonics (Knox, 1996).

The oldest Formation is that of the Thanet Sands, with outcrops occurring in southeast London and northern Kent. The formation sits unconformably on the eroded Upper Chalk surface (Knox, 1996). The Thanet Sand Formation in London is largely unfossiliferous and therefore we must still resolve how it correlates with the five lithological units identified using foraminifera assemblages by Haynes (1955) in the Thanet Sands in Kent. The basal unit, known as the Bullhead Beds, consists of up to 0.5 m of clayey, glauconitic silt or fine sand with abundant, relatively unrolled (hence like 'bulls' heads'), up to cobble-sized chalk-derived flints. The bulk of the formation as seen in the greater London area is a coarsening upward sequence of dominantly fine to medium grained glauconitic sand which can be silty or clayey in places (up to 30 m in thickness on the eastern side of the basin, Ellison et al., 2004). Leaching of heavy mineral suites at the top of the Thanet Sand Formation has been interpreted as evidence of uplift and emergence (Morton, 1982).

The succeeding Lambeth Group, although relatively thin (10–20 m in central London), lithology is highly variable both laterally and vertically and consists of three Formations (Table 3; after Ellison, 1983, Ellison et al., 1994 and Ellison et al., 2004). Skipper (1999)proposed a sequence stratigraphical interpretation which demonstrated that the Group had been deposited in three depositional sequences. This interpretation was used inPage and Skipper (2000) to invoke a predictive model and descriptive lithological stratigraphy (Table 3) which is now widely used.

Table 3: Paleogene, after King (1981) and Page and Skipper (2000).

Eocene	Thames Group	London Clay Formation	Divisions A to E Claygate Member,
		Harwich Formation	Swanscombe Member
			Oldhaven Member
			Blackheath Member
Paleocene	Lambeth Group	Woolwich Formation (lower)	Upper Shelly Clay
		Reading Formation (upper)	Upper Mottled Clay
		Woolwich Formation (lower)	Laminated Beds
			Lower Shelly Clay
		Reading Formation (lower)	Lower Mottled Clay
		Upnor Formation	
	Thanet Sand Formation	Thanet Sand	
		Bullhead Bed	

The Group is a complex sedimentary assemblage that was deposited in shallow marine, estuarine, lagoonal, and fluvial to terrestrial environments, the result of what has traditionally been considered as eustatically controlled marine transgressions (Knox, 1996 and Ellison et al., 2004). The shallow marine to estuarine Upnor Formation, deposited over a wide area of SE England, is succeeded by the lower part of the Reading Formation, which consists of terrestrial, pedogenically altered (colour mottled) and fluvial sediments. The upper surface of the lower Reading Formation is described as the Mid-Lambeth Hiatus (Page and Skipper, 2000), a useful marker horizon corresponding to an omission surface marking a basin wide fall in sea-level (Collingson et al., 2003).

In the London area the lower part of the Reading Formation is succeeded by the Woolwich Formation, consisting of the brackish marine Lower Shelly Clays (grey laminated clay, with occasional silts and sands and abundant layers of shells) and the Laminated Beds (thinly laminated clay-silt and silt-sand, with frequent organic remains and occasional shell beds). These sediments thin rapidly westwards across London and are succeeded by the upper Reading Formation (mottled clays, silts and sands with the rare deep sand channel deposits). In the far west of the London Basin, the lower Reading Formation is succeeded directly by the upper Reading Formation; both lower and upper Reading Formations thin eastwards across the London area (Ellison et al., 2004).

In south London (and rarely in east London), the upper Reading Formation sediments are succeeded by the upper part of the Woolwich Formation – the Upper Shelly Clays. Although in east London these manifest as similar sediments to the Lower Shelly Clays, in south London these sediments consist of up to 2.5 m of firm to very stiff firm organic clays and silts and laminated dark grey shelly clays and silts and mudstones. The*Paludina* Limestone and Oyster Limestone occur within this unit and each comprise up to 250 mm of moderately strong rock, the former having a freshwater monospecific fauna and the latter having an estuarine fauna (Dewey and Bromehead, 1921 and Ellison et al., 2004).

One of the most interesting aspects about the Lambeth Group is its ability, as a rapidly changing and colourful sequence of sediments, to show fault movements (Fig. 8) which are much more difficult to identify in most other sediments in the London basin. This unique attribute enhances the importance of this otherwise thin and seemingly enigmatic group of sediments to the London Basin Forum project. Ground investigations carried out for major engineering projects across London such as Crossrail and the Thames Tideway project provide evidence that faulting, possibly syn-sedimentary, played a significant part in controlling the distribution of the Lambeth Group. This implies that the variability of environments of deposition in the group is not just eustatically controlled.

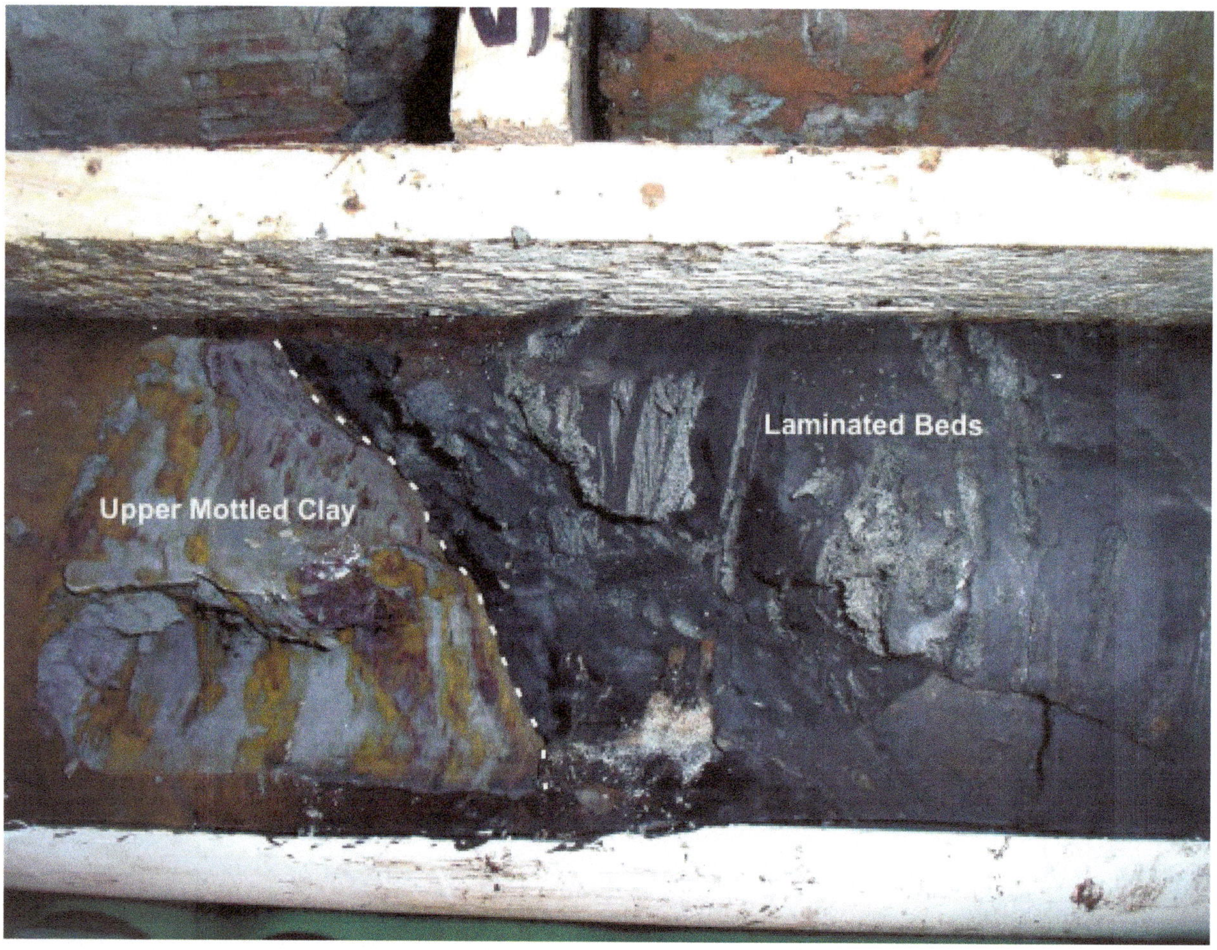

Figure 8: Photograph of a Thames Water rotary core from the Stoke Newington area (north London) showing lower Woolwich Formation (Laminated Beds) faulted against upper Reading Formation (Upper Mottled Clay).

As part of the work of the London Basin Forum project, the British Geological Survey utilised 1400 recent digital site investigation borehole logs to review and modify the original lithofacies maps produced for the Lambeth Group by Ellison et al. (1994). The modified lithofacies maps have been compared with structures within the London area, as revealed by recent 3D geological modelling work (Fig. 9, Ford et al., 2010). It is hoped that future work aimed at producing a 3D geological model for the Lambeth Group under London will provide a clearer understanding of the structural controls on the distribution of sediments within this group.

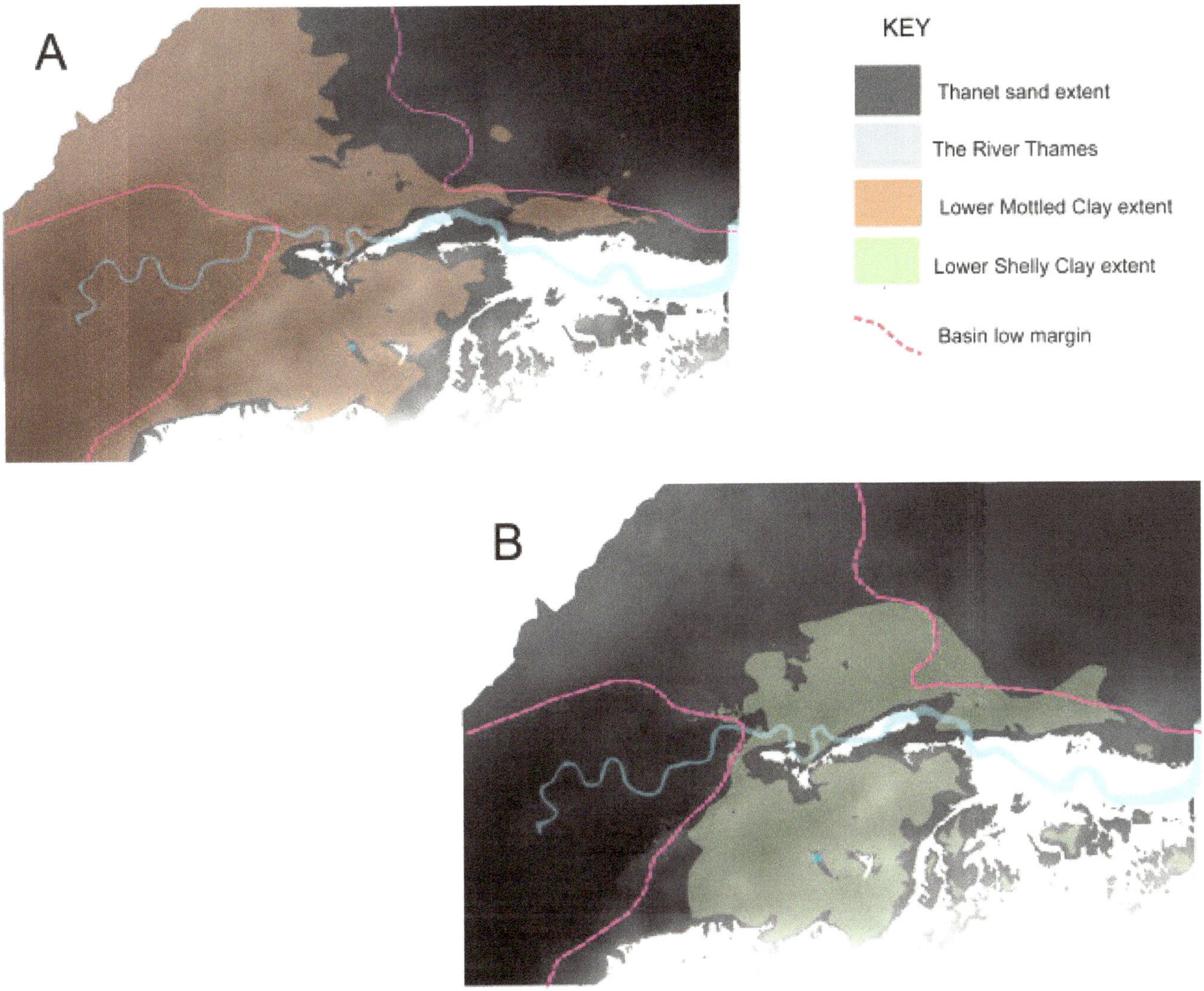

Figure 9: Distribution of (A) the Lower Mottled Clay of the Reading Formation (brown); (B) the Lower Shelly Clay of the Woolwich Formation (green); overlain on monochrome shaded relief map of the base of the Paleogene. Adapted from Ford et al. (2010).

THE THAMES GROUP

The Thames Group marks a return to marine conditions throughout the region, driven by pulsed global sea-level rise, with the coastline shifting far westwards from the present limits of the London Basin. It was preceded by an episode of uplift and erosion, with its initial sediments, included in the Blackheath Formation and the overlying Harwich Formation, deposited disconformably on the Lambeth Group. The Blackheath Formation includes estuarine pebble-gravels and sands, filling deep channels incised into the Lambeth Group and underlying sediments. The Harwich Formation comprises mainly thin glauconitic sands and sandy glauconitic clays. All these units have discontinuous distributions. Although often thin, they have great significance for engineering projects, particularly tunnelling, due to their often highly permeable lithology and the development of hard calcareous concretions at several levels. These features have caused significant problems in some recent construction projects, and detailed mapping of individual units is currently under way as part of the London Basin Forum project (Fig. 10).

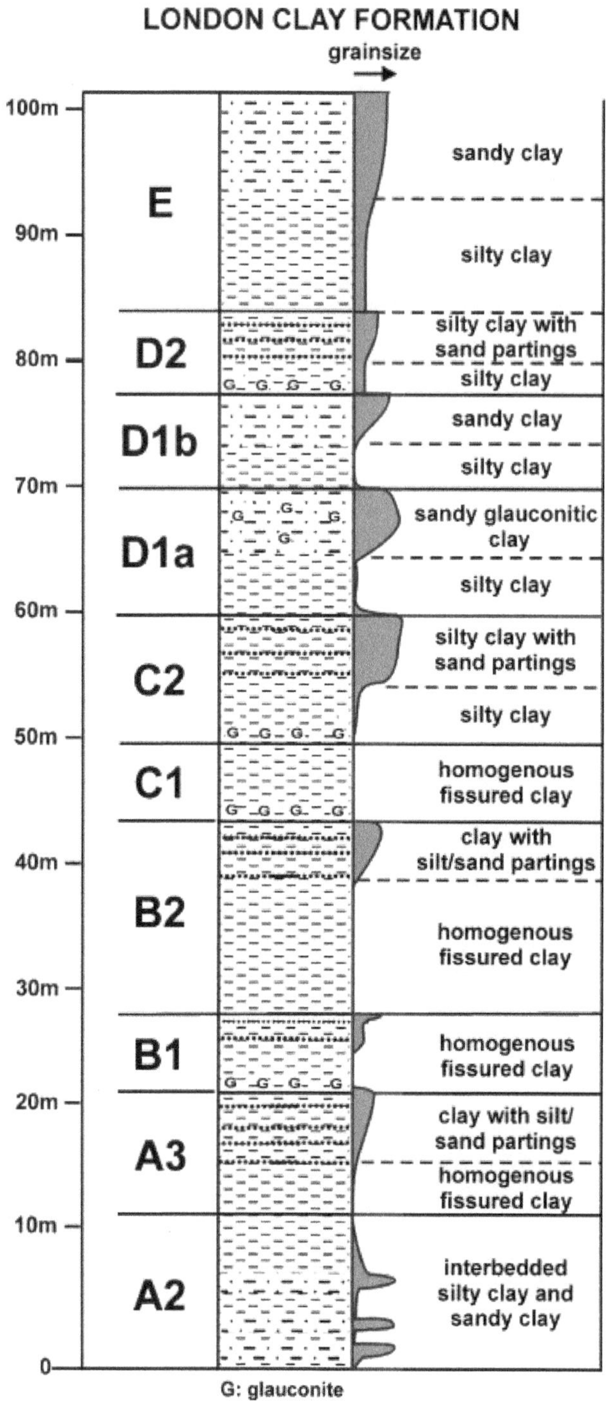

Figure 10: Distribution of the Blackheath Formation in London: a provisional version (2007), already modified by later information. The River Thames is shown in grey.

Overlying the Harwich Formation is the London Clay Formation, probably the most well-known of all the Formations in the London Basin and having a significant influence on the development of London's infrastructure. Its presence beneath much of central London, and its relatively homogeneous nature makes it a near perfect tunnelling medium, thus facilitating the development of the London Underground. The London Clay Formation, as its name suggests, is predominantly a bioturbated clay with silty and sandy clay intervals, with a maximum thickness of 130 m (Ellison et al., 2004).

It has long been acknowledged that the upper part of the London Clay is more sandy than the lower part (Whitaker, 1866), but early attempts to subdivide it were held back by a lack of exposure and the relatively monotonous lithology. However, King (1981) used a combination of biostratigraphy and lithological variation, and the identification of marine flooding events, to define five divisions (A–E) which have in general been found to be laterally and vertically consistent throughout the London Basin (Fig. 11). Improvements to this detailed understanding of the succession have been made from boreholes drilled throughout the London Basin, most significantly the stratigraphic boreholes drilled as part of the BGS mapping programme in the London area and Essex (Bristow, 1982, Lake et al., 1987 and Ellison et al., 2004).

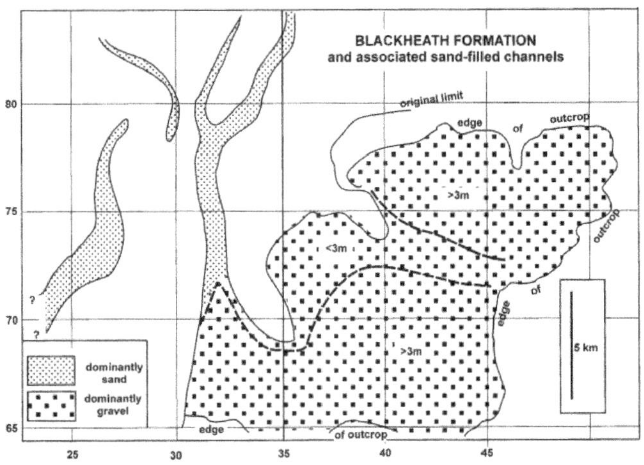

Figure 11: Stratigraphy of the London Clay Formation in London and the surrounding area (King, 1981).

The youngest Eocene sediments in the London Basin, the *Bracklesham Group* are predominantly sands, with some thinner clay units, largely removed by subsequent uplift and erosion, but preserved mainly west of London, forming the higher hills of Surrey and Berkshire. In north London these cap the hills of Highgate and Hampstead Heath. These are a complex series of sediments deposited in shallow marine and marginal marine environments, part of a formerly much more extensive sheet covering much of SE England.

QUATERNARY

By this time, the London Basin, as it would be recognised now, had been established for probably 20 Ma. The long history of weathering and erosion commenced, with the ice ages dominating events by the magnitude of the changes associated with them. These changes were not just erosional and depositional; glaciers of considerable thickness moved across the country to the north of the Basin imposing transient loads to the ground, causing the crust to deform. Those deformations would probably have been accommodated in the "basement" by movements on various faults. Much happened in the Quaternary but for the purposes of this review, the materials of two environments are considered: weathering producing Clay-with-Flints and deposition, most notably of river terraces.

Quaternary deposits formed in the London Basin over a period spanning approximately the last 1.65 Ma. They provide evidence of an ancient river system, a precursor to the River Thames, glaciations of Anglian age in the north of the district, and the development of the present River Thames valley. Understanding Quaternary history is difficult because of the complex and discontinuous nature of the sediments and the incompleteness of the stratigraphical record, and this is particularly so in the London Basin where considerable anthropological disturbance also occurs.

Clay-with-flints

Clay-with-Flints was originally described as 'stiff brown and red unctuous clay with large unworn flints and at the base a few inches of black clay with black coated flints' (Hull and Whitaker, 1861). It outcrops on the Chalk plateaux around the edges of the London Basin, between the Chalk scarp and the main outcrop of the Paleogene. It is a very extensive deposit, covering an area of the Basin, as defined by the limits of the Chalk outcrop, at least as great as that of the fluvial deposits. Paleogene deposits extend from their main outcrop to merge into Clay-with-Flints. The maps of the British Geological Survey record exposures of Chalk between the Clay-with-Flints and other deposits, both periglacial and fluvial (Fig. 12). This apparent gap arises because the Survey's mapping convention is not to show deposits less than 1 m in thickness; thus the thin veneer of Quaternary deposits overlying much of the Chalk outcrop is unrecorded.

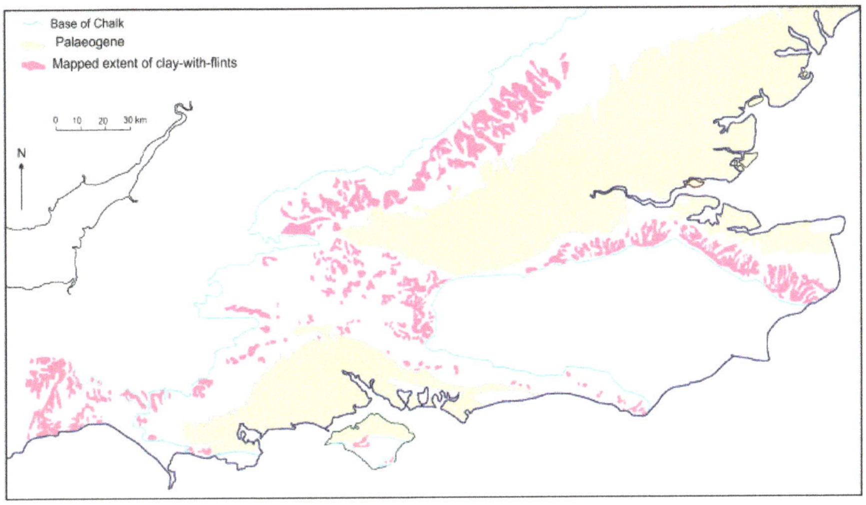

Figure 12: Distribution of the clay-with-flints in southern England based on BGS mapping (Klinck et al., 1998).

Clay-with-Flints is currently defined as an *in situ* residual soil derived from the Paleogene and the Chalk; it is underlain by intact Chalk. Remnants of weathered Chalk are confined to the clays immediately overlying the Chalk (Klinck et al., 1998) and consist predominantly of flint nodules along with a small percentage of clay and siliceous fossils. Pedological processes generated by infiltrating ground water redistribute the clay fractions within thc soil profile. The smectite clays, in particular, are concentrated immediately above the Chalk surface, possibly carried in suspension. The concentration of smectite clays is reflected in the variation of the engineering properties of the deposits with depth, especially by their ability to take up water, as seen in their liquid limit, which increases towards the surface of intact Chalk.

The location and orientation of faults and joints in the Chalk is reflected by the distribution of the Clay-with-Flints seen in dry valleys that dissect the Chalk outcrop. Many dry valleys follow the major directions of brittle fracture associated with deformation within the London Basin, usually the direction of tension parallel to dip and one of the conjugate shear directions, unless the outcrop has been modified by local drainage patterns. Detailed mapping of the Paleocene clay-with-flints has revealed that the Paleogene deposits on which they are found are preserved within secondary flexures within the Chalk surface. For example, the 1:50,000 scale Chatham Sheet (British Geological Survey, 1972) shows that the Paleogene deposits are associated with asymmetric folds running down and along the dip slope of the Chalk.

The Clay-with-Flints should be differentiated from the adjacent Quaternary deposits with which they are often and confusingly associated, by their different fabric and content, as these reflect their methods of formation. Much Valley Gravel, for example, consists of original Clay-with-Flints that has moved into dry valleys by periglacial slope processes and in so doing, has been mixed with materials that have also moved down-slope, having been originally derived from Paleogene deposits and the Chalk.

River deposits

The Thames system is the largest drainage basin in Britain. For convenience, it can be divided into three regions reflecting bedrock and river form, and referred to here as the Upper, Middle and Lower Thames. The Middle and Lower Thames occupy the London Basin. At this point the Thames is a broadly west to east aligned stream axial to the Basin, with tributaries entering from both the northern and southern margins (Fig. 13). In the extreme east, a large estuary occurs where the river and its drowned tributary valleys enter the North Sea. The deposits of the Thames and its tributaries occur from the tops of the highest hills on the Basin margin (180 m OD) to below sea level in the Thames Estuary (Fig. 13). The earliest Thames deposits, the so-called Pebble Gravel Formation, represent a fragmentary series of gravels composed predominantly of local materials, particularly flint. They postdate marine sands that are also found at up to 180 m OD on the margins of the London Basin, the occurrence of which indicate relative uplift of the western end of the Basin during the Pleistocene (Gibbard et al., 1988 and Mathers and Zalasiewicz, 1988). The Pebble Gravels therefore represent the Thames and tributaries established immediately following regression of the sea in the Early Pleistocene (Wooldridge and Linton, 1955).

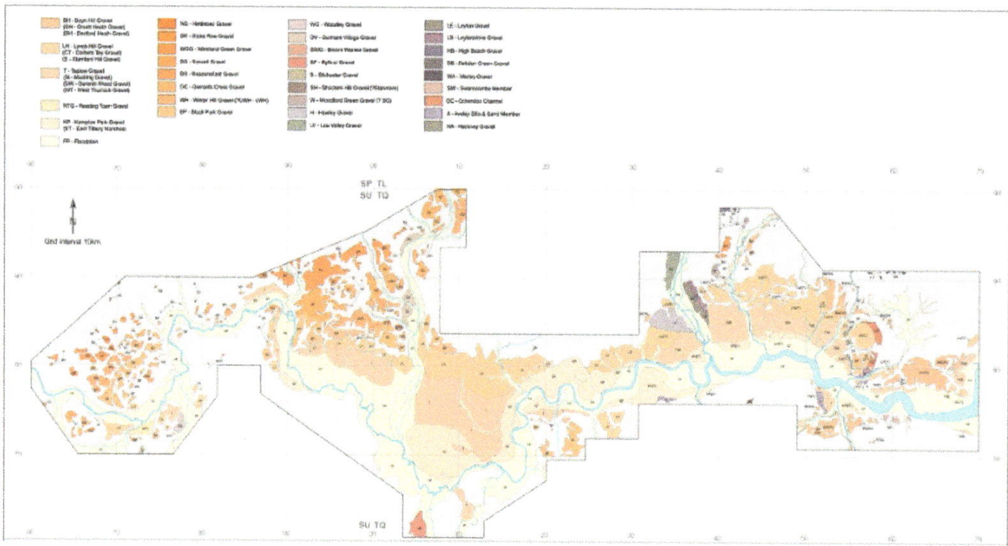

Figure 13: Distribution of terraces along the River Thames. Compiled from Gibbard, 1985 and Gibbard, 1994.

A profound change in gravel lithology is present in the next youngest units, which form a series of terrace remnants that are characterised by their content of rocks exotic to the present Thames catchment. These units can be traced from the Upper Thames, where they are aligned parallel to the modern Thames tributary, the Evenlode, downstream through the Middle Thames Valley. Here they diverge from the modern course and pass through Hertfordshire, parallel the shallow Vale of St Albans and enter East Anglia where they form a terrace-like system, mostly buried beneath tills of the Anglian glaciation (Hey, 1980, Rose et al., 1999 and Whiteman and Rose, 1992).

Glaciation in the Anglian overrode the drainage system of east and central England, damming the Thames and its southbank tributaries north of London. This resulted in the river adopting a new course through London (Gibbard, 1994, Gibbard, 1985 and Bridgland, 1988). Subsequent evolution of the Thames system has been marked by the cyclic development of a sequence of gravel and sand aggradations under periglacial climates during the Middle and Late Pleistocene. These aggradational members show a marked reduction in exotic material with time, accompanied by an equivalent increase in local lithologies as the deposits get younger.

The Thames Valley also includes many important interglacial fossiliferous sequences that provide both stratigraphical control and palaeo-environmental evidence. East of London, the valley was invaded repeatedly by the sea, so that, as today, a substantial estuary developed during periods of high eustatic sea level (interglacials). Submergence of the valley system has meant that offshore from south east Essex, a drowned course of the Thames and its tributaries occurs aligned towards the east and south east (Bridgland, 1988 and Bridgland et al., 1995). Upstream in the estuary, a thick wedge of Flandrian (Holocene) marsh and mud sediments accumulated (Devoy, 1979). The current subdivisions are largely based on morphological evidence, however the focus is now moving towards one based substantially on geological sequences and their three-dimensional relationships. This change reflects the explosion of developments in Quaternary science in general. Today, the outcomes of this approach are in a phase of consolidation and although substantial changes are still possible, they are less likely now than some 20 years ago. However, much remains to be elucidated, particularly in terms of dating and finer stratigraphical resolution of the unfossiliferous parts of the sequences.

Of the problems remaining, the most obvious is the disagreement over the stratigraphical status of individual terrace aggradations. Here the lithostratigraphical approach of Gibbard (Gibbard, 1985, Gibbard, 1994 and Gibbard et al., 1988) has been adopted. In this system, individual sediment bodies are assigned *member* status, because this is thought to be the most appropriate hierarchical level; it is compatible with neighboring areas and with other unit classifications. The term *formation* has been used to refer collectively to members with broadly unified lithological characteristics. This is contrary toBridgland, 1994 and Bridgland, 1988 who considers that terrace aggradations should be assigned formation status. Likewise, Bridgland favours *group* status for some of the formations. Bridgland considers this necessary, because the complexity of the Pleistocene sequence requires the use of all available hierarchical levels (Bridgland, personal communication). The term Group has also been used by (Whiteman and Rose, 1992) to refer collectively to deposits previously termed Kesgrave Formation (Eastern Essex, note 1). However, some consider this term is too large scale for use in Pleistocene stratigraphy, by comparison of group status units in other parts of the geological column, and so no units of this rank have so far been adopted. More problematic is the means by which chronostratigraphical correlation of individual temperate character deposits is achieved both between sites and between the conventional terrestrial and ocean isotope stages. Conflicting results have arisen from the application of conventional biostratigraphical techniques, particularly palynology, and geochronology, particularly amino-acid racemisation. This has led to the Thames' sequence being subdivided chronostratigraphically using systems that stress different elements of the sequence (*cf.* Gibbard, 1985, Gibbard, 1994 and Bridgland, 1994).

GEOLOGICAL STRUCTURES IN LONDON

In regional terms, London is located within the western area of the Anglo-Brabant Massif, north of the "Variscan front" and east of the Mesozoic North Sea rift system (Erratt et al., 1999 and Ellison et al., 2004). Traditionally, the geology in London has been considered to be 'relatively simple' (Ellison et al., 2004). However, observations and data collected from recent site investigations suggest that, in reality, the structure of the London Basin is more complex (e.g., Newman, 2009 and Skipper et al., 2008). The Chalk was deposited syntectonically over faulted basement blocks and it is these faults (Mortimore and Pomerol, 1997 and Mortimore et al., 2011) that controlled both its lithology and thickness. Modern alluvial deposits of the river Thames reveal multiple structurally controlled off-sets and flow patterns (de Freitas, 2009), implying that fault movements in London have occurred throughout the Cretaceous and Tertiary periods and remain modestly active at the present day.

Because the majority of bedrock is buried beneath thick Quaternary deposits related to the development of the River Thames (Royse, 2010) and/or the built environment of London, the nature and extent of faulting within London has been difficult to determine, but significant structural information can be inferred from an understanding of the structural controls imposed by these past tectonic events. We therefore need to consider each event in turn, identifying the brittle structures created by each event and the effects that subsequent tectonic events had on these structures. Three important events dominate the brittle structures found in the basement:

- The orogenies that formed Pangaea (Devonian – Carboniferous).
- The breakup of Pangaea (Jurassic – Cretaceous).
- The Alpine Orogeny (Cretaceous – Tertiary).

Deformation within NW Europe has been dominantly compressional in the last 50 Ma, a consequence of the Late Cretaceous collision between Africa and Europe that led to the formation of the Alpine orogenic belt. These compressive stresses have been transferred through the Variscan basement that underlies NW Europe. This basement is cut by major fracture zones: the Silurian NE–SW fractures of the Caledonian orogeny and the later E–W and NW–SE trending end-Carboniferous fractures of the Variscan orogeny. London lies at the boundary between these two major basement fracture sets (Fig. 14), on the southern margin of the Midland Craton (part of the Anglo-Brabant Massif), a stable Proterozoic crustal block that separates the Variscan fractures to the south and Caledonian fractures to the north. Variscan structures therefore provide the dominant structural control on the geology of London.

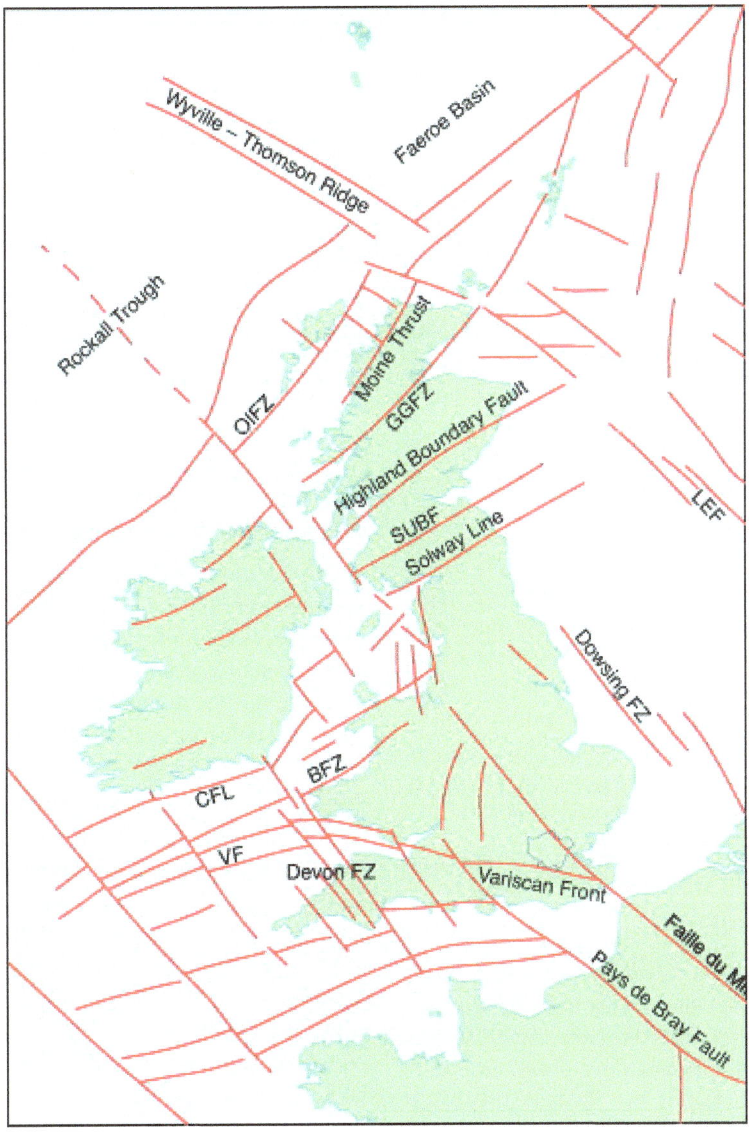

Figure 14: Palaeozoic basement structures of the British Isles. Adapted from Lee et al. (1993).

London lies immediately to the north of the Variscan Front (VF in Fig. 14), the origin and nature of which are still debated (Shackleton, 1984). In Pembrokeshire, South Wales, the basement rocks and their structures also crop out just north of the Variscan Front to reveal an approximately E–W trending Variscan fold/thrust belt, a tectonic regime compatible with their marginal position with respect to the main orogenic belt to the south. These folds and faults are now displaced by NW-SE trending wrench faults (Fig. 15) that are also related to the Variscan collision.

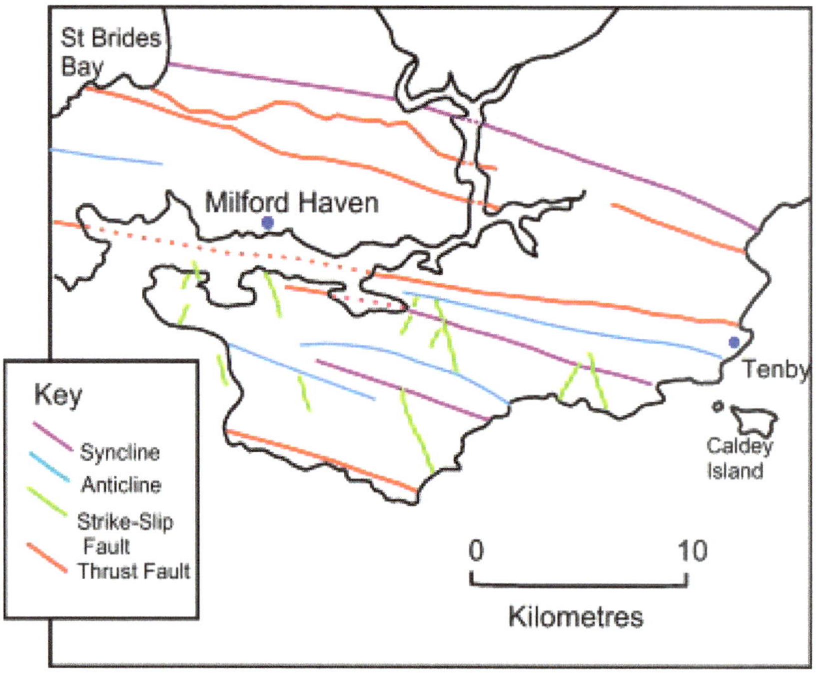

Figure 15: Approximately E–W trending Variscan fold/thrust belt as observed in Pembrokeshire (South Wales). These folds and faults are displaced by NW–SE trending wrench faults also related to the Variscan collision. Adapted from Hillier and Williams (2006).

This superposition of two fault regimes is predictable. A fold/thrust belt is characterised by the minimum principal stress being vertical, but as the belt develops, the overburden stress increases until it becomes the intermediate principal stress, a regime that results in wrench tectonics. These wrench faults, together with the ~E–W trending thrust faults, divide the basement into multiple blocks of different sizes. The response of these blocks to later tectonic stresses during the break-up of Pangea and the Alpine orogeny is critical to understanding the depositional environments of, and structures developed within, the basin sediments.

During the Cretaceous in London, nominally N–S crustal extension was focussed along the E–W fractures of these basement blocks, causing individual blocks to subside in an irregular pattern of horsts and grabens into which the Chalk was deposited. The fractured boundaries of these blocks propagated into the overlying Chalk and are now thought to influence the movement of groundwater in the Chalk aquifer, causing the observed grid-like pattern of water level drawdown and recovery across the central regions of the basin.

THE IMPACT OF GEOLOGY ON GROUND ENGINEERING IN LONDON

The intensity of development in London, coupled with its legacy of contaminated ground, vulnerability to rising sea level and size of population, all drive the need for establishing a geological model upon which planners, developers, engineers and insurers can rely. Examples of the geologically based problems to be solved are described below and illustrate the continuing need for such studies and the research currently being directed by the London Basin Forum.

Two aspects of geology that play a major role in the development of London are the success to which ground engineering can be accomplished in the sediments of the London Clay Formation and the management of the water resources of the Thames valley. These developments have been achieved as a consequence of long and continued research into the hydrogeology of the Chalk aquifer and the geology and mechanics of the London Clay Formation, through which many sections of the London Underground and utilities have been bored and in which major foundations have been sited.

In addition, all the studies to date in this recent phase of research show that the Formations in London have been affected by the reactivation of faults at depth, and this has significant implications for ground engineering. A major concern will be that faulting has divided the basin into compartments which have been able to move by different amounts vertically, relative to each other. It appears that neither the London Clay nor the aquifers beneath it can be relied on to have lateral continuity.

Much engineering has been completed within the London Clay Formation and as a consequence considerable banks of data have been accumulated in company and public records. A similar situation exists for the hydrology of the Chalk and water supplies derived from it. The value of such data for geotechnical engineering is considerable, because the subject relies so heavily on experience; however, the value of that experience depends on the confidence with which it can be used. That confidence is eroded if the data from one location cannot be used at another, because the body from which it comes is not laterally continuous between the two. Differences from place to place in a continuum can be explained as gradual change but that need not be so in a discontinuum; here, an abrupt change is expected, as can occur across faults. Thus the question arises, "*Is the London Basin divided into compartments by faults?*"

There are many lines of evidence relevant to answering this question; Fig. 16 illustrates some of the near-vertical structures known in the London Basin and should be compared with Fig. 17, which illustrates other possible locations based on the positions of tributaries of the Thames. At the very least, geotechnical engineering has good reason to consider the likely presence of faulting within the London Basin (de Freitas, 2009 and Mortimore et al., 2011). The sudden change in conditions which faults can create for sub-surface work, as revealed by the recent discovery of faulting at Plaistow (Fig. 18, Newman, 2008), and 3D modelling of the Chalk (Fig. 7, Royse, 2010) for the new hydrogeological model for the Chalk under London, are examples of what could be a widespread phenomenon.

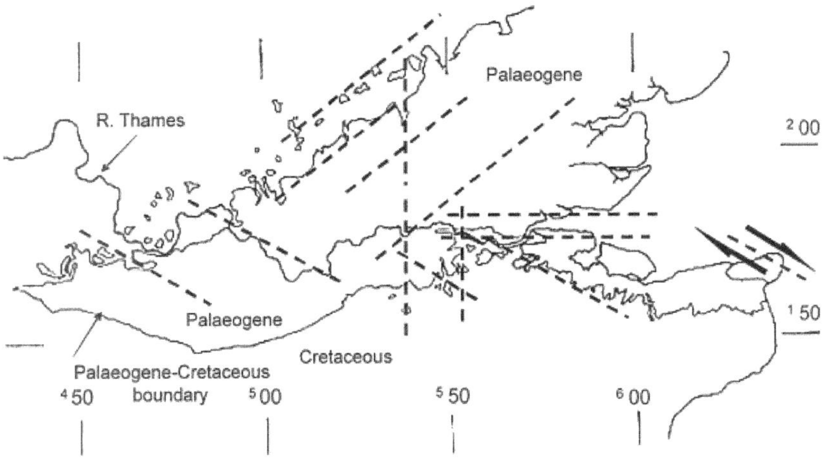

Figure 16: Known near-vertical structures in the London Basin. From Wooldridge, 1923 and Wooldridge, 1926,Wooldridge and Linton (1939), Water Resources Board, 1972a and Water Resources Board, 1972b, Carter and Hart (1977), Ameen (1995), Ellison et al. (2004), de Freitas (2009).

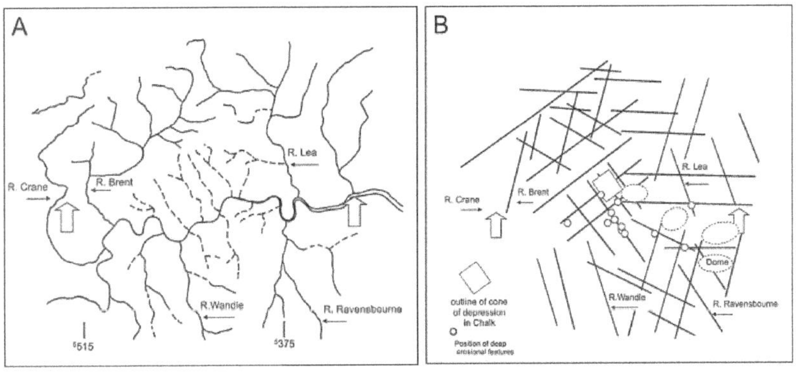

Figure 17: A fault pattern for London postulated from the trace of tributaries of the River Thames (Barton, 1992 and Thompson, 1997): (A) the tributaries, (B) the resulting postulated fault pattern. Arrows may be used for comparing A and B. Adapted from de Freitas (2009).

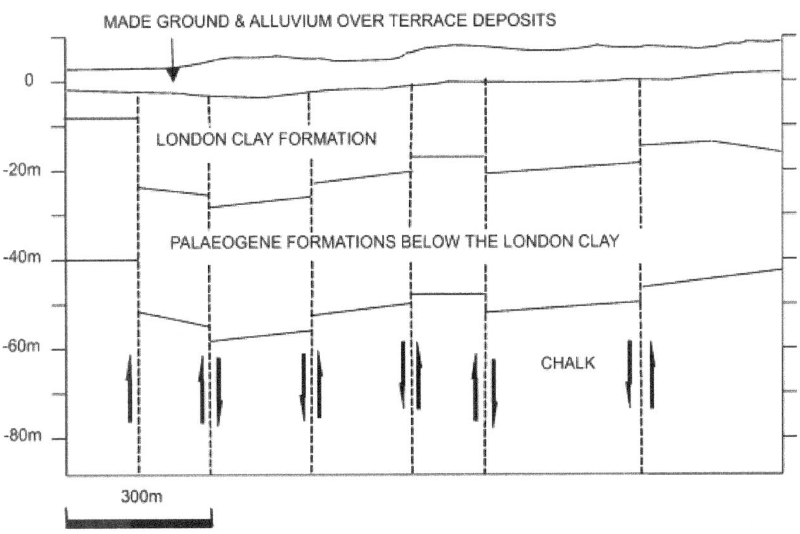

Figure 18: The Plaistow Graben. From Newman (2008), de Freitas (2009).

The implications of faulting for ground engineering are many. Data banks of geotechnical properties for the Basin (Hight et al., 2001), need to be used with care, as differences may reflect more than just local variations in lithology. The fault zones themselves may be thin in the clays but broader in gravels and quite wide in the Chalk, where their internal structure could be more permeable than their margins, especially if the latter have been reduced by displacement to a putty-like consistency. Such faults could divide the Chalk into compartments, yet may also be able to support a flow regime of their own if hydraulically insulated by their boundaries. The difficulties of dewatering three sites in Docklands, situated above the Chalk in an area where faulting in the Chalk is known, are recorded by Linney and Withers (1998). Although Linney and Withers (1998) did not consider faulting at the time to be a contributory factor, the fact that each site behaved quite differently, even though they were separated by a little over one kilometre, suggests that faulting had played a part.

The sedimentological history of a Formation influences how the ground will respond to engineering and can influence the geotechnical risk of working within it. However, the ability to assess accurately the impact of the ground conditions depends on the quality of the ground investigation, which is itself influenced by the geology (Clayton et al., 1995 and ICE, 1991). Ground investigations can introduce two sources of risk to a project: incomplete data, generated by either the sampling regime or result from drilling losses, and erroneous data. These contribute to the third and biggest ground risk, that of incorrectly interpreted data. Drilling losses are commonly experienced when harder materials occur within comparatively softer materials, and this can be a constant problem in the London Basin. A typical example is core loss due to pebble beds or concretions in clays, e.g., "claystones" in London Clay and flints in Chalk, and in pebble beds as found in the Harwich Formation, Lambeth Group and River Terrace Deposits.

The research carried out within the London Basin Forum has benefited from data generated by several large engineering projects that have procured high quality round investigations with detailed logging, closely spaced boreholes, sampling and testing to generate detailed profiles of both the geology and the engineering properties of considerable tracts of ground. The data from these profiles is used to model and predict ground movement in response to engineering work, and so link geological history to accurate models of ground conditions and reliable predictions of engineering performance.

The dewatering performance on the Jubilee Line was found to be influenced by the vertical variation in the permeability of the Chalk due to the (i) distribution of fissures, (ii) variation of infill on fissures and (iii) the presence of marl bands (Withers, 1996). Large engineering projects through the Chalk (e.g., CTRL and Crossrail) have encountered consistent properties associated with each of the Chalk Formations. Research on the engineering properties of Chalk has demonstrated a clear link with the diagenetic history, especially on compressibility and stiffness (Clayton, 1983). The Seaford Chalk which underlies much of London contains steeply dipping joint sets with consistent trends. The underlying Lewes Nodular Chalk contains inclined fractures which become more vertical and irregular as the Seaford Chalk is approached (Warren and Mortimore, 2003). Detailed lithostratigraphic logging has picked up many new faults which produce higher and more irregular groundwater flows. Similarly, the flints vary in size and character systematically through the lithology. They can form considerable obstructions to ground investigations, cause excessive wear to tunnelling equipment and have significant impact on rubber-tyred plant on the surface. Tabular sheet flints can significantly affect groundwater flows and frustrate dewatering if not accounted for (Lord et al., 2002). However, they are very difficult to recover during ground investigations and can often only be inferred from a zone of core loss.

The Thanet Sand Formation, being shallow marine in origin, consists of fine grained sand which initially appears to be fairly uniform in character. However, particle size distribution tests in the Thanet Sand show a coarsening up sequence. This has caused problems for vertical dewatering at several shaft sites along the Jubilee Line Extension. Care has to be taken with filter design so as to retain the fines in the lower sections but not unduly restrict flow in the upper sections. This variation in fines content is not readily apparent from samples. It was by careful profiling of grain size and downhole geophysics that the increase in fines was clearly demonstrated (Withers, 1996). It was also noted that the ability to profile the increasing fines content with depth is affected by difficulties in the recovery of representative samples (Linney and Withers, 1998). Cable percussion drilling in granular deposits below the water table consistently struggles to retain the finer fractions of samples. This can affect permeability values derived from correlations based on grain size, a problem not just confined to the Thanet Sand.

The intense variability of the Lambeth Group gives rise to complex and challenging ground conditions for civil engineering works encountering it (Page and Skipper, 2000). A better understanding of these deposits is critical for many current and future civil engineering projects, e.g., the Jubilee Line Extension (Bailey et al., 1999) the Channel Tunnel Rail Link (Dyke and Glover, 2007) and the development of the Crossrail network (Heath, 2001); therefore widespread acceptance of training in its stratigraphy and logging techniques (Skipper, 2008) is rapidly effecting an improvement in the understanding of this Group. The Lambeth Group contains sand channels reflecting the original tidal mudflats environment. These granular bodies can produce irregular groundwater flows when encountered particularly in tunnels and deep excavations, leading to instability at the tunnel face. The most famous case history of Lambeth Group instability is the construction of the Thames Tunnel by Brunel from Rotherhithe to Wapping (Hight et al., 2004). The initial ground investigation appeared to show significant thickness of clay. However, borings ahead of the face demonstrated that the clay was not continuous. There were many incidents during excavation of running sand and silt in the Laminated Beds. The most serious events caused the collapse of the overlying Upper Mottled Beds and inundation by river water. One collapse event was sufficient to irreparably damage the tunnelling shield. The collapses and inundations caused delays and resulted in the tunnel taking almost 20 years to complete.

Each of the lithological units within the London Clay has specific geotechnical properties of relevance to engineering within the Formation. In addition, the presence of silt and sand partings, claystones, concretions, pyrite and selenite have an influence on the behaviour of the London Clay and the design of structures within it. For instance, pyrite reacts with oxygen and water to produce acidic groundwater and a variety of minerals which have a larger volume than the original pyrite causing heave. The acidic groundwater can also attack cast iron and concrete structures in both tunnelling and highways environments (Hight et al., 2004). This is not just a problem with London geology, as it has been encountered in black shales especially where they were used as engineering fill above the water table (Steward and Cripps, 1983). It is therefore essential that in ground investigations the lithological units are clearly differentiated. Once this is achieved, it is then possible to start to make predictions as to how the ground might respond during major engineering works such as tunnelling. This type of study has been developed during recent major engineering projects and continues to be refined. One of the earliest engineering failures in the London area, the Highgate tunnel collapse of 1812, was the result of failure to appreciate that the London Clay was not quite as suitable for tunnelling as at lower levels in central London!

Drift deposits are not without their problems; for example, engineers have struggled with identifying, quantifying, describing and classifying Clay-with-Flints as a material, because of the apparently random distribution of flints within it. Sampling and testing regimes have often been conducted largely out of context of local geology, and this together with a reliance on traditional interpretation methods has resulted in the engineering parameters being poorly defined and unintentionally attributed to materials of different geological origin. Clay-with-Flints creates problems in engineering work for two other reasons: first, the finer fraction may have Atterberg limit values which are outside the specification for earthworks, and second, its highly irregular boundary with the underlying Chalk causes difficulty both for estimating earthwork volumes and for mixing materials from separate earthwork classes. Much remains to be known of this enigmatic material.

The thickness of drift deposits, e.g. river terrace deposits, is known to vary considerably within London (Berry, 1979). Some of the largest "anomalies" have been described as filling "Scour features" or "Drift Hollows" and are associated with an unexpected change in ground conditions where "host" strata such as the Chalk or the London Clay are eroded and replaced by collapsible granular materials with high groundwater flows (Ellison et al., 2004). They have been encountered in foundation excavations and in tunnelling projects both during the ground investigations and subsequent tunnelling, for example, for the Victoria Line (Berry, 1979). The most famous example is the scour feature encountered as part of the investigation and construction of the Blackwall Tunnels. This feature was sufficiently large to have eroded through the London Clay and into the underlying Lambeth Group. Hutchinson (1980) has suggested that many of these features could have originated as pingos.

HYDROGEOLOGY

The Upper Cretaceous – lower Tertiary stratigraphic sequence of the London Basin hosts the regionally important Chalk aquifer, which as an unconfined aquifer forms the northern and southern flanks of the basin and becomes confined by the London Clay across the central regions. In places, the Chalk aquifer is overlain by sands of the Lower Paleogene, with the Chalk and Lower London Tertiary sequences together forming a hydraulically connected, layered aquifer system. In London, Quaternary terrace gravels of the River Thames, overlying the London Clay, form a discrete shallow aquifer whose water table is much disturbed by near surface activity, the engineering of shallow drainage schemes, and by leakage from and to a multitude of service pipes (Gray and Foster, 1972, Price and Reed, 1989 and Price, 2004). Locally, in the vicinity of the structures at Greenwich and at the sites of discrete solution/collapse structures of disputed origin (Berry, 1979 and Hutchinson, 1980), direct contact between the Chalk aquifer and the shallow aquifer is established, which provides discrete routes for recharge (and potentially for contamination) of the aquifer at points throughout the confined region.

The earliest groundwater-sourced water supply for London from the Chalk aquifer was provided via an aqueduct, the 'New River', engineered at the beginning of the 17th Century to convey water by gravity from a Chalk spring on the northern limb of the Basin at Chadwell in Hertfordshire to Islington in north London (Ward, 2003). Boreholes in the Chalk aquifer situated directly at the points of demand in London contributed substantially to water supply from the middle of the 19th century, when the construction of deep wells and extensive adits in the Chalk became possible through the development of steam-powered drilling rigs and lifting machines. Excessive groundwater abstraction from central London between the mid-19th century and the mid-20th century led to the decline of the potentiometric surface of the Chalk aquifer by close to 100 m, the transition from confined to unconfined aquifer conditions occurring over much of central London, and under-drainage leading to partial dewatering of the London Clay. Minimum water levels were reached between 1950 and 1970, by which time large-scale expansion of the public water supply system based on surface water treatment and storage had led to abandonment of most private boreholes in the urban centre and the onset of water level recovery of the Chalk aquifer.

Groundwater conditions across the central regions of the Chalk aquifer in the London Basin have been substantially modified by this history of excessive groundwater abstraction across the capital. Regional potentiometric recovery since the 1970s has had deleterious consequences for groundwater quality, and potentially adverse implications for the stability of deep foundations engineered in the London Clay and underlying formations. It has also provided the impetus for development by Thames Water Utilities Ltd of one of the largest schemes in Europe for management of aquifer storage to augment groundwater recharge.

The strategic importance of the Chalk aquifer of London led to it becoming the focus for some of the earliest quantitative hydrogeological studies and analyses of regional groundwater flow in the UK. These studies, broadly accepted as demonstrating geological control on hydrogeological structure and response to pumping, included:

- The first use of groundwater contours in a regional map of water levels (Lucas, 1874 and Lucas, 1877 and Mather, 2001).
- The first analyses of spatial variability of aquifer transmissivity in the Chalk, related to geomorphological context (Ineson, 1962) and of the scale of karst development associated with acidic recharge from focussed surface run-off (Harold, 1937).

- Recognition of individual solution/collapse or pingo structures providing isolated routes for recharge and/or contamination of the aquifer throughout the confined region (Berry, 1979 and Hutchinson, 1980).
- The first regional descriptions of groundwater chemical trends (Ineson and Downing, 1963).
- The first regional mapping of groundwater O and H isotopes (Water Resources Board, 1972a, Water Resources Board, 1972b and Downing et al., 1979), indicating groundwater flow patterns and residence times.
- The first application of C isotopes to groundwater dating in the context of a regional groundwater flow system (Water Resources Board, 1972a, Water Resources Board, 1972b and Smith et al., 1976).
- The first measurements of groundwater tritium as an indicator of recent groundwater recharge and of rapid flow routes (Water Resources Board, 1972a, Water Resources Board, 1972b and Mather and Gray, 1973).

The 'rising groundwater levels' of the Chalk aquifer (Marsh and Davies, 1983), more properly the 'potentiometric recovery' (and the associated pore water pressure recovery of the London Clay) have been of interest, and concern, to the present day, with implications for groundwater quality in the aquifer and the bearing strength at the depths of building foundations in the London Clay and the formations of the Lambeth Group and Thanet Sands.

The original contributions by the Water Resources Board (1974) had been targeted at establishing the scope for 'artificial recharge' i.e. for strategic management/augmentation of storage in the Chalk aquifer beneath London. This was predicated on some of the earliest experiments in 'artificial recharge' (Boniface, 1959,Hawnt et al., 1981 and Flavin and Joseph, 1983). It underpinned development during the 1990s of the largest operational scheme in the UK for the augmentation and management of groundwater/aquifer storage (the North London Artificial Recharge Scheme (NLARS) – O'Shea et al., 1995 and O'Shea and Sage, 1999). The success of NLARS led to additional applications of 'artificial storage' south of the river Thames – South London Artificial recharge Scheme (SLARS) (Environment Agency, 2010).

Knowledge and understanding of the aquifer acquired through these fundamental investigations has been applied to predict the hydraulic and hydrochemical response of the aquifer to recovery from the hydraulic minimum established under stressed conditions by the late 1960s/1970s. These predictions have been based on:

- Long-term regional monitoring of the aquifer potentiometric recovery and groundwater quality (e.g., Environmental Agency annual reports).
- Development of numerical groundwater flow models (Wilkinson, 1985, Simpson et al., 1989, Lucas and Robinson, 1995 and Mott MacDonald, 2000), used for guiding the control of London's rising water levels by strategically positioned groundwater abstraction boreholes.
- Investigation of the hydrochemical implications of the rising groundwater levels for groundwater (artificial recharge) management schemes (Kinniburgh et al., 1994 and Mühlherr et al., 1998) and engineered structures (Rainey and Rosenbaum, 1989).
- Basin-wide determination of hydrochemical characteristics, including trace constituents, dissolved gases, environmental and radiogenic isotopes (Dennis et al., 1997 and Elliot et al., 1999).

The detailed monitoring of Chalk aquifer groundwater levels, and the requirements for improved calibration of the groundwater models, have indicated grid-like patterns in piezometric recovery and aquifer transmissivity, suggesting structural control not recognized in the early basin-wide studies. The emerging recognition of this possible structural control provides a basis for refining hydrogeological models of the basin (Buss and Daily, 2009) and a context within which observations made during geotechnical site investigations at sites of faults and other structures may profitably be interpreted. These new insights offer the prospect of a more detailed understanding of the styles of hydraulic functioning of fault structures (i.e. the nature of internal hydraulic boundaries within the London basin), the stratigraphical distribution of hydraulically active fractures, and the relative significance and the origins of deep solution enhancement.

THE USE OF 2D AND 3D MODELLING TECHNIQUES IN UNDERSTANDING THE GEOLOGY OF THE LONDON BASIN

If sound decisions are to be made, then organizations involved in planning and development need access to all relevant geo-environmental information (Royse et al., 2009). New and developing technologies allow geoscientists to present and communicate their information more effectively, especially to users of geology, many of whom are not geologists (e.g. planners, developers, financiers, insurers, engineers, Local Government officers etc.), so enabling development strategies, from regional to local, to include geoscientific information at an appropriate level for their purpose.

The application of geoscience has two major problems to overcome. Firstly, the geological map; although an excellent way of recording several sets of 2D information on a flat surface, it requires a significant amount of expert knowledge to interpret its meaning. For the non-geoscientist, the geological map presents a confusing array of colours and lines, which have little relevance to the users' everyday working lives (Royse et al., 2008). Secondly, through changes in planning policy; these have caused a significant change in the users of geoscience information. Culshaw (2003) suggested that academic users were no longer the majority users of geoscientific information, but those working in the land-use planning sector; and as a consequence, geoscientists have to change the way their data is presented and visualised.

Turner (2003) indicated that generic products (of which the traditional geological map is a classic example) are often insufficient to meet the needs of a specific user group such as planners, suggesting instead that geoscientists should also produce customized products. However, before the geoscientist can produce such outputs, three questions must be considered:

1. What geoscientific information do planners and developers need?

2. What types of geoscience data are required to meet these needs?
3. Why are geoscientific data not always fully utilized?

Urban areas require geological resources for construction and maintenance (Marker, 1998). They also require geological data to ensure that sterilization of resources or contaminative activities close to vulnerable aquifers do not occur. With these views firmly in mind, several authors such as: Brook and Marker (1987), Marker (1998), Bell and Culshaw (1998), Smith and Ellison (1999), Howland (2000), and Paul et al. (2002) have suggested the types of geoscientific data planners and developers require:

- Lithostratigraphical geology (at site, area and regional scale).
- Geomorphology.
- The nature and use of ground materials.
- The availability and quality of water.
- The susceptibility of aquifers to pollution.
- Natural and anthropogenic geohazards.
- The engineering behaviour of the ground.
- Land that may be contaminated.
- Identification of development potential (constraints on and resources for development).

Two key advances have enabled geoscientists to change the way they present data to planners and developers (Royse et al., 2008). Firstly, the availability of geoscientific data in computer-readable (i.e. digital) form (Bowie, 2005 and Jackson, 2004) and secondly, advances in GIS and 3D modelling software which allow geoscientists to take account of the 3rd dimension and in the future the 4th dimension (time). It is now possible to view and manipulate 3D models on a standard desktop computer and, more importantly, the model can be updated quickly and easily when new data become available. These are major steps forward from previous 3D urban modelling systems (Strange et al., 1998) which required a significant amount of specialist computer knowledge and access to large computing capabilities.

In essence, the 3D geological model provides either a framework within which or a platform on which the integration and visualisation of data from many different sub-disciplines can be achieved. This allows the model to portray some of the natural heterogeneity of real geological systems (Culshaw, 2005). The level of geological detail contained within the models will always be dependent on the amount and quality of the digital data available (Royse et al., 2010). Using this technology, a start can be made with predicting not only the type of rocks and soils that lie beneath the surface of London (Fig. 19), but also their physical and mechanical properties. For example, foundation conditions can be assessed by evaluating the ground at depth. In West Thurrock (Fig. 20), at 2 m below the surface, nearly half the area is underlain by soils whose compressibility falls within the 'high to very high' category. At 5 m below surface, only a small proportion of the same area contains highly compressible soils. Such data can be used in many ways, from predicting how long settlement might take, to choosing where to build. Robins et al. (2005) suggested that the 3D geological model could be improved, if it is combined with hydrogeological data, the two independent sources of data being used to complement and to check each other.

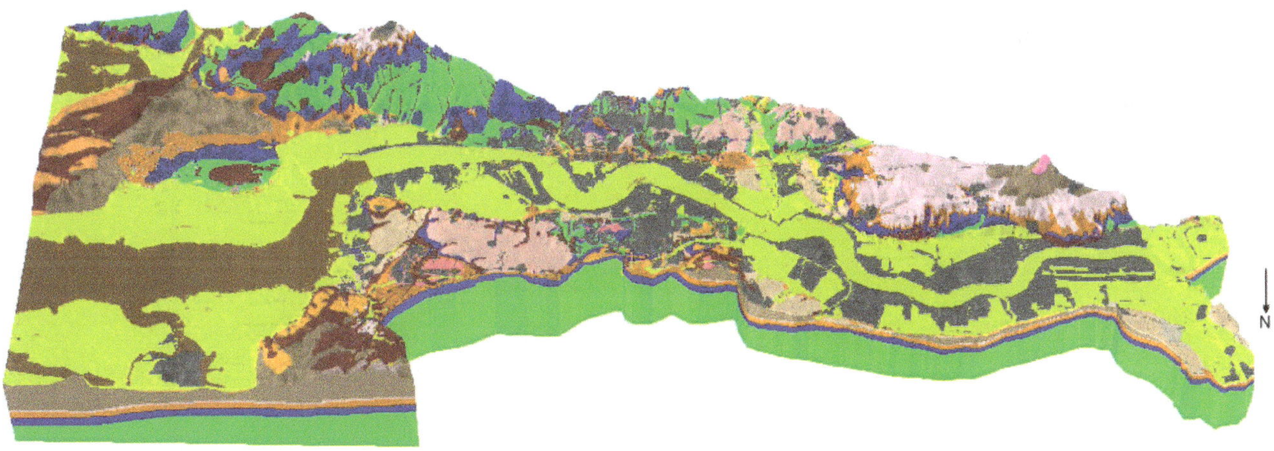

Figure 19: 3D geological model of the Thames Gateway Development Zone from Stratford in the west to Canvey Island in the east (Royse et al., 2008). Areas of peat (brown) are revealed beneath deposits of alluvium (yellow), river terrace deposits (orange) and anthropogenic deposits (grey). Bedrock is composed of Paleogene deposits (orange, blue and pink) underlain by Chalk (green).

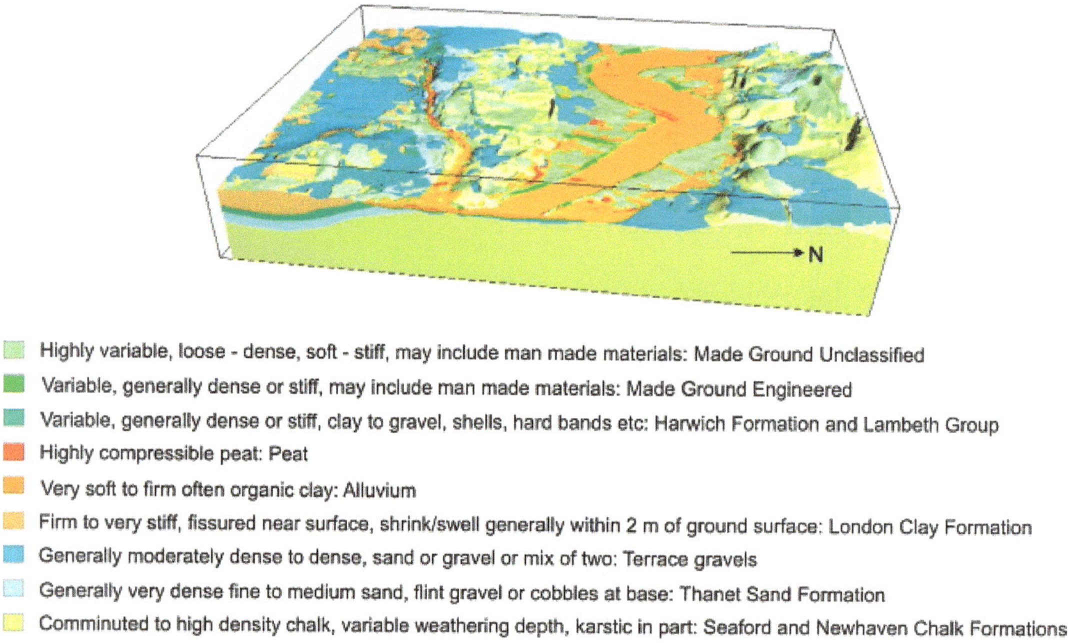

Figure 20: 3D block model of the engineering geological classification of the area between Dartford and Thurrock (Royse et al., 2009).

In this way, the 3D geological model can be used to show variations in hydrogeological properties for identifying the presence and location of high permeability units at depth. The 3D geological model constructed this way provides a means for assessing the potential hydrogeological performance of any lithological sequence modelled (Fig. 21). Thus it could be used to define either areas of likely recharge and discharge and to evaluate potential pollution pathways. If groundwater level data are added, the relationship between the potentiometric surfaces, geological units and the land surface can also be visualised (Royse et al., 2009). In summary, 3D attributed geological models are transforming the way geological maps are made and produced and changing the way groundwater modelling and ground investigations are carried out (Buss and Daily, 2009). In the future, it is probable that ground investigations will become more focused on areas where either the engineering and hydrogeological behaviour is known to be anomalous or data is sparse (Culshaw, 2005).

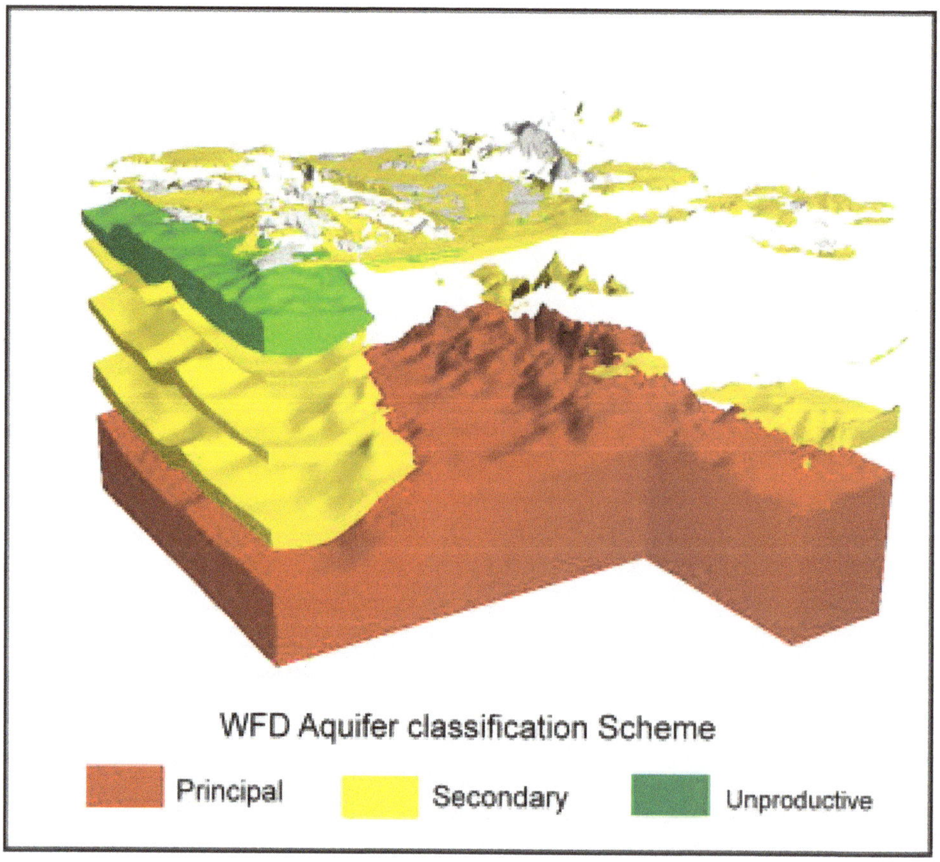

Figure 21: Exploded volume model illustrating the Environment Agency's Water Framework Directive aquifer classification scheme for the East End of London. Adapted from Royse et al. (2009).

A demonstration of this work will be the development of a new geological Atlas of the London basin by the London Basin Forum. Its task will be to distil the information now available for the London region and to synthesise it, in order to produce an holistic approach to the Basin, integrating its tectonic development, its record of sedimentation and its long history of uplift, weathering and erosion. This will be done using maps, sections, cartoons, 3D drawings and photographs.

Its data base will be integrated with that of the British Geological Survey, in a format that can be used by geologists, hydrogeologists, geotechnical and environmental engineers, and others involved with private industry and public works. In this way, the Atlas will become a geological basis for guiding local and national Government decisions concerning planning and development. To facilitate this, the text for the Atlas will address three levels of readership: geologists, users of geologists (mainly engineers) and users of geology (mainly planners and local government). In this respect, the Atlas is intended to be a key resource for science, industry and government.

Case histories will be presented that illustrate the practical consequences of geological controls on sedimentation and tectonics. These cases are drawn mainly from the construction industry but also include important examples from water supply, groundwater control and hydrogeology. Planning aspects are also to be considered, now that geo-data can be interrogated across the region, e.g. the propensity for swelling and shrinking of sediments beneath foundations and the incidence of insurance claims for structural damage resulting from these movements.

Such a resource is also of relevance to basic geological research, the region being a very well-documented example of the control that basement tectonics can exert upon sedimentation, stratigraphy, material properties, natural resources and neotectonics. The Atlas will not attempt to cover every aspect of the region's geology but will concentrate on areas where high quality data exist to illustrate the various aspects of geological control on materials and their properties, and to provide the appropriate interpretation of geological data.

A framework for interpretation is one of the major contributions the Atlas will make for its users; those who work in London are usually not short of data; quite the reverse, the problem for many is too much data and no way of knowing how best to interpret it. The Atlas will provide a basis for placing the data from local sites, even large local sites such as Crossrail, into a regional framework. In this way, features of London's geology that have up until now tended to be "anomalous" may become predictable and the severity of the consequences of encountering them unexpectedly much reduced.

SUMMARY AND CONCLUSIONS

There is as yet an untold history of the geology of the London Basin. Throughout the formation of its constituent strata, there is repeated evidence of tectonism, either by displacing strata that was once continuous, or by controlling rates and quantities of deposition, or by influencing the facies developed, or by subsidence, uplift and erosion, even up to the present day. There is much still to be learnt, but much can be achieved by using data already at hand and enabling one database to be compared with another over a common geographical area. This is the work of the London Basin Forum and the intended outcome of the Geological Atlas of the London Basin. Basic geological investigations are underway to re-evaluate the geological history recorded by the Gault and the Chalk, the Formations of the Paleogene and the Quaternary, and to compare this geology through time over the region of the basin, to reveal the existence and likely form of controlling structures at depth.

This basic research is therefore in sympathy with the pressures on urban space, above and below ground, as developers will need to make better use of the subsurface. In London, competing uses for underground space is requiring major new engineering projects to go deeper, thus associating themselves with higher stresses in geological formations whose strength is not easily able to carry them. Geological history, both sedimentological and structural, dominates the response of the ground and can influence geotechnical risk to a considerable degree. However, from the outset it should be stated that the ability to assess accurately the impact of ground conditions depends on the quality of the ground investigation. It has been said that you pay for a quality ground investigation whether you procure one or not (Terzaghi, 1943). Ground investigations do not inevitably reduce risk; they can introduce risk by supplying incomplete or erroneous data. These contribute to the third and biggest ground risk, i.e. incorrectly interpreted data. The challenge is now on to bring together locally held geodata sets so that the data are properly archived, collated and made publicly available.

The current failure to share ground investigation geodata and knowledge about anomalous ground conditions in London has prevented the geological model for the Basin from evolving and therefore failed to reduce the risk to engineering projects of unforeseen ground conditions. The current model oversimplifies the geological structure of the region (Royse, 2010). If nothing is done to improve this model, it will continue to result in unrealistic ground models and unreliable predictions (Wycisk et al., 2009). This is a situation that has already proved costly in terms of project overruns and is a continuing health and safety risk. An additional issue resulting in a failure to share geodata is that, currently, every major engineering project has to recreate a significant proportion of geodata, adding to the cost of each project (Hack, 2009). It is important to realize that the value in having geoscientific information is not in the possession of it, but in its amalgamation and interpretation. It is only when all geo-information is collected together that a realistic model can be generated. Data, even when collected and presented, as in 3D models, still has to be interpreted, and it is clear that London lacks a sound basis for the interpretation of its geological data; hence the anomalies and unexpected situations. Here the synthesis of geological evolution in place and time provided by the Atlas throughout what is now the London Basin offers a framework within which data can be interpreted.

With current advances in the ability of Geographical Information Systems (GIS) and 3D modelling technology to handle large datasets on nothing more than a regular desktop PC, coupled with a basis for their interpretation, at least for London, a revolution has occurred in the way geo-environmental information can be viewed, manipulated and interpreted. This is enabling the construction of the 'next generation' of geological models for London that will provide a platform for integrating and visualising data from many different sub-disciplines, so allowing a model to portray some of the natural heterogeneity of real geological systems. As with all geological models, the users must understand the limitations of the data on which they base their assessments. This is becoming more critical as technological improvements are allowing geoscientists to introduce a far greater level of realism into their models.

As this paper has shown, the accuracy of these new models will not only be dependent on the density and quality of the data input, but also on the theoretical understanding of the underlying geology. A good example of this is in the understanding of the impact of faulting on geotechnical properties of rocks and soils within the Basin. It is apparent from this paper, that re-evaluation of geological data from ground investigations (both new and old) is proving that faults are not only more numerous than previously thought, but have had a significant impact on the development and deposition of sediments within the Basin. With the amount and complexity of future engineering work planned in the London Basin, such as the tunnels for Crossrail, scheduled for 2010–2017, Thames Tideway, scheduled for 2012–2020 and the cable tunnels for National Grid, scheduled for 2009 – 2016, the presence or absence of faulting is becoming an ever more significant issue for which a new geological model is urgently required.

ACKNOWLEDGEMENTS

The authors would like to thank their many colleagues whose help and support has been invaluable throughout the production of this paper. Kate Royse would also like to acknowledge the support of BGS colleagues, particularly Matt Harrison for internally reviewing the manuscript. Phil Gibbard would like to thank David Watson for his expert drafting of the Thames Valley map. The two referees are thanked for their constructive comments on the original submission, which have helped to make significant improvements to the paper. Kate Royse has published with the permission of the Executive Director of the British Geological Survey (NERC).

REFERENCES

1. Ameen, M., 1995. Fracture characterisation in the Chalk and evolution of the Thanet monocline. In: Ameen, M. (Ed.), Fractography: Fracture Topography as a Tool in Fracture Mechanics and Stress Analysis, Geological Society of London, Special Publications 92, 149–174.
2. Bailey, R.P., Harris, D.L., Jenkins, M.M., 1999. Design and construction of Westminster station on the Jubilee Line Extension. Proceedings of the Institution of Civil Engineering 132, 36–46.
3. Barton, N., 1992. The Lost Rivers of London: A Study of Their Effects Upon London and Londoners and the Effects of London and Londoners Upon Them, 3rd edition. Historical Publications, Newbury, Berkshire.
4. Bell, F.G., Culshaw, M.G., 1998. Some geohazards caused by soil mineralogy, chemistry and mircrofabric: a review. In: Maund, J.G., Eddleston, M. (Eds.), Geohazards in Engineering Geology, Geological Society Engineering Geology Special Publication 15, 427–441.
5. Berry, F.G., 1979. Late Quaternary scour hollows and related features in central London. Quarterly Journal of Engineering Geology 12, 9–29.
6. British Geological Survey, 1972. England and Wales. Sheet 272 Chatham. Bedrock and Superficial Deposits in 1:50 000 Scale Geology Series. .
7. Boniface, E.S., 1959. Some experiments in artificial recharge in the lower Lee Valley. Proceedings of the Institute of Civil Engineers 14, 325–328.
8. Bowie, R., 2005. Creating a digital landslide. Geoscientist 15, 4–7.
9. Bridgland, D.R., 1988. The Pleistocene fluvial stratigraphy and palaeogeography of Essex. Proceedings of the Geologists' Association 99, 291–314.
10. Bridgland, D.R., 1994. The Quaternary of the Thames. Chapman & Hall, London.
11. Bridgland, D.R., Keen, D.H., Green, C.P., Bowen, D.Q., Sykes, G.A., 1995. Last interglacial deposits at Folkestone, Kent. Proceedings of the Geologists' Association 106, 183–193.
12. Bristow, C.R., 1982. The nomenclature of the Bagshot and the Claygate Beds of London and Essex. Tertiary Research 4, 7–8.
13. Bristow, C.R., Mortimore, R.N., Wood, C.J., 1997. Lithostratigraphy for mapping the Chalk of southern England. Proceedings of the Geologists' Association 108, 293–315.
14. Brook, D., Marker, B.R., 1987. Thematic geological mapping as an essential tool in land use planning. In: Culshaw, M.G., Bell, F.G., Cripps, J.C., O'Hara, M. (Eds.), Planning and Engineering Geology, Geological Society London Special Publication 4, 211–214.
15. Buss, S., Daily, P., 2009. London Basin Aquifer: Conceptual Model. ESI Ltd. Report 60121 R1D1 to the Environment Agency. .
16. Carter, D., Hart, M., 1977. Micropalaeontological investigations for the site of the Thames Barrior, London. Quarterly Journal of Engineering Geology 10, 321–338.
17. Clayton, C.R.I., 1983. The influence of diagenesis on some index properties of chalk in England. Geotechnique 33, 225–241.
18. Clayton, C.R.I., Matthews, M.C., Simons, N.E., 1995. Site Investigation. Blackwell.
19. Collingson, M.E., Hooker, J.J., Groke, D.R., 2003. Cobham Lignite Bed and penecontemporaneous macrofloras of southern England: a record of vegetation and fire across the Paleocene-Eocene Thermal Maximum. In: Wing, S.L.,
20. Gingerich, P.D., Schmitz, B., Thomas, E. (Eds.), Causes and Consequences of Globally Warm Climates on the Early Paleogene, Geological Society of America Special Papers 369, 339–349.
21. Culshaw, M.G., 2003. Bridging the gap between geoscience providers and the user community. In: Rosenbaum, M.S., Turner, A.K. (Eds.), New Paradigms in Subsurface Prediction. Springer-Verlag, Berlin, pp. 7–26.

22. Culshaw, M.G., 2005. From concept towards reality: developing the attributed 3D model of the shallow subsurface. Quarterly Journal of Engineering geology and Hydrogeology 39, 231–284. de Freitas, M.H., 2009. Geology; its principles, practice and potential for Geotechnics. Quarterly Journal of Engineering Geology and Hydrogeology 42, 397–441.
23. Dennis, F., Andrews, J.N., Parker, A., Poole, J., 1997. Isotopic and noble gas study of Chalk groundwater in the London Basin, England. Applied Geochemistry 12, 763–774.
24. Devoy, R.J.N., 1979. Flandrian sea level changes and vegetation history of the Lower Thames estuary. Philosophical Transactions of the Royal Society of London B285, 355–407.
25. Dewey, H., Bromehead, B.A., 1921. The Geology of South London. Memoirs of the Geological Survey, Sheet 270 (England and Wales) .
26. Downing, R.A., Pearson, F.J., Smith, D.B., 1979. The flow mechanism in the Chalk based on radio-isotope analyses of groundwater in the London Basin. Journal of Hydrology 40, 67–83.
27. Dyke, A., Glover, M., 2007. Channel tunnel rail link section 2: introduction. Proceedings of the Institution of Civil Engineering 160, 3–5.
28. Elliot, T., Andrews, J.N., Edmunds, W.M., 1999. Hydrochemical trends, palaeorecharge and groundwater ages in the fissured Chalk aquifer of the London and Berkshire Basins, UK. Applied Geochemistry 14, 333–363.
29. Ellison, R.A., 1983. Facies distribution in the Woolwich and Reading Beds of the London Basin, England. Proceedings of the Geologists' Association 94, 311–319.
30. Ellison, R.A., Knox, R.W.O., Jolley, D.W., King, C., 1994. A revision of the lithostratigraphical classification of the early Paleogene strata of the London Basin and East Anglia. Proceedings of the Geologists' Association 105, 187–197.
31. Ellison, R.A., Woods, M.A., Allen, D.J., Forster, A., Pharaoh, T.C., King, C., 2004. Geology of London. Special Memoir for 1:50 000 Geological sheets 256 (North London), 257 (Romford), 270 (Dartford) (England and Wales) British Geological Survey.
32. Environment Agency, 2010. Management of the London Basin Chalk Aquifer – Status Report. , 17 pp.
33. Erratt, D., Thomas, G., Wall, G., 1999. The evolution of the Central North Sea Rift. In: Fleet, A., Boldy, S. (Eds.), Petroleum Geology of Northwest Europe: Proceedings of the 5th Conference. Geological Society of London, pp. 63–82.
34. Flavin, R.J., Joseph, J.B., 1983. The hydrogeology of the Lee Valley and some effects of artificial recharge. Quarterly Journal of Engineering Geology and Hydrogeology 16, 65–82.
35. Ford, J.R., Mathers, S.J., Royse, K.R., Aldiss, D.T., Morgan, D., 2010. Scientific discovery and realisation through 3D geological modelling, with examples from the UK. Zeitschrift der Deutschen Gesellschaft fu¨ r Geowissenschaften 161, 205–218.
36. Gallois, R.W., 1994. Geology of the Country Around King's Lynn and The Wash. British Geological Survey Memoir for 1:50 000 Sheet 145 and Part of 129 (England and Wales) .
37. Gallois, R.W., Morter, A.A., 1982. The stratigraphy of the Gault of East Anglia. Proceedings of the Geologists' Association 93, 351–368.
38. Gibbard, P.L., 1985. The Pleistocene history of the Middle Thames Valley. Cambridge University Press, Cambridge.
39. Gibbard, P.L., 1994. Pleistocene history of the Lower Thames valley. Cambridge University Press.
40. Gibbard, P.L., Whiteman, C.A., Bridgland, D.R., 1988. A preliminary report on the stratigraphy of the Lower Thames valley. Quaternary Newsletter 56, 1–8.
41. Gray, D.A., 1965. The stratigraphical significance of electrical resistivity marker bands in the Cretaceous of the Leatherhead (Fetcham Mill) borehole, Surrey. Bulletin of the Geological Survey of Great Britain 23, 65–115 plates VIII, IX.
42. Gray, D.A., Foster, S.S.D., 1972. Urban influences upon groundwater conditions in the Thames flood plain deposits of central London. Philosophical Transactions of the Royal Society of London A 272, 245–257.
43. Greenough, G.B., 1820. A Geological Map of England and Wales [together with:] Memoir of a Geological Map of England. Longman, Hurst, Rees, Orme & Brown, London, Published under the direction of the Geological Society.
44. Hack, R., 2009. Advances in the use of geodata for the urban environment. In: Culshaw, M.G., Reeves, H.J., Jefferson, I., Spink, T.W. (Eds.), Engineering for Tomorrows Cities, Geological Society of London, Engineering Geology Special Publication 22, 201–207.
45. Hancock, J.M., 1975. The petrology of the Chalk. Proceedings of the Geologists' Association 86, 499–535.
46. Harold, C., 1937. Thirty-Second Annual Report on the Results of the Chemical and Bacteriological Examination of the London Waters for the Twelve Months Ended 31st December 1937. Metropolitan Water Board.
47. Hawnt, R.J.E., Joseph, J.B., Flavin, R.J., 1981. Experience with borehole recharge in the Lee Valley. Journal ofthe Institution of Water Engineers & Scientists 35, 437–451.
48. Haynes, J., 1955. Pelagic Foraminifera in the Thanet Beds, and the use of Thanetian as a stage name. Micropaleontology 1, 189.
49. Heath, D.L., 2001. Crossrail: introduction. Proceedings of the Institution of Civil Engineering 147, 59–60.
50. Hey, R., 1980. Equivalents of the Westland Green Gravels in Essex and East Anglia. Proceedings of the Geologists' Association 91, 279–290.
51. Hight, D.W., Ellison, R.A., Page, D.P., 2001. Engineering in the Lambeth Group. Construction Industry Research and Information Association (CIRIA), London.
52. Hight, D.W., Ellison, R.A., Page, D.P., 2004. Engineering in the Lambeth Group. Publication C583. Construction

Industry Research and Information Association (CIRIA), London.

53. Hillier, R.D., Williams, B.P.J., 2006. The alluvial Old Red Sandstone: fluvial basins. In: Brenchley, P.J., Rawson, P.F. (Eds.), The Geology of England and Wales. 2nd edition. Geological Society of London, pp. 155–171.

54. Howland, A.F., 2000. The history of the development of procedure for the rapid assessment of environmental conditions to aid the urban regeneration process at London Docklands. Engineering Geology 60, 117–125.

55. Hull, E., Whitaker, W., 1861. The geology of parts of Oxfordshire and Berkshire. In: Mem. Geological Survey of Great Britain. Geological Survey of Great Britain.

56. Hutchinson, J.N., 1980. Possible late Quaternary pingo remnants in central London. Nature 284, 253–255.

57. ICE, 1991. Inadequate Site Investigation. Thomas Telford.

58. Ineson, J., 1962. A hydrogeological study of the permeability of the Chalk. Journal of the Institution of Water Engineers 16, 449–463.

59. Ineson, J., Downing, R.A., 1963. Changes in the chemistry of ground waters of the Chalk passing beneath argillaceous strata. Bulletin of the Geological Survey of Great Britain (20), 519–541.

60. Jackson, I., 2004. Britain beneath our feet. British Geological Survey Occasional Publication 4, 114.

61. King, C., 1981. The stratigraphy of the London Clay and associated deposits. Tertiary Research Special Paper 6, 3–58.

62. King, C., 2006. Paleogene and Neogene: uplift and a cooling climate. In: Brenchley, P.J., Rawson, P.F. (Eds.), The Geology of England and Wales. The Geological Society, London, pp. 395–427.

63. Kinniburgh, D.G., Gale, I.N., Smedley, P.L., Darling, W.G., West, J.M., Kimblin, R.T., Parker, A., Rae, J.E., Aldous, P.J., O'Shea, M.J., 1994. The effects of historic abstraction of groundwater from the London Basin aquifers on groundwater quality. Applied Geochemistry 9, 175–195.

64. Klinck, B.A., Hopson, P.M., Lewis, M.A., Macdonald, D.M.J., Inglethorpe, S.D.J., Entwisle, D.C., Harrington, J.F., Williams, L., 1998. The Hydrogeological Behaviour of the Clay-With-Flints in Southern England. British Geological Survey, Keyworth, Nottingham.

65. Knox, R.W.O., 1996. Tectonic controls on sequence development in the Paleocene and earliest Eocene of southeast England: implications for North Sea stratigraphy. In: Hesselbro, S.P., Parkinson, D.N. (Eds.), Sequence Stratigraphy in British Geology, Geological Society of London Special Publication 103, 209–230.

66. Lake, R.D., Young, B., Wood, C.J., Mortimore, R.N., 1987. Geology of the country around Lewes. In: (sheet 319) Memoir of the British Geological Survey, HMSO, London.

67. Lee, M.K., Pharaoh, J.P., Williamson, J.P., Green, C.A., de Vos, W., 1993. Evidence on the deep structure of the Anglo-Brabant Massif from gravity and magnetic data. Geological Magazine 130, 575–582.

68. Linney, L.F., Withers, A.D., 1998. Dewatering the Thanet Beds in SE London: three case histories. Quarterly Journal of Engineering Geology 31, 115–122.

69. Lord, J.A., Clayton, C.R.I., Mortimore, R.N., 2002. Engineering in Chalk. Construction Industry Research and Information Association (CIRIA), 350 pp.

70. Lucas, J., 1874. Horizontal Wells. A new application of geological principles to effect the solution of the problem of supplying London with pure water. Edward Stanford, London.

71. Lucas, J., 1877. Hydrogeological Survey, Sheet 1. Edward Stanford, London.

72. Lucas, H.C., Robinson, V.K., 1995. Modelling of rising groundwater levels in the Chalk aquifer of the London Basin. Quarterly Journal of Engineering Geology and Hydrogeology 28, 51–62.

73. Marker, B.R., 1998. Incorporation of information on geohazards into the planning process. In: Maund, J.G.,Eddleston, M. (Eds.), Geohazards in Engineering Geology, Geological Society Engineering Geology Special Publication 385–389.

74. Marsh, T.J., Davies, P.A., 1983. The decline and partial recovery of groundwater levels below London. Proceedings of the Institution of Civil Engineers 74 (Pt 1), 263–276.

75. Mather, J.D., 2001. Joseph Lucas and the term "hydrogeology". Hydrogeology Journal 9, 413–415.

76. Mather, J.D., Gray, D.A., 1973. Groundwater recharge in the Lower Greensand of the London basin – results of tritium and carbon-14 determinations. Quarterly Journal of Engineering and Hydrogeology 6, 141–152.

77. Mathers, S., Zalasiewicz, J., 1988. The red Crag and Norwich Crag formations of southern East Anglia. Proceedings of the Geologist's Association 99, 261–278.

78. Mortimore, R.N., Newman, T., Royse, K.R., Scholes, H., Lawrence, U., 2011. Chalk: its stratigraphy, structure and engineering geology in east London and the Thames Gateway. Quarterly Journal of Engineering Geology and Hydrogeology.

79. Mortimore, R.N., Pomerol, B., 1997. Upper Cretaceous tectonic phases and end Cretaceous inversion in the chalk of the Anglo-Paris basin. Proceedings of the Geologists" Association 108, 231–255.

80. Morton, A.C., 1982. The provenance and digenesis of Paleogene sandstones of southeast England as indicated by heavy mineral analysis. Proceedings of the Geologists" Association 93, 263–274.

81. Mott MacDonald, 2000. London Basin Groundwater Modeling Study. .

82. Mu¨hlherr, I.H., Hiscock, K.M., Dennis, P.F., Feast, N.A., 1998. Changes in groundwater chemistry due to rising groundwater levels in the London Basin between 1963 and 1994. In: Robins, N.S. (Ed.), Groundwater Pollution, Aquifer Recharge and Vulnerability, Geological Society London Special Publications 130, 47–62.

83. Newman, T., 2008. Ground modelling for the Lee Tunnel. In: Engineering Geology and Experience in the East London and Thames Gateway Area. Paper

presented to the Engineering Group of the Geological Society of London Meeting, 3 September 2008. Disk available from the Geological Society, London.

84. Newman, T., 2009. The impact of adverse geological conditions on the design and construction of the Thames Water Ring Main in Greater London, UK. Quarterly Journal of Engineeging Geology and Hydrogeology 42, 5–21.

85. O'Shea, M.J., Baxter, K.M., Charalambous, A.N., 1995. The hydrogeology of the Enfield – Haringey artificial recharge scheme, north London. Quarterly Journal of Engineeging Geology 28, S115–S129.

86. O'Shea, M.J., Sage, R., 1999. Aquifer recharge: an operational drought-management strategy in north London. Water & Environmental Management 13, 400–405.

87. Owen, H.G., 1995. The upper part of the Carstone and the Hunstanton Red Chalk (Albian) of the Hunstanton Cliff, Norfolk. Proceedings of the Geologists" Association 106, 171–181.

88. Owen, H.G., 1971a. Middle Albian stratigraphy in the Anglo-Paris Basin. Bulletin of the British Museum (Natural History) Geology Series (Suppl. 8), 164 3 pls.

89. Owen, H.G., 1971b. The stratigraphy of the Gault in the Thames Estuary and its bearing on the Mesozoic tectonic history of the area. Proceedings of the Geologists' Association 82, 187–207.

90. Owen, H.G., 1976. The stratigraphy of the Gault and Upper Greensand of the Weald. Proceedings of the Geologists' Association 86, 475–498.

91. Owen, H.G., 1996. The Upper Gault and Upper Greensand of the M23/M25/M26 motorway systems and adjacent sections, Surrey & Kent. Proceedings of the Geologists' Association 107, 167–188.

92. Page, D., Skipper, J., 2000. Lithological characteristics of the Lambeth Group Ground Engineering, vol. 33, pp. 38–44.

93. Paul, T., Chow, F., Kjekstad, O., 2002. Hidden Aspects of Urban Planning – Surface and Underground Development. Thomas Telford Publishing, London.

94. Price, M., 2004. Dr John Snow and an early investigation of groundwater contamination. In: Mather, J.D. (Ed.), 200 Years of British Hydrogeology, Geological Society Special Publication 225, 31–49.

95. Price, M., Reed, D.W., 1989. The influence of mains leakage and urban drainage on groundwater levels beneath conurbations in the UK. Proceedings Institution of Civil Engineers 86 (1), 31–39.

96. Rainey, T.P., Rosenbaum, M.S., 1989. The adverse influence of geology and groundwater on the behaviour of London Underground railway tunnels near Old Streat Station. Procceedings of the Geolgists' Association 100, 123–134.

97. Robins, N.S., Rutter, H.K., Dumpleton, S., Peach, D.W., 2005. The role of 3D visualization as an analytical tool preparatory to numerical modelling. Journal of Hydrology 301, 287–295.

98. Rose, J., Whiteman, C.A., Allen, P., Kemp, R.A., 1999. The Kesgrave sands and gravels: &pre-glacial' Quaternary deposits of the River Thames in East Anglia and the Thames valley. Proceedings of the Geologists' Association 110, 93–116.

99. Royse, K.R., 2010. Combining numerical and cognitive 3D modelling approaches in order to determine the structure of the Chalk in the London Basin. Computers and Geosciences 36, 500–511.

100. Royse, K.R., Kessler, H., Robins, N.S., Hughes, A.G., Mathers, S., 2010. The use of 3D geological models in the development of the conceptual groundwater model. Zeitschrift der Deutschen Gesellschaft fu¨ r Geowissenschaften 161, 237–249.

101. Royse, K.R., Reeves, H.J., Gibson, A.R., 2008. The modelling and visualisation of digital geoscientific data as an aid to land-use planning in the urban environment, an example from the Thames Gateway. In: Liverman, D.G.E., Pereira, C., Marker, B. (Eds.), Communicating Environmental Geoscience. pp. 89–106.

102. Royse, K.R., Rutter, H.K., Entwisle, D.C., 2009. Property attribution of 3D geological models in the Thames Gateway: new ways of visualising geoscientific information. Bulletin of Engineering Geology and the Environment 68, 1–16.

103. Shackleton, R., 1984. Thin skinned tectonics, basement control and the Variscan front. Geological Society London Special Publication 14, 125–129.

104. Sheppard, T., 1917. William Smith: his maps and memoirs. Proceedings of the Yorkshire Geological Society 19, 75–253.

105. Sherlock, R.L., 1947. London and Thames Valley, 2nd edition. British Regional Geology, Geological Survey and Museum, His Majesty's Stationery Office, London.

106. Simpson, B., Blower, T., Craig, R.N., Wilkinson, W.B., 1989. The engineering implications of rising groundwater levels in the deep aquifer beneath London. CIRIA Special Publication 69 .

107. Skipper, J.A., 1999. The stratigraphy of the Lambeth Group (Paleocene) of South East England. PhD Thesis, Imperial College London.

108. Skipper, J.A., 2008. Project Specific Geological Training – A New Tool for Geotechnical Risk Remediation. Euroengeo International Association of Engineering Geology, Madrid.

109. Skipper, J.A., Newman, T., Mortimore, R.N., 2008. The Engineering Geology of the Lee Tunnel, East London, UK. Euroengeo International Association of Engineering Geology, Madrid.

110. Smith, D.B., Downing, R.A., Monkhouse, R.A., Otlet, R.L., Pearson, F.J., 1976. The age of groundwater in the Chalk of the London Basin. Water Resources Research 12, 392–404.

111. Smith, W., 1815. Map Entitled; A Delineation of the Strata of England and Wales with part of Scotland; Exhibiting the Collieries and Mines, the Marshes and Fen Lands Originally Overflowed by the Sea, and the Varieties of Soil According to the Variations in the

Substrata, Illustrated by the Most Descriptive Names by W. Smith. .

112. Smith, A., Ellison, R.A., 1999. Applied geological maps for planning and development: a review of examples from England and Wales, 1983 to 1996. Quaternary Journal of Engineering Geology 32, S1–S44.

113. Steward, H.E., Cripps, J.C., 1983. Some engineering implications of chemical weathering of pyritic shale. Quarterly Journal of Engineering Geology 16, 281–289.

114. Strange, P.J., Booth, S.J., Ellison, R.A., 1998. Development of 'rockhead' computergenerated geological models to assist geohazard prediction in London. In: Maund, J.G., Eddleston, M. (Eds.), Geohazards in Engineering Geology, Geological Society Engineering Geology Special Publication 15, 409–414.

115. Sumbler, M.G., 1996. British Regional Geology: London and the Thames Valley, 4th edition. British Geological Survey, London.

116. Terzaghi, K., 1943. Theoretical Soil Mechanics. John Wiley and Sons, New York.

117. Thompson, A., 1997. The relationship between rising groundwater levels and the geological structure of the London Basin. MSc Thesis, Imperial College London.

118. Turner, A.K., 2003. Putting the user first: implications for subsurface characterisation. In: Rosenbaum, M.S., Turner, A.K. (Eds.), New Paradigms in Subsurface Prediction. Springer-Verlag, Berlin, pp. 61–83.

119. Ward, R., 2003. London's New River. Historical Publications Ltd.

120. Warren, C.D., Mortimore, R.N., 2003. Chalk engineering geology – channel tunnel rail link and North Downs tunnel. Quarterly Journal of Engineering Geology and Hydrogeology 36, 17–34.

121. Water Resources Board (Downing, R.A., Davies, M.C., Young, C.P.), 1972. The Hydrogeology of the London Basin. Water Resource Board, Reading.

122. Water Resources Board, 1974. Artificial Recharge of the London Basin. Vol. 1: Hydrogeology. .

123. Water Resources Board, 1972b. The Hydrogeology of the London Basin. .

124. Whitaker, W., 1866. On the "Lower London Tertiaries" of Kent. Quarterly Journal of the Geological Society London 22, 404–434.

125. Whitaker, W., 1872. The Geology of the London Basin, Part 1. The Chalk and the Eocene Beds of the Southern and Western Tracts. Memoirs of the Geological

126. Survey, England and Wales. .

127. Whiteman, C.A., Rose, J., 1992. Thames river sediments of the British Early and Middle Pleistocene. Quaternary Science Reviews 11, 363–375.

128. Wilkinson, W.B., 1985. Rising groundwater levels in London and possible effects on engineering structures. In: Proceedings of the 18th Congress of the IAH, Cambridge, pp. 145–157.

129. Withers, A.D., 1996. Dewatering and depressurisation systems for the Thanet Sands and Chalk Aquifer, Case studies from the Jubilee line extension project. Unpublished MSc Thesis, University of London.

130. Wooldridge, S.W., 1923. The minor structures of the London Basin. Proceedings of the Geologists' Association 34, 175–193.

131. Wooldridge, S.W., 1926. The structural evolution of the London Basin. Proceedings of the Geologists' Association 37, 162–192.

132. Wooldridge, S.W., Linton, D.L., 1939. Structure, Surface and Drainage in South-East England. George Philip, London.

133. Wooldridge, S.W., Linton, D.L., 1955. Structure, Surface and Drainage in South-East England. George Philip, London.

134. Wycisk, P., Hubert, T., Gossel, W., Neumann, C., 2009. High-resolution 3D spatial modelling of complex geological structures for an environmental risk assessment of abundant mining and industrial megasites. Computers and Geosciences 35, 165–182.

INDEX

A

C

D

E

F

G

H

I

L

M

N

O

P

Q

R

S

T

U

V

W